Cambridge International AS Level

Information Technology

Second edition

Graham Brown
Brian Sargent

Endorsement indicates that a resource has passed Cambridge International Education's rigorous quality-assurance process and is suitable to support the delivery of a Cambridge syllabus. However, endorsed resources are not the only suitable materials available to support teaching and learning, and are not essential to achieve the qualification. Resource lists found on the Cambridge website will include this resource and other endorsed resources.

Any example answers to questions taken from past question papers, practice questions, accompanying marks and mark schemes included in this resource have been written by the authors and are for guidance only. They do not replicate examination papers. In examinations the way marks are awarded may be different. Any references to assessment and/or assessment preparation are the publisher's interpretation of the syllabus requirements. Examiners will not use endorsed resources as a source of material for any assessment set by Cambridge International Education.

While the publishers have made every attempt to ensure that advice on the qualification and its assessment is accurate, the official syllabus, specimen assessment materials and any associated assessment guidance materials produced by the awarding body are the only authoritative source of information and should always be referred to for definitive guidance.

Our approach is to provide teachers with access to a wide range of high-quality resources that suit different styles and types of teaching and learning.

For more information about the endorsement process, please visit www.cambridgeinternational.org/endorsed-resources.

Cambridge International Education copyright material in this publication is reproduced under licence and remains the intellectual property of Cambridge University Press & Assessment.

Third-party websites and resources referred to in this publication have not been endorsed by Cambridge International Education.

Answers to the practice questions and activities and all source files needed for the Student's Book can be downloaded from www.hoddereducation.com/cambridgeextras.

Computer hardware and software brand names mentioned in this book are protected by their respective trademarks and are acknowledged.

Photo credits: p50 [top] © Courtesy of International Business Machines Corporation, © International Business Machines Corporation; p50 [bottom] © Julian Herzog/CC BY (https://creativecommons.org/licenses/by/4.0); p93 [left] © Alvey & Towers Picture Library/Alamy Stock Photo; p93 [right] © David Jones/PA Images/Alamy Stock Photo; p93 [bottom] © Justin Kase zsixz/Alamy Stock Photo; p267 © MclittleStock/stock.adobe.com; p269 [left] © imageBROKER.com GmbH & Co. KG/Alamy Stock Photo; p269 [right] © Firas Nashed/stock.adobe.com; p271 © Designua/stock.adobe.com

Text credits: p46 Table 1.14 reproduced by permission of Lloyd's Register; p151 Figure 6.3 adapted from, 'Americans with lower incomes have lower levels of technology adoption' Pew Research Center, Washington, D.C. (21 June 2021) https://www.pewresearch.org/short-reads/2021/06/22/digital-divide-persists-even-as-americans-with-lower-incomes-make-gains-in-tech-adoption/ft_2021-06-22_digitaldivideincome_01/; Microsoft, (Access, Excel, Movie Maker, Photos, Windows and Word) are trademarks of the Microsoft group of companies; Apple Mac is a trademark of Apple Inc., registered in the U.S. and other countries and regions.

Although every effort has been made to ensure that website addresses are correct at time of going to press, Hodder Education cannot be held responsible for the content of any website mentioned in this book. It is sometimes possible to find a relocated web page by typing in the address of the home page for a website in the URL window of your browser.

Hachette UK's policy is to use papers that are natural, renewable and recyclable products and made from wood grown in well-managed forests and other controlled sources. The logging and manufacturing processes are expected to conform to the environmental regulations of the country of origin.

To order, please visit www.hoddereducation.com or contact Customer Service at education@hachette.co.uk/+44 (0)1235 827827.

ISBN: 978 1 0360 0560 3

© Graham Brown and Brian Sargent 2024

First published in 2021
This edition published in 2024 by
Hodder Education,
An Hachette UK Company
Carmelite House
50 Victoria Embankment
London EC4Y 0DZ
www.hoddereducation.com

Impression number   10 9 8 7 6 5 4 3 2 1

Year   2028 2027 2026 2025 2024

All rights reserved. Apart from any use permitted under UK copyright law, no part of this publication may be reproduced or transmitted in any form or by any means, electronic or mechanical, including photocopying and recording, or held within any information storage and retrieval system, without permission in writing from the publisher or under licence from the Copyright Licensing Agency Limited. Further details of such licences (for reprographic reproduction) may be obtained from the Copyright Licensing Agency Limited, www.cla.co.uk

Cover photo © cookiecutter – stock.adobe.com

Illustrations by Barking Dog Art

Typeset in India by Aptara Inc

Produced by DZS Grafik, Printed in Slovenia

A catalogue record for this title is available from the British Library.

Contents list

	Introduction	v
1	**Data processing and information**	**1**
	1.1 Data and information	1
	1.2 Quality of information	7
	1.3 Encryption	10
	1.4 Checking the accuracy of data	18
	1.5 Data processing	27
2	**Hardware and software**	**49**
	2.1 Mainframe computers and supercomputers	49
	2.2 System software	59
	2.3 Utility software	64
	2.4 Custom-written software and off-the-shelf software	70
	2.5 User interfaces	72
3	**Monitoring and control**	**76**
	3.1 Monitoring and measurement technologies	76
	3.2 Control technologies	82
4	**Algorithms and flowcharts**	**98**
	4.1 Algorithms	98
	4.2 Flowcharts	114
5	**eSecurity**	**120**
	5.1 Personal data	120
	5.2 Malware	130
6	**The digital divide**	**137**
	6.1 What is the digital divide?	137
	6.2 Causes of the digital divide	138
	6.3 The effects of the digital divide	140
	6.4 Groups affected by the digital divide	146
7	**Expert systems**	**155**
	7.1 What is an expert system?	155
	7.2 Different scenarios where expert systems are used	157
	7.3 Chaining	162

8 Spreadsheets — 168

- 8.1 Creating a spreadsheet — 169
- 8.2 Testing a spreadsheet — 230
- 8.3 Using a spreadsheet — 234
- 8.4 Graphs and charts — 244

9 Modelling — 254

- 9.1 Modelling — 254
- 9.2 Simulations — 268
- 9.3 Using what-if analysis — 272

10 Database and file concepts — 276

- 10.1 Database basics — 276
- 10.2 Normalising data — 325
- 10.3 Creating a data dictionary — 330
- 10.4 File and data management — 331

11 Video and audio editing — 340

- 11.1 Video editing — 340
- 11.2 Audio editing — 355

Glossary — 369

Index — 377

Answers can be found at www.hoddereducation.com/CambridgeExtras

Introduction

This textbook has been written to provide the knowledge, understanding and practical skills required by those studying the AS Level content (Topics 1–11) of the Cambridge International AS & A Level Information Technology syllabus (9626) for examination from 2025. Students studying the full A Level will also need to become familiar with the A Level content (Topics 12–21), covered in Hodder Education's *Cambridge International A Level Information Technology*.

How to use this book

This textbook, endorsed by Cambridge International Education, has been designed to make your study of Information Technology as successful and rewarding as possible.

Organisation

The book comprises 11 chapters, the titles of which correspond exactly with the topics in the syllabus. Each chapter is broken down into sections, which largely reflect the subtopics in the syllabus.

Features

Each chapter contains a number of features designed to help you effectively navigate the syllabus content.

At the start of each chapter, there is a blue box that provides a summary of the content to be covered in that topic.

> **In this chapter you will learn:**
> ★ about the sensors and calibration used in monitoring technologies
> ★ about the uses of monitoring technologies
> ★ about the sensors and actuators used in control technologies
> ★ how to write an algorithm and draw a flowchart.

There is also a box that lists the knowledge you should have before beginning to study the chapter.

> **Before starting this chapter you should:**
> ★ be familiar with the terms 'observation', 'Interviews', 'questionnaires', 'central processing unit (CPU)', 'chip and PIN', 'direct access', 'encryption', 'file', 'key field', 'RFID', 'sort', 'validation' and 'verification'.

Chapters that require you to do practical work also feature a list of source files that you will need to use. These can be found here: www.hoddereducation.com/cambridgeextras.

> **For this chapter you will need these source files:**
> ■ TuckShop.csv
> ■ Widget.csv

The practical chapters contain Tasks. The text demonstrates the techniques used to carry out the tasks. It provides easy-to-follow step-by-step instructions, so that practical skills are developed alongside the knowledge and understanding. Tasks often include the use of source files that you can download from www.hoddereducation.com/cambridgeextras.

> ### Task 8e
> Open and examine the file **Stock.csv**. Split this so that both types of stock can be viewed together. Save the spreadsheet as **Task_8e**.

Each chapter also includes Activities to allow you to check your understanding of the concepts covered and practise the skills demonstrated in the Tasks. In the practical chapters, these often require the use of source files from the website.

> ### Activity 1a
> Explain the difference between data and information.

Advice and shortcuts for improving your ICT skills are highlighted in Advice boxes.

> #### Advice
> A common error made by people writing algorithms with nested loops is not matching up the number of WHILE statements with the same number of ENDWHILEs. The same error can happen with REPEATs and UNTILs. You must always check this and make sure they are correctly indented.

Finally, each chapter ends with practice questions. These practice questions and their sample answers, as well as the activities throughout the book have been written by the authors.

Answers to the practice questions and activities and the source files needed for the Student's Book can be downloaded from www.hoddereducation.com/cambridgeextras.

Practice questions

1. A collection of data could be this: johan, Σ, $, <, AND
   Explain why they are regarded as just items of data. In your explanation give a possible context for each item of data and describe how the items would then become information. [5]

2. A company uses computers to process its payroll, which involves updating a master file.
   a State what processes must happen before the updating can begin. [2]
   b Describe how a master file is updated using a transaction file in a payroll system. You may assume that the only transaction being carried out is the calculation of the weekly pay before tax and other deductions. [6]

3. a Name and describe **three** validation checks other than a presence check. [3]
   b Explain why a presence check is not necessary for all fields. [3]

4. A space agency controls rockets to be sent to the moon.
   Describe how real-time processing would be used by the agency. [3]

5. Describe **three** different methods used to carry out verification. [3]

Text colours

Some words or phrases within the text are printed in red. Definitions of these terms can be found in the glossary at the back of the book. In the practical section, words that appear in blue indicate an action or location found within the software package, for example 'Select the Home tab.' In the database sections of the book, words in orange show fieldnames. Words in green show the functions or formulas entered into the cell of a spreadsheet, for example a cell may contain the function =SUM(B2:B12).

Assessment

The information in this section is taken from the Cambridge International Education syllabus. You should always refer to the appropriate syllabus document for the year of examination to confirm the details and for more information. The syllabus document is available on the Cambridge International Education website at www.cambridgeinternational.org.

If you are following the AS Level part of the course, you will take two examination papers: Paper 1 Theory (1 hour 45 minutes); Paper 2 Practical (2 hours 30 minutes).

Command words

The table below, taken from the syllabus, includes command words used in the assessment for this syllabus. The use of the command word will relate to the subject context. Make sure you are familiar with these.

Command word	What it means
Analyse	Examine in detail to show meaning, identify elements and the relationship between them
Assess	Make an informed judgement
Compare	Identify/comment on similarities and/or differences
Contrast	Identify/comment on differences
Define	Give precise meaning
Describe	State the points of a topic/give characteristics and main features
Discuss	Write about issue(s) or topic(s) in depth in a structured way
Evaluate	Judge or calculate the quality, importance, amount or value of something
Explain	Set out purposes or reasons/make the relationships between things clear/say why and/or how and support with relevant evidence
Identify	Name/select/recognise
Justify	Support a case with evidence/argument
State	Express in clear terms
Suggest	Apply knowledge and understanding to situations where there are a range of valid responses in order to make proposals/put forward considerations

Notes for teachers

Key concepts

These are the essential ideas that help learners to develop a deep understanding of the subject and to make links between the different topics. Although teachers are likely to have these in mind at all times when they are teaching the syllabus, the following icons are included in the textbook at points where the key concepts relate to the text (note that not all of these key concepts are relevant to the AS Level course and some will only feature in the A Level book).

Hardware and software

Hardware and software interact with each other in an IT system. It is important to understand how these work and how they work together with each other and with us in our environment.

Networks

Computer systems can be connected together to form networks allowing them to share data and resources. The central role networks play in the internet, mobile and wireless applications and cloud computing has rapidly increased the demand for network capacity and performance.

The internet

The internet is a global communications network. It uses standardised communications protocols to allow computers worldwide to connect and share information in many different forms. The impact of the internet on our lives is profound. While the services the internet supports can provide huge benefits to society they have also introduced issues, for example security of data.

System life cycle

Information systems are developed within a planned cycle of stages. They cover the initial development of the system and continue through to its scheduled updating or redevelopment.

New technologies

As the information industry changes so rapidly, it is important to keep track of new and emerging technologies and consider how they might affect everyday life.

Additional support

The *Cambridge International AS Level Information Technology Skills Workbook* is a write-in resource designed to be used throughout the course. It provides students with extra opportunities to test their understanding of the knowledge and skills required by the syllabus.

1 Data processing and information

In this chapter you will learn:
- what is meant by the terms 'data' and 'information' and about their use
- what is meant by the terms 'direct' and 'indirect' data and about their uses and sources
- what is meant by 'quality of information'
- what is meant by 'encryption', why it is needed, and about the methods and uses of encryption and protocols
- about the methods and uses of validation and verification
- about the methods and uses of different methods of processing (batch, online, real-time)
- how to write a simple algorithm.

Before starting this chapter you should:
- be familiar with the terms 'observation', 'interviews', 'questionnaires', 'central processing unit (CPU)', 'chip and PIN', 'direct access', 'encryption', 'file', 'key field', 'RFID', 'sort', 'validation' and 'verification'.

1.1 Data and information

Before we consider data processing, we need to define the term **data**. To be completely accurate, the word 'data' is the plural of 'datum', a single piece of data. Often, however, we use data in both the singular and the plural senses. It seems awkward to say 'the data are incorrect' so we tend to say 'the data is incorrect'. When we use the word 'data', it can mean many different things. A lot of people frequently confuse the terms 'data' and 'information'. For the purposes of this course we will consider data to be what is usually known as 'raw' data. Data can take several forms; it can be characters, symbols, images, audio clips, video clips and so on, none of which, on their own, have any meaning.

It is important for you to learn what the term **information** means when we use it in information technology. Information is data that has been given meaning, which often results from the processing of data, sometimes by a computer. The processed data can then be given a context and have meaning.

The difference between data and information is that data has no meaning, whereas information is data which has been given meaning.

Here are some examples:

Sets of data:

110053, 641609, 160012, 390072, 382397, 141186

01432 01223 01955 01384 01253 01284 01905 01227 01832 01902 01981 01926 01597

$\sigma \; \omega \; \rho \; F \; m \; a$

These are sets of data which do not have a meaning until they are put into context.

If we are told that 110053, 641609, 160012, 390072, 382397 and 141186 are all postal codes in India (a context), the first set of data becomes information as it now has meaning.

Similarly, if you are informed that 01432, 01223, 01955, 01384, 01253, 01284, 01905, 01227, 01832, 01902, 01981, 01926 and 01597 are telephone area dialling codes in the UK, they can now be read in context and we can understand them, as they now have a meaning.

The final set of data seems to be letters of the Greek alphabet apart from F, m and a, which are letters in the Latin alphabet. However, if we are told that the context is mathematical, scientific or engineering **formulas**, we can see that they all represent a different variable: σ represents standard deviation, ω represents angular velocity, ρ is density, F is force, m is mass and a is acceleration. They each now have meaning.

1.1.1 Data processing

On a computer, data is stored as a sequence of **b**inary dig**its** (**bits**) in the form of ones and zeros. We will discuss bits later in this chapter, when we look at **parity checks**. We can store data on a fixed or removeable media such as hard-disk drive, solid-state drive, DVDs, SD cards, memory sticks or in RAM. Data is usually processed for a particular purpose, often so that it can be analysed. The computer processing involved uses different operations to produce new data from the source data. You will, perhaps, have met this in previous practical work you have done, where you may have been given source files, including **.csv** files. You may have been asked to open these in a spreadsheet and add formulas. This is the processing of that data so that it then has meaning.

To sum up, data is input, stored, and processed by a computer, for output as usable information. Later in this chapter we will look at different types of processing.

> ### Activity 1a
> Explain the difference between data and information.

1.1.2 Direct and indirect data

Direct data is data that is collected for a specific purpose or task and is used for that purpose and that purpose only. It is often referred to as 'original source data'. Examples of sources of direct data are questionnaires, interviews, observation, and **data logging**.

Indirect data is data that is obtained from a third party and used for a different purpose to that which it was originally collected for and which is not necessarily related to the current task. Examples of sources of indirect data are the **electoral register** and businesses collecting personal information for use by other organisations (third parties).

Direct data sources

Direct data sources are sources that provide the data gatherer with original data. We will consider four such sources.

Questionnaires

A questionnaire consists of a set of questions, usually on a specific subject or issue. The questions are designed to gather data from those people being questioned. A questionnaire can be a mixture of what are called closed questions (where you have to choose one or more answers from those provided) and open-ended questions (questions where you can write in your answers in more detail). Questionnaires are easy to distribute, complete and collect as most people are familiar with the process. They can be completed on paper or on computer.

Interviews

An interview is a formal meeting, usually between two people, where one of them, the interviewer, asks questions, and the other person, the interviewee, answers those questions. Interviews are used to collect data about a topic and can be structured or unstructured. Structured interviews are similar to a questionnaire, whereby the same questions are asked in the same order for each interviewee and with a choice of answers. Unstructured interviews can be different for each interviewee, particularly as they give them the opportunity to expand on their answers. There is usually no pre-set list of answers in an unstructured interview.

Observation

Observation is a method of data collection in which the data collectors watch what happens in a given situation. The observer collects data by seeing for themselves what happens, rather than depending on the answers from interviewees or the accuracy of completed questionnaires. The collected data is then recorded and analysed. An activity could be for students to decide whether traffic lights are needed at a road junction to allow the smooth flow of traffic. Observation could consist of watching and counting the number of cars passing through the junction. It might also involve watching the traffic at the junction and timing how long it takes for a certain number of vehicles to pass.

Data logging

Data logging means using a computer and **sensors** to collect data. The data is then analysed, saved and the results are output, often in the form of graphs and charts. Data logging systems can gather and display data as an event happens. The data is usually collected over a period of time, either continuously or at regular intervals, in order to observe particular trends. It involves recording data from one or more sensors and the analysis usually requires special software. Data logging is commonly used in scientific experiments, in **monitoring** systems where there is the need to collect information faster than a human possibly could, in hazardous circumstances such as volcanoes and nuclear reactors, and in cases where accuracy is essential. Examples of the types of information a data logging system can collect include temperatures, sound frequencies, light intensities, electrical currents, and pressure.

Uses of direct data

An example of a use of direct data could be planning the alteration of a bus route. A committee of residents on a new housing development, just outside a local village, wants a bus company to re-route the bus service from the local village to the town centre so that residents on the new development are able to get to the town centre more easily. It will, however, involve the bus route running through open countryside near the village. In order to persuade the bus company to change the bus route, the committee will need to collect some original data to present to them.

This data will include:

- How long it takes to walk from the new development to the existing bus routes.
- The number of passengers who use the existing route.
- The number of passengers who would use the new route.
- The effect the villagers think the changed route would have on their daily lives.

Here are some examples of how data could be collected.

- How long it takes to walk from the new development to the existing bus routes.

Original data could be collected by actually walking from various points in the new development and timing how long it would take. This might not be practical if several points on the new development have to be considered, given the time it would take to measure all the possible walking times. This could be considered to be a form of observation.

- The number of passengers who use the existing route.

The suggested method to be used is a data-logger. Sensors fitted around the door of each bus could be used to count the numbers of passengers boarding and getting off at each stop. From these it can be calculated how many passengers are on the bus at any point along its route. The data would be fed back to a data-logger or a tablet computer.

- The number of passengers who would use the new route.

In order to save time, questionnaires could be used. People living on the new development would be asked to complete the questionnaires. The completed questionnaires could then be transferred to a computer. Provided that the questionnaires were completed honestly, an accurate assessment of how many passengers would use the new route could be obtained.

- The effect the villagers think the changed route would have on their daily lives.

In order to ensure completely honest responses, face-to-face interviews would be best. The disadvantages of interviews are the length of time the process would take and the potential difficulties of transferring the responses into a format that a computer could deal with. However, because the interviewer can add follow-up questions, the answers would be more accurate.

Indirect data sources

Indirect data sources are third-party sources that the data gatherer can obtain data from. We will consider five such sources.

Weather data

In Chapter 2, we will see how supercomputers are used to help with forecasting the weather. In the UK, the meteorological office (Met Office) has many weather stations around the country. Many different weather variables are recorded using computerised weather stations. This data is collected by the Met Office for use in weather forecasting. Because these weather stations belong to the Met Office and the data is collected solely for the purpose of weather forecasting, it can be considered to be direct data. If, however, a construction company purchases the data, it becomes indirect data. This is because designing the construction of a building is not the purpose the original weather data was collected for.

Electoral register

The electoral register, also referred to as the electoral roll, is an example of an indirect data source. It is a list of adults who are entitled to vote in an election. Some countries have an 'open' version of the register which can be purchased and used for any purpose. Electoral registers are used in countries such as the USA, the UK, Australia, New Zealand, and many others. They contain information such as name, address, age and other personal details, although individuals are often allowed to remove some details from the open version. In many countries, the full register can only be used for limited purposes specified by the law in that country. The personal data in the register must always be processed in a way that complies with the country's data protection laws.

Businesses collecting data from third parties

Businesses collect a great deal of personal information from third parties, such as their customers, when they sell them a product. Whenever they buy something online, customers have to enter personal information or they have already done this on a previous visit to that business's website. It is often the case that they agree to the business sharing this with other organisations. Another development, in this regard, has been the emergence of data brokers. These are companies that collect and analyse some of an individual's most sensitive personal information and sell it to each other and to advertisers and other organisations without the individual's knowledge. They usually gather it by following people's internet, social media, and mobile phone activity.

Research from textbooks, journals and websites

Indirect sources also include research from textbooks, journals and websites. These are the main sources that students tend to use when they are researching information on a particular subject. The students have not collected the information themselves but are relying on information others have collected and analysed. Because they did not collect the information directly from the authors' sources, these are regarded as indirect sources.

Census

We will learn more about censuses in Chapter 2. A population census is usually carried out by a government to determine the number of people in a country and information is collected about them. Everybody has to provide data to the government, usually in the form of a questionnaire. When the government uses the data to plan and run public services, it could be considered to be a direct source. However, the main uses of census data are for others who did not collect the information directly, such as businesses, voluntary organisations and academics. The census also provides a lot of information for genealogists, historians and family tree enthusiasts. Without the census, the lives and lifestyles of our ancestors would remain undocumented, making historical research much more difficult. When used by organisations and people other than the government, a census is considered to be an indirect source.

Uses of indirect data

A construction company can use weather data to design buildings capable of withstanding climate change. It is vital that buildings can withstand the effect of high winds and temperatures. Apart from elections and other government purposes, the electoral register can only be used to select individuals for jury service or by credit reference agencies. These agencies are allowed to buy the full register to help them check the names and addresses of people applying for credit. They are also allowed to use the register to carry out identity checks in an attempt to deal with money laundering. It is a criminal offence for anyone to supply or use the register for anything else. The open register, however, can have various uses; businesses can use it to perform identity checks on their customers, charities often use it for fundraising purposes, debt-collection agencies use it to track down people who owe money. Whenever the address of an individual is required, a business could use the open register to check it.

Businesses which collect personal information often use it to create mailing lists that they then sell to other organisations, which are then able to send emails or even brochures through the post.

Apart from the uses of census data to create family trees, there are many other uses of such data. Police forces can use census statistics to know where to concentrate their crime prevention efforts. For example, they can see the areas where there are lots of people over 65 years old. Older residents often experience a high number of burglars tricking homeowners into letting them in by claiming they are from utility companies. Police officers giving crime prevention advice can concentrate their efforts in these areas. Census statistics are also essential to local voluntary organisations to understand the local communities they are working in. They provide data such as age, race, gender, ethnicity, languages spoken and household structures.

These examples are not the only type of indirect data source. Any organisation that provides data or information to the general public for use by them can be said to be an indirect source. In the bus route example described above, an indirect source could be used to provide some of the required information. For instance, the timetable of the current bus service could be used by the committee to work out the number of passengers using the route by seeing how many times the bus runs during the day. However, this would not be very accurate, as the buses may not carry a full load of passengers each time and this is clearly not the purpose for which the data was intended.

Another scenario could be studying pollution in rivers. Direct data sources could be used, of course; questionnaires could be handed out to local landowners and residents in houses near to the river, asking about the effects on them of the pollution, and they could also be interviewed. Computers with sensors could be used to collect data from the river. However, indirect data sources could also be used; documents may have been published by government departments showing pollution data for the area and there may be environmental campaigners who have also published data related to pollution in the area.

1.1.3 Advantages and disadvantages of direct and indirect data

Here is a table showing the advantages and disadvantages of direct data when compared to indirect data. Notice how each paragraph contains comparisons:

▼ Table 1.1 Advantages and disadvantages of direct and indirect data

Advantages of direct data	Disadvantages of direct data
We know how reliable direct data is since we know where it originated. Where data is required from a whole group of people, we can ensure that a representative cross-section of that group is sampled. With indirect data sources we may not know where the data originated and it could be that the source is only a small section of that group, rather than a cross-section of the whole group. This is often referred to as sampling bias.	Because of time and cash restraints, the sample or group size may be small whereas indirect data sources tend to provide larger sets of data that would use up less time and money than using direct data collection with a larger sample size. The person collecting the data may not be able to gain physical access to particular groups of people (perhaps for geographical reasons), whereas the use of indirect data sources allows data from such groups to be gathered. In addition, using a direct data source could be problematic if the people being interviewed are not available, thus reducing sample size, whereas using indirect data sources allows the sample size to be greater, resulting in increased confidence in the results produced.
The person collecting the data can use methods to gather specific data even if the required data is obscure, whereas with indirect data sources this type of data may never have been collected before.	It may not be possible to gather original data due to the time of year; for example summer rainfall data may be needed but at the time of the data-gathering, it is winter. With indirect data, historical weather data is available irrespective of the time of year.
The data collector or gatherer only needs to collect as much or as little data as necessary compared to indirect data sources, where the original purpose for which data was collected may be quite different to the purpose for which it is needed now. Irrelevant data may need to be removed.	To gather data from a specific sample would take a lot longer than it would with indirect data. In addition, by the time all the required data has been collected it may possibly be out of date so an indirect data source could have been used. Indirect data may be of a higher quality as it might have already been collated and grouped into meaningful categories whereas with direct data sources, questionnaire answers can sometimes be difficult to read and the transcripts of interviews take time to read in order to create the data source.
Once the data has been collected it may be useful to other organisations and there may be opportunities to sell the data to them, reducing the expense of collection. With indirect data this opportunity will probably not arise as organisations can go direct to the source themselves.	Compared to indirect data sources, the collection of data may be more expensive than using an indirect data source as people may have to be paid to collect it. Extra cost may be incurred as special equipment has to be bought, such as data-loggers and computers with sensors, or purchasing the paper for questionnaires, whereas this would not be needed using an indirect source. There are, however, still costs involved when using indirect data sources, such as the travelling expenses and time taken to go to the source, which can be fairly expensive but not as expensive as using direct data sources.

> **Activity 1b**
> 1 Explain why observation is considered to be a direct data source.
> 2 Give **two** differences between indirect data and direct data.

1.2 Quality of information

Measuring the **quality of information** is sometimes based on the value which the user places on the information collected. As such it could be argued that the judgement regarding the quality of information is fairly subjective, that is, it depends on the user and such judgements can vary between users. However, many experts do suggest that these judgements can be objective if based on factors which are believed to affect the quality of information.

Poor quality data can lead to serious consequences. Poor data may give a distorted view of business dealings, which can then lead to a business making poor decisions. Customers can be put off dealing with businesses that give poor service due to inaccurate data, causing the business to get a poor reputation. With poor quality data it can be difficult for companies to have accurate knowledge of their current performance and sales trends, which makes it hard for them to identify worthwhile future opportunities.

One example can be seen in the data provided by a hospital in the UK, which resulted in it being temporarily closed down, until it was realised that the death rate data provided had been incorrect and it was actually significantly lower. Meanwhile, in the USA, incorrectly addressed mail costs the postal service a substantial amount of money and time to process correctly.

Some of the factors that affect the quality of information are described here.

1.2.1 Factors that affect the quality of information

Accuracy

As far as possible, information should be free from errors and mistakes. The **accuracy of information** often depends on the accuracy of the collected data before it is processed. If the original data is inaccurate then the resulting information will also be inaccurate. In order to make sure the information is accurate, a lot of time needs to be given to the collection and checking of the data. Mistakes can easily occur. Consider a simple stock check. If it is carried out manually, a quantity of 62 could easily be copied down as 26 if the digits were accidentally transposed. This information is now inaccurate. More careful checking of the data might have prevented this.

We will look at methods of checking the accuracy of data, such as **verification** and **validation**, later in this chapter. It is easy to see how errors might occur during the data collection process. When using a direct data source, if we have not made the questions clear then the people answering the questionnaires or being interviewed may not understand them. We need to make sure that questions are clearly phrased and are unambiguous, otherwise they might lead interviewees into providing the answers that they think the interviewer is expecting. This can lead to the same response being given by everyone, even though the question is open-ended. If the questions are too open-ended, it could be difficult to quantify the responses. It is often a good idea to include multiple-choice questions where the respondent chooses an answer from those provided. These can be quantified quite easily. It is important, however, to include a sufficient number of alternative answers.

Other reasons why the information derived from a study might be inaccurate are that the sample chosen is not representative of the whole group or that the data collector makes some errors when collecting or when entering the data into a computer. If sensors are being used, these must be calibrated before use and must be properly connected to the computer. In addition, the computer system needs to be set up correctly so that the readings are interpreted correctly.

Relevance

When judging the quality of information, we need to consider the data that is being collected. Relevance is an important factor because there has to be a good reason why that particular set of data is being collected. Data captured should be relevant to the purposes for which it is to be used. It must meet the requirements of the user. The question needs to be asked is the data really needed or is it being collected just for the sake of it. The relevance of data matters because the collection of irrelevant data will entail a waste of time as well as money.

There are a number of ways in which the data may or may not be relevant to the user's needs. It could be too detailed or concentrate too much on one aspect. On the other hand, it might be too general, covering more aspects of the task than is necessary. It may relate to geographical areas that are not really part of the study. Where the study is meant to be about pollution in a local area, for example, data from other parts of the country would not be relevant. When looking for relevant information, it is important to be clear about what the information needs are for each specific search.

It is also necessary to be clear about the search strategy: what the user wants and does not want to find and therefore what the user needs to look for. In an academic study, it is important to select academic sources. Business sources or sources which appear to have a vested interest should be ignored. Having selected the sources, it is important to select the relevant information within them. Consider a school situation. You need to study a tremendous amount of information to prepare for your exams. How would you feel if your teachers chose to spend several lessons talking about aspects of the subject that they found really interesting? You may find that it was very interesting, but it probably would not be very relevant to what you need to pass your course.

Age

How old information is can affect its quality. As well as being accurate and relevant, information needs to be up-to-date. Most information tends to change over time and inaccurate results can arise from information which has not been updated regularly. This could apply, for example, to personal information in a **database** being left unchanged. Someone could get married and have a baby. If the original information was used, which had the person as single with no dependants, this would produce inaccurate results if the person was applying for a loan. This is because people who are married with children tend to be viewed as being more responsible and more likely to keep up with repayments. This inaccurate information would also affect a retailer's targeted advertising if it wanted to sell baby products to such customers, as the person would not appear on its list of targets. The age of information is important, because information that is not up-to-date can lead to people making the wrong decisions. In turn, that costs organisations time, money, and therefore, profits.

Level of detail

For information to be useful, it needs to have the right amount of detail. Sometimes, it is possible for the information to have too much detail, making it difficult to extract the information you really want. Information should be in a form that is short enough to allow for its examination and use. There should be no extraneous information. For example, it is usual to summarise statistical data and produce this information either in the form of a **table** or using a chart. Most people would consider a chart to be more concise than data in tables, as there is little or no unnecessary information in a chart. A balance has to be struck between the level of detail and conciseness. Suppose a car company director wants to see a summary of the sales figures of all car models for the last year. The information with the correct level of detail would be a graph showing the overall figures for each month. If the director was given figures showing the sales of each model every day of the previous 12 months in the form of a large report, this would be seen as the wrong level of detail because it is not a summary. It is important to understand what the user needs when they ask you for specific information.

On the other hand, the information might not have enough detail, meaning that you do not get an overall view of the problem. This links closely to the issue of the completeness of the information, which we will look at next.

Completeness of the information

In order for information to be of high quality it needs to be complete. To be complete, information must deal with all the relevant parts of a problem. If it is incomplete, there will be gaps in the information and it will be very difficult to use to solve a particular problem or choose a certain course of action. Discovering and collecting the extra information in order to remove these gaps may result in improving the quality of the information, but can prove to be time-consuming. Therefore, if the information is not complete, a decision has to be made: either that it is complete enough to make a decision about a problem or that additional data needs to be collected to complete the information. Consider the car company director mentioned above who wants to see a summary of the sales figures for the last year. If the director was given figures showing the sales for the first six months, this would be incomplete. If the director was shown the figures for only the best-selling models, this would be incomplete. It is important to understand what the user needs when they ask you for specific information. To sum up, completeness is as necessary as accuracy when inputting data into a database.

> **Activity 1c**
> 1 List **two** factors that affect the quality of information.
> 2 Briefly describe what is meant by the quality of information.

1.3 Encryption

1.3.1 The need for encryption

Whenever you send personal information across the internet, whether it is credit card information or personal details, there is a risk that it can be intercepted. Once it is intercepted the information can be changed or used for purposes such as **identity theft**, **cyber-fraud**, or ransomed off. If it is information regarding a company's secrets, it could be sold by **hackers** to rival companies. If, however, the information is intercepted but it is unreadable or cannot be understood, it becomes useless to the hacker or interceptor. Too many companies or individuals become victims of hackers taking advantage of readily available usernames and **passwords**. No matter how vigilant we are regarding the security of our computer systems, hackers will always find a way of getting into them, but if they cannot decipher the information, it will mean the act of hacking is not worthwhile. This is where encryption comes in. Encryption keeps much of our personal data private and secure, often without us realising it. It prevents hackers from reading and understanding our personal communications and protects us when we bank and shop. Data is scrambled or jumbled up so that it is completely unreadable. This prevents hackers understanding the data, as all they see is a random selection of letters, numbers and symbols.

Encryption is a way of scrambling data so that only authorised people can understand the information. It is the process of converting information into a code which is impossible to understand. This process is used whether the data is being transmitted across the internet or is just being stored. It does not prevent cyber criminals intercepting sensitive information, but it does prevent them from understanding it. Technically, it is the process of converting **plaintext** to **ciphertext**.

It is not just personal computers that are affected; businesses and commercial organisations are also liable to be affected by hacking activities. Employing data encryption is a safe way for companies to protect their confidential information and their reputation with their clients, since the benefits of encryption do not just apply to the use of the internet.

Information should also be encrypted on computers, hard-disk drives, pen drives and portable devices, whether they be laptops, tablets, or smartphones. The misuse of the data on these devices will be prevented, should the device be hacked, lost, or stolen.

1.3.2 Methods of encryption

Encryption is the name given to converting data into a code by scrambling it, with the resulting symbols appearing to be all jumbled up. The **algorithms** (we will be looking at the topic of algorithms in much more detail in Chapter 4) which are used to convert the data are so complex that even the most dedicated hacker with plenty of time to spare and hacking software to help them would be extremely unlikely to discover the meaning of the data. Encrypted data is often called ciphertext, whereas data before it is encrypted is called plaintext.

The way that encryption works is that the computer sending the message uses an **encryption key** to encode the data. The receiving computer has a corresponding **decryption key** that can translate it back again. The process of decryption here is basically reversing the encryption. A key is just a collection of bits, often randomly generated by a computer. The greater the length of the key, the more effective the encryption. Many systems use 128-bit keys which gives 2^{128} different combinations. It has been estimated that it would take a really powerful computer 10^{18} (1 000 000 000 000 000 000 [one quintillion]) years to go through all the different combinations. Modern encryption uses 256-bit keys which would take very much longer to crack. As you can imagine, this makes this form of encryption virtually impossible to crack. The key is used in conjunction with an algorithm to create the ciphertext.

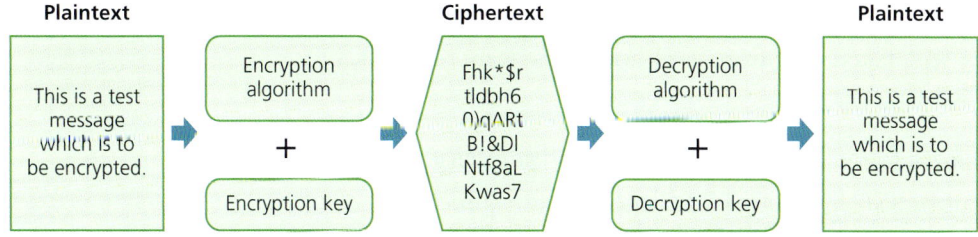

▲ **Figure 1.1** Encryption

There are two main types of encryption. One is called **symmetric encryption** and the other is **asymmetric encryption**, which is also referred to as **public-key encryption**.

Symmetric encryption

Symmetric encryption, often referred to as 'secret key encryption', involves the sending computer, or user, and the receiving computer, or user, having the same key to encrypt and decrypt a message. Although symmetric encryption is a much faster process than asymmetric encryption, there is the problem of the originator of the message making sure the person receiving the message has the same **private key**. The originator has to send the encryption key to the recipient before they can decrypt the message. This, however, leads to security problems, since this key could be intercepted by anybody and used to decrypt the message. Many companies overcome this problem by using asymmetric encryption to send the secret key but use symmetric encryption to encrypt data. So, with symmetric encryption both sender and recipient have the same secret, private, encryption key which scrambles the original data and unscrambles it back to an understandable format.

Asymmetric encryption

Asymmetric encryption, sometimes referred to as 'public-key encryption', uses two different keys, one public and one private. A **public key**, which is distributed among many users or computers, is used to encrypt the data. Essentially, this public key is published for anyone to use to encrypt messages. A private key, which is only available to the computers, or users, receiving the message, is used to decrypt the data. When a message is encrypted using the public key, it can be sent across a public channel such as the internet. This is not a problem as the public key cannot be used to decrypt a message that it was used to encrypt. It is incredibly complicated, if not impossible, to guess the private key using the public key and the encrypted message. Basically, any user who needs to send sensitive data over the internet securely, can do so by using the public key to encrypt the data, but the data can only be decrypted by the receiving computer if it has its own private key. Asymmetric encryption is often used to send emails and to digitally sign documents.

1.3.3 Encryption protocols

An encryption protocol is the set of rules setting out how the algorithms should be used to secure information. There are several encryption protocols. **IPsec (internet protocol security)** is one such protocol suite which allows the **authentication** of computers and encryption of packets of data in order to provide secure encrypted communication between two computers over an **internet protocol (IP)** network. It is often used in **VPNs (virtual private networks)**. **SSH (secure shell)** is another encryption protocol used to enable remote logging on to a computer network, securely. SSH is often used to login and perform operations on remote computers, but it can also be used for transferring data from one computer to another. The most popular protocol used when accessing web pages securely is **transport layer security (TLS)**. TLS is an improved version of the **secure socket layer (SSL)** protocol and has now, more or less, taken over from it, although the term SSL/TLS is still sometimes used to bracket the two protocols together.

The purpose of secure sockets layer (SSL)/transport layer security (TLS)

Because TLS is a development of SSL, the terms TLS and SSL are sometimes used interchangeably. We will use the term SSL/TLS in this book. The three main purposes of SSL/TLS are to:

» enable encryption in order to protect data
» make sure that the people/companies exchanging data are who they say they are (authentication)
» ensure the integrity of the data to make sure it has not been corrupted or altered.

Two other purposes are to:

» ensure that a website meets the **Payment Card Industry Data Security Standard (PCI DSS)** rules. The PCI DSS was set up so that company

websites could process bank card payments securely and to help reduce card fraud. This is achieved by setting standards for the storage, transmission and processing of bank card data that businesses deal with. Later versions of TLS are required to meet new standards which have been imposed
» improve customer trust. If customers know that a company is using the SSL/TLS protocol to protect its website, they are more inclined to do business with that company.

Many websites use SSL/TLS when encrypting data while it is being sent to and from them. This keeps attackers from accessing that data while it is being transferred. SSL/TLS should be used when storing or sending sensitive data over the internet, such as when completing tax returns, buying goods online, or renewing house and car insurance. Only going to websites which use SSL/TLS is good practice. The SSL/TLS protocol enables the creation of a secure connection between a web server and a browser. Data that is being transferred to the web server is protected from eavesdroppers (the name given to people who try to intercept internet communications).

The SSL/TLS protocol verifies the identity of the **server**. Any website with an **HTTPS** address uses SSL/TLS. In order to verify the identity of the server, the protocol makes use of **digital certificates**, which contain such information as the domain name that the certificate is issued for, which organisation, individual or device it was issued to, the **certificate authority (CA)** that issued it, the CA's digital signature, and the public key, as well as other items. Although SSL was replaced by TSL many years ago, these certificates are still referred to as SSL certificates today. As well as keeping the user's data secure, a website needs a digital certificate in order to verify ownership of the website and also to prevent fraudsters creating a fake version of the website. Valid SSL certificates can only be obtained from a CA. CAs can be private companies or even governments. Before allowing someone to have an SSL certificate, the CA will carry out a number of checks on an applicant and following that, it is the responsibility of the CA to make sure that the company or individual receives a unique certificate. Unfortunately, if hackers are able to break through a CA's security, they can start issuing bogus certificates to users and will then be in a strong position to crack the user's encryption.

The purpose of Internet Protocol Security (IPsec)

IPsec (Internet Protocol Security) is a suite of protocols used to secure data transmitted over a public network. It was added as an extension to the IP layer (explained in more detail in the A Level book). The IPsec protocols were developed in the mid-1990s to provide this security. Devices are authenticated to each other, thereby ensuring data integrity. IP network packets are encrypted so that the data remains confidential. In short, IPsec provides authentication, integrity and confidentiality of data.

IPsec commonly includes three protocols. These are the Authentication Header (AH) protocol, the Encapsulating Security Payload (ESP) protocol and the Internet Key Exchange (IKE) protocol. The AH protocol is used to authenticate data packets whereas the ESP protocol both authenticates and encrypts the data packets. The IKE generates security associations (SA) used to negotiate the encryption keys and algorithms which will be used in a session. The purpose of IPsec is to protect confidential data and to provide secure transfer of this data across a public network such as the internet. It is aimed at preventing interception of data by hackers as well as protecting the data being transmitted from being understood.

The use of SSL/TLS in client–server communication

Transport layer security (TLS) is used for applications that require data to be securely exchanged over a **client–server network**, such as web browsing sessions and file transfers. Just like IPsec it can enable VPN connections and Voice over IP (VoIP).

In order to open an SSL/TLS connection, a **client** needs to obtain the public key. For our purposes, we can consider the client to be a web user or a web browser and the server to be the website. The public key is found in the server's digital certificate. From this we can see that the SSL/TLS certificate proves that a client is communicating with the actual server that owns the domain, thereby proving the authenticity of the server.

When a browser (client) wants to access a website (server) that is secured by SSL/TLS, the client and the server must carry out an SSL/TLS handshake. A handshake, in IT terms, happens when two devices want to start communicating. One device sends a message to the other device telling it that it wants to set up a communications channel. The two devices then send several messages to each other so they can agree on the rules for communicating (a communications protocol). Handshaking occurs before the transfer of data can take place.

With an SSL/TLS handshake, the client sends a message to the server telling it what version of SSL/TLS it uses together with a list of the different ciphersuites (types of encryption) that the client can use. The list of ciphersuites has the client's preferred type at the top and its least favourite at the bottom. The server responds with a message which contains the ciphersuite it has chosen from the client's list. The server also shows the client its SSL certificate. The client then carries out a number of checks to make sure that the certificate was issued by a trusted CA and that it is in date and that the server is the legitimate owner of the public and private keys. The client now sends a random **string** of bits that is used by both the client and the server to calculate the private key. The string itself is encrypted using the server's public key. Authentication of the client is optional in the process. The client sends the server another message, encrypted using the secret key, telling the server that the client part of the handshake is complete. We will see in more detail in the section on HTTPS how any further transmitted data is encrypted.

The use of IPsec in client–server communication

IPsec is used for protecting confidential data transmitted across a network, for example, financial transactions, medical records and communications between, and within, businesses. The main use of IPsec, however, is in virtual private networks (VPNs). A VPN creates a secure connection between two computers over the public internet. It is almost as secure as a private internal network such as a LAN. It is used so that employees working from home on a client computer can access confidential files securely from the company's server as though they were working in the company's offices.

1.3.4 Uses of encryption

Data protection

There are many reasons to encrypt data. Companies often store **confidential data** about their employees, which could include medical records, **payroll** data, as well as personal data. These need to be encrypted to prevent them becoming public knowledge. An employee in a shared office may not want others to have access to their work which may be stored on a hard disk, so it needs to be encrypted. A company's head office may wish to share sensitive business plans with other offices using the internet. If the data is encrypted, they do not have to worry about what would happen if it were intercepted. Company employees and individuals may need to take their laptops or other portable devices with them when they travel for work or pleasure. If the device contains sensitive information which is not encrypted, it is possible that the information could be retrieved by a third party if the device is left unattended.

As recently as 2018, it was reported that over the previous four years, staff in five UK government departments had lost more than 600 laptops, mobile phones and memory sticks. Fortunately, the data had been encrypted and was therefore protected from prying eyes. Unfortunately, there have been occasions where laptops have been left on trains and the data was unencrypted, causing great embarrassment to the government when this was discovered.

There are other situations where encryption should be used. One example is when individuals are emailing each other with information they would want to remain confidential. They need to prevent anybody else from reading and understanding their mail. People use websites for online shopping and online banking. When doing so, the debit/credit card and other bank account details should be encrypted to prevent fraudulent activity taking place.

Let us now consider some specific uses of encryption.

Systems encryption

Systems encryption can be considered to be encryption of the device being used or the file system being used. Device encryption is when the whole device is encrypted such as a laptop. If a laptop is stolen after it has been encrypted the data on it will still be secure. The files can only be accessed if the device is turned on and the owner has logged in to it. Otherwise the encrypted data cannot be understood by anyone without the password or recovery key. Filesystem-level encryption or file-based encryption (FBE), is when individual files or directories are encrypted by the file system itself. This is different to full disk encryption which involves encrypting the whole disk. A description of which follows below. FBE can allow different files to be encrypted using different keys, thereby making it even more difficult for hackers to unencrypt all the data in a system. It does not encrypt metadata such as the time the file was modified or its size.

Hard-disk encryption

The principle of hard-disk encryption is fairly straightforward. When a file is written to the disk, it is automatically encrypted by specialised software. When a file is read from the disk, the software automatically decrypts it while leaving all other data on the disk encrypted. The encryption and decryption processes are understood by the most frequently used application software such as spreadsheets, databases and word processors. The whole disk is encrypted, including data files, the OS and any other software on the disk. Full (or whole) disk encryption is your protection should the disk be stolen, or just left unattended. So, even if the disk is still in the original computer, or removed and put into another computer, the disk remains encrypted and only the keyholder can make use of its contents.

Another benefit of full disk encryption is that it automatically encrypts the data as soon as it is saved to the hard disk. You do not have to do anything, unlike the encryption of files and folders, where you have to individually encrypt them as you go.

There are, however, drawbacks to encrypting the whole disk. If an encrypted disk crashes or the OS becomes corrupted, you can lose all your data permanently or, at the very least, disk data recovery becomes problematic. It is also important to store encryption keys in a safe place, because as soon as a disk is fully encrypted, no one can make use of any of the data or software without the key. Another drawback can be that booting up the computer can be a slower process.

Email encryption

When sending emails, it is good practice to encrypt messages so that their content cannot be read by anyone other than the person they are being sent to. Many people might think that having a password to login to their **email account** is sufficient protection. Unfortunately, emails tend to be susceptible to interception and, if they are not encrypted, sensitive information can become readily available to hackers. In the early days of email communication, most messages were sent as plaintext, which meant hackers could easily read their content. Fortunately, most email providers now provide encryption by default. There are three parts to email encryption.

1. The first is to encrypt the actual connection from the email provider, because this prevents hackers intercepting and acquiring login details and reading any messages sent (or received) as they leave (or arrive at) the email provider's server.
2. Then, messages should be encrypted before sending them so that even if a hacker intercepts the message, they will not be able to understand it. They could still delete it on interception, but this is unlikely.
3. Finally, since hackers could bypass your computer's security settings, it is important to encrypt all your saved or archived messages.

Asymmetric encryption is the preferred method of email encryption. The email sender uses the public key to encrypt the message while the recipient uses the private key to decrypt it. It is considered good practice to encrypt all email messages. If only the ones that are considered to be important are encrypted, the hacker will know which emails contain sensitive data and will therefore spend more time and energy trying to decode those particular ones. Encryption only scrambles the message contents, not the sender's email address, making it very difficult to send messages anonymously.

Most types of email encryption require the sending of some form of digital certificates to prove authenticity. The management of digital certificates, though time-consuming, is crucial, as users would not want them to fall into the hands of hackers.

Encryption in HTTPS websites

HTTP (Hypertext Transfer Protocol) is the basic protocol used by web browsers and web servers. Unfortunately, it is not encrypted and so can cause internet traffic to be intercepted, read and understood. Hackers could intercept any private information including bank details and then use these to commit fraud. HTTPS (Hypertext Transfer Protocol Secure), however, enables users to browse the world wide web securely. To do this, it uses the HTTP protocol but with SSL/TSL encryption overlaid. HTTPS websites have a digital certificate issued by a trusted CA, which means that users know the website is who it says it is; as we mentioned in the section on SSL/TLS, it proves it is authentic. It again ensures the integrity of the data by showing that web pages have not been changed by a hacker while being transferred and by encrypting any information transferred from the client to the server, and vice versa. As a result of the use of HTTPS, users can securely transmit confidential information such as credit card numbers, social security numbers and login credentials over the internet. If they used an ordinary HTTP website, data is sent as plain text, which could easily be intercepted by a hacker or fraudster.

Indicators that you are using a secure site are the inclusion of the HTTPS:// prefix as the starting part of the URL. There should also be a **padlock** icon next to the URL. Depending on your browser and the type of certificate the website has installed, the padlock may be green. The way HTTPS works is that the web browser on the client computer performs a handshake with the web server, as described earlier in the section on SSL/TLS. Then, what is sometimes called a session key is created randomly by the web browser and is encrypted using the public key, then sent to the server. The server then decrypts the session key using its private key. All data sent between the two from then on is encrypted using this session key. So, the generation of the session key is done through asymmetric encryption, but symmetric encryption is used to encrypt all further communications. Asymmetric encryption requires a lot of processing power and time, so HTTPS uses a combination of asymmetric and symmetric encryption. Once the session is finished, each client and the server discard the symmetric key used for that session, so each separate session requires a new session key to be created.

One of the benefits of HTTPS is security of data, with information remaining confidential because only the client browser and the server can decrypt it. Another benefit of HTTPS is that search engine results tend to rank HTTPS sites higher than HTTP sites. However, the time required to load an HTTPS website tends to be greater. Websites have to ensure that their SSL certificate has not expired and this creates extra work for the host as it has to keep on top of certificate management.

1.3.5 Advantages and disadvantages of different protocols and methods of encryption

There are a number of advantages of using encryption. As was said earlier, if personal information is sent across the internet, whether it is credit card information or personal details, once it is encrypted the information can no longer be changed or used for purposes such as identity theft or cyber-fraud or ransomed off. Company secrets would not be able to be sold by hackers to rival companies. One drawback with encryption, however, is that it takes more time to load encrypted data as well as requiring additional processing power. When browsing, the client and server must send messages to each other several times before any data is transmitted. This increases the time it takes to load a webpage by several milliseconds. It also uses up valuable memory for both the client and the server. Encryption involves the use of keys and while a larger key size means more effective encryption, it also increases the computational power required to perform the encryption. Encryption is meant to protect data, but it can also be the source of great inconvenience to a computer user. **Ransomware** can be used against individual computer users; hackers can encrypt computers and servers and then demand a ransom. If the ransom is paid, they provide a key to decrypt the encrypted data. Another problem with encryption is that if the private key is lost, it is extremely difficult to recover the data and, in certain circumstances, the data may well be lost permanently. It is possible for the data to be recovered by the reissuing of the digital certificate, but this can take time and, in the meantime, if a hacker has managed to get hold of the key, they will have full access to the encrypted data. In addition, users can get careless and forget that decrypted data should not be left in the decrypted state for too long, as it then becomes susceptible to further attack from hackers.

Regarding the different forms of encryption, asymmetric is much slower compared to symmetric due to its mathematical complexity, and it is therefore not suitable for computing vast amounts of data. It also requires greater computational power.

While it is difficult to give the advantages and disadvantages of individual protocols since they all do a similar job, it is possible to compare the advantages and disadvantages of SSL/TLS with IPsec for setting up a VPN, for example. With SSL/TLS, digital certificates are only essential with the server (client digital certificates are optional), whereas with IPsec both client and server have to be authenticated, which makes it more difficult to manage an IPsec system. However, as user authentication is optional, this means that security is weakened with SSL/TLS compared to IPsec.

Extra software has to be downloaded when using SSL/TLS if non-web-based applications are used, which may be a problem if a **firewall** prevents or slows down access to these downloads. VPN tunnels using SSL/TLS are not supported by certain operating systems which do support IPsec. In conclusion, the time-consuming management of digital certificates is less of a problem with SSL compared to IPsec, which could lead to saving of money. Another cost implication is that, unlike most uses of IPsec, you do not need to buy client software and the process of setting up and managing such a system tends to be easier.

> ### Activity 1d
> 1 Briefly describe what is meant by symmetric encryption.
> 2 Write a sentence about each of the **three** uses of encryption.
> 3 Explain the difference between HTTP and HTTPS.

1.4 Checking the accuracy of data

1.4.1 Validation and verification

It is vital that data is input accurately to ensure that the processing of that data produces accurate results. When using a computer system, data entry is probably the most time-consuming part of data processing. Consequently, it is important to try and ensure that the number of errors which occur when entering data directly or transferring from another medium is very small, otherwise more time will need to be spent correcting the data or even re-entering it. To try and ensure that data entry is accurate, we use two methods called validation and verification. Neither method ensures that the data being entered is correct or is the value that the user intended to enter. Verification tries to ensure that the actual process of data entry is accurate, whereas validation ensures the data values entered are reasonable. It may well be that the original data when collected was incorrect; here we are just ensuring that no further errors occur during the transfer process.

Methods of validation

As stated above, data validation simply ensures that data is reasonable and sensible. In the early days of computerisation there were some horror stories of people getting utility bills for 1 million dollars! This was usually the result of poor checking being carried out and nobody noticing that the decimal point had been put in the wrong place. A validation check that ensured nobody could have a bill of more than $1000 would have stopped this. However, it would not prevent somebody getting an incorrect bill of $321 instead of $231, as the amount being charged would still be regarded as sensible or reasonable. To emphasise then, validation ensures data is reasonable or sensible but not necessarily correct.

In order to ensure that data input to a system is valid, it is essential to incorporate as many validation routines as possible. The number and methods of validation routines or checks will obviously depend on the form or type of input to the system. Not every **field** can have a validation check. For example, there are so many variations of an individual's name that this would be very difficult to validate. Some have letters that might not be recognised by a computer using an alphabet based on the English language and some contain punctuation marks such as apostrophes and hyphens.

In order to illustrate the types of validation, let us consider a school library database which has one table for books and another table for upper-school students/borrowers. Let us look at an extract showing some of the typical books and borrowers from the complete database. We shall assume that these **records** are representative of the whole database.

▼ Table 1.2 Books database table

ISBN	Title	Author	Published	Cost
9781474606189	The Labyrinth of the Spirits	Carlos Ruiz Zafón	2016	$10
9780751572858	Lethal White	Robert Galbraith	2018	$18
9781780899329	18th Abduction	James Patterson	2019	$29
9781408711095	The Colours of All the Cattle	Alexander McCall Smith	1997	$26

▼ Table 1.3 Years 10 and 11 upper-school borrowers database table

Borrower_ID	Name	Date_of_birth	Class	Book_borrowed
0205	Chew Ming	21/12/04	11D	9781474606189
1016	Gurvinder Sidhu	19/11/05	10A	9781408711095
0628	Gary Goh	18/04/06	10C	9781474606189
1014	Jasmine Reeves	13/02/05	11A	9780751572858

Presence check

When important data has to be included in a database and must not be left out, we use a **presence check** to make sure data has been entered in certain fields. A common mistake made by many students when asked which validation check should be used on a field in a database, is to always say 'presence check'. It is the easiest to remember but is rarely used on any field except the key field. With many of the fields illustrated above, it is not necessary to use a presence check. The data can be updated or entered at a later date. We would, however, probably use a presence check on the ISBN field in the Books table and on the Borrower_ID field in the Borrowers table. These are key fields and, if data were not entered, it would be very difficult to identify unique records if that data were missing. You may well have come across this when completing online data-capture forms. Fields sometimes have a red asterisk next to them, along with a message instructing users that fields marked with an asterisk must be completed. Presence checks are frequently used to check that you have entered data into these fields and, if you have not, then there is usually a warning message saying you must enter the data before you can move on to the next page. While the presence check prevents you from missing out a field or record, it is otherwise a fairly inefficient method, as it does not prevent you from entering incorrect or unreasonable data.

Range check

Fields which contain numeric data tend to have **range checks** designed for them. Using the extract from the Books table above, we can see that the maximum cost of a book is $29 and the minimum cost is $10. We can make sure that the person entering the data does not enter a cost less than $10 or more than $29 by using a range check. A range check always has an upper value and a lower value which form the range of acceptable values. If a value is entered that is outside this range, an error message is produced. We will deal with the limit validation check later in this section. Suffice to say at this point that a **limit check** only has one boundary, which is either upper or lower.

So, to sum up, a range check checks that the data is within a specified range of values. In our example, the cost of a book should lie between 10 and 29. A range check could be carried out on each part of the Date_of_birth field in the Borrowers table to ensure that 32 or 0 was not entered in the day part or that 13 or 0 was not entered for the month part. However, this would not prevent impossible dates such as 31/2/05 being entered (February can only have day values of 29 or less).

Type check

A **type check** ensures that data is of a particular data type. In the above example, Borrower_ID in the Borrowers table could be set up so that a validation rule would ensure that every character was **numeric**. Most students think that you only have to set the field data type to numeric to prevent text from being entered. While this is true, it would not be sufficient with the Borrower_ID field, as any leading zeros, such as with 0205, would be removed in such a field. This would also not be acceptable in most parts of the world in fields containing telephone numbers. The data type of Borrower_ID would have to be set to **alphanumeric** and a validation routine would need to be created to only allow the digits 0 to 9 to be entered. Fields in some databases would contain letters of the alphabet only. Setting the field type to alphanumeric does not prevent numbers from being entered. A separate validation routine would need to be set up. A type check can be performed on most fields to make sure that no invalid characters are entered. For that reason, it is often referred to as an **invalid character check**, or by its shortened name, **character check**. One of the shortcomings of this check is that it would not alert you if you had not typed in the correct number of characters in a particular field.

Length check

A **length check** is performed on alphanumeric fields to ensure they have the correct number of characters, but it is not used with numeric fields. Generally, it is carried out on fields that have a fixed number of characters, for example the length of a telephone number in France tends to be 10 digits. As the leading digit is 0, the phone numbers would be stored in the alphanumeric data type, so it is fairly straightforward to apply a length check.

Again, the application of this type of validation check is not to be confused with setting up the data structure of a file so that the field length is a fixed number of characters, as this does not prevent a user from typing in a string of less than that fixed number of characters. In this instance no error message would be produced, but with a length check, if you did not type in a correct number of characters, an error message would result.

In our database, we would apply a length check to the ISBN field in the Books table. All ISBNs in our table are 13 characters long, so we need to have a check to make sure we are not typing in fewer or more than 13 characters. We could also apply this type of check to the Borrower_ID field in the Borrowers table as these all appear to be four characters in length. This would not give us full validation on the field, however, as letters of the alphabet, if entered, would not be flagged up as an error, as the check only counts the number of characters.

It is also worth noting that length checks can be set to a range of lengths. For example, phone numbers in Ireland tend to vary between 8 and 10 characters in length. In that instance, you would need a routine that would not allow you to type in fewer than eight characters or more than ten characters. This is not to be confused with the range check described above.

Format check

New vehicle registration or licence plates in the UK follow an identical pattern. The format is two letters followed by two digits followed by a space and then three letters, for example, XX21 YYY. Any new registration number when entered into a database can therefore be validated using a **format check** or **picture check**. If a combination of characters is entered which does not follow this pattern, it should produce an error message.

In a format check, you can specify any specific combination of characters that must be followed. In our example database, we could use a format or picture check on the Class field in the Borrowers table. We could set it so that it must be two digits followed by one letter. If somebody did not realise that it was the Upper-school borrowers table and typed in 9B, for example, this would cause an error message to be output saying that the field must have two numbers followed by a letter. While a format check is very useful in this scenario, it would not prevent somebody mistyping an entry, and entering 19A for example, by mistake, as this would be accepted by the system. This form of validation is very useful for checking dates if they are in a specific format, such as Date_of_birth in the Borrowers table, which consists of two digits followed by a slash, followed by two digits followed by a slash, followed by two digits.

Many students mistakenly think that you can put a format check on a **currency** field. This is not the case. The data is stored as a number and the currency symbol is added by the software. The user is only entering numbers not the currency symbol.

Check digit

For our purposes we will consider the use of a **check digit** as a means of validating data as it is being input (some people would argue that it can be used as a verification check in certain other circumstances, but that is outside the scope of this book). It is used on numerical data, which is often stored as a string of alphanumeric data-type. For example, the last digit of an ISBN for a book is a check digit calculated using simple arithmetic. There are a number of ways of calculating the check digit but one of the most frequently used methods is described here. Using the first 12 digits in the ISBN, each individual digit in the string is multiplied by 1 if it is in an odd-numbered position (such as 1st, 3rd and so on) or 3 if it is in an even-numbered position (such as 2nd, 4th and so on). The resulting numbers are added together and divided by 10. If the remainder is 0 that becomes the check digit, otherwise the remainder is subtracted from 10 and that becomes the check digit. This is then added to the end of the string, in this case as the 13th digit of the ISBN.

This happens at a stage before the data is entered, for example when the ISBN is allocated before a book is published. When this data comes to be entered into a database in a library, the computer recalculates the check digit to check whether it gives the same check digit. If it does not, then an error message is produced. This usually happens when the person entering the data has transposed two digits. In our example if we typed in 9781447606189 instead of 9781474606189, the check digit would be recalculated from the first 12 digits and would produce the check digit 5 instead of 9. This would produce an error message.

Lookup check

A lookup check compares the data that has been entered with a limited number of valid entries. If it matches one of these then it is allowed, but if it does not then an error message is produced. It can only be used efficiently if there are a limited number of lookup values, such as the days of the week, where there are only seven. A lookup check is not to be confused with setting up a lookup list for the user to select data items from when entering data. A lookup list does not produce error messages and in certain circumstances users can still enter data which overwrites the given list. In our database, we could use the Class field in the Borrowers table, which would probably only have the classes 10A, 10B, 10C, 10D, 11A, 11B, 11C, 11D to choose from. If, using the example above, somebody entered 9B, the computer would compare this to the values stored in a separate table and would not be able to find a match and would produce an error message.

Consistency check

This is sometimes called an integrity check, but for our purposes, so that we do not get confused with referential integrity (which we will meet later in the book), we will refer to it as a **consistency check**. It checks that data across two fields is consistent. A good example would be to ensure data consistency between the Class field in the Borrowers table and the Date_of_birth field. We will assume that each student is allocated to a class according to their age when they join the school. We will assume that, in our example, students in year 11 were born between 1 September 2004 and 31 August 2005. A consistency check can be applied here so that if the first two digits of the class are 11 then the student's Date_of_birth must be between 01/09/04 and 31/08/05. If this is not the case, then this validation check would output an error message. It is often applied so that a field that contains a person's age must be consistent with a field that contains their date of birth, though it should be stressed that storing the age of a person is considered to be bad practice as it changes regularly and needs updating often.

Limit check

We have already considered the range check; now we will look at a limit check. A limit check is similar to a range check, but the check is only applied to one boundary. For example, in the UK you are only allowed to drive from the age of 17, but there is no upper limit. If somebody enters a number lower than 17 when asked to enter their age when applying for a driving licence, for example, this will generate an error message. In our database it is difficult to apply a limit check given the data provided.

1.4.2 The need for both validation and verification

The two methods of checking the accuracy of data are complementary. We have seen that verification can report on errors that validation cannot and, similarly, validation will pick up errors that verification cannot. Both are needed to ensure that data is sensible and reasonable in the first place and also transferred accurately.

1.4.3 The difference between validation and verification

As we saw earlier both these methods are essential to ensuring the entry of data is accurate. It is important to emphasise that we are checking that the data was entered accurately; we are not checking that the data itself is accurate. There are a number of differences between verification and validation. Validation is always carried out by a computer whereas verification can be carried out by a computer or by a human. Validation is checking that the data is reasonable and sensible. Verification is checking that the data has been copied or entered correctly but cannot tell you whether the data is sensible or not. Similarly, validation does not help if you have copied the data incorrectly. If you type in FD236CS instead of DF236CS, a format check would accept this as valid input (as it is still two letters followed by three **numbers** followed by two letters) even though it has been copied incorrectly. Verification would alert the user to this error. Data may have been invalid when collected but verification only helps you to know that the data has been transferred accurately to another medium. It does not help if the original data is incorrect.

Consider an electricity company which employs meter readers to read customers' electricity meters. Suppose the meter reader accidentally writes down that, for one customer, the number of units used was 4866 instead of the actual reading, which was 4860. When the readings for all the customers have been collected, they are entered into the computer, including the incorrect reading of 4866. At this stage all verification would do is check that the number entered was the same as that in the source document, 4866, so incorrect data would pass the verification test. The validation check might be that readings must be between 2000 and 6000. Again, incorrect data would pass this test as well. This shows how important it is that the correct data is collected in the first place, since verification and validation might still allow the data to pass through the system undetected. Verification is a way of ensuring that the user does not make a mistake when inputting data whereas validation is checking that the data input conforms with what the system considers to be sensible and reasonable.

Verification

As already stated, verification simply ensures that data has either been entered accurately by a human or that it has been transferred accurately from one storage medium to another. There are a number of methods of verification, some related to manual entry of data and some related to data transfer.

Visual checking

Visual checking is carried out by the person who enters the data, who visually compares the data they have entered with that on the source document. They can see the differences and then correct the mistakes. This is the simplest form of verification and can be done by reading the data on the screen to make sure it is the same as the source document. An alternative method is to print out the data entered and compare the printout side by side with the source document. Visual checking can be rather time-consuming and possibly costly as a result. Another problem is that the person who is checking that the data has been entered correctly may be the same person who entered it. It is very easy for them to overlook their own mistakes. A possible way around this is to get somebody else to do the check.

Double data entry

Double data entry, as the name suggests, involves the entry of data twice. The first version is stored. The second entry is compared to the first by a computer, and the person entering the data is alerted by the computer to any differences. The user then checks to see if the second attempt is correct, corrects the error if necessary, and continues entering the data.

The alternative to this way of entering data twice is for two different people to enter the data, which is temporarily saved to the same hard disk. The computer compares the two versions on the disk and alerts both operators to any differences, which are then checked to see which version is correct. Some systems cause the keyboards to freeze so that the people entering the data cannot continue until the mistake is corrected.

Double data entry is similar to visual verification in that both methods compare two versions of data and check that data is copied accurately, not checking that the data collected in the first place was accurate or correct. The essential difference is that visual verification is carried out by the user, whereas with double data entry it is the computer that compares the two versions.

Parity check

As was mentioned in Section 1.1.1 the computer stores data in the form of bits. Each string of bits is called a byte, with a byte normally consisting of 8 bits. Each bit is either 1 or 0, for example 10001101. A byte represents a number between 0 and 255. Most computers use the **American Standard Code for Information Interchange (ASCII)**. It is a code which uses numbers to represent 96 English-language characters, with each character being given a number between 32 and 127. The first 32 codes (0–31) in ASCII are unprintable control codes and are used to control peripherals such as printers. For example, 0 represents the **null string**, 8 is equivalent to the backspace key on the keyboard and 13 is equivalent to the return key. The codes 32–127 represent letters of the alphabet, numbers, and other symbols such as $, %, &.

So, for example, the ASCII code for uppercase I is 73, which is 01001001 in binary.

If we wanted to represent the word BROWN, we would use the ASCII codes 66, 82, 79, 87 and 78, which would in turn be represented by the following bytes: 01000010 01010010 01001111 01010111 01001110

In the early days of computers, there were only 7 bits used to contain information with the extra bit used for a parity check. It soon became apparent that extra characters needed to be represented in the system, such as the Spanish ñ, the French è, the symbol ©. These form what is called **extended ASCII** and provide characters represented by the codes 128–255. This provided 128 additional characters. Now, the **parity bit** was added to make 9 bits. It is possible to use the seven bits of the byte to represent data and have a parity bit to make it 8 bits, but this would give us a limited set of characters to work with. In this section, we will only be considering 9-bit parity checking.

Most computers use ASCII codes to represent text, which makes it possible to transfer data from one computer to another. There has to be, however, a way to check that data has been transmitted accurately. This is called parity checking, which involves the use of parity bits. The parity bit is added to every byte (8 bits) that is transmitted. The parity bit is added to the end of the byte so that there are an even number of 1s. Some systems use an odd number of 1s but we will be looking at the most commonly used, which is even parity.

When data is being transmitted from one device to another, the sending device counts the number of 1s in each byte. If the number of 1s is even, it sets the parity bit to 0 and adds this on to the end of the byte. However, if the number of 1s is odd, it sets the parity bit to 1 and adds it on. The result is that every byte of transmitted data consists of an even number of 1s. When the other device receives the data, it checks each byte to make sure that it has an even number of 1s. If there is an odd number of 1s, then this means there has been an error during the transfer of data. So how does this all work?

Consider the example given above: the word BROWN

B – 01000010 there are two 1s (even) so we add 0; it now becomes 010000100 (two 1s [even])

R – 01010010 there are three 1s (odd) so we add 1; it now becomes 010100101 (four 1s [even])

O – 01001111 there are five 1s (odd) so we add 1; it now becomes 010011111 (six 1s [even])

W – 01010111 there are five 1s (odd) so we add 1; it now becomes 010101111 (six 1s [even])

N – 01001110 there are four 1s (even) so we add 0; it now becomes 010011100 (four 1s [even])

This is a very effective verification method. It makes sure that all bytes have an even number of 1s. If a 1 within the byte is transmitted as a 0, then the error will be trapped by the system. However, there are still errors which can go undetected. If two 1s within the byte get transmitted as 0s, the byte will still have an even number of 1s and so the system will not report an error. For example, if 010101111 (W with parity bit added) is transmitted as 010001101 (F with parity bit added) the parity check will not notice this, as there is still an even number of 1s. Also, if, somehow, a 1 and a 0 are transposed, such as 010000100 (B with a parity bit added) being transmitted as 010000010 (A with a parity bit added), again the parity check will not report an error as there is still an even number of 1s. More complex error-checking methods have had to be developed, but parity checking is still very common because it is such a simple method for detecting errors.

Checksum

Checksums are a follow-on from the use of parity checks in that they are used to check that data has been transmitted accurately from one device to another. A checksum is used for whole files of data, as opposed to a parity check which is performed byte by byte. They are used when data is transmitted, whether it be from one computer to another in a network, or across the internet, in an attempt to ensure that the file which has been received is exactly the same as the file which was sent.

A checksum can be calculated in many different ways, using different algorithms, for example a simple checksum could simply be the number of bytes in a file. Just as we saw with the problem with transposition of bits deceiving a parity check, this type of checksum would not be able to notice if two or more bytes were swapped; the data would be different, but the checksum would be the same. Sometimes, encryption algorithms are used to verify data; the checksum is calculated using an algorithm called a hash function (not to be confused with a **hash total**, which we will be looking at next) and is transmitted at the end of the file. The receiving device recalculates the checksum, and then compares it to the one it received, to make sure they are identical.

Two common checksum algorithms are MD5 and SHA-1 but both have been found to have weaknesses. It is possible for two different files to have the same calculated checksum, so because of this a newer SHA-2 and even newer SHA-3 have been developed which are much more reliable.

The actual checksum is produced in hexadecimal format. This is a counting system that is based on the number 16, whereas we typically count numbers based on 10. You can see what each hexadecimal value represents in this table.

▼ **Table 1.4** Hexadecimal values

Hexadecimal	0	1	2	3	4	5	6	7	8	9	A	B	C	D	E	F
Base 10	0	1	2	3	4	5	6	7	8	9	10	11	12	13	14	15

MD5 checksums consist of 32 hexadecimal characters, such as 591a23eacc5d55a528e22ec7b99705cc. These are added to the end of the file. After the file is transmitted, the checksum is recalculated by the receiving device and compared with the original checksum. If the checksum is different, then the file has probably been corrupted during transmission and must be sent again.

Hash total

This is similar to the previous two methods in that a calculation is performed using the data before it is sent, then it is recalculated, and if the data has transmitted successfully with no errors, the result of the calculation will be the same. However, this time the calculation takes a different form; a hash total is usually found by adding up all the numbers in a specific field or fields in a file. It is usually performed on data not normally used in calculations, such as an employee code number. After the data is transmitted, the hash total is recalculated and compared with the original value. If it has not been transmitted properly or data has been lost or corrupted, the totals will be different. Data will have to be sent again or the data will have to be visually checked to detect the error.

This type of check is normally performed on large files but, for demonstration purposes, we will just consider a simple example. Sometimes, school examinations secretaries are asked to do a statistical analysis of exam results. Here we have a small extract from the data that might have been collected.

▼ **Table 1.5** Sample data

Student ID	Number of exam passes
4762	6
0153	8
2539	7
4651	3

Normally, the Student ID would be stored as an alphanumeric type, so for the purpose of a hash check, it would be converted to a number. The hash check involves adding all the Student IDs together. In this example it would perform the calculation 4762 + 153 + 2539 + 4651 giving us a hash total of 12105. The data would be transmitted along with the hash total and then the hash total would be recalculated and compared with the original to make sure it was the same and that the data had been transmitted correctly. We would use a hash total here because there is no other point to adding the Student IDs together. Apart from verification purposes, the hash total produced is meaningless and is not used for any other purpose.

Control total

A **control total** is calculated in exactly the same way as a hash total, but is only carried out on numeric fields. There is no need to convert alphanumeric data to numeric. The value produced is a meaningful one which has a use. In our example above, we can see that it would be useful for the head teacher to know what the average pass rate was each year. The control total can be used to calculate this average by dividing it by the number of students. The calculation is 6 + 8 + 7 + 3 giving us a control total of 24. If that is divided by 4, the number of students, we find that the average number of passes per student is 6. Obviously, the control total check is usually carried out on much larger volumes of data than our small extract.

The use of a control total is the same as for a hash total in that the control total is added to the file, the file is transmitted and the control total is recalculated. Just as with the hash total, if the values are different, it is an indication that the data has not been transmitted or entered correctly. However, both types of check do have their shortcomings. If two numbers were transposed, say student 4762 was entered as having 8 passes and 0153 with 6 passes, this would obviously be an error but would not be picked up by either a control or hash total check.

> **Activity 1e**
> 1 Briefly describe what is meant by verification.
> 2 Write a brief description for each of **three** different validation checks.

1.4.4 Advantages and disadvantages of validation and verification

This section should be read in conjunction with Section 1.4.3.

Advantages and disadvantages of individual validation and verification checks have been included within the descriptions given above, where appropriate.

▼ Table 1.6 Advantages and disadvantages of validation and verification checks

Advantages	Disadvantages
Both make the data entered into a system more accurate, with verification ensuring that the data is copied accurately whereas validation tends to ensure that data is sensible and not ridiculous.	Both tend to slow down the processing of data with neither method checking that data is correct. For example, validation may not alert the user if an incorrect number has been input providing it is within an acceptable range and is numeric.

1.5 Data processing

As we saw in Section 1.1, data must be processed so that it can become information. Data can include personal data, transaction data, sensor data and much more. Data processing is when data is collected and translated into usable information. Data processing starts with data in its raw form and translates it into a more readable format such as diagrams, graphs, and reports. The processing is required to give data the structure and context necessary so it can be understood by other computers and then used by employees throughout an organisation. There are several different methods of data processing, but the three most popular ones are batch, online, and real-time. We will now consider each of these in turn.

1.5.1 Batch processing

In business, a transaction occurs when someone buys or sells something, but in IT the term '**transaction**' can mean much more, such as adding, deleting or changing values in a database.

Batch processing is still used today as it is an effective method of processing large volumes of data; sometimes millions of transactions are collected over a period of time. The data is entered and processed altogether in one batch by the computer, which then produces the required results. The processing of one batch is often called a job. Batch processing allows computers to process data when computing resources are not being fully utilised, such as overnight, and requires very little, and often no, human interaction. Examples of batch processing include payrolls and customer billing systems.

Batch processing provides a number of benefits. One of these is that it allows a business to process jobs at the times of day when computer resources are not being used fully, thereby saving the cost of the computer doing very little.

Companies are able to make sure that vital tasks which need immediate attention, such as online services, can use the computer resources as and when needed. They can then timetable batch-processing jobs for those times of day when online processing tasks are fewer. Compared to **real-time processing**, batch processing requires a simpler computer system without complex hardware or software. Once the system is up and running, it does not need as much maintenance as a real-time system and, because there is less human interaction, data entry methods used in batch processing tend to be more accurate. The actual processing, unlike the data collection, is very fast with batch processing, with several jobs being processed at the same time. Many utility companies such as water, gas and electricity companies handle huge amounts of data in their billing systems. Batch processing the collection of data, calculating and printing the bills can be done mainly at night, so that the computers can be used to help control the actual delivery of the utility during the busy part of the day. Before we look at the use of batch processing in the production of utility bills, we need to study the use of master files in batch processing.

Master and transaction files

In the next section, we will look at how batch processing is involved in producing a company payroll. First, we need to be sure we understand that two separate files are involved in this process. Let us consider the payroll system of a company that pays its workers weekly. One file is called the **master file**, which contains all the important data that does not change often, such as name, works number, department, hourly rate, and so on. The other file, called the **transaction file**, contains the data that changes each week, such as hours worked. The master file would already be sorted in order of the key field so that processing is easier to perform. The transaction file would probably be in the order that the transactions were collected. Before these two files can be processed together to produce the payroll, the transaction file needs to be sorted in the same order as the master file and will also need to be checked for errors using validation checks.

A master file would be used together with a transaction file in most batch processing applications, including a computerised customer orders system (which we will also be looking at shortly). When data is stored in order of a key field, it often forms what is called a 'sequential file' (the data is in a predetermined sequence). These used to be stored on magnetic tapes, but magnetic disks and solid state drives are now used much more than magnetic tapes, which tend, in modern systems, only to be used for backing up systems. The transaction file will therefore be stored on disk, even though it holds data in sequential order. When data is searched for in batch processing, each record is looked at one by one until the computer finds the record it is looking for. This is called **sequential access**.

The steps involved in updating a master file using a transaction file

There are occasions when the data in a master file has to be updated, for example so that a new worker will get paid. We will assume that the changes will happen on a weekly basis. It is likely that the transaction file would contain any needed updates as well as the payroll data and there would only be one transaction file, but to make it simpler to understand we will assume that the updating of the master file happens first and then the payroll runs immediately afterwards.

The three types of transaction involved in updating a master file in the scenario we have outlined are when:

» a worker moves to different department: their record must be amended or changed

- a worker leaves the company: their record needs to be removed or deleted
- a new worker starts with the company: their record needs to be added.

We can give each type of update a letter. Changing an existing record in the master file will be C, whereas deleting a record from the master file will be D and adding a new record to the master file will be A. At the end of each week, the computer system will process the data stored in the transaction file and make any changes that are necessary to the master file, thereby producing an updated master file.

We can demonstrate this, using the following small sample of data.

▼ Table 1.7 Transaction file

ID	Transaction	Employee name	Department
2	D	Julia Bolero	Sales
4	C	Nigel Ndlovu	Buying
7	D	Adrienne Pascal	IT
11	A	John Ward	Stores
12	A	Paolo Miserere	IT
EOF	End of file marker		

▼ Table 1.8 Master file

ID	Employee name	Department
1	Jose Fernandez	Buying
2	Julia Bolero	Sales
3	Louis Cordoba	Sales
4	Nigel Ndlovu	Stores
5	Bertrand Couture	Buying
6	Lionel Sucio	Stores
7	Adrienne Pascal	IT
8	Gurjit Mandare	Stores
9	Iqbal Sadiq	IT
10	Tyler Lewis	Buying
EOF	End of file marker	

In order to update the master file, a new blank file will be created and act as the new master file. The following very basic algorithm will be followed.

> **Advice**
>
> This algorithm and the subsequent algorithms in this chapter are simplified versions of what an algorithm might look like. More efficient ways of writing these and other algorithms will be covered in Chapter 4.
>
> We will look at REPEAT…UNTIL in more detail in Chapter 4.

```
1   First record in the transaction file is read
2   First record in the old master file is read
3   REPEAT
4       IDs are compared
5       IF IDs do not match, old master file record is written to new
        master file
6       IF IDs match transaction is carried out
7           IF transaction is D, old master file record is not written to
            new master file
8           IF transaction is C, data in transaction file is written to new
            master file
9       IF IDs match, next record from transaction file is read
10      Next record from master file is read
11  UNTIL end of old master file
12  Data in transaction file record is written to new master file
13  Any remaining records of the transaction file are written to the
    master file
```

What is happening in this algorithm is that the computer reads the first record in the transaction file and the first record in the old master file. If the ID does not match, as it does not in this case (ID is 2 in the transaction file but the ID in the master file is 1), there is no change necessary. The computer simply writes the old master file record to the new master file and misses out the next instructions 6 to 9 (IDs do not match so they can be ignored) then the next record from the master file is read.

After the last record, an End of file marker would be stored. As it has not been read yet we cannot be at the end of the file so the UNTIL statement tells the algorithm to go back to the REPEAT instruction. It starts again at the 'IF IDs do not match' in instruction 5. The IDs match in this example because the ID is 2 in the transaction file and the ID in the master file record is now 2. Instruction 5 is therefore ignored and instruction 6 indicates that the transaction is carried out and moves on to instruction 7. In this case the transaction is D so the record has to be deleted, so the old master file record is not written to the new master file. Instruction 8 is ignored as the transaction is not C. Instruction 9 causes the next record from the transaction file to be read. Instruction 10 means the next record from the master file is read. We again meet the 'UNTIL end of old master file' instruction.

As the algorithm has yet to meet the end of file marker, we go back to the REPEAT which is instruction 3. We are now looking at the second record of the transaction file (ID 4) and the third record of the old master file (ID 3). If they do not match (instruction 5), which they do not, the old master file record is written to the new master file and we jump to instruction 10 and the next record (the fourth, ID 4) of the old master file is read.

We are not at the end of the file, so the algorithm takes us back to instruction 3 and then on to instruction 4 then instruction 5: the 'IF IDs do not match' instruction. However, this master file record matches with the second record in the transaction file so the transaction is carried out. It is C, so the master file record is not copied across, instead the record from the transaction file (apart from the C) is copied into the new master file. The next record from the transaction file is read and then the next record from the master file is read.

This carries on until the EOF marker is met in the old master file. The UNTIL instruction is now true, so the algorithm moves on to instruction 12 so the current transaction record is written to the master file. Then instruction 13 is followed and the remaining records of the transaction file are added to the master file, in this case two records.

The steps are carried out as shown, making two assumptions. One is that all additional records will be at the end of the transaction file. This is usually the case as new employees would be given the next available ID and the transaction file would be sorted so these new workers would appear at the end of the file. The other assumption, which is not really likely in a real situation, is that the transaction file records will have fields identical to all those in the master file. To make it easier to follow, we have not included, for example, the rate of pay field in those records which need to be added, but this field would have to be included in the master file for every employee.

▼ Table 1.9 The new master file

ID	Employee name	Department
1	Jose Fernandez	Buying
3	Louis Cordoba	Sales
4	Nigel Ndlovu	Buying
5	Bertrand Couture	Buying
6	Lionel Sucio	Stores
8	Gurjit Mandare	Stores
9	Iqbal Sadiq	IT
10	Tyler Lewis	Buying
11	John Ward	Stores
12	Paolo Miserere	IT
EOF	**End of file marker**	

Records 2 and 7 have been deleted.

Record 4 has been changed.

Records 11 and 12 have been added.

Use of batch processing in utility bills

The bills that are sent to customers by utility companies are produced using batch processing, usually on a monthly basis. Meter readings are taken on a specific date and then processed. A transaction file of meter readings and customer account numbers is created. The master file will contain customer account numbers, the previous month's reading as well as other details of the customer such as contact and bank details. We only need to concern ourselves with their account number and previous meter reading. We will assume that the transaction file is sorted using Customer_Account_number and has been validated. A very simple algorithm describing the process would be as follows:

```
1   First record in the transaction file is read
2   First record in the old master file is read
3   REPEAT
4   Customer_Account_numbers are compared
5   IF Customer_Account_numbers do not match, old master file record is
    written to new master file
6   IF Customer_Account_numbers match transaction/calculation is
    carried out
7   Computer subtracts old_reading (from master file) from new_reading
    (from transaction file) and calculates Bill_amount
8   Bill_amount is written to new master file with rest of customer
    record
9   IF Customer_Account_numbers match, next record from transaction file
    is read
10  Next record from master file is read
11  UNTIL end of transaction file
12  Remaining records of the master file are written to the new master file
```

Use of batch processing in credit card and debit card accounts

Batch credit card processing is an example of how batch processing is used. It can be used in restaurants, bars, stores or anywhere goods are traded. When a transaction takes place, the point-of-sale terminal (POS) sends the customer's credit card information and the amount to be paid to the card issuer. The card is then checked that it exists and that it has not been reported as stolen. A check also takes place that the card owner has enough credit to cover the amount. If all is satisfactory an authorisation code is sent to the trader and the amount is temporarily blocked on the customer's credit card account. These authorisation codes are gathered during the day and are transmitted in a batch to a credit card processing company so they can be processed. This is done at the end of the day. The company charges a fee for this and then pays the trader for the transactions having, in turn, been paid by the card issuer. It then lets the card issuers know that money has been paid and the card issuer can then add the transaction to the customer's account. It can take more than one day for the trader to receive payments for a batch and consequently the same applies for the transactions to be added to customers' accounts. Credit card batch processing is popular among traders because instead of being charged a fee for each transaction they are charged a fee for each batch of transactions, which reduces costs. Batch processing is sometimes used with debit card transactions. The amount of the transaction is often debited immediately from the customer's account although the payment may not be credited to the trader until after a period of time.

> ### Activity 1f
> Amend the algorithm on page 29 so that it would work if there were no records to be added, that is if there were no records in the original transaction file on page 29 after 7, D, Adrienne Pascal, IT.

Use of batch processing in payroll

As mentioned earlier, batch processing is used to calculate wages in a payroll. Let us look at a typical master file and transaction file which might be used in a payroll system. We will only consider a very small company but, in real life, it is not unusual for payroll systems to cater for thousands of employees.

▼ Table 1.11 Master file

ID	Department	Rate ($)	Wages_to_date
036	Sales	20	1280
047	Buying	25	1475
165	Buying	25	1525
469	Sales	20	1160
512	Stores	15	825
545	Sales	20	1220
578	IT	30	1860
682	Sales	20	1080
778	IT	30	1920
786	Buying	25	1575
789	IT	30	1830
861	Stores	15	795
EOF	End of file		

▼ Table 1.10 Transaction file

ID	Hours_worked
036	40
469	40
578	38
778	40
789	40
EOF	End of file

We can assume that the transaction file has been sorted and validated. The system would have to go through each transaction file record and using the hourly rate (Rate) from the matching master file record, calculate that employee's wages for the week. This would be added to the employee's wages paid so far this year and replace the Wages_to_date value. We shall assume that the workers pay no tax and have no other deductions.

```
1   First record in the transaction file is read
2   First record in the old master file is read
3   REPEAT
4       IDs are compared
5       IF IDs do not match, old master file record is written to new master
        file
6       IF IDs match transaction/calculation is carried out
7           Computer calculates the pay, Rate (from master file) multiplied by hours
            worked (from transaction file)
8           Wages _ to _ date is updated and record is written to new master file
9       IF IDs match, next record from transaction file is read
10      Next record from master file is read
11  UNTIL end of transaction file
12  Remaining records of the master file are written to the new master file
```

The basic outline of the algorithm, to show how the payroll is processed, is quite similar to the updating algorithm, but the middle part is different:

Notice that because no additional records need to be added to the master file, we stop the processing when we get to the end of the transaction file.

> ### Activity 1g
> Work through this algorithm using the two files shown on the previous page and write down the new master file.

Use of batch processing with customer orders

We have already mentioned that customer orders can be dealt with using batch processing. When a company receives an order from a customer, it is added to a transaction file. At the end of the day, the transaction file is used together with a master file to check the items are in stock and to update the master file with the new number of items in stock after making deductions arising from the orders. If items are not in stock, they must be ordered from a supplier. If they are in stock, then they are added to what is called a picking list which is sent to the warehouse. The warehouse staff then find the goods that have been ordered and package them ready for shipment to the customers. The next step is to allocate the packaged goods to delivery vehicles. At the same time, invoices are produced and sent to customers for immediate payment or added to their account (if customers make payments every month). Let us look at how account payments are processed.

Use of batch processing with customer accounts

The transaction file will contain details of the orders a customer has made in the last month, together with any payments the customer has made. We will consider a simple master file which just contains the money owed by each customer, which is called 'the balance'.

Again, we can assume that the transaction file is sorted in Cust_no order and has been validated. The algorithm describing the process would be similar to the payroll algorithm:

▼ **Table 1.12** Transaction file

Cust_no	New_orders	Payment_made
219	320	200
451	870	1500
523	190	340
834	520	250

▼ **Table 1.13** Master file

Cust_no	Balance
138	0
187	0
219	0
451	-800
487	-260
523	-340
764	0
802	-920
834	0
869	-540

```
1   First record in the transaction file is read
2   First record in the old master file is read
3   REPEAT
4       Cust_nos are compared
5       IF Cust_nos do not match, old master file record
        is written to new master file
6       IF Cust_nos match transaction/calculation is
        carried out
7           Computer calculates New_orders minus Payment_
            made and subtracts from Balance
8           Balance is updated and record is written to new
            master file
9       IF Cust_nos match, next record from transaction
        file is read
10      Next record from master file is read
11  UNTIL end of transaction file
12  Remaining records of the master file are written to
    the new master file
```

> **Activity 1h**
> Work through this algorithm using the two files shown on the previous page and write down the new master file.

1.5.2 Online processing

We have seen how batch processing involves gathering data together ready for processing at a later date. This has an obvious drawback in that the processing is delayed. In some cases, this is not a problem, for example with applications such as payroll, which only need to be processed weekly or monthly, or utility billing systems which are often processed every three months. Some processing, however, has to be done almost immediately, such as at supermarket checkouts or interrogating a database for an employee's details. The original definition of online processing was that the user was in direct communication with a central computer. This has now evolved to include any aspect of IT which takes place over the internet. In this section, we shall be looking at how applications such as **electronic funds transfer (EFT)** and automatic stock control, among others, take place. One of the differences between batch processing and online processing, is that in batch processing data is searched using sequential access, whereas direct access tends to be used in online processing. Direct access is simply the ability to go straight to the record required without having to read all the previous records.

Uses of online processing

When data is input into an online system, processing takes place almost immediately with just a short delay, so short that the user believes they are in direct communication with the computer. Each transaction is processed before the next transaction is dealt with. This means that online processing can be used in a variety of ways. We shall look at some of these here.

Electronic funds transfer

One definition of electronic funds transfer (EFT) is that it is the electronic transfer of money from one bank account to another using computer-based systems, without the direct intervention of bank staff. Examples include the use of an **automated teller machine (ATM)**, a direct payment of money to another person, and direct debits, when a company debits the customer's bank account for payment for goods or services. EFTs can be transfers resulting from credit or debit card transactions at a supermarket, a store or online. They usually involve one bank's computer communicating with another bank's computer, though not always, such as when the ATM being used belongs to the customer's bank.

Most people receive their wages as a result of an EFT. Money from the employer's bank account is transferred electronically to the employee's bank account.

EFT has become a common way of paying bills. For example, you may decide that your house needs redecorating, so you ask a painter to come and paint your house. When the painter has finished, he or she will require payment. One of the easiest ways of doing this is to take the painter's bank details and transfer the money from your account to the painter's account.

The following steps describe what happens after you have logged in to your online bank account, although the process may differ slightly from bank to bank, and assumes you are paying someone new:

```
1   Select transfer money
2   Select the account you wish to transfer money from
3   Select new payee
4   Type in sort code, account number and payee name
5   Type in amount to transfer
6   Computer checks available balance
7   If you have sufficient funds, the transaction is authorised
8   Your bank's computer contacts payee's bank's computer which searches
    for the payee's record
9   Amount is subtracted from your account balance
10  Amount is added to payee's account balance
```

The most common form of electronic funds transfer is purchasing goods in a store or supermarket when paying at a checkout. This is called EFTPOS and stands for Electronic Funds Transfer at Point Of Sale. Checkouts at supermarkets are called point-of-sale terminals. When a customer goes to a checkout to pay for their goods, they insert their bank card and the following steps are carried out. (This assumes it is not a contactless transaction, in which case steps 3 to 8 are omitted. Most countries only allow contactless transactions if the value of the goods is less than a certain amount.)

```
1   Card chip is read and checked to make sure it is in date and it is a
    valid card number
2   If not, card is rejected and transaction terminated
3   PIN is entered by customer into PIN pad
4   Chip reader determines PIN from the chip
5   The two PINs are compared
6   If they are identical the transaction is authorised
7   If they are not identical, error message appears on the chip reader
    and two more attempts are allowed
8   If the two PINs are still not identical, the transaction is rejected
    and error message issued
9   Customer's bank is contacted by supermarket's computer
10  Customer's bank retrieves customer's record
11  Customer's bank checks if sufficient funds in account
12  If there are insufficient funds then transaction is rejected
13  If there are sufficient funds then transaction is authorised
14  The amount of the bill is deducted from customer's account
15  The amount of the bill is credited to supermarket's account
```

Automatic stock control

There are many ways in which IT can be used in stock control. In this section we are going to concentrate on automatic stock control. This involves the use of an automated system where stock is controlled by a computer with little human input.

We have already met the use of EFTPOS terminals. These also serve another purpose in stores and supermarkets because they are used for stock control. The checkout operator swipes the barcode of an item and the computer uses this to update the stock. The terminal in a supermarket or store consists of a screen (which can be a touchscreen), a barcode reader to input the barcode of the product, a number pad to enter the barcode in case the barcode label is damaged, and electronic scales. Each terminal is connected to a computer network. The hard disk on the network server stores a file (we shall call it the product file) containing the records of each product that is sold. Each record consists of different fields containing data, for example:

- barcode number: the number which identifies each different product; this is the key field because it is different for each product
- product details: a description, such as tin of beans, packet of teabags and so on
- price of the product
- size: weight or volume of the product
- number in stock: the current total of that product in stock; this changes every time a product is sold or new stock arrives
- re-order level: the number which the computer will use to see if more of that product needs re-ordering. If the number in stock falls to this level, the supermarket or store must re-order
- re-order quantity: when the product needs re-ordering, this is the number of products which are automatically reordered
- supplier number: the identification number of the supplier which will be used to look for the details on the supplier file.

There is also a file (the supplier file), which contains details of the supplier of each product, including their contact details. It is more than likely that these two files would be stored as separate tables in a **relational database**.

The processing involved in automatic stock control is as follows:

```
1   The product's barcode is input from the barcode reader
2   The computer searches for this barcode number in the product file and
    finds it using direct access
3   The number in stock is reduced by one
4   The computer then compares the number in stock with the re-order
    level
5   If the number in stock is not equal to the re-order level then go
    back to step 1 and repeat
6   If the number in stock is equal to the re-order level then the
    computer creates an automatic order
7    It looks up the re-order quantity of that product
8    It looks up the supplier number of that product
9    It searches the supplier file for the record corresponding to the
     supplier number found in the product file
10   It sends the order automatically to the supplier using the
     supplier's contact details
11  Go back to step 1 and repeat
```

When new goods are delivered, the computer automatically updates the product file by following steps 1 and 2 and then increasing the number in stock by the re-order quantity.

Electronic data exchange

Electronic data exchange is often referred to as Electronic Data Interchange (EDI). For the purposes of this book we will call it EDI, though the terms are interchangeable. This is a method of exchanging data and documents without using paper. The documents can take any form such as orders and invoices, with the electronic exchange between computers using a standard format. An invoice is a type of bill sent to a customer containing a list of goods sent or services that have been provided, including a statement of the sum of money due for these.

Most companies create invoices using a computer system. Many then print a paper copy of the invoice and post it to the customer. If the customer is a business, they will often type the details into their computer, meaning that the whole activity of sending and receiving invoices is actually the transfer of information from the seller's computer to the customer's computer. EDI replaces this activity with an electronic method. The physical exchange of documents could take between three and five days. EDI often occurs overnight and can take less than an hour.

The old paper-based method was usually made up of these steps:

- A company decides to buy some goods, creates an order and prints it.
- Company posts the order to the supplier.
- Supplier receives the order and enters it into their computer system.
- Company calls supplier to make sure the order has been received, or supplier posts a letter to the company to say it has received the order.

An EDI system generally has these steps:

- A company decides to buy some goods, creates an order and does not print it.
- EDI software creates an electronic version of the order and sends it automatically to the supplier.
- Supplier's computer system receives the order and updates its system.
- Supplier's computer system automatically sends a message back to the company, confirming receipt of the order.

EDI systems save companies money by providing an alternative to systems that require humans to operate them, thereby saving wages that would be paid to people who would sort and search paper documents. There is no need to pay operators to manually enter the data. EDI also reduces human error during data entry, since there would be no need to re-enter data that had been sent originally. Productivity is improved as more documents are processed in less time.

EDI systems are often used because of the security aspects of the system. As alternative systems using the internet have grown, EDI has had to innovate and this has been achieved largely by increased security in the transmission of data. EDI is also used by some examination boards to allow exam entries to be made and for issuing results. It is also used by hospitals to send and receive documents to and from doctors, again due to the increased security of this system.

Business-to-business buying and selling

Business-to-business – often termed B2B – refers to buying and selling between two businesses, rather than between a business and an individual customer (B2C). The value of B2B transactions is noticeably higher than that of B2C, as businesses are more likely to buy higher-priced goods and services and to buy more of them than individual customers are. A car manufacturer, for example, will buy thousands of tyres, whereas a customer is only likely to buy four tyres at most when replacements are needed.

Many companies still use EDI for sending orders and invoices, but there are other aspects of B2B which require online processing. Businesses can buy and sell using online marketplaces, but many B2B sellers do not take advantage of these. There is sometimes little difference between B2C and B2B marketplaces, for example Amazon® has B2C and B2B versions of its site. However, you are advised to read the syllabus regarding the use of brand names. B2B marketplaces work just like a B2C marketplace in that they connect many sellers to buyers. Buyers have the opportunity to compare and buy products from many different sellers all on one site. However, a B2B marketplace is different to a B2C marketplace in that bulk orders can be placed, discounts can be received for ordering large quantities and orders can be edited online. Sellers have benefits too, such as lower costs, since they do not have to spend as much money on marketing or setting up a larger website, as the marketplace is responsible for that part of marketing, although the business has less control over the look of its products on the website. Sellers can save the time that would be spent on setting up the sales aspect of a website. The audience is now global and to a certain extent, a captive audience. Marketplaces can also be used to test out new products by putting a few up for sale. If they sell well, the volume can be increased and if not, they can be easily withdrawn.

B2C online transactions are fairly straightforward, whereas B2B transactions tend to be more complicated. B2B selling prices can vary a great deal with discounts needing to be taken into account. The quantity of products being sold is much greater, resulting in more complicated shipping requirements. In addition, B2B tends to have more government regulation and complex taxation. However, the failure of companies to invest in online buying and selling can lead to being left behind by competitors who sell more and make greater profits.

Although EDI is still used by many companies, there are other methods of buying and selling. Companies can have their own selling website that can be used by other companies to buy their goods. E-procurement is the term used to describe the process of obtaining goods and services through the internet. E-procurement software can be used by sellers and buyers to link directly to each other's computer systems.

Online shopping

Internet shopping began in the 1990s. At that time, however, the vast majority of the world's population was not even aware of it. Today, many people shop online, a development that has arisen due to the emergence of online shopping, from both traditional high-street big names and new online-only companies. Online shopping has become many people's preferred method of shopping for many reasons.

- Customers are not rushed by store assistants into hurriedly comparing products and prices.
- Customers can shop at a convenient time for them.
- Customers do not have to spend time and money travelling around different stores to find the best bargains; it is much faster and cheaper to do it online.
- If a local branch of a chain of stores has closed, customers can still shop with that chain online.
- Customers can look at a wide range of stores all around the world.
- Items are usually cheaper online because warehouse and staff costs are lower than maintaining main street stores.
- Stores can deliver goods at a time to suit the customer.
- Supermarkets can remember the customer's shopping list and favourite brands, making it easier for the customer to enter their shopping list.
- There is a greater choice of manufacturers. Many main street stores can only stock items from a few manufacturers.

A typical online shopping website opens with a home page which contains different categories, on tabs across the top or down the side or on a drop-down menu. Customers are able to click on a category tab, which takes them to a different web page on the site. They can browse the products within that category to get to the one they want. After opening the store website and browsing product categories, the customer then decides what they want to buy. At this point, some stores may ask for a postcode or zip code to check that they actually deliver to that area. In a real store or supermarket, the customer would place products in a shopping trolley or basket; similarly, with an online shop, they place them in a virtual shopping basket. This is usually done by clicking on an 'add to basket' or simply 'add' icon. As with a real store, items can be removed from, as well as added to, a customer's shopping basket and, when the customer has decided that they have finished shopping and they want to pay, they go to the checkout.

Here is an example of the steps, in the form of a simple algorithm, required at the checkout, although the actual sequence of steps varies depending on the online shopping website.

```
1   IF user is new customer, they must register
2       Enter a username
3       Enter and verify (by double entry) password
4       Enter phone number and email address
5       Enter delivery/shipping address
6       Enter billing address if different to shipping address
                                                        [some sites only]
7       Choose speed/cost of delivery                   [some sites only]
8       Enter type of card and credit/debit card number
9       Enter date of expiry
10  IF user is existing customer
11      Log on by entering username and password
12      Confirm delivery address
13      Choose speed/cost of delivery                   [some sites only]
14      Select credit card/debit card account to be debited if stored,
        otherwise enter type of card and credit/debit card number
15  Enter card security code            [some sites may not require this]
16  Confirm order
```

An email address is nearly always needed, so the store can notify the customer when the order has been received. Some sites inform you of the progress of the order's delivery.

The delivery address is needed so the store knows where the goods will be sent to.

The billing address is required because the store wants to know where to send the bill as this is sometimes different to where the goods will be delivered. As payments are made electronically, this piece of information is largely academic, but it can be used for additional credit checks.

The customer is often able to choose how quickly the goods should be delivered or choose a delivery time slot, if the store has its own delivery vehicles. Some stores offer same-day or next-day delivery, although usually the quicker the delivery, the higher the cost.

Often, the customer has to pay a delivery charge as well as the price of the goods.

1.5.3 Real-time processing systems

Real-time processing is an example of online processing in that it requires the inputs to go directly to the CPU of the computer, but the response time of the computer must be immediate, with no delay whatsoever. This type of processing is usually found in systems that use sensors, for example computer-controlled **greenhouses**, often referred to as glasshouses. Temperature, light and moisture sensors are all used to monitor **physical variables** and send these as input data to the computer so that it can take immediate action. For example if the temperature falls below a certain value, the heater is automatically switched on. If a batch-processing system was used, the temperature might have been below the required value for a long period of time, damaging or even killing the plants inside, so it is essential that the response from the computer is immediate. Real-time processing is continuous so the process is never ending unless the user switches the system off. We will now look at some systems which use real-time processing. These will be examined in more detail in Chapter 3.

Uses of real-time processing

Greenhouses

Real-time systems are often systems where the output affects the input. Consider the glasshouse example above, where the input to the computer or **microprocessor** is the temperature. If the temperature is below the required level, the computer turns the heater on. How this is achieved will be described in greater detail in Chapter 3. The temperature will now rise and because this process is continuous the temperature sensor will feed the new temperature back to the computer. Here, the output, which is the switching on of the heater, has affected the input, the temperature. This cycle continues until the desired temperature is reached. Unlike other online systems, any output is produced quickly enough so that it affects the system before the next input is received. The output happens immediately. We will now look at three uses of real-time systems.

Central-heating systems

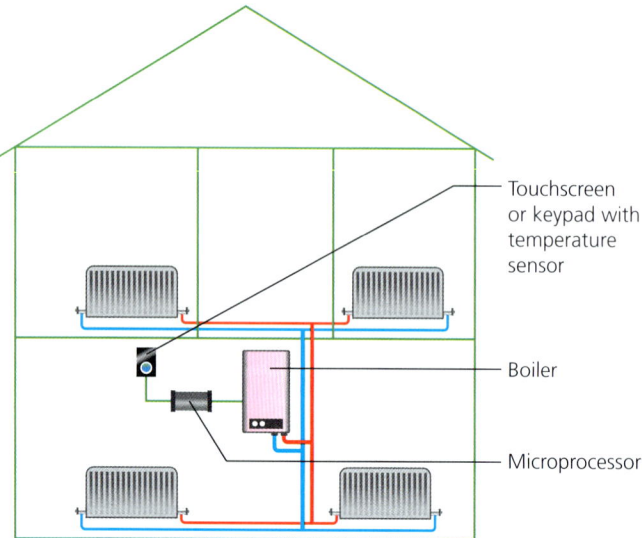

▲ Figure 1.2 A typical central-heating system

In the system shown in Figure 1.2, there is what is called a combination or combi boiler. The boiler contains the pump as well as the means of heating the water. A central-heating system is a real-time system since it involves the use of sensors; in this case temperature sensor(s) are used in central heating to continuously monitor the physical variable, the temperature of the house. As with any real-time system, it involves the use of a feedback loop. This is called a 'closed system' in that the temperature is fed into the system. The boiler heats the water, which causes the temperature to rise. This in turn, eventually, causes the boiler to be switched off by the microprocessor, which then results in a drop in temperature and the boiler has to come back on again and so the sequence is repeated. We know it is a real-time system since the output affects the input.

In a microprocessor-controlled central-heating system, users select their required temperature using a touchscreen or keypad. The microprocessor reads the data from a temperature sensor on the wall and compares it with the value the user has requested. If it is lower, then the microprocessor switches the boiler and the pump on. If it is higher, the microprocessor switches both off. In order for the microprocessor to process the data from the temperature sensor (which is an analogue sensor), it uses an analogue-to-digital converter to convert this data into a digital form that it can understand. As a result of the input, the microprocessor may or may not send signals to **actuators** which open the gas valves in the boiler and/or switch the pump on.

The microprocessor is also used to control the times at which the system switches itself on and off. For example, users can set the system to come on before they get up in the morning and set it to switch off just before they go out to work. When the system is off, the microprocessor ignores all readings.

A simple algorithm shows how the system works:

```
1  User enters required temperature using keypad/touchscreen
2  Microprocessor stores required temperature as a pre-set value
3  Microprocessor receives temperature from sensor
4  Microprocessor compares temperature from sensor to pre-set value
5  If temperature from the sensor is lower than the pre-set value, the
   microprocessor sends a signal to an actuator to open the gas valves
6  If temperature from the sensor is lower than the pre-set value, the
   microprocessor sends a signal to an actuator to switch the pump on
7  If temperature is higher than or equal to the pre-set value, the
   microprocessor sends a signal to switch the pump off and close the valves
8  This sequence is repeated until the system is switched off
```

Air-conditioning systems

Air-conditioning systems are more sophisticated than central-heating systems and involve units such as valves, compressors, condensing units and evaporating units, which together form a system that, basically, feeds cold air into a room via a unit containing enclosed fans to circulate it around the room. We shall just concern ourselves with what happens in one room in a house. Each room has a temperature sensor and the system uses this to determine whether it needs to switch the fans on or, in the case of more complex systems, change the speed of the fans (we will only be considering the simpler version). The user, as with a central-heating system, enters the required temperature using a keypad or touchscreen. The output of the system (the switching on of the fans) results in the cooling of a room thereby lowering the temperature, which is the input to the system.

Again, we can use a basic algorithm to describe the process:

```
1   User enters required temperature using keypad/
    touchscreen
2   Microprocessor stores required temperature as a
    pre-set value
3   Microprocessor receives temperature from sensor
4   Microprocessor compares temperature from sensor to
    pre-set value
5   If temperature from the sensor is higher than the
    pre-set value, the microprocessor sends a signal to an
    actuator to switch the fans on
6   If temperature is lower than or equal to the
    pre-set value the microprocessor sends a signal to
    an actuator to switch the fans off
7   This sequence is repeated until the system is
    switched off
```

Burglar alarm systems

Burglar alarm systems use a number of sensors to detect the presence of intruders. For example, a pressure sensor detects the weight of an object placed on it. The output is the sounding of an alarm or the beeping of the control box where the user inputs their passcode. These systems will be discussed in detail in Section 3.2.

Control of traffic/pedestrian flow/smart motorways

Computers are used to control the flow of traffic and pedestrians. With smart motorways the output is in the form of changing speed limits. The input is effectively a count of the number of vehicles passing a given point at a specific point in time.

Car park barriers

Car park barrier systems are computer or microprocessor-controlled. They often consist of input devices such as light sensors and induction loops. The output is the raising and lowering of the barrier using motors. The system has to be real time as it is essential that the barrier rises when a vehicle presents itself without any delay.

Traffic lights

Computer-controlled traffic light systems often use induction loops as input. The output is the changing of the traffic lights in sequence which will be explained in greater detail in Chapter 3. The processing has to be immediate or unnecessary traffic jams might develop.

Wireless sensor and actuator networks

Wireless sensor and actuator networks use a variety of input devices in the form of sensors as well as several different actuators as output devices. We will look at these in greater detail in Chapter 3 as well as examining smart homes more thoroughly in that chapter.

Guidance systems

A guidance system can be used to control the movement of different types of vessel or moving object, such as a ship, aircraft, missile, rocket or satellite. It involves the process of calculating the changes in position and velocity and other more complex variables, as well as controlling the object's course and speed.

A guidance system, like all computer systems, has inputs, processing, and outputs. The inputs include data from sensors, the course set by the controller and data from radio and satellite links. The processing involves using all the input data to decide what actions, if any, are necessary to maintain or achieve the required course. The outputs are the actions decided upon as a result of the processing and use devices such as turbines, fuel pumps and rudders, among others, to change or maintain the course.

Missiles have a high-precision, real-time guidance system built into their nose. The guidance system has a radar system, which consists of a radar which looks forwards and one that looks downwards. There is a navigation system and a 'divert-and-attitude-control' system, which is able to increase or decrease the amount of thrust in different engines driving the missile and thereby control the direction and speed of the missile. Basically, the radar scans the surrounding terrain and feeds the information into the navigation system, which then uses the data from the radar together with its stored maps of the area to calculate the required flight path, avoiding obstacles. The new flight path is immediately fed in to the divert-and-attitude-control system so that it can adjust the direction and speed of the missile to achieve this flight path. The flight path is constantly adjusted to ensure the missile does not drift off its projected flight path. A missile guidance system is an example of a true real-time system, because there must be no delay between the navigation system realising that the flight path has been deviated from and the flight path being adjusted.

Autonomous vehicles

An autonomous vehicle is a vehicle that is able to operate itself without a driver or human intervention. It uses technology to sense its surroundings and respond to external conditions. This is not the same as an automated vehicle, which requires a human driver or pilot to resume the steering and navigating tasks under certain situations. An autonomous vehicle can be an aircraft, a car, a drone, or a ship. It makes use of a fully automated driving/piloting system in order to allow the vehicle to respond to external conditions that a human driver or pilot would normally be able to react to.

Autonomous aircraft

An autonomous aircraft or unmanned aerial vehicle (UAV) is an aircraft which flies under the control of automatic systems requiring no intervention from a human pilot. The overall term 'autonomous aircraft' can be considered to be the same as the term 'drones' which we will look at in detail further on. With passenger airlines, although autonomous features such as autopilot systems and automated navigation systems are currently relieving the human pilot of ever more duties, the human pilot still remains essential. However, as UAV technology improves it is possible that carrying passengers in UAVs might become a possibility. There is a desire to increase the use of UAVs, in particular, to discourage hijacking as the aircraft would not allow a stranger to take over the controls.

There is a range of levels of autonomy. At the lowest level, a computer system has no part to play with the human pilot being in total control. In the middle of the range, the system has some control, such as the autopilot, but is mainly an adviser to the human pilot. At the highest level, the computer system is fully in control with human pilots having no input.

UAVs are fitted with a range of sensors. They are used within the following devices:

- Accelerometers to measure acceleration, i.e. the rate of change in velocity per unit of time. By measuring acceleration in three dimensions, accelerometers can detect changes in speed and direction. This data is transmitted to the drone's flight controller, enabling it to maintain stability and control during flight.
- Angle of attack (AOA) indicators to measure the angle between the oncoming air and the airplane or wing. An AOA indicator gives an indication of the amount of lift the wing is generating.
- Barometers to measure air pressure. Their readings are used to help the system to stabilise the altitude of the aircraft.
- Gyroscopes to measure the rotation of the nose of the aircraft. In simple terms they measure how far the nose is up and down or left to right as well as the amount of tilt in the wings. It allows the control system to automatically maintain the aircraft's required compass heading, its altitude and to stabilise its wings to maintain level flight.
- GPS helps determine the position of the UAV based on signals received from GPS satellites. The readings help the system maintain the correct course.
- Magnetometers indicate the strength and direction of the magnetic field around the aircraft. They are used by the system to confirm or make adjustments to the aircraft's heading or course.
- Collision avoidance systems are intended to make sure a UAV or drone does not crash into other obstacles or aircraft. They make use of cameras which see obstacles or aircraft, RADAR to detect them, LiDAR which sends laser pulses that reflect back to a sensor to determine the distance an object is from the UAV, ultrasonic sensors which detect ultrasound and infrared sensors which detect infrared signals. These last three make use of a transmitter to send signals and a sensor to detect the reflected signals.
- Pitot tubes to measure pressure from the incoming airflow around an aircraft. It can then be used to calculate air speed and increase or decrease if necessary.
- A velocity sensor is used to measure the change in distance over time.
- An air speed indicator (ASI) is the device for measuring the forward speed of the aircraft. The ASI uses the aircraft pitot-static system to compare pitot and static pressure and thus determine forward speed.

In addition to the sensors, an autonomous aircraft needs computers that can process enormous amounts of data in real time, and complex software to provide the artificial intelligence (AI) to become autonomous rather than just highly automated. The aircraft will also need to be able to communicate with external systems, such as air traffic control.

Autonomous cars/automobiles

A fully autonomous or driverless car can sense its surroundings and operate without the need of a driver. A human is not required to take control of the vehicle at any time. An autonomous car is one that can navigate to a destination, avoid obstacles, and park without any human intervention. To achieve this a driverless car must have an artificial intelligence (AI) system. Such a system must be able to sense its surroundings, detect objects so as to avoid collisions, control different aspects of driving such as steering and braking, and use GPS to track the car's current location and navigate to its destination. So it can recognise its surroundings and detect possible collisions, a self-driving car usually has two or more of cameras, RADAR and LiDAR.

The Society of Automotive Engineers (SAE), based in the USA, currently defines six levels of driving automation:

Levels 0–2 have some driver support features:

- Level 0: No Driving Automation. The human performs all driving tasks although the on-board computer can provide driver assistance in the form of warnings of blind spots or leaving a lane on a motorway.
- Level 1: Driver Assistance. The computer system controls one aspect of driving such as the speed of the car or the steering.
- Level 2: Partial Driving Automation. The system controls both steering and acceleration. The human still needs to supervise all aspects of driving.

Levels 3–5 have some automated driving features.

- Level 3: Conditional Driving Automation. The system controls most driving tasks but the human needs to be in a position to take over the driving if required.
- Level 4: High Driving Automation. The system performs all driving tasks as long as the car is within a specific geographical area. It cannot leave that area.
- Level 5: Full Driving Automation. The car can drive itself in all locations and in all conditions. A human is not required. As of the current time this is still a dream and does not exist as yet.

Drones

Technically any UAV can be called a drone and as we saw above, large aircraft which are unmanned could be referred to as drones. A drone can receive commands from a pilot using remote control or, as with the description of aircraft, can be controlled by onboard software. Originally drones were developed by the military and aerospace industries. More recently they have been developed to cover small distances as well as large. Because drones can be controlled remotely and can be flown at varying distances and heights, they can be used for a variety of tasks. They are used in the search for survivors of natural disasters. They are also used in warfare for observing enemy positions as well as being able to help in research of extreme climactic conditions. Many of the smaller drones are used by hobbyists and photographers as they have a camera attached. These have the ability to take off and land vertically using four rotors although, in an effort to save battery life, some drones have been developed which use just two propellors. Single rotor drones tend to be much larger and more difficult to control. Four rotor drones can make deliveries from warehouses to people's front doors and are being tested for use by a number of large delivery companies. They are also used by farmers to observe their crops and to track their livestock.

Larger, fixed-wing, drones can be used by armies to carry out attacks on enemy positions. They look like normal airplanes using wings to provide the lift instead of rotors. They can glide in the air for more than 16 hours. As they are usually much larger they need to take off and land on runways. Drones use a variety of sensors that have been described above in the section on aircraft.

Autonomous ships

These are often referred to as autonomous unmanned surface vessels (USVs). They are vessels without a crew which transport containers or other large cargos across seas or along rivers with little or no human interaction. In recent times ships have been built which are automated to some extent. Software has been produced which helps to control speed and steering. Some manufacturers have even developed remote ships. The ships are controlled by a human operator but there is help from people ashore in making decisions. Some ships have the ability to make decisions without involving the crew of the ship. Although there are currently no fully autonomous ships many companies are developing these and they are likely to be in operation within the next few years. As with other types of autonomous vehicles there are different levels of autonomy. Lloyd's Register has produced a table defining these levels. Lloyd's Register is a global professional services group specialising in marine engineering and technology. Created more than 260 years ago as the world's first marine classification society, to improve and set standards for the safety of ships.

▼ Table 1.14 Levels of autonomy for autonomous vehicles

AL 0) Manual – no autonomous function	All action and decision making is performed manually, i.e. a human controls all actions at the ship level.
AL 1) On-ship decision support	All actions are taken by a human operator, but a decision support tool can present options or otherwise influence the actions chosen.
AL 2) On and off-ship decision support	All actions are taken by a human operator on board the vessel, but a decision support tool can present options or otherwise influence the actions chosen. Data may be provided by systems on or off the ship.
AL 3) 'Active' human in the loop	Decisions and actions are performed with human supervision. Some decisions can give human operators the opportunity to intervene and over-ride them. Data may be provided by systems on or off the ship.
AL 4) Human on the loop – operator/ supervisory	Decisions and actions are performed autonomously with human supervision. Some decisions can give human operators the opportunity to intervene and over-ride them.
AL 5) Fully autonomous	Unsupervised or rarely supervised operation where decisions are made and carried out by the system.
AL 6) Fully autonomous	Totally unsupervised operation where decisions are made and carried out by the system.

Autonomous ships also use the technologies found in autonomous cars and drones. They must also be equipped with a suitable collision warning system. These include a range of sensors, RADAR, LiDAR and cameras, as well as AIS (automatic ship identification). The data is processed by AI systems on the ship or at an onshore location. These then suggest the best route. The possible benefits relate to safety and cost. Many accidents at sea are caused by human error as a result of fatigue and, of course, computers do not get tired and make mistakes. With a minimal or even no crew, much less money would be spent on salaries. Ships could be constructed so that it would be difficult to board and control them, meaning that cases of piracy would be reduced.

There would, however, be the risk of cyber attacks. The computer systems could be hacked and ships hijacked.

1.5.4 Advantages and disadvantages of different methods of processing

It is important to give both sides of the argument and to take into account the specific context and the user's needs when comparing methods.

While performing real-time processing, there is no significant delay in response whereas with batch processing the processing can take place well after the initial inputs have been entered. In real-time processing, information is always up to date, so the computer or microprocessor is able to take immediate action. With batch processing, the information is only up to date after the master file has been updated by the transaction file. Because real-time processing occupies the CPU constantly it can be very expensive, unlike batch processing which only uses the CPU at less busy times. Real-time processing needs expensive, complex computer systems whereas lower specification computers can suffice for batch-processing systems. Data is collected instantaneously with real-time systems whereas gathering of data for batch-processing systems can take a long time occupying more people so leading to greater expense in wages.

Comparing real-time systems to online systems, it is easier to maintain and upgrade online processing systems as computers are less busy overnight, so shutting down banking or shopping systems is less of a problem at less busy times, unlike real-time systems which do not have less busy times.

Comparing online systems to batch-processing systems, the extra hardware requirements, input devices, workstations and so on of a large online processing system can make it more expensive than batch processing. In a company using an online processing system, each transaction requires the entry of information immediately. This means that salespeople and other employees must be connected to the system at all times, unlike batch processing which only requires a limited number of employees to enter all the data at once. This leads to savings in wages. An advantage of batch processing compared to online processing is that it is less expensive than online input as it uses very little computer processing time to prepare a batch of data or transaction file, and this can be done at a time convenient to the employees responsible for entering the data. In online processing errors are revealed, and can be acted upon, immediately, compared to a batch processing system, where if there is an error it is only revealed when processing takes place. This can be overnight and as there may be no human involvement at this stage, management may not be aware of it until some time after the error actually occurred. Batch processing can be carried out overnight when the computer would not normally be used. This means that a company can get more work out of its computer hardware.

Interrogative databases are not well suited to batch processing as details may be required. Consider the case where a company uses batch processing for employee details, payroll and so on. An employee may have an accident and so a manager needs to contact the employee's next of kin immediately. It would be pointless running this query overnight. The system must be available whenever necessary so that the business can function properly.

Most modern business systems use a mixture of batch and online processing, in order to overcome some of the disadvantages described and benefit from the advantages.

Practice questions

1. A collection of data could be this: johan, Σ, $, <, AND

   Explain why these are regarded as just items of data. In your explanation give a possible context for each item of data and describe how the items would then become information. [5]

2. A company uses computers to process its payroll, which involves updating a master file.

   a. State what processes must happen before the updating can begin. [2]

   b. Describe how a master file is updated using a transaction file in a payroll system. You may assume that the only transaction being carried out is the calculation of the weekly pay before tax and other deductions. [6]

3. a. Name and describe the purpose of **three** validation checks other than a presence check. [3]

   b. Explain why a presence check is not necessary for all fields. [3]

4. A space agency controls rockets to be sent to the moon.

   Describe how real-time processing would be used by the agency. [5]

5. Describe **three** different methods used to carry out verification. [3]

6. L12345 is an example of a student identification code.

   Describe **two** appropriate validation checks which could be applied to this data. [2]

7. Describe **three** drawbacks of gathering data from direct data sources. [3]

2 Hardware and software

In this chapter you will learn:
- the characteristics and uses of mainframe computers and supercomputers
- the types and uses of system software
- the need for, types and uses of utility software
- the use of custom-written software and off-the-shelf software
- the types and uses of user interfaces.

Before starting this chapter you should:
- be familiar with the terms 'application software', 'CPU', 'command line interface', 'device', 'graphical user interface', 'medium', 'operating system', 'processor' and 'utilities'.

2.1 Mainframe computers and supercomputers

Mainframe computers are often referred to simply as mainframes. They are used mainly by large organisations for bulk data-processing applications such as censuses, industry and consumer statistics, and **transaction processing**. Most individuals tend to use personal computers, laptops or tablets, but mainframes are much larger and have more processing power than these and cost considerably more to buy. In 2020, the cheapest mainframe would cost at least $75 000. In the early days of computers, the central processing unit was very large compared to modern-day computers and used to be housed in a steel cabinet. This was often referred to as the 'main frame' and sometimes as the 'big iron'. Although many people thought that with the advent of PCs, mainframes would die out, they have continued to evolve and are still in major use today, mainly because of their features such as **RAS**, which we will go into in more detail later in this section.

Most PCs and laptops used to have a single processor, but today they tend to have a CPU with many cores which gives the effect of having many processors. This allows these computers to carry out **parallel processing** rather than the serial processing of their predecessors. Serial processing is when the PC performs tasks one at a time, whereas parallel processing allows several tasks to be carried out simultaneously. Currently, the best performing PCs have a processor with 18 cores, which allows the computer to carry out 18 tasks at the same time, resulting in much faster performance.

A mainframe computer can have hundreds of processor cores and can process a large number of small tasks at the same time very quickly. A mainframe is a multitasking, multi-user computer, meaning it is designed so that many different people can work on many different problems, all at the same time. Mainframe computers are now the size of a large cupboard, but between 1950 and 1990 a mainframe was big enough to fill a large room, so you can see how much the average mainframe has decreased in size while also improving its performance. Mainframes are not the only computers within large companies as many organisations own hundreds of personal computers (PCs) and laptops. Despite their popularity amongst commercial organisations, most other people are completely unaware of their existence. Mainframe computers have almost total reliability, being very resistant to **viruses** and **Trojan horses**.

Even more powerful are modern **supercomputers**, which can have in excess of 100 000 processing cores. In Chapter 1, batch processing of payroll was described. A supercomputer can multiply different hourly wage rates from a master file by a list of hours in a transaction file for hundreds of workers in roughly the same time that it would take a personal computer to calculate the wages of just one employee. The Oak Ridge National Laboratory in the USA launched its **Summit** supercomputer in 2018. It claimed, 'if every person on Earth completed one calculation per second, it would take the world population 305 days to do what Summit can do in 1 second'. The Summit supercomputer, however, fills a room the size of two tennis courts.

▲ Figure 2.1 The IBM z15 mainframe computer (© IBM Corporation)

2.1.1 Mainframe computers

The most advanced mainframe computer at the time of publication was the IBM z15, shown in Figure 2.1. The cabinet itself is taller than the average person.

The IBM z15 with up to 190 cores and its predecessor, the Z14 with 170 cores, are used by large banking organisations, government departments and large insurance organisations.

2.1.2 Supercomputers

▲ Figure 2.2 The Cray XC40 supercomputer

This is a picture of the **Cray XC40 supercomputer**, which can have up to 172 000 processor cores. Several countries have one of these and they use them in the fields of weather forecasting and scientific research, such as the study of astrophysics and mathematical and computational **modelling**.

2.1.3 Characteristics of mainframe computers and supercomputers

Longevity

Mainframe computers have great longevity, or lifespans. This is because they can run continuously for very long periods of time and provide businesses with security in the shape of extensive encryption in all aspects of their operation. Governments, banking organisations and telecommunications companies still base their business dealings on mainframes. Many of these computer systems have existed for decades and are still working well. To shut them down then dispose of the hardware is very expensive, as is the hiring of companies to securely remove their data.

Although they tend not to cause problems, there are several factors threatening their continued existence. One is the lack of experienced IT professionals who can maintain or program mainframes. Many of the older computer systems still work on the COBOL programming language, which is being taught at fewer and fewer universities in recent times. It has been gradually replaced by courses in Java, C and Python. Yet as recently as 2017, the news agency Reuters stated that 95 per cent of ATM transactions were carried out by computers using COBOL code. Another threat to mainframes comes from new technological developments, specifically Cloud computing, which is accessible from anywhere, thereby reducing the need to maintain expensive hardware within an organisation. The solution may well be to combine the use of mainframes with Cloud computing, thus giving organisations the flexibility and accessibility of the Cloud, while at the same time providing the processing power and security of the mainframe.

Despite these threats, the mainframe has remained popular for a long time, largely as a result of its efficiency and dependability. The mainframe is still able to process more transactions and calculations in a set period of time when compared with alternatives and it continues to operate with a minimum of downtime, which means that companies can operate 24 hours a day, every day.

In contrast, supercomputers have a lifespan of about five years. Research institutions and meteorology organisations are always looking for faster ways to process their data and so, unlike companies using mainframes, will tend to look at replacing their existing systems whenever much faster supercomputers come on to the market. Companies using mainframes are more inclined to modernise them using different software tools.

RAS

The term 'RAS' is frequently used when referring to mainframe computers and stands for reliability, availability and serviceability. RAS is not a term that is used, on the whole, with supercomputers.

Reliability

Mainframes are the most reliable computers because their processors are able to check themselves for errors and are able to recover without any undue effects on the mainframe's operation. The system's software is also very reliable, as it is thoroughly tested and updates are made quickly to overcome any errors.

Availability

This refers to the fact that a mainframe is available at all times and for extended periods. **Mean time between failures (MTBF)** is a common measure of systems, not just those involving computers. It is the average period of time that exists between failures (or downtimes) of a system during its normal operation. Mainframes give months or even years of system availability between system downtimes. In addition to that, even if the mainframe becomes unavailable due to failure, the length of time it is unavailable is very short. It is possible for a mainframe to recover quickly, even if one of its components fails, by automatically replacing failed components with spares. Spare CPUs are often included in mainframes so that when errors are found with one, the mainframe is programmed to switch to the other automatically. The operator is then alerted and the faulty CPU is replaced, but all the time the system continues to work.

Serviceability

This is the ability of a mainframe to discover why a failure occurred and means that hardware and software components can be replaced without having too great an effect on the mainframe's operations.

Security

In addition to their other characteristics, mainframes have greater security than other types of computer systems. Data security is considered to be the protection of data from intentional or accidental destruction, modification or disclosure. As has already been mentioned, mainframes are used to handle large volumes of data. Most of this is personal data and, especially in the banking sector, it has to be shared by the banks with customers. This means that the data has to be extremely secure so that only those users entitled to see the data can do so; in other words, customers must be able to see their own data but not each other's. This applies to other uses of mainframes. Many companies store the personal information of their employees, customers, and so on. Fortunately, the mainframe computer has wide-ranging security that enables it to share a company's data among several users but still be in a position to protect it. A mainframe has many layers of security including:

- user identification and authentication, although more and more systems are using multi-factor authentication, which is a combination of two or more of the following: a password, a physical token, a biometric identifier or a time-restricted randomised PIN
- **levels of access**, which means that it depends on a user's level of security as to which sets of data they can access
- encryption of transmitted data and data within the system
- secure operating systems
- continual monitoring by the system for unauthorised access attempts.

In addition to the use of supercomputers to perform massive calculations, they may also be used to store sensitive data such as DNA profiles. This obviously requires a very high level of security. Most supercomputers use end-to-end encryption, which means that only the sender or recipient is able to decrypt and understand the data.

Performance metrics

The **performance metrics** of a computer are basically the measures used to determine how well, or how fast, the processor deals with data. The speed of a mainframe's CPU is measured in millions of instructions per second (**MIPS**). However, it is not always the best measure, because not all instructions are the same. Mainframes use a very large number of different instructions, with some being straightforward and easy to carry out, while others can be more complex and slower to process. An application using five million simple instructions will take a lot less time than one using five million complex ones. In addition, the number of available instructions is increasing as time goes by and mainframes improve. It is important that the comparison between the performance of one mainframe and another is made by measuring how fast the CPUs are when carrying out the same task. This is referred to as a benchmark test. This measure is usually obtained when processing application software. MIPS are often linked to cost by calculating how much a mainframe costs per one million instructions per second.

Supercomputers use a different set of metrics. As they are used mainly with scientific calculations, their performance is measured by how many FLoating point Operations can be carried out Per Second (**FLOPS**). Since the original supercomputers were developed, speeds have increased incredibly and are now measured in petaflops. One petaflop is 1 000 000 000 000 floating point operations per second. Experts are already using the term exaflops, which are 1000 times faster than petaflops, and are expecting the first supercomputer to attain this speed sometime in the current decade. The speed of the current fastest supercomputer, at the time of publication, is 148 petaflops and even the tenth fastest operates at 18 petaflops.

Volume of input, output and throughput

The volume of input, output and throughput has to be considered when describing computers.

Mainframes have specialised hardware, called peripheral processors, that deal specifically with all input and output operations, leaving the CPU to concentrate on the processing of data. This enables mainframes to deal with very large amounts of data being input (terabytes or more), records being accessed, and subsequently very large volumes of output being produced. Modern mainframes can carry out many billions of transactions every day. This large number of simultaneous transactions and extremely large volumes of input and output in a given period of time is referred to as 'throughput'.

A supercomputer is designed for maximum processing power and speed, whereas throughput is a distinct mainframe characteristic.

Fault tolerance

A computer with **fault tolerance** means that it can continue to operate even if one or more of its components have failed. It may have to operate at a reduced level, but does not fail completely. Mainframe computers have the characteristic of being fault-tolerant in terms of their hardware. While in operation, if a processor fails to function, the system is able to switch to another processor without disrupting the processing of data. The system is also able to deal with software problems by having two different versions of the software. If the first version produces errors, the other version is automatically run.

Supercomputers have far more components than a mainframe, with up to a thousand times more processors alone. This means that statistically, a failure is more likely to occur and consequently interrupt the operation of the system. The approaches to fault tolerance are much the same as those for mainframe computers, but with millions of components, the system can go down at any time, even though it tends to be up and running again quite quickly.

Operating system

Most mainframes run more than one operating system (OS) at any given time and the use of z/OS, z/VM®, and Linux® (which are all different operating systems) at the same time often occurs. The OS on a mainframe divides a task into various sub-tasks, assigning each one to a different processor core. When each sub-task has been processed, the results are recombined to provide meaningful output. This is called parallel processing and it is what makes a mainframe far more efficient than a PC, which, despite having more than one core these days, has a very limited capability regarding parallel processing.

Supercomputers tend to have just one OS, Linux, but most supercomputers utilise **massively parallel processing** in that they have many processor cores, each one with its own OS. Linux is the most popular, mainly because it is open-source software, that is, it is free to use.

Number of processors

Early mainframes had just one processor (the CPU), but as they evolved more and more processors were included in the mainframe system and the distinction between the terms 'CPU' and 'processor' became confused. One major mainframe manufacturer called them 'CPU complexes', which contained many processors. The number of processor cores found in a mainframe is now measured in the hundreds.

By contrast, supercomputers have hundreds of thousands of processor cores. Unlike mainframes, modern supercomputers use more than one **GPU** or **graphics processing unit**.

Heat maintenance

Because of the large number of processors in both mainframes and supercomputers, overheating becomes a major problem and **heat maintenance** or heat management, as it is often called, has to be implemented. In the early days of mainframe computing, the heat produced by the computer could not be controlled using cooling by fans. Liquid-cooling systems had to be used. When integrated circuits were developed in the 1970s and became universal in the 1980s, mainframes were developed which produced less heat and so fans were able to dissipate the heat. However, recent developments in mainframe technology involving more powerful hardware mean the overheating issue has resurfaced. More powerful systems produce more heat. What were considered to be a relatively cheap option – air cooling systems – are becoming more complex and more expensive to use in more powerful systems. At the same time, water-cooling solutions have become more cost-effective. We now appear to have come full circle and mainframe manufacturers are once again recommending water-cooled systems for larger machines.

The overheating problem has always been present with the use of supercomputers. The large amount of heat produced by a system also has an effect on the lifetime of components other than the processors. Some supercomputers draw four or more megawatts of power to keep them operating at high efficiency. That is enough to power several thousand homes! This, together with having so many processors very close together, results in a great deal of heat being produced and requires the use of direct liquid cooling to remove any excess heat.

> **Activity 2a**
>
> Explain what is meant by these terms:
> 1. MIPS
> 2. FLOPS
> 3. Fault tolerance

2.1.4 Mainframe computer uses

Mainframe computers play a vital role in the daily operations of many organisations. Finance companies, health care providers, insurance companies, energy providers, travel agencies, and airlines all make use of mainframes. By far and away, however, the greatest use of mainframes is in the banking sector, with banks throughout the world using mainframes to process billions of transactions. The key benefit of mainframes is their ability to process many terabytes of data, which is very useful when carrying out batch processing. Batches of transactions are processed by the mainframe without user interaction. Large volumes of data are read and processed, then customers' bank statements, for example, are output. During batch processing, which usually takes place overnight, other jobs can be carried out such as back-ups. Mainframes are also used in other areas such as censuses, industry statistics, consumer statistics, and transaction processing.

Census

Census is a term that, when used alone, normally refers to a population census. A population census is an official survey of the people and households in a country that is carried out in order to find out how many people live there. It is used to obtain details of such things as people's ages and types of employment. The amount of data that has to be processed is enormous. The processing of census data has long been associated with computers. The **UNIVAC 1**, which was the first computer on general sale to organisations or businesses, was purchased by the United States Bureau of the Census in 1951. After that, technological innovations have seen the production of different generations of mainframes and other developments have enabled census agencies to become more knowledgeable in the way they process and manipulate data. A census usually takes place every ten years. Not surprisingly, as populations increased, each census produced more data and so the use of mainframes to process the data was crucial. However, by the year 2000, the increase in processing power and storage capacities of PCs provided many countries with the opportunity to take a new approach and many are deciding against buying expensive mainframe computers.

Industry statistics

Industry statistics are statistics that are recorded regarding trends in different industries, such as those that process raw materials, make goods in factories, or provide services. They can include the number and names of businesses, the number of employees and wages paid. Some businesses in certain sectors of industry need mainframes to process the vast amount of data which helps to identify their major competitors. It shows their competitors' share of the market and the trends in their sales. This helps to identify those products which could compete profitably with other businesses. Where companies do not feel the need to process their own data, they can obtain reports from organisations that collect the data for them and those companies could use mainframes for this purpose.

Consumer statistics

Consumer statistics allow businesses to assess the demand for their product, that is how many people need or want that type of product. They can inform them about the range of household incomes and employment status of those consumers who might be interested in the product so that a price can be set. This data will also inform businesses of where the consumers live for local sales or how inclined they are to use the internet for shopping. It may also allow businesses to know how many similar products are already available to consumers and what price they pay for them. These statistics produce an incredible amount of data and the organisations that produce these statistics are likely to use mainframes.

Transaction processing

Transaction processing can consist of more than one computer-processing operation, but these operations must combine to make a single transaction. Each of the operations that constitute the transaction must be carried out without errors occurring, otherwise the transaction will be deemed to have failed. A transaction-processing system, in the event of an error, will remove all traces of the operations and the system will continue as though the transaction never happened. However, if there are no errors, the transaction is completed and the relevant database is updated before the system continues on to the next transaction. All this, of course, has to happen in milliseconds, as it must seem to the user that their transaction has been carried out immediately. Transaction processing involves the use of online processing, which was looked at in Chapter 1.

One example of transaction processing is the transfer of a sum of money from one bank account to another, for example when you pay for goods at a supermarket checkout. The bill might be $100. The customer needs to transfer $100 from their bank account to the supermarket's bank account. To make it simpler, consider that the customer and the supermarket use the same bank. There are two operations involved, one is that $100 must be subtracted from the customer's account balance and the other is that $100 must be added to the supermarket's account balance. If either of these operations fails then the transaction will be cancelled by the mainframe, or the bank would not be able to balance the books at the end of the day.

Why is the mainframe computer so well suited to transaction processing? First, the effectiveness of a transaction-processing system is often measured by the number of transactions it can process in a given period of time. We know a mainframe can perform hundreds of MIPS. The system must be continuously available during the times when users are entering transactions and the system must be able to deal with hardware or software problems without affecting the integrity of the data. It must also be possible to add, replace, or update hardware and software components without shutting down the system. With its ability to transfer processing from one core to another, the mainframe is more than suitable for the task.

> **Activity 2b**
>
> Give **two** reasons why the use of mainframes for carrying out censuses is reducing.

2.1.5 Supercomputer uses

The first use of supercomputers was in national defence, for example, designing nuclear weapons and data encryption. Supercomputers have become important in the field of scientific research, particularly quantum mechanics. They have also become essential for weather forecasting. There are now many other uses of supercomputers, one of which is drug research. Before a drug is tested, it needs to be developed, but it does not start with a blank sheet of paper. Quite often the results of previous research are stored and then compared with the results of the new drug that is being developed. The computer is used to monitor the amount of ingredients that are being used. It is important that the researchers can change these amounts by the smallest fractions. Modern drug research involves the use of complex computer **models** to see how changing the structure of the drug affects the way the body reacts. Computer models can also help to predict any side effects. As the changes to the model are so tiny and the number of times the model must run is so high, a computer with great computational power is required. This makes the use of supercomputers indispensable to drug manufacturers.

Another field where supercomputers are used is genetic analysis. Finding the genes that make humans susceptible to disease has always been very difficult. There are so many genes, and variations of them, that humans were unable to carry out the calculations required to identify those genes responsible. With the advent of supercomputers, however, this task has become more manageable, reducing the time taken to perform the calculations from months to minutes.

Here, we will focus on three uses of supercomputers: quantum mechanics, weather forecasting and climate research.

Quantum mechanics

Quantum mechanics is the study of the behaviour of matter and light on the atomic and subatomic scale. It attempts to describe the properties of the constituent parts of an atom, such as electrons, protons and neutrons, and how they interact with each other. They do not behave in the same way as larger bodies, which obey the laws of physics. You do not need to understand quantum mechanics, but it is important to realise that there are a very large number of calculations which require great accuracy and thus require the use of a supercomputer. The Juqueen supercomputer in Germany, now decommissioned, was used by a team of physicists to calculate the difference in mass between a neutron and a proton as well as predicting the make-up of dark matter. The Juqueen had a maximum performance of 5 petaflops with 459 000 cores. It has been replaced by the Juwels supercomputer with a performance of 12 petaflops (although the developers are currently working to increase that to 70 petaflops!).

Weather forecasting

Weather forecasting is based on the use of very complex computer models. Data from the sensors at **weather stations** around the world is input to the model and then many calculations are performed. Records of previous weather conditions have also been collected over a very long period. Using the past weather readings, the computer examines similar patterns of weather to those being experienced at the moment and is able to predict the resulting weather. Variables such as atmospheric pressure, humidity, rainfall, temperature, wind speed and wind direction are recorded using computerised weather stations around the world. These readings, together with observations from radar, satellites, soundings from space, and information from ships and aircraft, help the supercomputer to produce a 3-D model of the Earth's atmosphere. Because of the complexity of the calculations, and the very large number of them that need to be carried out, they can only run effectively on supercomputers.

In 2017, the UK Meteorological Office commissioned three Cray XC40 supercomputers to help with formulating its weather forecasts. Together they are capable of running at 14 petaflops and have a total of 460 000 cores, each of which is slightly faster than those found in a quad (4) core laptop!

Climate research

This is an extension of the use of IT in weather monitoring. Climate is measured over a much longer timescale. The data collected over several decades can be used to show the trends of different variables over time. For example, the levels of nitrogen dioxide, sulphur dioxide and ozone are monitored to determine air quality. Rivers are monitored, measuring different variables such as temperature, pH, dissolved oxygen, turbidity (the cloudiness of the water caused by minute particles being present), and water level.

Climate is now being looked at from a global point of view. The planet Earth can be regarded as a complex system which has many different constituent parts, such as heat from the Sun, ocean currents, vegetation, land ice as found in Greenland and the Antarctic, sea ice as found in the Arctic, weather and climate, volcanoes and, of course, humans. The interactions of each of these components within the Earth system can be described in mathematical terms. However, just as with weather forecasting, there are very many variables which have to be collected and complex calculations which need to be carried out. We have to rely on computer models to help us understand how the system works. Climate is one component of this planet's system and only supercomputers are able to run the models that represent the interactions between the components of this system. CESM®, the Community Earth System Model, is the most widely used climate model in the world today. It includes modules which allow for atmosphere, land, ocean, sea ice and land ice, and uses various equations to mimic the process of climate change using a virtual environment created on the supercomputer.

Many people and organisations are becoming involved in developing supercomputer climate models as climate change is developing into a potentially catastrophic situation. In 2019, a team of students from the Sun Yat-sen University in Guangzhou, China, won the e-Prize at the 2019 ASC Student Supercomputer Challenge (ASC19) finals. They reduced the full-mode operation time of CESM to 1.83 hours.

> **Activity 2c**
> Give **two** reasons why supercomputers are used in weather forecasting.

2.1.6 Advantages and disadvantages of mainframes and supercomputers

Advantages of mainframe computers

Mainframes are very reliable and rarely have any system downtime. This is one reason why organisations such as banks use them. Most other systems do fail at some point and then have to be rebooted or restarted, as most people who use laptops or PCs will know. In addition, hardware and software upgrades can occur while the mainframe system is still up and running.

Mainframes are getting faster and more powerful every year and are completely outperforming PCs, laptops and other devices.

Mainframes can deal with the huge amounts of data that some organisations need to store and process. Most organisations that use mainframes would find it exceptionally difficult to transfer all the data they have stored on their current mainframe to an alternative system. Because of a mainframe's ability to run different operating systems, it can cope with data coming in a variety of database formats which other platforms would find problematic.

Mainframes have stronger security than other systems with complex encryption systems and authorisation procedures in place.

Disadvantages of mainframe computers

Mainframes are very expensive to buy and can only be afforded by large organisations such as multinational banks. There is also a high cost for the personnel required to run and maintain them. Large rooms are required to house the system, which is not needed with other systems. As mainframes become more advanced, the cooling systems needed become more expensive to install and run.

Many organisations are migrating to Cloud-based services so they do not have to buy their own system or hire the expertise required. The software required to run mainframe systems is more expensive to buy than using the Cloud.

Advantages and disadvantages of supercomputers

Supercomputers are the fastest data processing computers but are also the most expensive to buy and install. The Summit supercomputer, which at the beginning of 2020 was the world's fastest supercomputer, cost $200 million to build compared with IBM's latest mainframe which costs between $250 000 and $4 million depending on the organisation's requirements.

Most supercomputers have one operating system, whereas mainframes can have more than one. Supercomputers are less fault tolerant than mainframes meaning they are less likely to recover as quickly in the event of the failure of one component, and are down more often than mainframes, although not as often as some other systems.

Supercomputers use massively parallel processing, which makes them more powerful compared to the parallel processing of mainframes, and much more powerful than PCs which have far fewer processor cores than mainframes or supercomputers.

2.2 System software

System software refers to the programs that run and control a computer's hardware and application software.

2.2.1 Types and functions of system software

Examples of system software are **compilers**, **interpreters**, **linkers**, **device drivers**, operating systems and utilities.

Compilers

Most software that runs on computers is in machine code, which is stored in binary form within the computer. Machine code consists of the instructions that computers understand and each instruction is actually a number written in binary. Unfortunately, programmers who write the software for computer users find it difficult to use machine code for programming purposes. In the early days of computing it was felt necessary to develop languages which programmers could understand and was close to their use of everyday language.

Some of the early high-level languages developed included:

» **FORTRAN (Formula Translation)** was the first high-level language to be developed. It was very technical and mathematical in nature. FORTRAN is rarely used these days, but still remains popular for simulating large physical systems, such as the modelling of stars and galaxies. The fact that it is still widely used by physicists today often mystifies computer scientists, who regard it as being out of date.

- » **COBOL (Common Business-Oriented Language)** was developed not long after FORTRAN to help businesses. Unlike FORTRAN, which relied on mathematical equations, COBOL tended to use simple expressions, such as SUBTRACT money FROM account GIVING balance, rather than an equation like bal=acc-mon.
- » **LISP (List Processor)** was also an early high-level language. In LISP, simple statements often start with the mathematical operator, such as (+ 2 4) which would calculate 2+4. LISP was used extensively in the early years of AI but, apart from its use in association with CAD, it is rarely used today.

Most high-level language programming for modern-day systems is done using languages such as Visual C++®, Visual C#®, Visual Basic® or Pascal®, among others. These have largely replaced the other languages mentioned above. However, the development of all these high-level languages led to the need for software which could translate a program written in high-level language into the machine code that computers could understand.

A compiler is software (a program) that processes statements written in a high-level programming language and converts them into machine language or code that a computer's processor understands and can execute. In other words, it translates a high-level language program called source code into an executable file called object code. The compiled program is then run directly without the need for the compiler to be present.

Although compilers translate the whole program as one unit, they often need to make more than one pass through the program to do this. This is because after translating, it may become apparent that line 10 of the program refers to a statement at line 5. The first pass picks up this type of information and then the second pass completes the translation. The whole program is still effectively translated in one go before it is executed. However, some programming languages, such as Pascal, have been designed so that they would only require one pass. The object code is machine code that the processor can execute one instruction at a time. A compiler produces a list of error messages after it has translated the program. These cannot be corrected without going back to the source code.

Interpreters

An interpreter translates the high-level language program one statement, or line, at a time into an intermediate form, which it then executes. It continues translating the program until the first error is met, at which point it stops. This means that errors are located by the programmer, exactly where they occur. An interpreter has to do this conversion every time a statement is executed.

Unlike with a compiler, translation occurs at the same time as the program is being executed. The interpreter has to be resident in memory in order for the program to be executed. Interpreters just run through a program line by line and execute each command. As a result, another benefit of using an interpreter is that only a few lines of the program need to be in memory at any one time, saving memory space.

There are other benefits of using interpreters, such as they help programmers when they are developing programs. Interpreters are able to execute each statement as it is entered and are able to generate helpful error reports. This means that interpreters can be used during program development allowing a programmer to add a few lines at a time and test them quickly.

Some high-level language programs such as Python® can be translated using a combination of a compiler and an interpreter. It is possible for the program to be translated or compiled into what is called 'bytecode'. This is then processed by a program called a 'virtual machine', which is often an interpreter, instead of by the processor within the computer.

> **Activity 2d**
> Describe the purpose of compilers and interpreters.

Advantages and disadvantages of interpreters compared with compilers

▼ Table 2.1 Advantages and disadvantages of interpreters compared with compilers

Advantages of interpreters	Disadvantages of interpreters
While a program is being compiled, if it is a major application and takes a long time, the programmer has to wait, doing nothing, before they can correct any errors. With an interpreted program the programmer can correct errors as they are found.	The execution process is slower for an interpreted program as each statement must be translated before execution, whereas once a program has been compiled it executes all in one go. The time it takes to compile a program can be long but once it is compiled it does not have to be translated, which compares favourably with an interpreted program which has to be translated every time it runs.
Debugging is easier with an interpreter as error messages are output as soon as an error in a statement is encountered, which gives the programmer the opportunity to correct the error there and then. With a compiler all the error messages are output at the end of the process and errors can be difficult to locate and correct.	With an interpreted program, the source code must always be available, which can lead to software copyright infringement or intellectual property rights being at risk. With a compiled program, the source code has been translated and machine code is difficult to understand and alter.
A compiled program will only run on a computer with the same operating system as the computer it was originally compiled on, whereas an interpreted program will still be in its original source code and so it will work on any system with the appropriate interpreter.	
Compiling a program uses more memory than interpreting it as the whole program must be loaded before translation, whereas with an interpreter only a few statements of the program have to be in memory at any given time. This means that small pieces of code can be tested to make sure they work before continuing with the rest of the program.	

Linkers

A linker, or to give it its full name a **link editor**, is a system program that combines object files or modules that have been created using a compiler into one single executable file. Most programs are written in modular form. That is to say, a number of programmers write separate pieces of code or modules that form the required program when combined, which has the advantage of saving time than if one person wrote the whole code, although it is still possible for one person to write all the modules. If there is an error only that module has to be corrected.

In short, a linker is used to combine different modules of object code into one single executable code program. It could be that a large program can only be compiled in small parts because there may not be enough RAM to hold the whole program and the compiler program. The parts of the program can be stored on backing storage and then, one at a time, each part is brought into RAM and compiled. The resulting object code is then saved to the backing storage. When all the parts have been compiled, the compiler is no longer required to be in RAM, all the pieces of object code can be brought back into RAM and the linker can be used to combine them into the complete program. An advantage of using linkers, therefore, is that programs can be written in modules which requires less RAM so saving money. The disadvantages are that there can be problems with variable names (this term will be explained in Chapter 4) and also documentation has to be more detailed and takes longer to write, or read when completed.

Device drivers

A device driver is a small program that enables the operating system (OS) and application software to communicate with a hardware device. One example is a printer driver which acts as an interface between the OS (or any application software that is running) and the printer. The user, through the application, might want to print information. The application tells the printer driver and the printer driver tells the printer. This, in effect, allows the user to have control of the device. Other devices which need drivers are sound cards, monitors, mice, SSDs, network cards, keyboards, disk drives and many other items of hardware.

The software used by a computer tends to be created by different companies to those that manufacture hardware. This results in software which uses a different type of language to the hardware. There has to be some means of converting the languages so that the software is able to communicate with the hardware and vice versa. Think of a Spanish person who knows no French trying to communicate with a French person who speaks no Spanish. They would not be able to understand each other. If, however, they managed to find some translating software, that is a program that could convert Spanish to French and French to Spanish, they could run that software on a computer and be able to communicate. The device driver, which is, after all, a piece of system software, performs the same function as translating software. So, application software such as a word processor sends information to a driver saying what it wants the hardware to do; the device driver understands this and then tells the hardware what it needs to do.

If the appropriate driver is not installed, the computer is unable to send to or receive data from the hardware devices. This means, in effect, the device will not work. Modern operating systems are supplied with many drivers that allow hardware to work at a basic level. However, if the OS of a computer does not recognise certain features of the device, it will not work without drivers. It is possible to connect any keyboard or mouse into a computer and it may work. However, if that keyboard has any special keys or features or the mouse has extra buttons, they will not work until the drivers are installed. In addition, drivers that are produced for a specific OS will rarely work with an alternative OS.

Operating systems

Before an operating system (OS) is loaded the computer has to boot up. Booting up a computer is starting it up by loading the BIOS. BIOS is stored in ROM and is the Basic Input/Output System for the computer which executes during boot-up. It checks various devices are present. It then loads the OS.

An OS is system software that manages computer hardware and software resources as well as interacting with device drivers. Some of the device drivers are separate to the OS, but are often included by the provider of the OS. The OS acts an interface between the user and the computer, as well as supplying important utilities for managing the computer. A utility program is a type of system software that assists users in controlling or maintaining the operation of a computer, its devices or its software. The OS also acts as an interface between an application program and the computer hardware, so that an application program can communicate with the hardware. The part of the operating system that facilitates this is called the kernel. To sum up, an operating system interacts with application software, device drivers and hardware to manage a computer's resources, such as the processor, RAM, storage space, and peripherals.

For nearly all types of computer, the operating system program is large and would occupy too much ROM, so most of it is stored on a hard disk. However, the instructions for loading the operating system are stored in ROM and are executed every time the computer is switched on.

One of the major functions of an operating system is to manage the computer's memory. The OS allocates a particular part of RAM for each program, whether it is an application, system software or a utility that is running. It needs to make sure that instructions and data from one program do not spread into another program's memory allocation, otherwise data can get corrupted and the computer could crash.

Another of the main functions of an OS is to manage data input and output. To do this, it needs to be able to respond to input devices in order to receive and manage that data. It does this by communicating with the device driver so that it can receive data from the input device. It also uses device drivers when it sends data or instructions to the printer.

The OS manages the storing of files on, as well as the loading of them from, backing storage. It knows the names of each file and exactly where they are stored on the hard disk, tape, pen drive or SSD. It also keeps a record of any empty spaces on the medium so that it knows where new files can be stored. Using a disk drive as an example, it requests from the disk drive the position of the first available free storage location on the disk. It then records on the disk the position of the start of file and the end of file, as well as other details of the file.

Strictly speaking, multitasking and multi-programming systems are not the same thing. However, the OS has the same responsibility in both, in that it must allocate time to each task or program fairly, so that all tasks or programs get a reasonable amount of time. Most computers these days are able to multitask, so we will concentrate on those here. The OS loads the software in RAM for each task and the computer gives each application a tiny amount of time before moving to the next task. This process is repeated for however many tasks or programs are running at the same time.

Another responsibility of the OS is to display error messages to the user should an error occur which requires the user to intervene. A typical error might be that a user, when they are trying to save their work, types in a symbol that is not allowed in the file name, such as /. The OS will output a message saying that it is an invalid file name and will not resume until the user takes action.

When a user logs in to a system, it is the OS that deals with this. Passwords are no longer stored as plaintext, but are encrypted. A calculation is performed on the password and the result is stored. When a password is entered by a user, the calculation is performed again by the OS. If the result is the same as the previous calculation result stored for that user, then the OS allows the user to access the system. Even when a user has successfully logged in to a system, they may still only have permission to access certain files. These are often referred to as file permissions or access rights. Some of the files may be particularly sensitive and only certain people will be allowed to look at them. In general, it is the operating system's responsibility to handle the security of the system.

When a user wishes to shut down the computer, the OS has to safely close all software running on the computer. It then shuts itself down bit by bit before finally sending a signal to the power management hardware to turn off the power.

> **Activity 2e**
> List the different functions of an operating system.

2.2.2 Uses of system software

How a high-level language is translated to run on different computer systems using interpreters

There may be times when programmers wish to transfer their program to another computer which has a different operating system and processor. In this case they tend to write their programs in high level languages which use interpreters. An interpreted program can be transferred between computers with different operating systems because it remains in the form of source code, but it needs to be translated in each computer it is moved to. This takes more time than with a compiler, but it means that the program can be distributed regardless of the processor or operating system (OS) of the computer. The computer that it is transferred to must have a suitable interpreter. However, if the interpreter needs to be supplied together with the program, the overall process is more complicated than with a compiler.

How a high-level language is translated to run on different computer systems using cross compilers

For compiled languages, it is possible to use a cross compiler, although this can be complicated. The computer the program is written on is called the host computer. The computer which the program is to be run on, after compilation, is called the target computer. The cross-compiler is responsible for translating the source code into machine code suitable for the target computer. This machine code interacts with the kernel of the target computer to access systems services and hardware resources. This enables the development of software for computers with different architectures and characteristics than the host computer.

One problem with using a cross compiler is that the compiled program will then no longer run on the host computer.

Another problem comes as a result of a cross compiler tending to be a slimmed down version of a 'native' compiler (the compiler normally found on the host computer so that the compiled program can be run on that computer). The cross compiler produces more errors and mistakes than a native compiler. In addition, the compiled code can run slower on the target machine than if it had been produced on that machine.

2.3 Utility software

2.3.1 The need for utility software and its uses

Utility software is a type of system software that is needed to help maintain a computer system. A basic set of utility programs is often supplied with the OS, most of which help to manage files and their associated storage devices. However, users can sometimes feel the need to get additional utilities which perhaps the

OS does not possess. Much of the work a computer does revolves around the use of files of data so it is no surprise that it needs a number of programs that deal with file handling. Without utility software, the computer would not function properly. Utility software is required to manage the allocation of computer memory in order to improve the computer's performance and so that users can customise the appearance of their desktop.

The structure of hard disk storage

A hard-disk drive consists of several **platters** (individual disks) with each side (surface) of the platter (top and bottom) having its own read–write head. The read–write heads move across the platters, floating on a film of air and, when they stop, they either read data from or write data to the surface. They never touch the **disk surface** and each surface is used to store data. The platters are stacked one above the other and spin together at the same speed. Each surface is divided into several **tracks** and each track is divided into **sectors**. The tracks are in the same position on each disk. The track on the top platter together with the tracks exactly below it form a cylinder. Data is written to the platters starting with the outside track. When a cylinder has been filled with data, the read–write heads move toward the centre of the disk. They write data on each of the second tracks until the second cylinder is full. The diagram just shows one track out of the many that would be on the disk. There are normally 512 bytes of data in a sector. Operating systems normally deal with data in **blocks**. A block can consist of one, two, four, eight or more sectors.

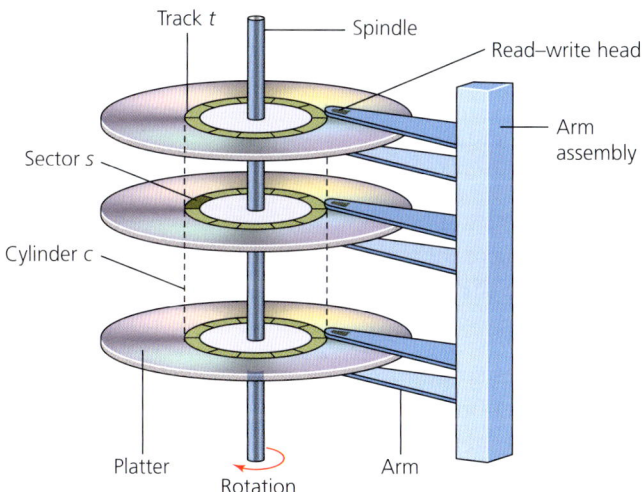

▲ **Figure 2.3** The structure of hard-disk storage

Formatting

When a hard disk is first manufactured, it is not suitable for use so manufacturers format the disk so it can be used with a computer, otherwise it would be completely unusable. When a customer receives the disk it is ready to use. Other reasons for formatting disks is to remove data if the disk is to be used to replace a hard disk in another computer. Sometimes a user may wish to change the operating system on a computer so they need to remove all existing system files and settings. **Disk formatting** is the configuring of a data storage medium such as a hard disk or SSD for initial use. It can be performed on a disk that already has files on it, but all those files would be erased. Disk formatting is usually carried out on a new disk or on an existing disk if a new OS has to be installed. There are two levels of formatting: **low-level formatting** and **high-level formatting**.

Low-level formatting prepares the actual structure of the disk by dividing the disk into cylinders and tracks and then dividing the tracks into sectors. This type of formatting is usually carried out by manufacturers rather than individual users. If users were to attempt this, there are at least two drawbacks: after erasing all the files, it would be almost impossible to restore them and if done repeatedly, it would shorten the life of the medium. When high-level formatting is carried out, it does not permanently erase data files but deletes the pointers on the disk that tell the OS where to find them. These pointers are in a File Allocation Table which is stored on the disk and contains where the file starts and the length of the file. High-level formatting is typically done to erase the hard disk and reinstall the OS back onto the disk drive. Unlike low-level formatting, the files are retrievable. One benefit of disk formatting is that it can remove viruses.

Often users want to split their hard disk into two sections. They may want to separate the operating system from their data files. This is useful so that if the operating system has to be reinstalled it is only done in that one partition while the users' data files remain intact on another partition. Partitions can be setup so that sensitive data which needs to be encrypted exist on one partition while user data which is not confidential or sensitive is stored on another partition. A partition table is created which stores information such as where about on the disk the partition is as well as the amount of memory it is taking up. This table is read by the operating system before any other part of the disk is read. It appears to the operating system that each partition is a disk in its own right and is treated as such. System administrators are able to use a program called a partition editor which enables them to create or change the partitions.

Disk defragmentation

This is needed so that the time taken to retrieve data from a fragmented file is reduced as well as the need to increase the available storage space on a disk.

When a file of data is stored on a disk, it may consist of several blocks. There may not be enough blank or empty sectors next to each other to store all the blocks together which means they have to be spread out across the disk. The file is now said to be fragmented.

When data is no longer needed it is deleted from the disk, which leaves some empty sectors. **Defragmentation** software is used to organise the data on the disk by moving the data blocks around to bring all the parts of a file together so they are contiguous. As a result, data retrieval is made easier and quicker. With fragmented files it takes longer for the read–write heads to move over the surfaces to find all the different fragments of files than if the data is held in sequence. In addition, the software provides additional areas of free space and more storage capacity. Sometimes it is impossible to bring all the blocks of data belonging to a file together, but the software will move as many blocks together as possible. It can also attempt to keep smaller files which belong in the same folder or directory together by reorganising other files.

Data compression

This is often needed when disks are getting full. It is also needed when large files are being transmitted over a network or as email attachments so that the speed of transmission is increased.

Data compression is the modifying of data so that it occupies less storage space on a disk. It can be lossless or lossy. **Lossless compression** is where, after compression, the file is converted back into its original state, without the loss of a single bit (binary digit) of data. When the lossless compression software sees a repeated sequence of bits it replaces the repeated sequences with a special character which indicates what is being repeated and how many times. This type of compression is normally used with spreadsheets, databases and word-processed files, where the loss of just one bit could change the meaning completely. **Lossy compression** permanently deletes data bits that are unnecessary, such as the background in a **frame** of a video which might not change over several frames. Lossy compression is commonly used with images and sound, where the loss of some data bits would have little effect. **JPEG** is an image file format that supports lossy image compression. Formats such as **GIF** and **PNG** use lossless compression. Other methods of **file compression** can be found in Chapter 11.

An advantage of data compression is that it means data can be transmitted more quickly over a network or the internet. Another advantage is that it saves storage space on a disk or SSD. Consequently, administrators spend less money and less time on storage. It also allows the streaming of high definition (HD) videos which would ordinarily occupy a great deal of bandwidth and memory.

There are disadvantages however, such as the fact that data compression software uses a lot of computer memory during the compression process. Another disadvantage is that the process of loading or opening a compressed file takes a lot longer than opening the original. Also, lossy compression causes a slight lowering of the quality of the resulting sound and video files.

Back-up

Back-up software is a program that is used to keep copies of files from a computer or copy the content of a server's backing storage. The back-up is an exact duplicate of the files. It can be used for restoring the original files, should the original files become corrupted or deleted, accidentally or deliberately. Back-up software allows the user to select the time and type of back-up they want and how regularly the back-up is to take place.

When tapes or external disk drives are used to store server back-ups, they are normally stored in a room separate to the one where the server is housed. Should a disaster occur with the server, such as fire or malicious damage, the back-up tapes or disks are still safe. This is also the case where the Cloud is used; damage to the server will not affect the state of the back-ups.

Most back-up software allows different types of back-up, not just a full back-up. An incremental back-up is one where only the data that has been added or changed since a specific date and time is backed up. Back-ups can take a long time to carry out and require a certain amount of storage space, so this situation is improved if back-ups are done incrementally. However, this can lead to difficulties when restoring what may well be several back-ups. An alternative is a differential back-up which only backs up the data which has changed since the last full back-up. Restoring the system only requires the use of two back-ups in this case. Users can also opt to verify their back-up; this is when checksums (see Chapter 1) are used to make sure the data matches. Back-up software can also allow the user to choose whether they want the back-up encrypted or not.

File copying

A copy of a file can be needed when a file needs to be worked on so a temporary copy can be made for amendments leaving the original intact.

File copying is creating a duplicate copy of an existing file. The copy will have exactly the same content as the original. There are a number of ways a file-copying utility works. In disk operating systems, they tend to involve the use of a **command line interface (CLI)** where the user types some form of 'copy' command which contains the name of the original file as well as the destination directory or folder. With a **graphical user interface (GUI)** the user can have a variety of ways of doing this. One of the simplest is to have both folder windows present on the screen and, using the mouse, right click over the original file, select the Copy option from the drop-down menu, right click over the destination folder, then select the Paste option from the drop-down menu. There are now two versions of the same file in different folders.

Deleting

The need to delete files often occurs when a user is running out of disk space. Files which are no longer needed can be deleted to increase the available disk space.

The **delete utility** is a piece of software which deletes the pointers that tell the OS where to find the file. The OS now considers these sectors to be available for the writing of fresh data. Until new data is written to the space the file occupies, users can still retrieve the data. Data recovery software is available which allows users to retrieve deleted data. It can identify the start and end of file markers on the disk and reset the pointers accordingly. If, however, data has already overwritten part of the file it will be extremely difficult, if not impossible, to retrieve any of the data from the file.

The delete utility is not to be confused with pressing the delete button on the keyboard. With some systems there is a trash folder and depending on the OS, it may be called the recycle bin, trash or, simply, bin. When the delete button is pressed the file is sent to this folder and files are stored there in case they need to be retrieved.

Anti-virus

As the name suggests, **anti-virus software** can be a program or set of programs whose function is to detect and remove viruses. It monitors a computer in a bid to prevent attacks from many types of **malware** (malicious software) such as viruses, **worms**, Trojan horses, **adware** and many more. (The different types of malware are described in detail in Chapter 5.) It is important to keep anti-virus software up to date because new viruses and other types of malware appear at frequent intervals and older versions of anti-virus will not detect these.

Anti-virus software will either remove the virus or malware or will quarantine it, asking the user if they want to delete the file or 'clean' it. It can be set up so that it scans specific files or folders for threats, as well as having the option to scan the whole disk or the whole computer. Scans can be set up so that they take place automatically and can be scheduled to take place at a specific time.

There are different methods employed by the anti-virus software to detect viruses. One such method is **signature-based detection**, which is based on recognising existing viruses. When a virus is identified, anti-virus manufacturers add its signature to a database of known viruses. This virus signature is a sequence of bytes (remember, a byte is usually a string of 8 bits) contained within the virus and this sequence will exist within any similar virus. The anti-virus checks all files for this sequence of bytes and when discovered within a file, that file is either deleted or quarantined. In essence, signature-based detection compares the contents of a file with its database of known malware signatures. The main drawback with this method is that it is only capable of dealing with known threats. If a new virus is created, it will do untold damage to the software or data stored on a hard disk because it is not within the known database. It was for this reason that the **heuristic-based detection** method, sometimes referred to as static heuristic, was devised whereby a program is decompiled, its source code compared to that of known viruses, and if a specific percentage matches, it is flagged as a potential threat and quarantined. The drawback with this method is that a **false positive** can occur. This happens when the detection algorithm is so general that matches can be made with files that do not contain viruses, but just happen to contain a small part of the sequence of bytes which make up the virus.

Behavioural-based malware detection, sometimes referred to as the dynamic heuristic method, looks for abnormal or suspicious behaviour, for example sending a large number of emails or attempting to alter certain files. It too can generate false positives. Sometimes **behavioural-based detection** can be used in a **sandbox** environment. This is a virtual environment within the computer whereby the suspected virus infected code is executed in the sandbox, so that it can do no real harm.

2.3.2 Other types of utility software

File management systems

This is the collective name for the software that manages data files in a computer system. As well as allowing the use of utilities such as copying and deleting of files which were described above, it allows files to be named so that each separate file can be identified by the system. It allows you to name files using combinations of the majority of characters on a keyboard. Certain characters, however, are not allowed in file names such as characters used to indicate directory prefixes or file path separators. It also allows directories to be created so that the user can store groups of similar types of files within different directories. A file management system is used to arrange the files, and move them from directory to directory. It manages how the files are organised as well as how they are stored. The user interface displays these files. The system lets you browse, move, and sort the files using different criteria such as date of last modification, date of creation, file type and size, for example. The system may contain features that enable users to generate reports as well as copy files, delete files, edit metadata and more. The metadata associated with a file consists of the file name, the size of the file, the location of the file on the disk, when it was created, when it was last edited and other properties.

Disk Management systems

Disk Management systems software is a collection of utilities which, as the name suggests, help users to manage their hard disk(s). Many of these utilities have been described above. These utilities are used to format disks, partition hard disks, compress the data on hard disks, defragment disks and back up disks. There are other utilities within disk management systems, which have not been previously described, that allow users to assign different names to disks, set up or initialise a new disk drive, extend a disk volume, shrink a partition and change the letter of the disk drive. When a user looks at a PC running Microsoft® Windows®, they can see the disk drive labels such as Windows (C:) and Data (D:). These are actually partitions of the main hard disk. It is possible to rename these partitions to names which the user feels more familiar with. Similarly, new partitions can be made for storing photos or personal documents, etc. These can be named or renamed **Photos** and **Personal**, for example. Extending a disk volume is actually just increasing the storage space allocated to a particular partition. The utility does this by reallocating any unallocated space on the hard disk. One way of doing this is to delete another partition. The storage space previously occupied by the now deleted partition becomes 'unallocated' and can be used by the disk management system to make the particular partition larger. An alternative approach is to shrink a partition, i.e. make it smaller in size. The extra storage space now available on the hard disk is unallocated and can be used to extend another partition, making it larger. Another feature of disk management systems is they allow the user to change the letter of a drive although it is inadvisable to change the C: and D: drives. It is possible to change the drive letters of partitions. It may be that the user likes the drive letters of their partitions to be in order; E:, F:, G: and so on. It could be that the user wants the partitions to be instantly recognisable, e.g. P: for photos, G: for games, etc. Whatever the reason, Disk Management systems have utilities which allow the letters of the drives to be changed quite easily.

Data compression utilities have been discussed above.

> **Activity 2f**
> Write a sentence about each different type of utility software.

2.4 Custom-written software and off-the-shelf software

When a company installs a new computer system, it will need to obtain application software, whether it is for producing word-processed documents, keeping databases of employee records or producing invoices. However, many companies discover that finding the most appropriate software for them can prove difficult. If a wrong decision is made, they may need to keep modifying their system or even have to abandon the software and start again. Generally, there are two choices: either buy **off-the-shelf software** or ask someone to create **custom-written software**.

2.4.1 Custom-written software

This is software which has to be specially written for a particular task and is developed for a specific organisation such as a company or business or it could be an individual. If the organisation with the new computer system is a large company, it might want to employ a programmer to write software specifically to solve a particular problem or provide for a specific need. Once the software is written, the company will own it and may be able to sell it to other companies.

Uses by individuals and organisations

Organisations often use databases in many shapes and forms. They are needed for employee records, stock control and customer account records. These have to be designed and this requires specialist programming skills. In the current age, organisations need a website. This has to be created and it must be tailored to the organisation's own needs. Websites are used by banks, online stores and sports clubs, to name just a few. Each organisation has its own specific needs and often requires custom-written software to meet these needs. Individuals do not usually require custom-written software as computer users tend to have their needs catered for by off-the-shelf software at lower cost. Some businesses can just involve one person and the reason they need to use custom-written software are that their specific needs are complex and possibly unique, such as an integrated website with database and spreadsheet features may be required. These may not be catered for by off-the-shelf software.

2.4.2 Off-the-shelf software

This is software which already exists and is available straight away, ready for use. The software may need to be adapted for the organisation's own specific purpose, which can mean that it occupies a large amount of storage space. Unlike custom-written software, it will not be owned by the purchasing organisation and even if adapted for it, it will not be able to sell it to other companies.

Uses by individuals and organisations

If the organisation with a new computer system is a smaller company, for example, it may turn to one of the large software companies that already produce business software packages, for example invoicing software, accounting software, payroll software, web design packages and other programs that are available to all organisations and individuals. Such programs may provide features that many organisations want, such as text editing, accounting functions and mail merge. Individual users tend to use only basic word processing and spreadsheet features. The drawback is that in trying to cater for a wide range of organisations and individuals there may be a substantial number of features that the purchasing organisation or individual does not need, such as trigonometric functions or engineering functions.

2.4.3 Advantages and disadvantages of custom-written and off-the-shelf software

Here is a table showing custom-written software compared to off-the-shelf software. Notice how each paragraph contains comparisons.

▼ **Table 2.2** Advantages and disadvantages of custom-written software and off-the-shelf software

Advantages of custom-written software	Disadvantages of custom-written software
Custom-written software is designed specifically for the task and will meet all the customer's requirements. It will not need to be adapted, unlike off-the-shelf software, which may be difficult to adapt to the particular use the customer requires.	The customer will have to pay the programmers to write the programs which have to be written specifically for the task and this will cost more than if they were buying off-the-shelf software. Off-the-shelf software is cheaper to buy because it is mass-produced. Instead of one customer paying for the development there are several customers paying the software company and thereby indirectly contributing to the development costs.
If the software does not quite meet what the customer wants, alterations to the software can be made by the programmer who is readily available.	Because there will only be one programmer or a small team of programmers, customers may find it difficult to get support if anything goes wrong. With off-the-shelf software, there are likely to be internet forums or websites to help users, as well as telephone helplines with operators who will be experienced with all sorts of queries other customers have made.
Off-the-shelf software may have some features which are not necessary for the customer, but custom-written software will not have any unnecessary features. Off-the-shelf software may not have all the functions the customer needs, but the programmer will make sure these are included in custom-written software.	It can take a long time to develop the software since the programmers will be starting from scratch, whereas off-the-shelf software is available immediately because it has already been written. There is also the likelihood that a lot of time will be spent having meetings with the programmers to tell them what will be required of the system.
The programmers will know what the current computer system is and will be able to make sure that the software is compatible with it, unlike off-the-shelf software which may not necessarily be compatible with the hardware or operating system currently being used. There may be settings within the off-the-shelf software that cannot be changed.	There may be more bugs in custom-written software as it may not have been tested as thoroughly. Often the tests that are carried out are those which the programmer thinks are necessary based on how they think the software will be used, which may not be accurate. Off-the-shelf software is usually tested rigorously so it is highly unlikely that there will be any bugs in it and it will have been used many times already, so any bugs that were present will have been discovered and removed.
The customer will own the copyright of the custom-written software and so be able to sell it to others and have extra income. With off-the-shelf software, even if they adapt it, customers cannot sell it as this would infringe the software company's copyright.	

Proprietary and open source software will be described in Chapter 10, Section 10.4.

> ### Activity 2g
> Describe custom-written software and off-the-shelf software in terms of:
> a availability
> b number of bugs.

2.5 User interfaces

A **user interface** is the means by which the computer system interacts with the user. It enables the user, with the help of input devices, to communicate with the computer and then, via the OS, communicate with a piece of software or any output device. A good user interface is one which allows the user to perform this communication without encountering any problems; it needs to be user-friendly. It should also be intuitive; users should be able to predict what will happen if they take a certain action, for example if a button on a user interface looks like a button in real life (such as the on/off button on a computer), a user should be able to press it and get a response. The four major interfaces are the command line interface, graphical user interface, **dialogue interface** and **gesture-based interface**.

2.5.1 Command line interface

The command line interface (CLI) is a means of interacting with a computer using commands in the form of successive lines of text. Normally, a prompt appears on the screen to which the user responds by typing a command. The output from the computer could be to produce a list or take some other action. In the early days of computing, the CLI was the only way a user could get the computer to run software or carry out tasks. This type of user interface is rarely used today except by software developers, system administrators and more advanced users. Most individuals use a graphical user interface (GUI) when communicating with their computers. However, even within a GUI it is still possible to access a CLI.

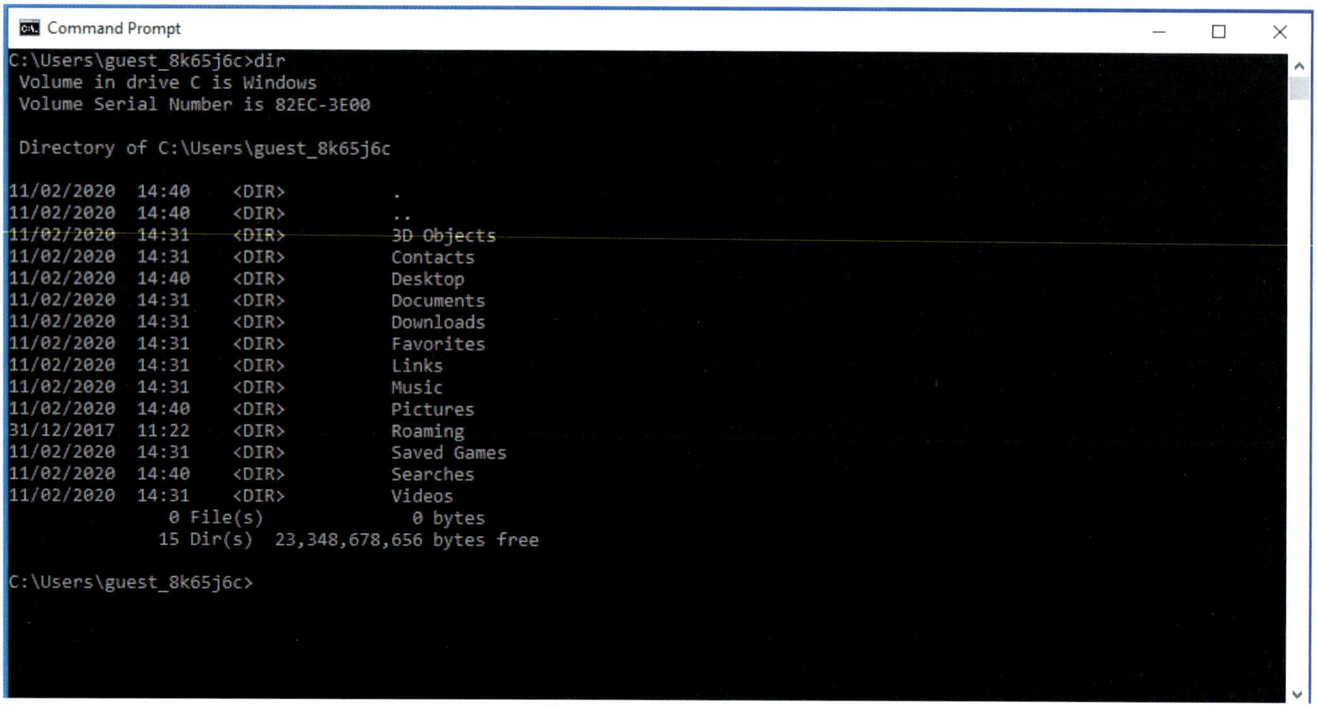

```
Command Prompt
C:\Users\guest_8k65j6c>dir
 Volume in drive C is Windows
 Volume Serial Number is 82EC-3E00

 Directory of C:\Users\guest_8k65j6c

11/02/2020  14:40    <DIR>          .
11/02/2020  14:40    <DIR>          ..
11/02/2020  14:31    <DIR>          3D Objects
11/02/2020  14:31    <DIR>          Contacts
11/02/2020  14:40    <DIR>          Desktop
11/02/2020  14:31    <DIR>          Documents
11/02/2020  14:31    <DIR>          Downloads
11/02/2020  14:31    <DIR>          Favorites
11/02/2020  14:31    <DIR>          Links
11/02/2020  14:31    <DIR>          Music
11/02/2020  14:40    <DIR>          Pictures
31/12/2017  11:22    <DIR>          Roaming
11/02/2020  14:31    <DIR>          Saved Games
11/02/2020  14:40    <DIR>          Searches
11/02/2020  14:31    <DIR>          Videos
               0 File(s)              0 bytes
              15 Dir(s)  23,348,678,656 bytes free

C:\Users\guest_8k65j6c>
```

▲ **Figure 2.4** A command line interface screen

As we can see in Figure 2.4, the CLI usually consists of a black box with white text. In the early days of computing the text was green. In Figure 2.4, on entering the interface, a prompt C:\Users\guest_8k65j6c> appears. The > symbol tells the user that they need to type in a command. Here, the command 'dir' was entered, which asked the computer to list the files and directories (folders) that are in the directory (folder) belonging to the user guest_8k65j6c. Notice that after executing the command, the original prompt is shown again. In this example there are no files in this directory, only other directories. The 'dir' command is only one of hundreds of commands which are available in a CLI. It is possible to do any action with a CLI that could be achieved with a GUI. In fact, it could take several clicks of the mouse and negotiation through a number of dialogue boxes and menus in a GUI to achieve the same outcome as a single line of text in a CLI.

2.5.2 Graphical user interface

When CLIs were introduced, commands had to be typed in correctly with any misspellings potentially causing the system to fail to perform as desired, and this made the interface clumsy and confusing. There was a need for a less inefficient means of communicating with the computer, which resulted in the creation of the GUI. Instead of typing in commands, the GUI used windows, icons, menus and pointers, collectively known as a 'WIMP' interface, to carry out commands, such as opening, deleting, and moving files. Technically speaking, a WIMP interface is only a subset of GUI and requires input devices such as a keyboard and mouse, whereas other types of GUI use different input devices such as a touchscreen. Figure 2.5 shows a window, icons and a menu. A user can double-click on an icon, which could represent a file, a folder or an application, and open it. A right click on a two-button mouse opens a menu. By moving the mouse, the pointer can be moved up and down through the menu and then an option can be selected by clicking on it.

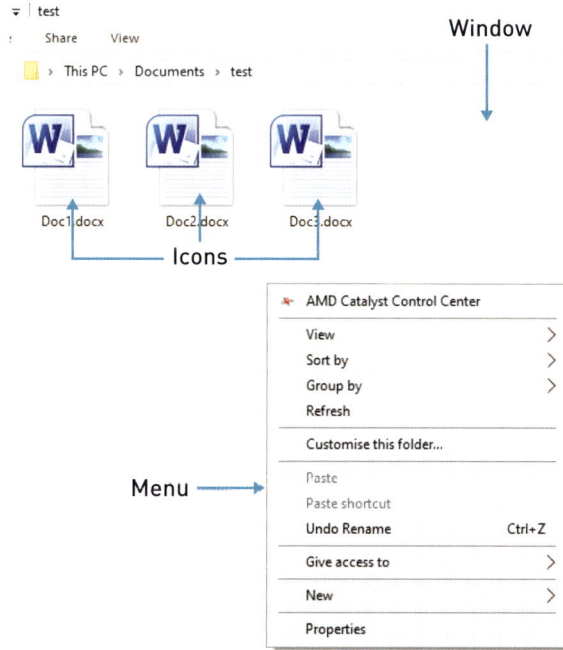

▲ Figure 2.5 A graphical user interface screen

2.5.3 Dialogue interface

A dialogue interface allows a user to communicate with a computer or device using their voice. The computer is able to use speech-recognition software to convert the spoken word into commands it can understand. Many cars have such a system whereby the driver is able to control their phone or radio without touching them. For example the driver is able to initiate a phone call by saying 'phone Graham' or switch a particular radio channel on by saying the name of the channel. Laptops and PCs often come with voice control these days. The user is able to load and run software packages and files by speaking into a microphone and saying the commands. The computer or device responds with spoken words after the required text has been converted into speech. It requires the device to learn the way the speaker talks by asking the user to repeat certain sentences until it has 'learnt' the way they speak. It can, however, become quite capable of understanding simple commands. Noise in the background while the user is speaking and the ability to recognise only a limited vocabulary can cause problems.

2.5.4 Gesture-based interface

A gesture-based interface is designed to interpret human gestures and convert these into commands. Gestures can be made with any part of the body, but it is usually the face or hand that makes the gestures the computer can interpret. An example of where this type of interface is used is in 'smart' homes where a gesture can turn on the lights, for example. In this area of IT, a gesture can be said to be any physical movement, large or small, that can be interpreted by a computer. Examples of these are the pointing of a finger, nodding the head, or a wave of the hand. These types of gestures can be understood by a computer. First of all, a camera in conjunction with an infrared sensor detects the movements being made in front of it. The computer, using a special type of software, searches through all the gestures it has stored in a database to match it with the input. Each stored gesture is linked to a specific command which is then executed after the gesture has been matched.

2.5.5 Advantages and disadvantages of different types of user interface

Users with disabilities may not be able to hold a mouse or click it, so a GUI may not be suitable for them. They may not be able to type using a keyboard, so using a CLI would also be inappropriate. They may not be able to control the movement of their limbs accurately, so a gesture-based interface would be unsuitable. In this instance, then, a dialogue interface would be the best type of interface for them to use, although people with different disabilities may favour a different type of user interface.

Despite this potential advantage of a dialogue interface, if there is any background noise when the user is speaking, the computer might misunderstand what is being said, which is not a problem when using the other types of interface.

Some types of user interface would be inappropriate for hygienic reasons. If a user was, for example, involved in the food service industry, then touching a mouse, keyboard or touchscreen may be inadvisable while working, so a GUI or CLI would not be suitable for them. However, they would be able to make gestures and speak, so a gesture-based or dialogue interface would be a sensible choice for them. On the other hand, certain gestures could be misunderstood by the computer, particularly if the user has made them without realising, and some gestures in certain cultures might be judged to be inappropriate.

In terms of accuracy, a GUI is better than gesture-based or dialogue interfaces. A gesture may not point exactly at an icon whereas a mouse can be more accurately controlled, and voice inputs can be misunderstood by the speech-recognition software.

CLIs and, to an extent, dialogue interfaces require the user to learn a lot of commands, whereas GUIs and gesture-based interfaces are more user-friendly and reasonably intuitive. CLIs do tend to be used by IT specialists and require a certain amount of IT knowledge but this is not the case with GUIs or gesture-based interfaces. Compared with a GUI, commands entered into a CLI are far more difficult to correct. There is a degree of editing allowed using the arrow keys on a keyboard to manoeuvre to the line with the error, but this can be far more awkward to do than using a mouse, for example. However, CLIs tend not to change over time and once a user is familiar with one, they do not have to relearn how to use a changed version. GUIs do tend to develop and consequently the user has to learn how to use the new version, which can take time.

Generally, the more advanced the type of interface, and this not only means GUIs but also includes gesture-based and dialogue interfaces, the faster the processing is and the greater the storage space required to store the interface software compared with a CLI. Gesture-based and dialogue interfaces also tend to be more expensive to develop than CLIs or even GUIs. Some situations require the use of dialogue interfaces for safety reasons. A driver in a vehicle may wish to play a particular piece of music using their in-car entertainment system. To select it using a GUI or a gesture would require taking a hand off the steering wheel, which could be dangerous, whereas using their voice through a dialogue interface would not.

Activity 2h
1 Write down **two** advantages of using a GUI rather than a CLI.
2 Write down **two** disadvantages of using a gesture-based interface rather than a dialogue interface.

Practice questions

1 Mainframe computer manufacturers often refer to RAS.
  Explain what is meant by the term 'RAS'. [3]
2 Explain how high-level language is translated to run on different computer systems. [5]
3 Describe the terms: [4]
  a sector
  b block
  c track
  d cylinder.
4 Explain what is meant by custom-written software. [3]
5 Describe the features of a command line interface. [4]

3 Monitoring and control

In this chapter you will learn:
- ★ about the sensors and calibration used in monitoring technologies
- ★ about the uses of monitoring technologies
- ★ about the sensors and actuators used in control technologies
- ★ how to write an algorithm and draw a flowchart.

Before starting this chapter you should:
- ★ be familiar with the terms 'sensor', 'physical variable' and 'microprocessor'.

3.1 Monitoring and measurement technologies

Monitoring, or **measurement** as it is sometimes called, involves the use of a computer or microprocessor-based device to monitor or measure physical variables over a period of time. It is important to know which sensors would be appropriate in a given situation to measure physical variables such as light, temperature, atmospheric pressure, humidity, moisture, sound, blood pressure and pH, among others.

In Chapter 1 in the section on data logging, we saw that a sensor is a device that is used to collect (input) data. The data usually relates to physical changes in the environment that are being monitored. A sensor converts the physical characteristic, such as temperature, light or pressure, into a signal which can be measured electrically. Monitoring systems are different to **control systems** regarding their outputs. Control systems process the data and then make decisions based on the data regarding which actions to take, as we saw in Chapter 1 in the section on real-time processing. Monitoring systems only process and record data so that people can see trends in the changes that are taking place in the environment being monitored.

Sensors continually send data back to the computer or microprocessor. Students often misunderstand this process, with some thinking that sensors send data every few seconds or even minutes or that sensors only send data when there is a change in the environment. The process is actually continuous and never ends. Most sensors used in monitoring systems are analogue, which means the data sent to the computer is in analogue form. However, the computer can only process data in digital form, so the data has to be converted using an analogue-to-digital converter (ADC). This is so the computer, which can only understand digital data, is able to process it.

The output from the system is usually on a screen or printed out, but can be a warning sound if the monitoring that is taking place is critical, for example an overheating nuclear reactor.

3.1.1 Sensors

Below is a table showing the uses of some of the more commonly occurring sensors. In reality there are very few basic sensors, with several sensors actually being combinations of the others.

▼ Table 3.1 Uses of common sensors

Type of sensor	Description and use
Light/ultraviolet (UV)	Light sensors, as the name suggests, measure the amount of light. There are many types of light sensor, but they all follow the same principle. When light falls on this type of sensor, it generates electrical energy. The amount of energy is then converted to give a value for the amount of light. These sensors can be used in weather stations to measure the amount of sunshine.
	Turbidity sensors measure the cloudiness of water in a river that is affected by pollution. A turbidity sensor is actually a light sensor usually placed at right angles to a light emitter. The greater the number of particles in the water, the greater the amount of light reflected off them on to the sensor.
	UV sensors are used to measure the amount of UV (specifically UVB radiation, which can be dangerous to humans, sometimes causing skin cancer).
Temperature	The components of different types of temperature sensor either change their electrical resistance or generate a voltage according to temperature. Whichever type is used, electrical signals are generated which are converted into values to represent temperature.
	These sensors are used to feed data back to the computer about, for example, the **ambient temperature** in a weather station, the temperature in a river or air quality.
Pressure	A pressure sensor converts the force applied to its surface to generate electrical energy which is then converted into values to represent the applied pressure.
	These sensors are used to measure atmospheric pressure in weather stations.
Humidity/ moisture	Humidity sensors are often a combination of a moisture sensor and a temperature sensor in one unit. This is because humidity can only be calculated by knowing how much water there is in the atmosphere together with the temperature. A moisture sensor is actually a combination of a light sensor and a light emitter, as the amount of light transmitted depends on the moisture content of the air.
	Humidity sensors are used to measure the air humidity in a weather station. Moisture sensors are also used when monitoring soil quality.
pH	pH sensors measure the acidity of soil and acidity in rivers, lakes, etc. They are very similar to a simple battery and generate electricity depending on the number of hydrogen ions in the solution, which causes an electrode to generate a voltage.
Gas	These are sensors used in environmental monitoring and water pollution monitoring.
	Oxygen (O_2) sensors measure the level of oxygen in soil and water.
	Carbon dioxide (CO_2) sensors measure the level of carbon dioxide in the air or in water and are basically an adaptation of an infrared sensor.
	Carbon monoxide (CO) sensors measure the level of carbon monoxide in the air, usually in homes to detect carbon monoxide gas leaks. One type uses stannic oxide (SnO_2 – also called tin dioxide) whose resistance reduces when exposed to carbon monoxide.
	There are a number of oxides of nitrogen. The most common of which are nitrogen dioxide (NO_2), nitric oxide (NO) and nitrous oxide (N_2O – also known as laughing gas). These are commonly emitted from both petrol and diesel car engines and are a big contributor to air pollution with NO being by far the largest contributor, hence the need for sensors to measure car exhaust fumes. The sensor consists of an electrode containing a catalyst which reacts with the nitrogen oxides. The sensor produces a voltage proportional to the amount of nitrogen oxides that the car is producing.
Sound	These sensors convert sound waves into voltages or electrical signals which are converted by the computer into values to represent sound.
	Sound sensors can be used in environmental monitoring systems to measure noise pollution.

Type of sensor	Description and use
Infrared	All bodies possess thermal energy and therefore emit infrared radiation. This radiation is converted into electrical signals as a result.
	These sensors can be used in environmental monitoring, for example the Earth's surface temperature can be monitored by satellites.
Oxygen, carbon dioxide, pH, turbidity	These are sensors used in environmental monitoring and water pollution monitoring.
	Oxygen (O_2) sensors measure the level of oxygen in soil and water.
	Carbon dioxide (CO_2) sensors measure the level of carbon dioxide in the air or in water and are basically an adaptation of an infrared sensor.
	pH sensors measure the acidity of soil and acidity in rivers, lakes, etc. They are very similar to a simple battery and generate electricity depending on the number of hydrogen ions in the solution, which causes an electrode to generate a voltage.
Reed switch	Strictly speaking, a reed switch is not a sensor, but it is used to help measure rainfall (see the next section about the uses of monitoring technologies in weather stations).
Touch sensors, (electro)magnetic field sensors and proximity sensors	These sensors can also be used in monitoring and measurement technologies but are more commonly used in control systems. These will be described more fully in that section.

3.1.2 Uses of monitoring and measurement technologies

Environmental monitoring

Environmental monitoring is the collection of data relating to our environment. It can include:

» monitoring sound in cities, in addition to other pollution produced by motor vehicles
» monitoring soil quality in outdoor gardens and also in greenhouses/glasshouses, using pH, moisture and temperature sensors, among others.

Environmental monitoring is closely linked to climate study, so can also include:

» detecting abnormally high temperatures using temperature sensors, so that people can be warned of the dangers
» monitoring the level of air pollution using O_2 sensors and CO_2 sensors; pH sensors can also help in this regard as they can provide acidity readings
» monitoring ultraviolet levels; excessive amounts can cause skin cancer, so these are often monitored by governments in high-risk countries to judge whether people should be advised to wear skin protection cream.

Monitoring water pollution

Studies of water pollution usually happen with reference to bodies of water such as rivers, lakes and, sometimes, seas. There are basically two ways of carrying out the study.

One is to compare the readings with those that would normally be expected. This requires the lowering of one set of sensors into the river or lake.

The other relates usually to industrial pollution, though it can also be used to measure pollution from a farmer's field. This involves inserting two sets of sensors, one upstream from the suspected site of pollution and the other downstream, immediately after the site, whatever that may be, farm or factory. The readings from the two sets of sensors are compared to see if there are any differences so that a conclusion can be reached as to whether the site is causing pollution.

The system operates the same as the weather station, with sensors feeding data to an ADC and then the computer processing the digital data. The sensors involved are temperature sensors, pH sensors, turbidity sensors, O_2 and CO_2 sensors. The processing carried out is a comparison of the readings with normal values if it is the first method, or a comparison of the readings from the two sets of sensors if the second method is being used.

Weather stations

There are many types of weather station, ranging from very sophisticated and complex systems to a basic home-based system. We will be considering the type of station that might exist in a school. This type of weather station could be used to monitor the weather in terms of temperature, rainfall, hours of sunlight, atmospheric pressure, humidity, and UV radiation. For this purpose, it would need the following:

» **Temperature sensors** to measure the ambient temperature. When referring to the weather, ambient temperature means the temperature of the surrounding air of the weather station.
» **Pressure sensors** to measure atmospheric pressure, which is the pressure of the air above us. Weather forecasters use pressure readings to help them formulate weather forecasts. High pressure is usually an indication of fine sunny weather, whereas low pressure tends to be associated with wet and windy weather.
» **Humidity sensors** to measure absolute and relative humidity. Absolute humidity is the amount of moisture in the air, measured without taking temperature into account. It is usually measured as the number of grams of water per cubic metre of air. Relative humidity is also a measure of moisture but does consider the temperature of the air and is actually a percentage value. It is the fraction of the amount of moisture in the air compared to the maximum amount of water that could be held in air at that temperature. It is used by weather forecasters to predict the likelihood of rain or snow occurring.
» **Light sensors** to measure the number of hours of sunlight. Measuring sunlight requires an array of light sensors which collectively measure the intensity of the light radiation.
» **A tipping bucket and a reed switch** to measure rainfall. Most non-automated weather stations use a bucket into which the rain falls. When the bucket reaches a certain weight (usually after a very small amount of rain) the bucket mechanism causes it to tip over and empty the water. It then tips back to collect further rainfall. Originally, the tipping bucket was connected to a rotating, mechanical, graph plotter that would make a mark every time the bucket was tipped. By counting the number of marks on the graph human operators could work out the amount of rain by multiplying this count by the amount it took to tip the bucket. In modern automated systems, the tipping of the bucket activates a reed switch which sends a signal back to the microprocessor. The microprocessor, counting the number of times the bucket tips, performs the same calculations that human operators used to do.
» Wind speed and direction can also be measured using sensors or combinations of sensors.

When the weather station is operating, the readings from the sensors are fed back to an ADC and then sent to the computer. The ADC converts the data from analogue to digital so that the computer can understand and process it. On receiving the digital data, the computer stores the data in the form of a table, which could be done using a spreadsheet or database package, so that it can be processed. The processing might consist of calculating, for each day, the highest, lowest and mean temperature, the level of UV radiation, total rainfall, hours of sunshine, highest and lowest value of atmospheric pressure, and wind speed and direction. These values can also be calculated for the month and year to date. Results can be output in the form of graphs, either to a monitor or printed out. This all happens automatically without the need for human intervention.

Monitoring patients

Intensive care units (ICUs) in hospitals use a range of sensors to monitor their patients. The sensors are used to feed back data about the patient to a computer. The data relates to blood pressure, pulse rate, rate of breathing, body temperature and the level of oxygen in the blood. The computer is programmed with a range of pre-set or normal values. It constantly compares the data from the sensors to these pre-set values. If any value in the data is outside the pre-set range, the computer sounds an alarm. However, this type of system is not a computer control system as we understand it since nurses are required to intervene in the event of an alarm sounding. Apart from the use of sensors and computers making the process more accurate, the computer can monitor the health of several patients simultaneously, thus enabling nurses to carry out other tasks.

Increasingly, wearable sensors are being used to monitor patients enabling them to return home where their vital signs are monitored remotely. This means that the risk of infection is reduced as well as freeing up hospital beds for other patients.

> **Activity 3a**
>
> Explain how you would set up a study to monitor pollution in your school or college grounds.

3.1.3 Calibration

When sensors measure physical variables, people believe that the results are always going to be accurate because a computer is involved. However, for this to be the case, sensors have to be calibrated. **Calibration** is making sure that when, for example, a temperature sensor is used to measure the temperature of boiling water, it actually causes the computer to output a value of 100 °C. It is done by comparing the value a sensor produces to a known measurement.

The importance of calibration

The accuracy of all sensors reduces after a period of time. This is often caused by constant use and exposure to the atmosphere or liquids that are being used. Slight erosion of the material the sensor is made out of is also bound to occur. It depends on the type of sensor and how it is being used as to how quickly this occurs but whatever the cause, regular calibration helps to maintain accuracy. The sensor is only one component in the monitoring system and devices like the ADC may also deteriorate over time, so calibration is important for that reason too. If a sensor is being used as part of a sensing system, as with a humidity sensor, then if that sensor loses accuracy, the whole system will need calibrating.

Methods used to calibrate devices

Let us take a look at how to calibrate a temperature sensor as an example. In order to calibrate the sensor, there needs to be some way of creating a known temperature. This can be done by using a heat source that generates an accurate temperature or using a mixture of ice and water which can give an accurate reading for 0 °C. The sensor, when activated, generates a voltage which is converted into a temperature reading. The algorithm to do this can be refined so that the temperature reading matches the actual temperature. The alternative is to take a reading from a sensor that has already been calibrated and use this to compare the temperature reading of the sensor that is being calibrated.

One-point calibration

This type of calibration is the easiest to perform. Only one measurement point is needed and this makes it a particularly appropriate method to use with sensors which have to measure a set temperature or temperature range that is constant, that is, it never changes. The sensor need not be calibrated against any other range of temperatures since it is not going to be required to measure these. This type of calibration is often needed for sensors which are constantly used at very high temperatures and after some time lose accuracy. Accuracy can be checked by doing a one-point calibration every so often and comparing the result with the previous calibration. In order to carry out a one-point calibration, a reading is taken from the sensor in the range being measured and it is compared with either a pre-calibrated sensor or a known value. The sensor reading is subtracted from the known value which gives the 'offset'. In the algorithm, this offset is then added to every reading in the temperature range being measured.

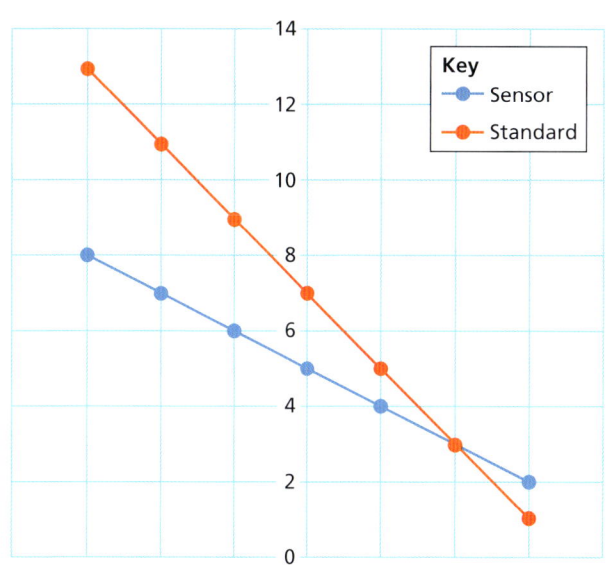

▲ **Figure 3.1** Two-point calibration

Two-point calibration

Figure 3.1 shows a graph of a sensor's readings plotted against a standardised (pre-calibrated) sensor for a range of pH values. A number of solutions with different pH values are being used. The readings shown in the graph are exaggerated as it is most unlikely that even the most inaccurate sensor would produce readings this far out. The sensor being calibrated is different to the standardised sensor by +1 at the lower end and -5 at the higher end. The calibration cannot be carried out by just adding an offset value in this case, since the offset values are different for every pH value. There is however a linear relationship between the two and to calibrate the sensor the algorithm would need to multiply each reading by 2 and subtract 3. This calibration compensates for both offset errors and what are slope errors, so called because the slope of the graph has to be considered.

To arrive at this conclusion, obviously we need to compare more than one reading. With a pH sensor, it is recommended to place sensors in solutions which are neutral (pH 7) and either one which is acidic (say, 4.0) or one that is alkaline (say, 10.0). The use of the two values will enable the relationship between the sensor and the standard to be established.

Multipoint calibration

Figure 3.2 shows a graph of a sensor's readings plotted against a standardised (pre-calibrated) sensor, but this time for a range of temperatures. The temperature readings are taken from a hot liquid as it cools down. Although the sensor being calibrated in Figure 3.2 has the same reading as the standardised sensor at the lower end, it is markedly different for other readings. The calibration cannot be carried out by just adding an offset

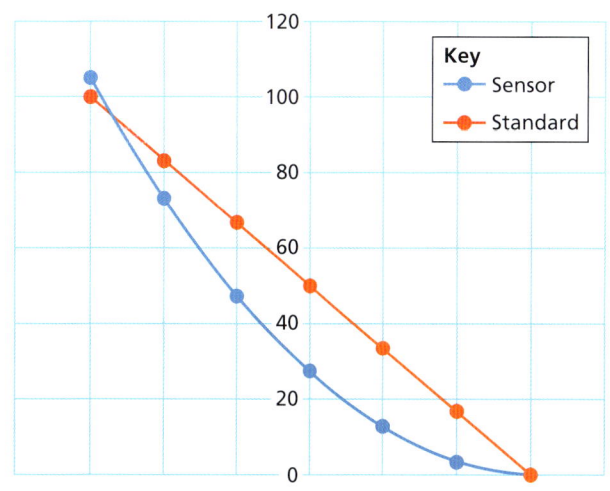

▲ **Figure 3.2** Multipoint calibration

value or by allowing for the slope. There is no longer a linear relationship between the two and to calibrate the sensor, the algorithm would need to include some form of what is called 'curve fitting'. This is because it is a nonlinear relationship and needs to be described using a quadratic function of the form $y = ax^2 + bx + c$, where y is the standardised reading and x is the reading from the sensor that needs calibration. Do not worry if you find this complicated as all you really need to understand is that the values from the sensor are a curve, whereas they should be a straight line and the algorithm needs to be amended to deal with this.

To carry out the calibration, at least three known temperatures need to be created and the sensor readings at those temperatures need to be taken. Good practice would be to use temperatures of 100 °C, 0 °C and say, 50 °C. This should provide the relationship between the sensor and the standardised sensor.

3.1.4 Advantages and disadvantages of monitoring technologies

Although computers are now used in all aspects of monitoring, there is still the need for humans to be involved. Here is a table showing the advantages and disadvantages of computer monitoring compared with people taking readings.

▼ Table 3.2 Advantages and disadvantages of monitoring technologies

Advantages of computer monitoring	Disadvantages of computer monitoring
Humans are unable to take readings at very frequent intervals as they need to make a note of each reading. This takes time, during which they cannot take another reading. Computers are able to take readings at more frequent intervals and are capable of reading more than one variable simultaneously. Humans can only do one thing at a time, so take longer.	
It is very difficult for humans to keep taking readings for sustained periods, whereas computers can be left on to take readings at any time, day or night. The readings are always taken at regular intervals unlike with a human who might forget to take them.	Computers can be expensive to buy, whereas humans would be expected to do the monitoring as part of their job. Computers are also expensive to maintain.
It takes time for people to draw accurate graphs, whereas computers can produce them automatically after processing the data.	It is quite difficult to program computers to interpret the results, but humans can interpret results and are also needed to program the computers in the first place.
Results are produced automatically after the readings are received by the microprocessor or computer, unlike a human who would take a lot of time to write them down.	
Readings taken by computers tend to be more accurate than those by humans as computers are not subject to 'human error'.	Sensors can deteriorate after a period of time, whereas humans will tend to be more consistent.

3.2 Control technologies

A control system is one that uses microprocessors or computers to control certain physical variables. Computers can do this by maintaining certain physical conditions at the same level for a period of time or by controlling certain devices which cause the variables to change. Physical variables that can be controlled by computers and microprocessors include temperature, pressure, humidity, light, and moisture.

Control systems use real-time processing, which was described in Chapter 1. They make use of actuators to control devices, although some devices are actuators in their own right, such as a motor. Unlike in monitoring systems, in control systems the output affects the input. For example, think about a room's temperature as being controlled by a microprocessor connected to a temperature sensor and a heater. The temperature is *input* by the sensor to the microprocessor. If the temperature is below a certain value, the microprocessor sends a signal to the heater to switch on, which is the *output*. The heater being on causes the temperature to rise which means the input value has now changed, so the output has obviously affected the input. Control systems involve continuous processes.

3.2.1 Sensors and their uses

All of the sensors described in Section 3.1 on monitoring technologies can also be used in control systems, but there are some sensors that are more likely to be found in control systems than monitoring systems. Table 3.3 shows some of these.

▼ Table 3.3 Uses of sensors in control systems

Type of sensor	Description and use
Electromagnetic field sensor	Often referred to simply as a magnetic field sensor, this sensor measures the change in the Earth's natural magnetic field caused by the presence of a ferromagnetic object. When a vehicle, for example, is above the sensor, the metal in the body of the vehicle distorts the Earth's magnetic field and so its presence is detected. The sensor is small, so its installation and maintenance are easier and cheaper than an induction loop which performs more or less the same function. Magnetic field sensors and induction loops are used at the entrances to car parks to control barriers but magnetic field sensors can be used to detect the number of spaces available. They are also used in some automated car parking systems to help drivers park their car, in a similar way to ultrasonic sensors (see below).
Ultrasonic sensors	An ultrasonic sensor is actually made up of a device that sends out sound waves with a frequency greater than that of the human audible range (so that a human cannot hear it) and a sensor that receives the sound waves which are reflected back. It can be used to measure how far away an object is. It measures the amount of time taken for the sound to be sent and received which, combined with knowledge of the speed of these sound waves, can be used by the microprocessor to calculate the distance. It is used in automated car parking systems which let the driver know when they are close to another vehicle or other object so they can park their car without hitting that obstacle.
Proximity sensor	A proximity sensor can be a mixture of sensors but usually comprises a device that sends out a signal and sensor which receives the reflected signal back. This can be an infrared beam, ultrasound or a magnetic field. One use is in smartphones to switch off the screen display when the phone is held near to the ear.
Touch sensors	One type of touch sensor is used for measuring fluid levels. A capacitive touch sensor measures the capacitance between two conductors separated by an insulating plate. One of the conductors will be the fluid whose level is being measured. When the fluid is touching the sensor, it detects that it is at that level. This type of sensor is often used in detection devices used to measure fluid levels such as the cooling water level in nuclear power plants to ensure that there is sufficient water to cool the reactors.

3.2.2 Actuators and their uses

Just as sensors provide the input to a control system, so actuators provide the output. An actuator controls a device, such as the valve which allows water to flow through heaters or sprinklers in a greenhouse. It can actually be a motor which controls the opening or shutting of windows in a greenhouse or a switch which turns a heater on. It is essentially a device that turns an electrical signal from a microprocessor into movement. It is an output device that carries out an action or causes movement. Actuators can be classified according to the type of motion they produce, either linear (the device moves in a straight line) or rotary (the device rotates) actuators. They can also be classified according to the type of power they use such as hydraulic, pneumatic, electric, thermal or magnetic actuators. To begin with, let us look at the actuators in terms of the type of motion they produce.

Linear actuator

A common actuator is the linear solenoid actuator, which is basically an electrical coil wound around a cylindrical tube encasing a plunger, which can move in and out of the coil. They can be used to open doors, open or close valves and activate electrical switches.

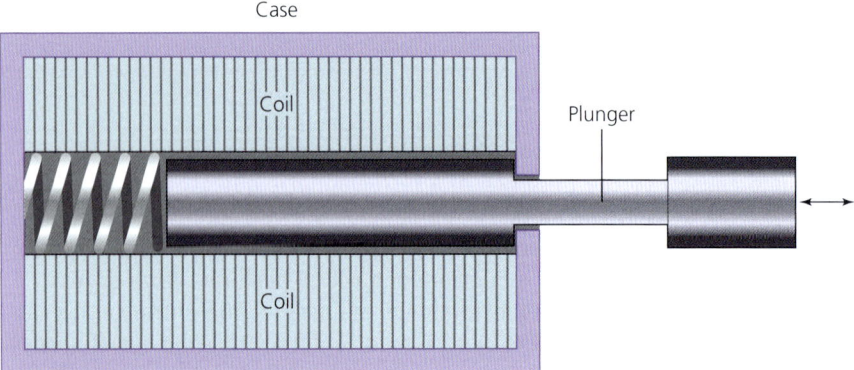

▲ **Figure 3.3** A linear solenoid actuator

Rotary actuator

A rotary actuator, as its name suggests, produces movement that is rotary. In other words it causes a device to rotate. The resulting motion can cause the device to continually rotate or it can be moved to a fixed position at a specific angle. There are different types of rotary actuator. Rotary solenoid actuators can be used where the angular movement is very small with the angle of rotation being 90° or less. Typical rotational angles are 25°, 35°, 45°, 60° and 90° but there can be a number of movements such as, for example, moving forward 90° and then returning back to the start position. In the linear solenoid actuator shown in the diagram above, depending on its polarity (north-south or vice versa) the magnetic field either attracts the plunger (or rotor) or repels it to or from the centre of the coil. Rotary solenoids use the magnetic field to produce a rotational movement of the rotor when current passes through the coil.

Another type of rotary actuator is the rack and pinion type (Figure 3.4). The rack and pinion mechanism is just an extension of the idea of a linear actuator. The rack is equivalent to the plunger in a linear actuator. It has teeth in it and as it moves in a linear manner, the teeth of the rack engage with those in the pinion (these are similar to cogs on a bicycle) causing the pinion to rotate. The pinion is connected to the device which the user requires to rotate. It can be used in the movement of robotic arms, for example, which need to move through an arc or particular angle.

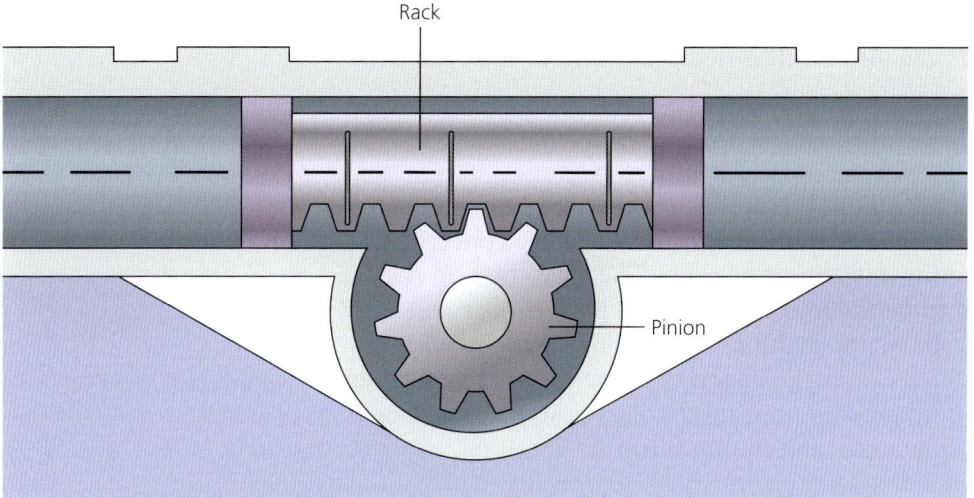

▲ **Figure 3.4** A rack and pinion

Vane rotary actuators are more mechanical in nature. They consist of vanes inside a cylinder. Pressure is applied to the vanes using air or fluids which causes the vanes to rotate causing the rotor, which the vanes are attached to, to rotate.

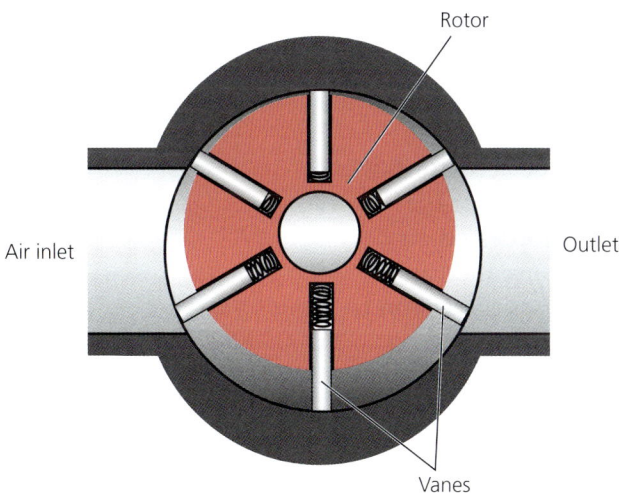

▲ **Figure 3.5** Vane rotary actuator

Soft actuator

This is a special type of actuator that responds to different types of stimuli. The stimuli can be thermal, magnetic, mechanical, moisture, electrical, humidity or even pH. It is able to change shape in response to these various stimuli. The development of this type of sensor came about as a result of studying the behaviour of human muscles and soft-bodied animals which have various properties such as suppleness or flexibility. These properties needed to be mimicked in this type of actuator. They are found in robotic arms or grippers and are used, for example, in the picking of fruit from trees. They are also used in devices that help support the functions of the human heart or in artificial muscles.

As mentioned above, actuators can also be classified according to the type of power they use.

Hydraulic actuator

This type of actuator uses a fluid or liquid to apply pressure to a motor causing a linear movement or causing a rotational movement to the device concerned. It consists of a cylinder containing a piston or plunger. The space behind the piston head can be considered to be the lower chamber of the cylinder and the space ahead of the piston, the upper chamber. When fluid enters the lower chamber pressure builds up which causes the piston to move forward in a linear fashion, thereby converting liquid pressure into mechanical power. Rotational movement can be produced using a rack and pinion setup.

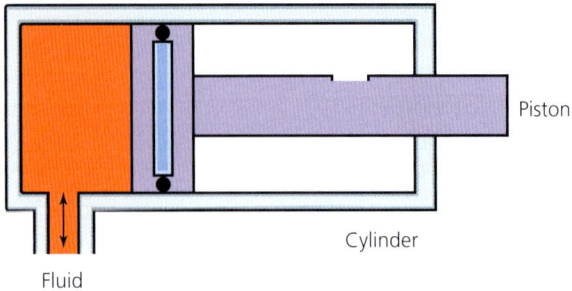

▲ **Figure 3.6** A hydraulic actuator

Pneumatic actuator

This type of actuator operates in a similar fashion to an hydraulic actuator but uses compressed air or pressurised gas instead of a fluid. The energy of the compressed air is converted into mechanical motion. As the compressed air enters the lower chamber the pressure increases. Once the pressure increases above that of the atmosphere surrounding the cylinder this causes the piston to move forward.

Electric actuator

This type of actuator uses electricity, often in the form of a battery, as its energy source or power supply. The components of an electric actuator are an electrical signal (which can possibly come from a computer), a power supply and a cylinder-housed piston or shaft. Electrical energy is converted into mechanical movement. Electrical rotary actuators are used in robotic arms on industrial production lines.

Thermal actuator

A thermal actuator is one that responds to temperature changes by creating linear movement. It is made up of a piston or plunger within a cylinder and thermally sensitive material. One type of thermal actuator contains a diaphragm next to the piston within a cylinder as shown in Figure 3.7.

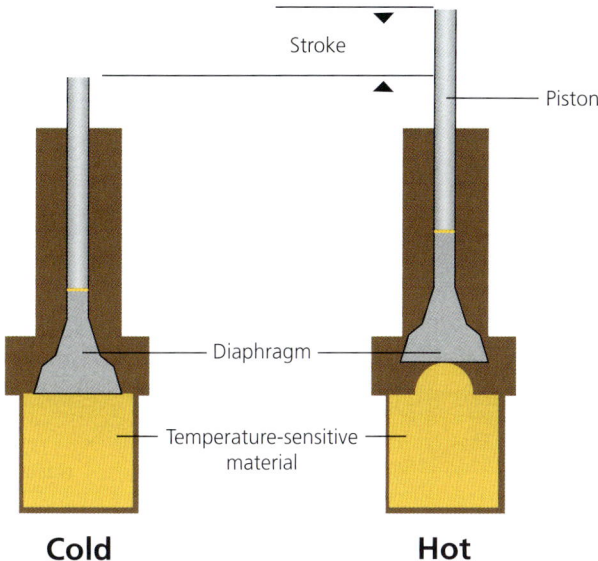
▲ Figure 3.7 A thermal actuator

When the temperature increases, this causes the temperature-sensitive material to expand. This pushes the diaphragm upwards making the piston rise out of the cylinder. Different waxes melt at different temperatures but most do so in the range 40° to 70 °C. They expand as they melt and so are often used in thermal actuators. When the temperature falls these substances contract, causing the piston to retract. Thermal actuators convert thermal energy into mechanical motion. They have many uses including maintaining the engine coolant temperature in UAVs which we met in Chapter 1. As the coolant temperature rises the actuator causes a flap to lift, letting in more cooling air and when the temperature falls the flap closes.

Magnetic actuator

We have seen that magnetism is used in actuators in the descriptions of both linear and rotary actuators. The use of magnetic fields in actuators is based on the polarity (north-south or south-north) either attracting or repelling the components causing motion. Magnetic actuators use magnetic fields to produce mechanical motion, either rotary or linear.

Mechanical actuator

Most actuators could be considered to be mechanical in the sense that they all produce mechanical motion. This type of actuator is a device that converts one type of motion like rotary into another, linear. They operate by making use of gears and pulleys. A typical example is a rack and pinion setup. If the description we saw above of the workings of a rack and pinion device is considered in reverse, it is easy to see how rotary motion can be converted into linear motion. The rotating of the pinion by an external force would cause movement of the rack in a linear fashion. A non-IT example sometimes used by car owners is a very basic car jack, where the turning of a handle in a circular motion raises the saddle of the jack, thereby lifting the car.

3.2.3 Uses of control-technology systems

All of the following examples are microprocessor-controlled. The term 'computer' can be substituted for microprocessor, so all these examples could be called computer-controlled.

Greenhouses

Greenhouses are used in countries with cooler weather so that people can grow plants that are normally grown in warmer countries. The sensors needed in a greenhouse are:

- a temperature sensor to measure the air temperature in the greenhouse
- a moisture sensor to measure the water content of the soil
- a light sensor to measure the light level inside the greenhouse.

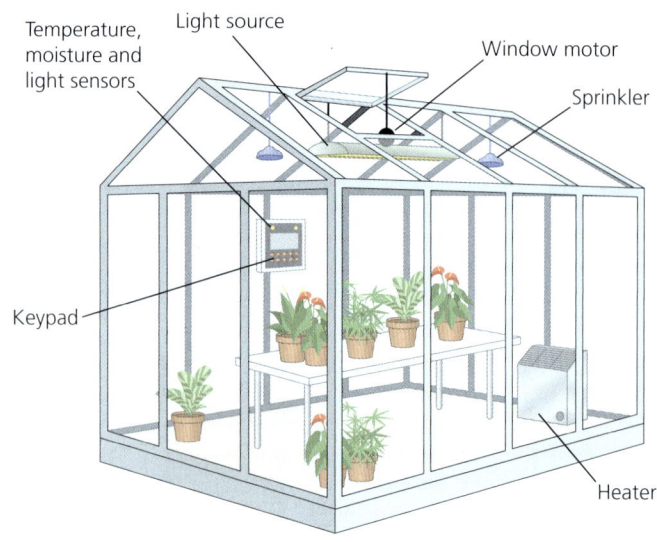

▲ Figure 3.8 A greenhouse with sensors

Maintaining the required temperature

At the start of the process, the user inputs the required temperature (pre-set value) using a keypad, number pad or touchscreen.

The computer receives data about the temperature of the greenhouse from the temperature sensor. It needs an ADC to change the analogue temperature data to a digital value the computer can understand. The computer compares the sensor data to the pre-set value input to the system earlier by the user. If the temperature is above the pre-set value, it sends a signal to the window motor to open or leave it open if it already is. If the temperature is below or equal to the pre-set value, the computer sends a signal to the motor to close the window or leave it closed if it already is closed. It also sends a signal to an actuator to activate the heater or leave it on if it already is on. This whole process is continuous as long as the system is switched on.

The layout of algorithmic constructs will be explained in detail in Chapter 4. The algorithms shown below are not as detailed as they might be. In Chapter 4 you will be asked to refine them and full answers will be provided. Here is a possible algorithm which describes the processing carried out by the microprocessor (computer), assuming the pre-set (required) temperature has already been input to the system:

```
1   WHILE system switched on
2       INPUT temperature
3           IF temperature > pre-set
4               THEN
5                   send signal to window motor to open window or leave it open
6                   send signal to actuator to switch off heater or leave it off
7               ELSE
8                   send signal to window motor to close window or leave it closed
9                   send signal to actuator to switch on heater or leave it on
10          ENDIF
11  ENDWHILE
```

Notice that this is just a representation of how the microprocessor (computer) deals with the data. It does not contain all the steps which are involved in this system. The input of a pre-set value needs to be done beforehand and the analogue to digital conversion has not been mentioned.

Maintaining the required soil moisture

At the start of the process, the user inputs the required moisture level (pre-set value) using a keypad, number pad or touchscreen.

The microprocessor (computer) receives data about the amount of moisture in the soil in the greenhouse from the moisture sensor. It needs an ADC to change the analogue moisture data to a digital value the computer can understand. The computer compares the sensor data to the pre-set value input earlier by the user. If the level of moisture is below the pre-set value, it sends a signal to an actuator which activates the sprinkler valve to open or leave it open if it already is open. If the moisture level is above or equal to the pre-set value, the computer sends a signal to the actuator to close the sprinkler valve or leave it closed if it already is closed. This whole process is continuous as long as the system is switched on. Some greenhouses have sensors which measure the humidity of the air as well, because some plants require high humidity as well as warmth. These have separate watering systems, one to spray water into the air and another to put more moisture into the soil. The principle for these systems is much the same as for the light one described later.

Here is an algorithm which describes the processing carried out by the microprocessor, assuming the pre-set (required) moisture level has already been input to the system:

```
1   WHILE system switched on
2       INPUT moisture
3       IF moisture < pre-set
4           THEN
5               send signal to actuator to open sprinkler valve or remain
                open
6           ELSE
7               send signal to actuator to close sprinkler valve or remain
                closed
8       ENDIF
9   ENDWHILE
```

Maintaining the required level of light

With regard to the level of light in the greenhouse, things are a little more complicated. When the light is not very good, a light source needs to come on so that the plants have good growing conditions. A pre-set value still needs to be entered. If the light value is less than the pre-set value, the light source needs to be activated. However, when night comes, the light source needs to be switched off. A second pre-set value therefore needs to be entered, so that when the amount of light falls below that value, as at night, the light source is switched off.

The steps required in this process are similar to the two outlined above, except that two pre-set values need to be input. A light sensor is needed to measure the amount of light coming into the greenhouse and the analogue to digital conversion is still required. Here is an algorithm which could describe the processing carried out by the microprocessor, assuming the pre-set (required) light levels have already been input to the system. Note that pre-set1 is the light level below which it is dull during the day, and pre-set2 is the light level as it gets dark at night time.

```
1   WHILE system switched on
2       INPUT light_level
3       IF light_level < pre-set1
4           THEN
5               IF light_level < pre-set2
6                   THEN
7                       send signal to actuator to switch off light source
                        or leave off
8                   ELSE
9                       send signal to actuator to switch on light source
                        or leave on
10              ENDIF
11          ELSE
12              send signal to actuator to switch off light source or leave off
13      ENDIF
14  ENDWHILE
```

Maintaining pH

Some sophisticated greenhouse systems are able to control the acidity of the soil. The computer will have been pre-programmed with the required level of acidity of the soil. pH sensors are used to measure the pH of the soil. Different plants often require a different level of acidity or pH. A pH of 7 is neutral, a pH above 7 means the soil is basic or alkaline whereas below 7 it is acidic. Chemicals can be added to the water supply to make the soil more acidic whilst different chemicals can be added to make it more alkaline. The computer can take readings from the sensors and compare it to the pre-set value. Depending on the difference the computer can control the addition of the chemicals and therefore the soil acidity or alkalinity.

Central-heating systems

Central-heating systems and microprocessor control were discussed in Chapter 1 in the section on real-time processing. The steps involved are outlined below.

```
1   User enters required temperature using keypad/touchscreen
2   Microprocessor stores required temperature as a pre-set value
3   Microprocessor receives temperature from sensor
4   Microprocessor compares temperature from sensor to pre-set value
5   If temperature from the sensor is lower than the pre-set value, the
    microprocessor sends a signal to an actuator to open the gas valves
6   If temperature from the sensor is lower than the pre-set value, the
    microprocessor sends a signal to an actuator to switch the pump on
7   If temperature is higher than or equal to the pre-set value the
    microprocessor sends a signal to switch the pump off and close the
    valves
8   This sequence is repeated until the system is switched off
```

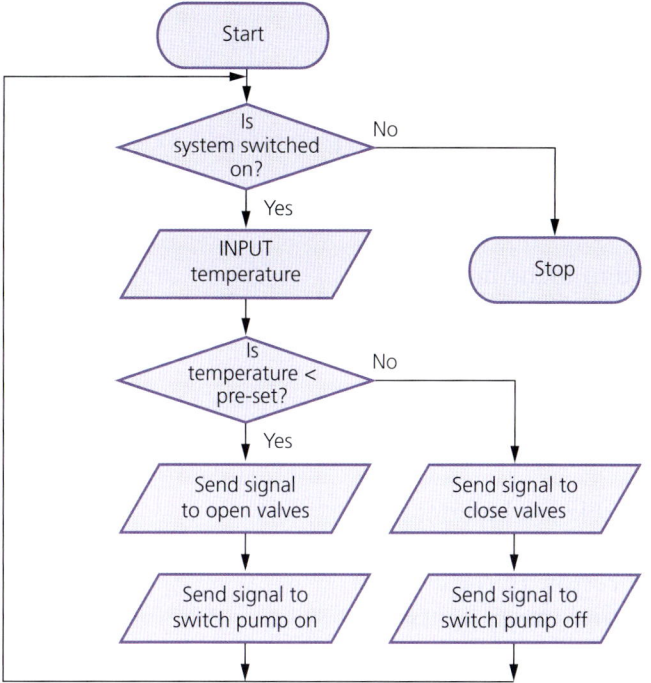

▲ Figure 3.9 Flowchart describing the processing carried out by the microprocessor in a central heating system

Notice that the flowchart has been drawn assuming the pre-set (required) temperature has already been input to the system. The flowchart does not take into account whether the valves are already open/closed or whether the pump is already on/off. This will be addressed in Chapter 4.

Air-conditioning systems

Like central-heating systems, air-conditioning systems and microprocessor control were discussed in Chapter 1 in the section on real-time processing. The steps involved were outlined as shown below.

```
1  User enters required temperature using keypad/touchscreen
2  Microprocessor stores required temperature as a pre-set value
3  Microprocessor receives temperature from sensor
4  Microprocessor compares temperature from sensor to pre-set value
5  If temperature from the sensor is higher than the pre-set value, the
   microprocessor sends a signal to an actuator to switch on the fans
6  If temperature is lower than, or equal to, the pre-set value, the
   microprocessor sends a signal to switch the fans off
7  This sequence is repeated until the system is switched off
```

It is important to be aware that the algorithms which have been written up to this stage are fairly limited. We have not taken into account that when a signal is sent to switch on a device such as a heater, light source, valve or fan, the device may already be switched on. Similarly, we have not considered the fact that if a signal is sent to switch off a device, it may already be switched off. This will be dealt with in Chapter 4.

> **Activity 3b**
> Draw a flowchart for an air-conditioning system which could describe the processing carried out by the microprocessor, assuming the pre-set (required) temperature has already been input to the system.

Burglar alarms

Microprocessor-controlled burglar alarm systems are used in many houses to protect against intruders. The sensors needed in such a system are:

» infrared sensors to detect movement of human bodies, which emit heat
» sound sensors to detect the level of sound an intruder might make
» pressure sensors that are placed under a carpet or rug to detect an increase in weight caused by a burglar treading on it.

The microprocessor is programmed to have certain acceptable levels and it only acts if the sensor readings are greater than these. An ADC is required so that data from the sensors can be understood by the microprocessor. In the event of detecting an intruder, the burglar alarm sounds an alarm and causes lights to flash and probably also sends a signal to the police to alert them to the presence of a possible intruder.

Of course, at the start of the process the user would need to switch the system on! This is achieved by the user inputting a passcode into the control box. The computer compares this entry with the passcode it has been programmed to accept. If it is the same the user then has a few seconds to leave the house through the door. A beeping noise is emitted from the control box. When the door is closed, a contact switch is closed, the system is activated and the beeping noise ceases. When the door is opened, upon the user's return, the user has a few seconds to enter the passcode. When the front door is opened a beeping sound is output from the control box warning the user that they only have a few seconds to enter the passcode. Again, the computer compares the entered passcode with the stored passcode. If it is the same the beeping stops and the system is deactivated.

> **Activity 3c**
> Write out all the steps in the operation of a burglar alarm system, from the moment that the system is activated.

Control of traffic/pedestrian flow

Apart from traffic lights, the main use of computer control regarding managing traffic flow is found on motorways (known as freeways or expressways in some countries). Many countries around the world are introducing **smart motorways**, but this is particularly the case in the UK, Australia and New Zealand. A smart motorway is a section of motorway that uses **active traffic management (ATM)** techniques. The system involves the use of variable speed limits and/or being able to drive on the hard shoulder ('shoulder' in the USA and elsewhere; this is the area at the side of a motorway or other road where you are allowed to stop if your car breaks down) at certain times. Computers constantly monitor the road and can change the speed limit or open the hard shoulder to traffic. Normally, traffic is not allowed on the hard shoulder and a red cross is displayed above it (see Figure 3.10). If the traffic becomes congested, the computer opens up the hard shoulder by removing the red cross (Figure 3.11).

▲ **Figure 3.10** A red cross shows that the hard shoulder should not be used

▲ **Figure 3.11** The red cross is removed and so traffic can use the hard shoulder

There are two main types of devices computers use to monitor the volume of traffic. One is rather like an induction loop (which will be explained in the next section) and is positioned just beneath the road surface. There are several of these sited at intervals of about 500 metres. The other uses a 'side-fire radar', which involves the firing of a radar beam at an angle to the roadside. The device measures the speed of the vehicles it is pointed at by detecting a change in frequency of the radar signal when it is reflected back off an oncoming vehicle. This has advantages over the induction loop system because it is not susceptible to damage caused by potholes or erosion in the road.

All this data is constantly fed back to computers, which then process the data and decide what action, if any, needs to be taken; if the road becomes very busy, the system can automatically lower the speed limit or open the hard shoulder to traffic.

The computer can reduce the speed limit from 70 miles per hour to, for example, 50 mph. This slows traffic down and reduces the chance of traffic congestion. The decision made by the computer will often remain free of human intervention unless the decision appears to be unusual. However, computers are not programmed to react to a crash and are not allowed to close lanes. Figure 3.12 shows the result of action taken by people in a control room which makes use of CCTV to monitor traffic.

▲ **Figure 3.12** Drivers are alerted to an accident on the motorway ahead

Traffic lights have been used to manage pedestrian flow for many years. The use of computers to control traffic lights has been the case since the 1990s. Their use will be explained in much more detail in the section on traffic lights.

Car-park barriers

One of the most common ways to detect vehicles in a microprocessor-controlled car-park barrier system is by using an induction (sometimes called inductive) loop buried just below the surface of the road in front of the barrier. As a vehicle passes over the loop, it causes a change in inductance which is detected by the loop. The metal in the vehicle causes a change in the magnetic field. This in turn causes a current to flow. The loop sends back data which is converted to digital and if the computer detects any change compared to a pre-set value, it sends a signal to the actuator. In this case, the actuator is the motor which, when activated, causes the barrier to rise. There is usually a second sensor, often a light sensor, which is used to detect when the vehicle has passed beyond the barrier. A light beam passes across the space occupied by the vehicle. If the vehicle prevents the light beam from reaching the sensor, then the microprocessor will keep the barrier raised. When the vehicle is clear of the barrier, the microprocessor detects that the light beam has resumed and so can send a signal to the motor to retract and allow the barrier to lower. This makes sure the barrier stays up until the vehicle has passed through the beam.

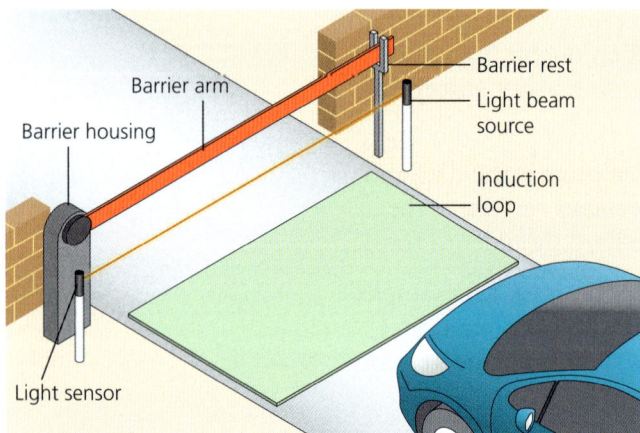

▲ Figure 3.13 Car-park barrier using an induction loop system

> ### Activity 3d
> Write out the steps in the operation of a car-park barrier system, putting each distinct step on a separate line.

Some car-park barrier systems issue a ticket on entry which contains a barcode. When the driver wishes to leave the car park, he or she enters the ticket into a machine which has a barcode reader, and then pays for the time parked. The barcode reader feeds the information to a computer and registers that the barcode belongs to a driver that has paid. As the vehicle leaves the car park, the driver inserts the ticket into a machine at the barrier which contains another barcode reader. The computer checks that the barcode matches the list of those barcodes belonging to drivers that have paid and sends a signal to the motor to lift the barrier.

Some systems use a magnetic stripe instead of a barcode, but the process is the same. Some systems use automatic number plate recognition (ANPR) instead of a ticket with a barcode. The customer types in details of their car's number plate at the ticket machine before paying. As the car approaches the exit barrier, a camera sends the details of the number plate to the computer which uses optical character recognition (OCR) to read the number plate. If the number plate matches a plate which is recorded as having paid, it sends a signal to the motor to lift the barrier.

Ultrasonic sensors are sometimes used in car-park barrier systems instead of light sensors to prevent the barrier descending if the car has not moved beyond the barrier.

Traffic lights

There are two types of traffic light systems. One is called a fixed-time traffic light control system, which is controlled mechanically and turns the traffic lights green or red after a certain amount of time. This is usually about one minute, though some systems have less time and others have more. With this system, drivers often get very annoyed when they see that there is no traffic going across them, yet their light stays red. A system which can respond to the volume of traffic using the road is much more preferable and, to this end, many traffic lights at road intersections are now controlled by computers.

The computer is programmed to react to different volumes of traffic during the day and often uses the same method as in car-park barriers (induction loops) to detect these. The computer receives data from the induction loops by way of the ADC and counts the number of vehicles travelling in each direction. These counts are then compared with pre-set values and the computer changes the traffic light timings/sequences as required by sending signals back to the control box in the traffic lights, which then operates the new sequence or timings. The whole process is continuous.

So, for example if a line of cars is coming from one direction and none from the other, the computer will decide to keep the light on red for the road which has no traffic. When a sufficient number of vehicles have stopped at the red light, the computer will cause the red light to turn green and the other one red.

These systems sometimes need to allow for pedestrians crossing at the junction. If this is the case, the pedestrian presses a button at the side of the road. The computer registers this input and after a predefined delay sends a signal to the actuator to change the traffic lights from green to red, in addition to lighting up the sign telling the pedestrians they can cross (usually a green man). There can be a reasonable delay if the lights have only recently changed, but most modern systems, using induction loops, can tell if traffic is light and thereby reduce the time delay.

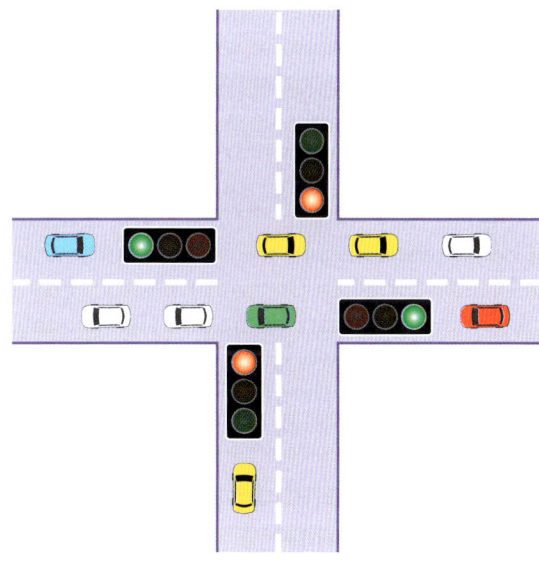

▲ Figure 3.14 A computer-controlled traffic light junction

Wireless sensor and actuator networks

A **wireless sensor and actuator network (WSAN)**, as the name suggests, is a networked group of sensors and actuators that communicate wirelessly. They are sometimes called 'wireless sensor and actor networks', because the actuators might be grouped together to perform a collaborative action and are thus referred to as an **actor**. For example, a robot is classified as an actor because it contains several actuators in one unit. An actor works as if it has a microprocessor included within it because it is capable of making decisions. Other examples of WSANs are smart parking systems, which alert drivers to where there are car parking spaces available, and combined heating, ventilation and air-conditioning (HVAC) systems. Networks such as wireless sensor networks (WSN) are slightly different, in that they take no action but merely monitor the environment. Examples are wearable computer devices used to monitor an athlete's performance, and air and water pollution monitoring where these are performed remotely from the computer without the use of cabled connections.

Smart homes

A **smart home** is a home in which devices and appliances are connected so that they can communicate with each other and can be controlled using commands by anyone living in that home. A smart home makes use of the home computer network and router and is an example of a WSAN. The commands can be given by voice, remote control, tablet or smartphone. The most common systems controlled in this way are televisions, music centres, lights, burglar alarms, central heating and air-conditioning units. Smart homes have developed due to the increase in the use of smartphones and tablet computers. These are continuously connected to the internet and can thus be used to control any number of online devices.

The ability to control such devices remotely – it is possible to switch on the heating while the user is in an office miles away, for example – is called the Internet of Things (IoT). IoT is a term used to describe the remote control of appliances and devices that are interconnected through digital networks and includes refrigerators, cookers and others. One of the benefits of a smart home is preventing the possible panic which can occur when a homeowner is on their way to work. Did they lock the front door? Was the cooker switched off? Was the burglar alarm switched on? A smart home or IoT allows the homeowner to do all these things and more, remotely. For example, when still in their office at work, the homeowner can switch the oven on so that dinner will be ready as soon as they walk in the door; the central heating or air-conditioning unit can be switched on and the temperature set so that the house will be nice and comfortable for when they return home. However, there is at least one drawback to having a smart home and that is being vulnerable to hackers, who could access a home network and turn off the burglar alarm, making it easy for someone to break in, or they could just cause a nuisance by turning lights off, changing channels on the television, and so on.

3.2.4 Advantages and disadvantages of using control technology

The use of computer-controlled car-park barriers instead of human parking attendants, for example, has increased unemployment but, on the other hand, IT technicians are needed to maintain the computers, and programmers are required to program the systems, so increasing employment. However, the number of new jobs created tends to be far fewer than the number of old jobs lost.

Microprocessor-controlled burglar alarms can give people a greater sense of security as they feel free from the risk of being burgled.

Smart homes can reduce the amount of energy needed to heat and provide light within a home. But they can lead to people becoming lazy since they can become over-reliant on microprocessor-controlled devices in the home. They have also caused a loss of manual household skills and prevent people from performing simple exercise such as walking around or using their hands and arms as much as they used to.

The use of computer-controlled traffic lights has led to there being fewer traffic jams than when they are mechanically or time controlled.

Air-conditioning units in shopping malls tend to make shopping a more comfortable experience when the weather is warm but can lead to increased costs for shops, which in turn leads to an increase in prices for the consumer.

Most control systems, generally, can help people with disabilities who may find it difficult to get around and use devices in the home. The use of smart-home technology can also lead to savings in household costs such as electricity as it tends to ensure that more economic use is made of power, switching off appliances automatically when not required. However, it may cost a lot of money to buy the system in the first place. Smart devices used in a smart home are much more expensive than non-smart devices.

Computer-controlled systems process data more quickly than a human could which, in turn, leads to almost immediate reactions to changes in the inputs to the system. A computer-controlled system will not be able to function if there is a problem with the computer or there is a power outage without backup power supplies.

The use of such systems does leave people free to do other things such as pursuing leisure activities since they do not have to be in the house to do the cooking or washing clothes or dishes.

With a computer-controlled greenhouse it is possible that a human could forget to take readings or be so busy that they are unable to take readings, whereas a computer can take readings at regular intervals. This means it can take action almost immediately, whereas human action might be delayed if they forget to take a reading. Computers can monitor more than one variable at a time, whereas a human would have to spend more time taking the readings since they would have to do them one at a time. The readings taken by computers are more reliable and more accurate than human readings as humans can make mistakes. Computers can carry out readings more frequently so in a greenhouse, any lack of water in the soil, for example, can be dealt with far sooner than if a human was taking readings every couple of hours, say. In addition, readings can be taken and control can be carried out at any time, such as at night or when people go on holiday.

Practice questions

1 A school has a computerised weather station which monitors a number of atmospheric conditions or variables. Describe how data is collected and processed by the computer. [6]

2 Give **three** advantages of using computers rather than humans to monitor soil quality. [3]

3 Give **three** disadvantages of using computers in a smart home. [3]

4 Draw a flowchart to represent the computer control of a car-park barrier. [6]

4 Algorithms and flowcharts

In this chapter you will learn how to:
- write a basic algorithm that demonstrates a decision-making process
- use conditional branching within an algorithm
- use loops within an algorithm
- use nested loops within an algorithm
- include procedures/subroutines within an algorithm
- edit a given algorithm
- write an algorithm to solve a given problem
- edit a given flowchart
- draw a basic program flowchart that demonstrates a decision-making process
- draw a program flowchart to solve a given problem
- identify errors in an algorithm or program flowchart for a given scenario.

Before starting this chapter you should understand the:
- comparison operators >, <, =
- arithmetic operators +, -, *, /
- order of arithmetical operations in an equation.

4.1 Algorithms

Before we look at how to write or edit an algorithm, let us consider what is meant by an algorithm. In computer science, information technology and mathematics, an algorithm is a set of instructions sequenced to solve a problem or to represent a calculation. Some would say that an algorithm is basically a data or **program flowchart** without the boxes. It is actually a list of precise steps. The order in which these steps are carried out is always crucial to the way an algorithm works. We start at the first line of the algorithm and work downwards to the final instruction. When an algorithm is created to produce information, data is input, processed, then the result is output. An algorithm must be carefully designed. It must be written in a way that caters for all possible scenarios. Any steps that rely on decisions being made must be dealt with in sequence and the conditions must be clear and unambiguous. Depending on the way the algorithm is written, not all instructions are carried out for a particular scenario, but they still have to be written in a way that allows for all possible scenarios.

Algorithms can be written in many different ways, including everyday natural language, **pseudocode**, **flowcharts** or programming languages. Natural language can sometimes lead to instructions which are open to interpretation and so tends not to be used in very complex algorithms. Pseudocode and flowcharts are structured ways of writing algorithms and we will concentrate on these approaches. Programming languages are primarily intended for converting algorithms to a form that can be understood and executed by a computer.

When solving problems in this chapter, the algorithms will be written in pseudocode. Pseudocode is independent of any programming language. Once the pseudocode is written and checked to make sure that it solves the problem, it would be fairly straightforward to translate it into any programming language we were familiar with.

Any algorithm consists of statements which have **linear progression**, that is the result of one statement is used in the statements that follow. It may also contain **conditional branching**, such as **IF…THEN…ELSE** or **CASE…ENDCASE**, which results in a decision being made between two or more courses of actions. It will probably also involve the use of **loops** such as **WHILE…ENDWHILE** or **REPEAT…UNTIL**. A loop is a sequence of statements that are repeated a number of times.

Here are some key actions that are performed and the common terms used in pseudocode:

- inputting data: **INPUT** or **READ**
- outputting data: **WRITE** or **PRINT**
- calculation: +, -, *, /
- comparison: >, <, =, <>
- setting values: ←

Do not worry if you have not met some of these before as we will be going into much greater detail when we begin to write our own algorithms.

4.1.1 Writing an algorithm

By the end of this chapter, you will be able to write a basic algorithm that demonstrates a decision-making process. In order for you to be able to do this you will need to become familiar with a number of pseudocode terms. One of the most basic units of an algorithm is a variable. It is best to think of a variable as an area of the computer's memory that stores one item of data, such as a number. The algorithm designer is able to choose the names of the variables, making writing an algorithm easier, and also it is possible to write an algorithm where we can change the values each time we work the algorithm. Assigning a value to a variable name can be done using an arrow symbol, ←. The variable name is to the left of the sign and the value is to the right, for example:

```
X ← 42
```

means we want to store the number 42 in X.

We can also assign calculations and other types of processing this way, for example:

```
Z ← W + X + Y
```

This assigns the result of adding the contents of W, X and Y together and stores the result of this calculation in Z.

Input and output

Consider your use of calculators. When you want to do a calculation such as 42 × 36, three things have to happen. First of all, you have to type in the numbers, then the calculator's internal computer calculates the answer and finally it outputs the answer on the screen. The last two occur so quickly that you probably do not realise that the calculation occurs before the answer appears on the screen.

If we were writing an algorithm to describe this calculation it would simply be:

```
INPUT X, Y        // input two numbers, one is X, the other is Y
Z ← X*Y           // multiply the two numbers and call the answer Z
PRINT Z           // output the answer Z
```

It is written like this so that X and Y represent any two numbers.

We can think of Z as somewhere we store the result of X multiplied by Y.

The last line causes the answer (Z) to be output.

Notice we have used // to make comments. This just helps us to see what is going on but is not part of an instruction to be carried out.

If we use 42 and 36 as our numbers, Z would be 42*36, which is 1512.

We could use any letters or words instead of X, Y and Z.

We could have written:

```
INPUT firstnumber, secondnumber
answer ← firstnumber*secondnumber
PRINT answer
```

This is an example of how writing an algorithm can be made simpler just by using variable names that make sense and result in the algorithm being easier to follow. Always try to use variable names which give a clue to what they are storing.

The last statement could also be written as:

```
PRINT "The answer is", answer
```

This would result, using our calculation, in the following output:

```
The answer is 1512
```

Conditional branching

This is sometimes referred to as selection. Conditional branching means that certain statements are carried out depending on the value stored within a particular variable name. There are generally considered to be two types, which are IF...THEN...ELSE and CASE...ENDCASE.

IF...THEN...ELSE

From now on, whenever you see words within angle brackets, this is just an explanation of what you will be typing in and not the exact words themselves, for example <statement> would mean that a statement must go here, such as:

```
A ← B+C
```

Sometimes, IF statements do not have an ELSE part, such as in this sequence of instructions:

```
IF <condition>
   THEN
      <statement or list of statements to be carried out>
ENDIF
```

An example of this type is:

```
INPUT number1, number2
X ← number1/number2
   IF number2 > number1
      THEN
         X ← number2/number1
   ENDIF
PRINT X
```

Here, two numbers are input and the first number is divided by the second number, with the result being stored in X. The IF condition checks to see if the second number is bigger than the first. If it is, then the second number is divided by the first number and that result is stored in X, overwriting the previous result. If it is not, then X remains unchanged. The value in X is then output.

It is important to note that IF statements are slightly indented. It does not matter how much the indentation is, as long as you are consistent and use the same amount of indentation for each IF.

THEN statements are further indented and again, you just need to be consistent. The statement after THEN is further indented. ENDIF must always be in line with the IF statement. Any statements that come after ENDIF are not indented.

IF statements with an ELSE clause are written as follows:

```
IF <condition>
   THEN
      <statement or list of statements to be carried out>
   ELSE
      <alternative statement or list of alternative statements to be
      carried out>
ENDIF
```

Using our example above, a better way of writing the algorithm would be:

```
INPUT number1, number2
   IF number2 > number1
      THEN
         X ← number2/number1
      ELSE
         X ← number1/number2
   ENDIF
PRINT X
```

Here, as before, the value of X is the bigger of the two numbers divided by the smaller number.

The algorithm inputs two numbers. It checks to see if the second number is bigger than the first and if it is then it divides the second number by the first number, storing the answer in X. If it is not bigger (ELSE) it divides the first number by the second number, storing the answer in X.

Again, it is important to note the indents and the fact that ELSE is immediately in line with THEN. The statement below THEN is in line with the statement below ELSE.

Sometimes it is necessary to have what are called nested IF statements. This is an IF condition within another IF. You will have met these when using spreadsheet software during your practical lessons.

Let us consider an exam system where if a student gains 75 or more marks they are awarded a distinction; if they get between 60 and 74 marks they are awarded a merit; between 40 and 59 marks the student is awarded a pass; otherwise they fail the course.

This is the type of problem which requires nested IF statements to solve it. One solution is as follows:

```
1    INPUT mark
2       IF mark <75
3          THEN
4             IF mark <60
5                THEN
6                   IF mark <40
7                      THEN
8                         PRINT "Sorry you have failed"
9                      ELSE
10                        PRINT "You have passed"
11                   ENDIF
12                ELSE
13                   PRINT "Well done. You have been awarded a merit"
14             ENDIF
15          ELSE
16             PRINT "Congratulations you have been awarded a distinction"
17       ENDIF
```

Each line of pseudocode is numbered. This does not normally tend to be the case but it has been done here to make the explanation of this algorithm easier to follow.

Consider marks of 34, 45, 62 and 78. We will work through the algorithm for each mark.

With an IF statement we follow the THEN part when the IF statement is true but we only follow the ELSE part if the IF statement is false.

For the first mark of 34:

Statement 1 will store 34 in the variable name 'mark'.

Statement 2 is true so we go on to statements 3 and 4.

Statement 4 is true so we go on to statements 5 and 6.

Statement 6 is also true so we move on to statements 7 and 8, which cause the message 'Sorry you have failed' to be printed out.

We can go on to statement 11 as we can ignore statements 9 and 10 which are only followed if statement 6 was false.

Statement 11 is just there to tell us that statement 6 is now closed.

We can ignore statements 12 and 13 which would only be followed if statement 4 were false.

Statement 14 tells us we have reached the end of the IF in statement 4.

Again, we can ignore statements 15 and 16 as statement 2 is not false, which brings us to statement 17, which shows us that we have reached the end of the whole sequence of the nested IF.

Briefly then, for the mark of 45, statements 2 and 4 are true, so we reach statement 6 which is false. As a result, we jump to the ELSE statement 9 and reach statement 10 which causes the message 'You have passed' to be printed out. We then ignore all the remaining ELSE statements.

For the mark of 62, statement 2 is true but statement 4 is false, so we jump to the corresponding ELSE statement to the line 5 THEN statement, which is line 12. That causes us to go on to line 13, which causes the message 'Well done. You have been awarded a merit' to be printed out.

For the mark of 78, line 2 is false so we jump straight to the corresponding ELSE statement in line 15 which leads to line 16 being printed out 'Congratulations you have been awarded a distinction' and then we reach line 17.

Always make sure that the number of ENDIFs matches the number of IFs and that they are indented to match. Always make sure the ELSEs are in line with the THENs.

CASE...ENDCASE

With this type of condition, depending on the value input, the algorithm carries out one of a number of statements. The condition can be phrased differently but we shall use CASE <identifier> OF, though some would use CASE OF <identifier>. Whichever you choose to use, just be consistent.

The value input is usually a number which then determines which statement will be carried out.

```
CASE <identifier> OF
    <value 1> : <statement or list of statements>
    <value 2> : <statement or list of statements>
    OTHERWISE : <statement or list of statements>
ENDCASE
```

Depending on the value entered, either one statement or several statements can be carried out. For example:

```
INPUT grade
CASE grade OF
    'distinction':    X ← 75
                      Y ← 100
    'merit':          X ← 60
                      Y ← 74
    'pass':           X ← 40
                      Y ← 59
    'fail':           X ← 0
                      Y ← 39
ENDCASE
PRINT "Your mark must have been between ",X, " and ",Y
```

This algorithm allows you to type in your grade as a word. The values of X and Y are set accordingly depending on the word you type in. The range of marks that your mark must have been within is then printed out after the phrase 'Your mark must have been between'. Note the comma after the quotation marks. Commas are used to separate variable names from the words that are printed out.

For example, if 'distinction' is typed in, then X is set to 75 and Y is set to 100 and the algorithm jumps immediately to ENDCASE and then prints out the message: 'Your mark must have been between 75 and 100'.

Notice neither the quotes nor the commas are printed out.

If the word 'merit' had been typed in then the first case ('distinction') is ignored and the second case is carried out, then it jumps to ENDCASE and prints out the appropriate message. With 'pass', it goes straight to the third case then ENDCASE before printing the message, and with 'fail', it goes to the fourth case then ENDCASE and prints out the message.

It is quite common for CASE…ENDCASE to have numbers input and the appropriately numbered CASE statement to be carried out. Here is an example which includes an OTHERWISE statement.

```
INPUT number
CASE number OF
      1 : PRINT "This is the number 1"
      2 : PRINT "This is the number 2"
      OTHERWISE: PRINT "You have entered an unacceptable number"
ENDCASE
```

The OTHERWISE case must always be the last case statement.

There are a number of variations of the CASE…ENDCASE statement but you will need to be consistent with your use of any variation to the method given above.

> ## Activity 4a
> 1. Write an algorithm using IF…THEN which will input a number, then print out the number multiplied by 8 if it is less than 3, multiplied by 4 if it is between 3 and 5, and multiplied by 2 if it is any other number.
> 2. Write an algorithm using CASE…ENDCASE which will input a number, then print out the number divided by 2 if it is less than 6, divided by 5 if it is between 6 and 10, and divided by 10 if it is any other number.

Loops

When referring to the writing of computer programs or algorithms, a loop is a sequence of repeated statements that are carried out a number of times. This number may be just one or it can be a very large number, depending on the problem which needs solving. There are three ways a loop operates.

» One way is to use a count-controlled loop. The first statement in the loop specifies the size of the loop (how many times the subsequent statements are to be carried out). The final statement in the loop is just a marker so that the first statement is then returned to. The statements in the loop are repeatedly carried out until the number of times the statements within the loop are to be carried out (as shown in the first statement) has been met.

» Another way is sometimes referred to as a pre-condition loop, meaning the conditional statement is checked before (pre) any of the sequence of statements is followed. This is when a condition is checked to see if it is met and if it is, the sequence of statements is carried out. When the last statement is reached, the conditional statement is returned to. The condition is checked again and the whole process is repeated. Once the condition has not been met, it jumps to the last statement and exits the loop.

» The third type of loop is sometimes referred to as a post-condition loop. As its name suggests, the conditional statement is only checked after the sequence has been carried out for the first time. If the condition is not met, then the first statement is returned to. This is repeated until the condition is met, in which case the next statement after the close of the loop is carried out. The loops which will be described here are the WHILE…ENDWHILE loop (a pre-condition loop) and the REPEAT…UNTIL loop (a post-condition loop).

Before describing the different types of loop, the concept of counting within an algorithm has to be considered. If a loop is to be repeated a certain number of times, it is usual to set up a counter. This usually involves the initialising of a variable (setting it to 0). The variable name used is often just 'count'. It is quite common for the first statement in any algorithm to be:

```
count ← 0
```

This count is then incremented (increased by 1) every time a sequence of statements is carried out.

```
count ← count + 1
```

The loop can then be set up to be executed an exact number of times. This is done by basing the conditional statement on whether the count has reached the number of times we want the loop to be carried out.

Although, as we shall see, it is possible to have WHILE…ENDWHILE and REPEAT…UNTIL loops which use counts, these loops can be controlled using other means. The condition that has to be met doesn't necessarily have to involve a count. The FOR…NEXT loop on the other hand is purely a count-controlled loop.

FOR…NEXT

The FOR…NEXT loop looks like this

```
FOR <identifier> ← <lowvalue> TO <highvalue>
    <a statement or a list of statements to be carried out>
NEXT <identifier>
```

The identifier, lowvalue and highvalue must all be integers.

If lowvalue = highvalue then the statements are carried out once only. If lowvalue is greater than highvalue then the statements are not carried out. The NEXT instruction adds one to the count and goes back to the FOR statement.

When highvalue is greater than lowvalue (which is normally the case), the value of the identifier is set to that of the lowvalue and the statement or statements are carried out. The identifier's value then increases by one (it increments) and then the statements are carried out again. This process is repeated until the value of the identifier is the same as highvalue and the statement(s) are carried out for the final time.

An example of a FOR…NEXT algorithm would be.

```
INPUT number
FOR count ← 1 TO number
    PRINT count, "x 8 = ", count * 8
NEXT count
```

This algorithm takes as input the number of times the loop is to be carried out – 'number'. As long as the value of 'number' is a whole number greater than 0 it carries out the PRINT statement. To begin with, providing the number which has been input is a whole number greater than one, the identifier 'count' is set to 1 and the NEXT statement increases it by 1. When the count is equal to the value of 'number' the statement is carried out one more time before leaving the loop except, of course, if the number input is 1. This algorithm is printing out the multiplication table for 8, often called the 8 times table, up to the number that is input. If the number input was 5 it would calculate 1*8, 2*8, 3*8, 4*8 and finally 5*8 before exiting the loop.

FOR...NEXT...STEP

It is possible to write a counting algorithm where the count does not increase by one every time but another number.

This looks like:

```
FOR <identifier> ← <lowvalue> TO <highvalue> STEP <incremental value>
    <a statement or a list of statements to be carried out>
NEXT <identifier>
```

The incremental value must be an integer such as 2, 3, 4, etc. The identifier will be given the values from lowvalue in successive increments of incremental value until it reaches highvalue. If identifier becomes greater than the highvalue, the loop stops without carrying out any more statements. For example, FOR count ← 1 TO 12 STEP 3 would result in count taking the values 1, 4, 7, 10, 13 on successive runs of the algorithm but when it got to count being 13 it would stop without carrying out the statements.

The incremental value can be negative. For example,

```
INPUT number
FOR count ← number TO 1 STEP -1
PRINT count, "x 8 = ", count * 8
NEXT count
```

If we input the number as 5, this would produce the 8 times table but written backwards. It would calculate 5*8, 4*8, 3*8, 2*8 and finally 1*8 before exiting the loop.

WHILE...ENDWHILE

The WHILE...ENDWHILE loop looks like this:

```
WHILE <condition>
    <a statement or a list of statements to be carried out>
ENDWHILE
```

Some variations of this loop have the word 'DO' after the condition:

```
WHILE <condition> DO
```

We will not be using this in this book.

While the condition is met, the statements between the WHILE and ENDWHILE statements will be carried out. As soon as the conditional statement stops being true, the loop ends.

An example of a WHILE…ENDWHILE algorithm using a counter would be:

```
count ← 0
INPUT number
WHILE count < number
    count ← count + 1
    PRINT count, "x 10 = ", count*10
ENDWHILE
```

This algorithm takes as input the number of times the loop is to be executed – 'number'. It checks whether the count is less than this number before proceeding with the sequence of statements to be carried out. The first statement increments the counter so that it goes up by one before the calculation is performed. This algorithm is printing out the multiplication table for 10, often called the 10 times table, up to the number that is input. If the number input was 6 it would calculate 1*10, 2*10, 3*10, 4*10, 5*10 and finally 6*10 before exiting.

REPEAT…UNTIL

Unlike the WHILE…ENDWHILE loop, which tests the condition at the start of the loop, the REPEAT…UNTIL loop checks the condition at the end of the loop. It is similar to a WHILE…ENDWHILE loop, except that a REPEAT…UNTIL loop is executed at least once.

The REPEAT…UNTIL loop looks like this:

```
REPEAT
    <a statement or a list of statements to be carried out>
UNTIL <condition>
```

The statements between the REPEAT and UNTIL are carried out and from then on, for as long as the UNTIL condition is not met, they will be repeated. After the first time, they will only be carried out while the conditional statement is false. As soon as the condition becomes true, the loop ends.

We can convert the example above which uses the WHILE…ENDWHILE loop into the following:

```
count ← 0
INPUT number
REPEAT
    count ← count + 1
    PRINT count, "x 10 = ", count*10
UNTIL count = number
```

The only difference is that if you did not want anything printed out, you could type in 0 for the number. With the WHILE…ENDWHILE loop it would print nothing out. With the REPEAT…UNTIL loop it would print out:

```
1 x 10 = 10
```

despite the fact we might not want it to. This is a big disadvantage of the REPEAT…UNTIL loop, but as long as you want the loop to be executed at least once, it works well.

You may use another technique for finishing a loop. You can come out of a loop by inputting a rogue value (i.e. a value that is out of the ordinary). For example, supposing you needed to write an algorithm which added up a set of numbers. If all the numbers to be added consisted of numbers less than 100 you could end the loop by inputting a rogue value such as 999 as follows.

```
Total ← 0
number ← 0
WHILE number <> 999
total ← total + number
INPUT number
ENDWHILE
PRINT total
```

When the WHILE condition is met for the first time the value of number is 0, so the statements within the loop are executed. It will only terminate when number is 999. It will carry on increasing the total by adding the number that is input each time. As soon as 999 is input the loop stops and the total is output. The equivalent REPEAT…UNTIL algorithm would be:

```
Total ← 0
number ← 0
REPEAT
total ← total + number
INPUT number
UNTIL number = 999
PRINT total
```

Notice that in order to avoid adding 999 to the total, the input occurs *after* the total has been increased. The total has to be increased by the value of the number even when the number is equal to 0, which it is on the first pass.

Nested loops

A nested loop, as its name suggests, is a loop within a loop. Let us start with an example. We have already seen an algorithm which can output a multiplication table. Now we will look at an algorithm which will output several multiplication tables.

```
count1 ← 0
INPUT number1              // this is the number of multiplication tables
                              printed out
INPUT number2              // this is the number we want to go up to in
                              each table
WHILE count1 < number2
   count1 ← count1 + 1     // count1 is incremented so we do not
                              get 0 x 0
   count2 ← 0
   WHILE count2 < number1
      count2 ← count2 + 1  // count2 is incremented so we do not get 1 x 0
      PRINT count1, "x ", count2, "= ", count1*count2
   ENDWHILE
ENDWHILE
```

The variable number1 is the number of multiplication tables we want and number2 is how far we want the table to go up to. For example, if 6 is input for number1, then the 1, 2, 3, 4, 5 and 6 multiplication tables will be printed out, and if 12 is input for number2 each table will go up to 12×.

Notice that both count1 and count2 are incremented immediately after the loop starts to prevent the algorithm from printing out 0×1 or 1×0. The WHILE… statement only checks whether the condition is met before the loop is allowed to start. When count2 gets to 6, it will have already printed out 1×6 and the count1 loop will increment then count2 will be set back to 0. This is repeated until count1 is 12 and count2 is 6.

The final output would be:

1 × 1 = 1	1 × 2 = 2	1 × 3 = 3	1 × 4 = 4	1 × 5 = 5	1 × 6 = 6
2 × 1 = 2	2 × 2 = 4	2 × 3 = 6	2 × 4 = 8	2 × 5 = 10	2 × 6 = 12
3 × 1 = 3	3 × 2 = 6	3 × 3 = 9	3 × 4 = 12	3 × 5 = 15	3 × 6 = 18
4 × 1 = 4	4 × 2 = 8	4 × 3 = 12	4 × 4 = 16	4 × 5 = 20	4 × 6 = 24
5 × 1 = 5	5 × 2 = 10	5 × 3 = 15	5 × 4 = 20	5 × 5 = 25	5 × 6 = 30
6 × 1 = 6	6 × 2 = 12	6 × 3 = 18	6 × 4 = 24	6 × 5 = 30	6 × 6 = 36
7 × 1 = 7	7 × 2 = 14	7 × 3 = 21	7 × 4 = 28	7 × 5 = 35	7 × 6 = 42
8 × 1 = 8	8 × 2 = 16	8 × 3 = 24	8 × 4 = 32	8 × 5 = 40	8 × 6 = 48
9 × 1 = 9	9 × 2 = 18	9 × 3 = 27	9 × 4 = 36	9 × 5 = 45	9 × 6 = 54
10 × 1 = 10	10 × 2 = 20	10 × 3 = 30	10 × 4 = 40	10 × 5 = 50	10 × 6 = 60
11 × 1 = 11	11 × 2 = 22	11 × 3 = 33	11 × 4 = 44	11 × 5 = 55	11 × 6 = 66
12 × 1 = 12	12 × 2 = 24	12 × 3 = 36	12 × 4 = 48	12 × 5 = 60	12 × 6 = 72

The same output would be produced using the following nested REPEAT… UNTIL loop:

```
count1 ← 0
INPUT number1
INPUT number2
REPEAT
    count1 ← count1 + 1
    count2 ← 0
    REPEAT
        count2 ← count2 + 1
        PRINT count1, "x ", count2, "= ", count1*count2
    UNTIL count2 = number1
UNTIL count1 = number2
```

The equivalent nested FOR…NEXT loop could be:

```
INPUT number1, number2
FOR count1 ← 1 TO number1
    FOR count2 ← 1 TO number2
        PRINT count1, "x", count2, " = ", count1*count2
    NEXT count2
NEXT count1
```

Advice

Common errors

An error sometimes made by people writing algorithms with nested loops is not matching up the number of WHILE statements with the same number of ENDWHILEs. The same error can happen with REPEATs and UNTILs. You must always check this and make sure they are correctly indented.

An IF statement checks if something is true and if it is, it carries out a matching statement. As we have seen above, if it is not true, the algorithm will either carry on or have an alternative statement to carry out which requires an ELSE statement. Sometimes, people forget to include a matching ELSE statement.

Procedures/subroutines

When writing algorithms, it is often good practice to use subroutines, which are sometimes called procedures, although technically a procedure is a subset of subroutines. A subroutine is a sequence of algorithm statements that perform a specific task. This subroutine can then be used in other algorithms as and when needed.

The subroutines or procedures follow the following pattern:

```
PROCEDURE <value>
    <a statement or list of statements>
ENDPROCEDURE
```

Let us consider a house builder who needs to know the area of the top part of a brick wall underneath the roof. He or she wants to calculate the number of bricks to be used. Such a house is shown in the figure below, with the area which needs to be calculated indicated as a red triangle. The other areas which would need to be calculated are rectangle shapes. An algorithm can be written to work out these areas and this could then be used to calculate the number of bricks needed. This could all be done by using simple subroutines.

The subroutines or procedures could be simply written as:

```
PROCEDURE triangle(width, height)
    area ← width*height*0.5
    PRINT area
ENDPROCEDURE
```

```
PROCEDURE rectangle(length, width)
    area ← length*width
    PRINT area
ENDPROCEDURE
```

The main algorithm could be:

```
INPUT width, height                    // type in the height and width of
                                       //   the triangle

CALL triangle (width, height)          // this would print out the area
INPUT number                           // type in the number of walls
count ← 0
WHILE count < number
    INPUT length, height
    count ← count + 1
    CALL rectangle (length, height)    // this would print out the area of
                                       //   each wall
ENDWHILE
```

Notice that although we used length and width as the variable names when we set up the rectangle procedure, we can use different variable names when we 'CALL' it. As long as we 'pass' two numbers to the procedure, it will perform the calculation.

Once the procedure or subroutine is set up, it can be called as many times as we want.

File handling

There are a number of file-handling functions used in pseudocode, but they are really outside the scope of this book. However, you will need to demonstrate an understanding of how file handling works, without necessarily knowing the technical terms. Let us consider the very basic algorithm that was constructed in Chapter 1.

```
1   First record in the transaction file is read
2   First record in the old master file is read
3   REPEAT
4   IDs are compared
5   IF IDs do not match, old master file record is written to new master
    file
6       IF IDs match transaction is carried out
7           IF transaction is D or C, old master file record is not written
            to new master file
8           IF transaction is C, data in transaction file is written to new
            master file
9       IF IDs match, next record from transaction file is read
10  Next record from master file is read
11  UNTIL end of old master file
12      Data in transaction file record is written to new master file
13      Any remaining records of the transaction file are written to the
        master file
```

You should now be able to convert it into a properly structured algorithm such as this (notice we are using command words for the file input – READ and output – WRITE):

```
READ first record from the transaction file
READ first record from the old master file
REPEAT
    IF ID of old master file record <> ID of transaction file record
      THEN
          WRITE old master file record to new master file
      ELSE
          IF transaction = C
            THEN
                WRITE data in transaction file record to new master file
          ENDIF
          READ next record from transaction file
    ENDIF
    READ next record from master file
UNTIL end of old master file
WRITE data in transaction file record to new master file
WRITE any remaining transaction file records to the master file
```

4.2 Flowcharts

There is another way of writing an algorithm and that is by using a program flowchart. A program flowchart is a diagrammatic way of representing an algorithm used to produce the solution to a problem. A flowchart consists of symbols, connected to each other by arrows which indicate the path to be followed when working through the flowchart. Each different shape symbol represents a different stage in the process. From now on when we mention the word flowchart we are referring to program flowchart. This is not the same as a system flowchart, which you will meet in the A2 text book. Below is a list of the more common symbols.

▼ **Table 4.1** Common flowchart symbols

Input/output	*parallelogram*	This shape symbol represents input or output. It is equivalent to either the INPUT or PRINT statement.
Decision	*diamond*	This is a decision box and is equivalent to the IF statement but is also used in loops.
Terminator (Start/Stop)	*rounded rectangle*	This is the symbol used to show where the flowchart begins and also where it ends. It is also used at the start and end of a subroutine.
Process box	*rectangle*	This symbol represents any calculation or assigning of variables, usually contains the ← symbol.
Subroutine	*rectangle with double vertical lines*	This is the symbol used to call a subroutine from the main flowchart. It is equivalent to CALL.
Flow line	*arrow*	This is the symbol that shows which direction you should follow when working through the flowchart.
Connector	*circle with A*	Sometimes a flowchart can extend over many pages. This symbol indicates the continuation of the flowchart.

Sometimes flowcharts are so complex that they occupy more than one page. To illustrate this, let us imagine that the central heating system flowchart shown in Chapter 3 has been extended. Here it is, labelled.

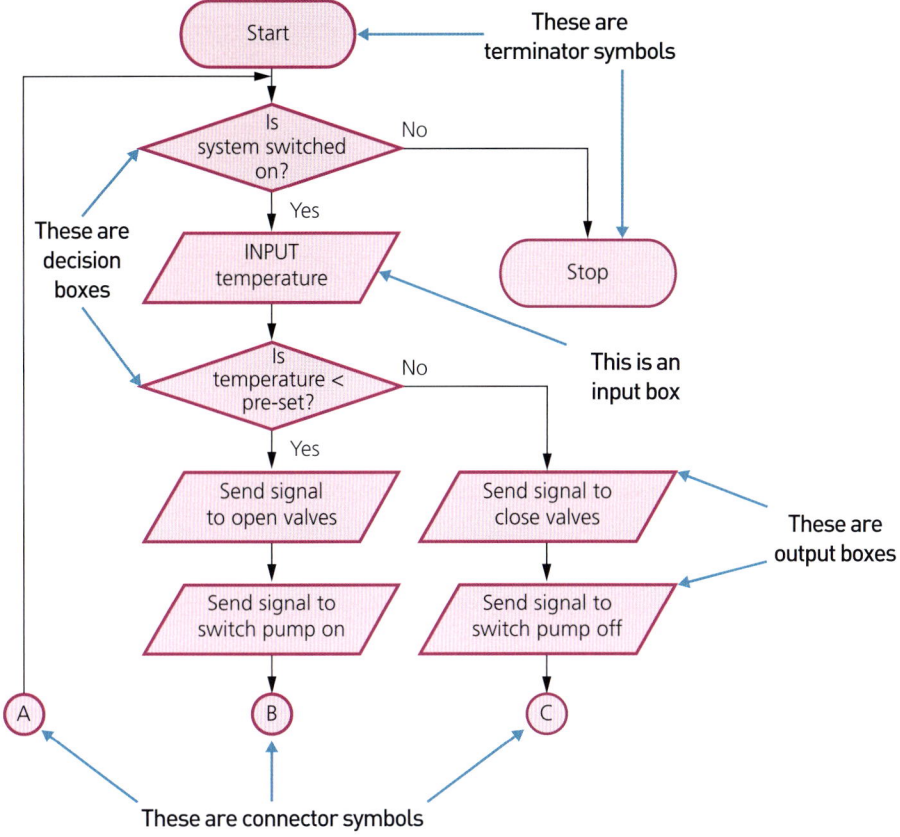

▲ Figure 4.1 Labelled flowchart from Chapter 3

It is possible that this flowchart might have continued on to another page. The connector symbols have been added to the ends of the flow lines to show how they would be positioned, if this were the case. Connector symbols A, B and C would be drawn at the top of the next page with the appropriate flow lines flowing down to the next symbol(s) if the flowchart continued.

To illustrate how to draw a flowchart, let us consider the earlier algorithms:

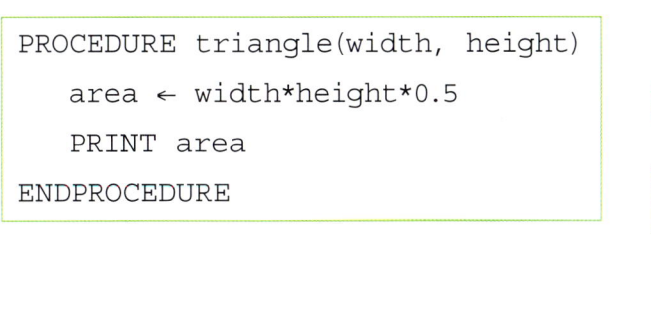

This becomes:

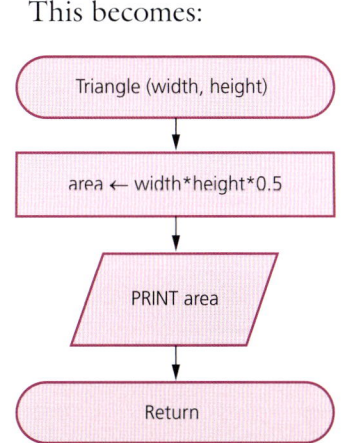

```
PROCEDURE rectangle(length, width)
    area ← length*width
    PRINT area
ENDPROCEDURE
```

This becomes:

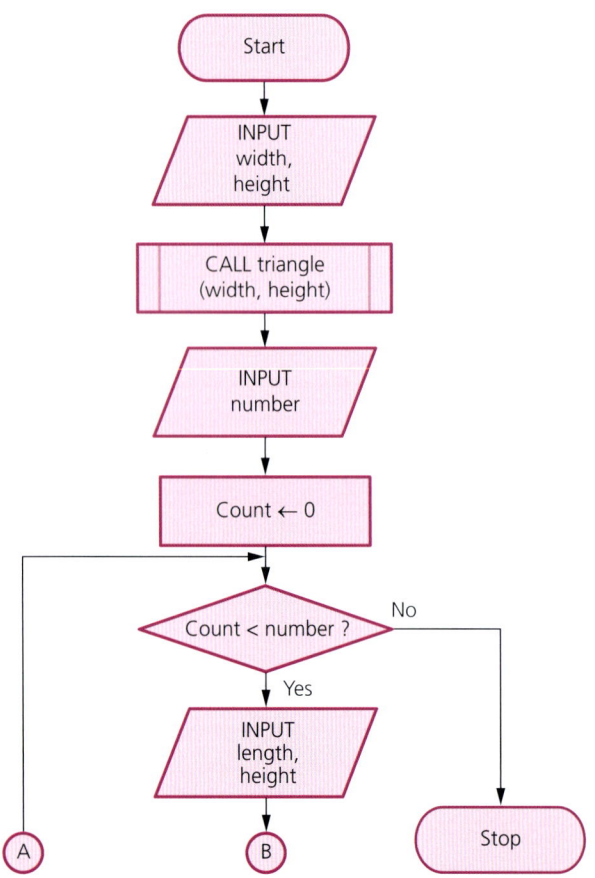

```
INPUT width, height            // type in the height and width of the
                                  triangle
CALL triangle (width, height)  // this would print out the area
INPUT number                   // type in the number of walls
count ← 0
WHILE count <number
    INPUT length, height
    count ← count + 1
    CALL rectangle (length, height)
ENDWHILE
```

This becomes:

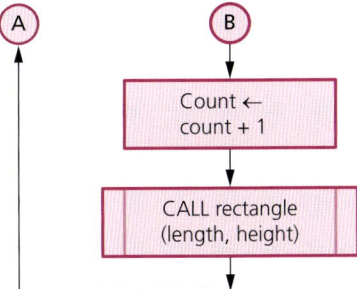

Notice the use of connector symbols because the flowchart goes over two pages.

As well as being able to increase the value of a counter, it is possible to add up a set of numbers in the same way, using:

```
total ← total + number
```

but you need to initialise the total as you would initialise a count. The flowchart to add up five numbers would look like this.

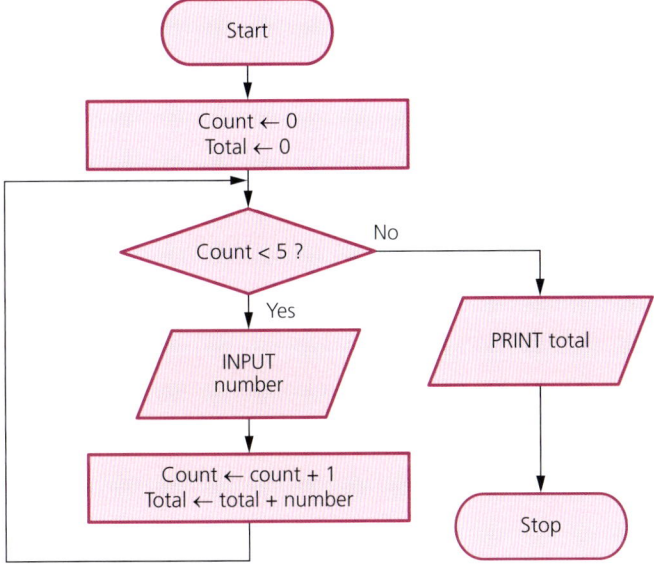

Common errors people make when drawing flowcharts and writing algorithms are:

» failing to initialise a count (leaving out count ← 0)
» failing to initialise a running total (leaving out total ← 0)
» getting the greater than (>) confused with the less than (<) symbol
» In flowcharts, having the yes and no the wrong way around.

Activity 4b

1. Look back at the section on greenhouses in Chapter 3. Refine the algorithms provided for temperature, moisture and light control to make them more efficient. Consider what should happen if the device is already switched off or switched on. Include extra statements to replace where the original algorithm has phrases like 'or leave on' and 'or leave off'.

The sequence of statements needed for this will be similar to the following:

```
WHILE system switched on
    INPUT value
        IF value < pre-set
            THEN
                IF device off
                    THEN
                        send signal to device to switch on
                ENDIF
            ELSE
                IF device on
                    THEN
                        send signal to device to switch off
                ENDIF
        ENDIF
ENDWHILE
```

You will need to change the word 'device' to the name of the device being used. You will need to change 'value' to the physical variable being measured. Draw program flowcharts for each algorithm so that each is a subroutine with an appropriate name. You can assume the system is switched on.

2 Draw a simple program flowchart which would check if the system was on and would then call the three subroutines.

3 Here is a program flowchart to output the sum of eight numbers. Identify the errors in it and suggest improvements (the yes and no flow lines are positioned correctly).

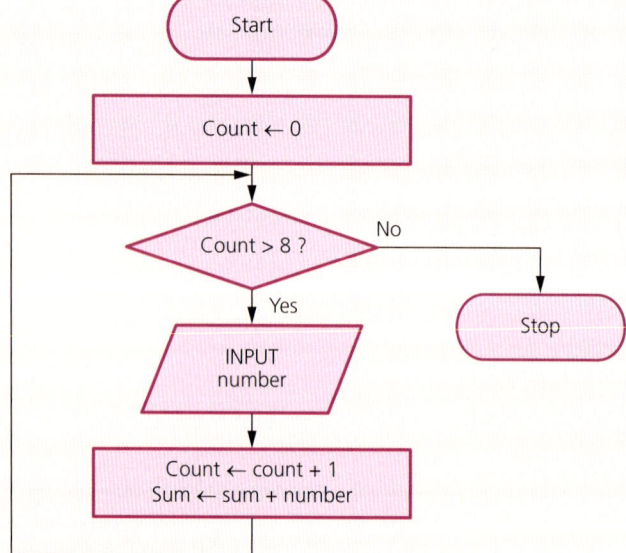

Practice questions

1 Write an algorithm using WHILE...ENDWHILE conditions, which would input 10 exam marks and if the mark was greater than 40 the output would be 'You have passed'. If it was not, then the output would be 'You have not reached the pass mark'. [6]

2 Write an algorithm using a REPEAT...UNTIL loop which would input five numbers and would output the largest number. [4]

3 Draw a flowchart which will allow the average of six numbers to be calculated. [6]

4 A student wants to output the 10 multiplication (times) table up to a certain number, which will be input at the start of the algorithm. The student wishes to make sure that an invalid number is not input.

Identify the errors in this algorithm and suggest improvements. [5]

```
count ← 0
WHILE count = 0
    INPUT number
    IF number < 1
        THEN
            PRINT "number is invalid"
ENDWHILE
WHILE count < number
    PRINT count, "x 10 = ", count*10
ENDWHILE
```

5 This flowchart inputs an examination mark and then outputs an appropriate message. If the mark is 80 or more, the message is distinction; if it is 60 or more, it is a merit; if it is 40 or more, it is a pass; otherwise, it is a fail. Identify the errors in this flowchart and suggest how it might be edited to produce the required output. [2]

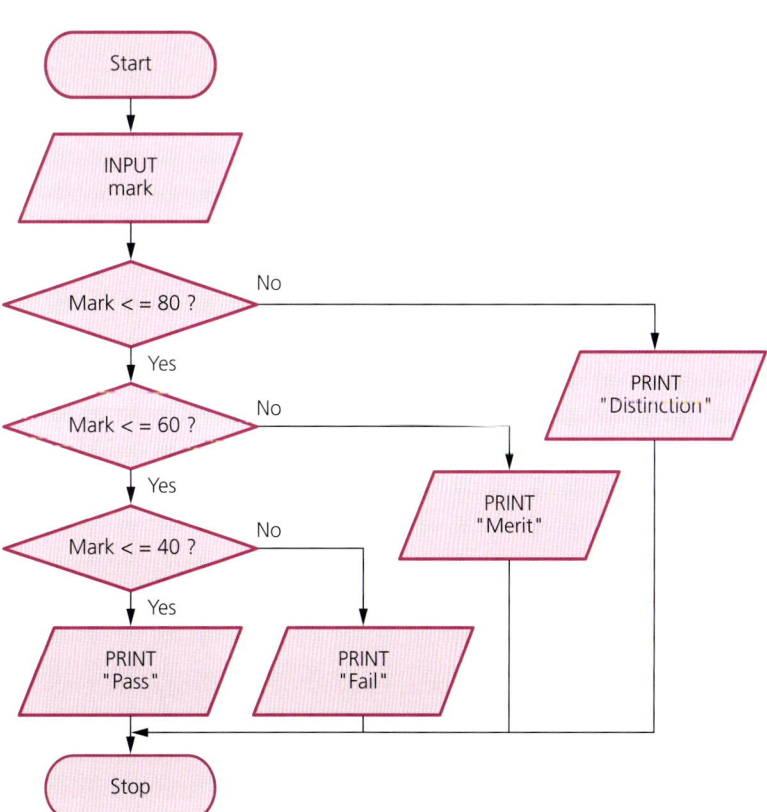

5 eSecurity

In this chapter you will learn:
- what personal data is
- why personal data should be kept confidential
- how personal data can be kept confidential
- the methods used by unauthorised persons to gather personal data
- how to evaluate the methods of prevention
- the types and uses of malware
- the consequences of malware for organisations and individuals
- how to prevent malware from entering computers

Before starting this chapter you should:
- be familiar with the terms 'encryption', 'data packet', 'anti-virus software', 'firewall', 'hacking', 'Uniform Resource Locator (URL)' and 'internet service provider (ISP)'.

5.1 Personal data

One useful definition of personal data is provided by the European Union's website. It defines personal data as follows:

> Personal data is any information that relates to an identified or identifiable living individual. Different pieces of information, which collected together can lead to the identification of a particular person, also constitute personal data.

It means that any data that can be used to identify or recognise somebody is classed as personal data. Sometimes the data has been manipulated so that it does not allow an individual to be identified. The EU's definition goes on to suggest that any data which can be reconstituted so that it does enable an individual to be identified can also be classed as personal data. So, even if personal data has been de-identified, encrypted or pseudonymised, it is still classed as personal data.

De-identification is a common strategy when trying to prevent a person's identity from being revealed. Items of personal data might be removed from a record, such as the individual's name. If the person could still be recognised from the remaining data it would be possible to re-identify the data and add the removed data, for example the name, back in.

Encryption was covered in Chapter 1; it is used to make data unidentifiable. However, the problem with encryption is that data can be decrypted and so become personal data again.

Pseudonymised data is when, instead of removing the personal items of data, they are replaced with a temporary ID. This means instead of seeing the person's name, you would see an ID which would mean nothing to you. The problem is that if, similar to de-identification, somebody can recognise that individual from the rest of the record, they can replace the ID with the individual's name.

If, however, the data has been amended to make it appear anonymous in such a way that it is impossible to recognise the individual, then it is no longer classed as personal data. For this to be the case the anonymisation of the record must be irreversible.

The EU, like many areas of the world, has a set of rules governing the protection of data. These are together called the **General Data Protection Regulation (GDPR)**. The GDPR promotes both the pseudonymisation and anonymisation of personal data. Organisations with personal data should use one or other of these methods to reduce the risk.

Examples of personal data as outlined by the EU are:

- a name and surname
- a home address
- an email address, such as name.surname@company.com
- an identification card number
- location data (for example from the location data function on a mobile phone)
- an IP address
- a cookie ID
- the advertising identifier of your phone
- data held by a hospital or doctor, which could be a symbol that uniquely identifies a person.

5.1.1 Keeping personal data confidential

Computers are used by organisations and companies to store large amounts of personal information. The question might arise: why do we need to keep data confidential? The answer is, if it were to fall into the wrong hands, the data could be used for identity theft or to withdraw huge sums of money from bank accounts. Identity theft is when a fraudster pretends to be another individual online by using that individual's personal information. Fraudsters who have accessed an individual's personal data can use their login details to access their bank accounts or commit other types of fraud, while pretending to be that individual. They can take your banking information and make unauthorised withdrawals and purchases, and transfer money between accounts. If burglars obtain personal data such as addresses and information about when a person is at work, then they can burgle that person's house. Therefore, people should not post information about their holiday or vacation plans on social media.

Organisations and businesses can take certain measures to ensure the confidentiality of data. It is essential that personal information should only be seen by those people who are authorised to see it. Keeping data confidential is an essential part of an organisation's responsibilities. Encryption is the main IT technique used to ensure the confidentiality of data in online systems. Anybody illegally accessing this data will not be able to understand it. They can still perform malicious actions, like deleting the data, but they cannot gain any information from it. Data protection acts, which are the rules organisations must stick to in order to protect data, exist in most countries. A number of workers in organisations need to look at the personal data of other individuals as part of their job. It is important that these workers maintain the confidentiality of the personal data. Organisations can encourage workers to be aware of their responsibilities. Workers who deal with confidential information about other individuals have a **duty of confidence**, both to the individual and their employer. They must not tell anybody or use the information for any reason except with the permission of the person who gave it. Should they attempt to do so, the person whose data it is can take out a legal injuction preventing them.

So that a duty of confidence can exist, the worker must be asked to treat the information as confidential or it must be obvious to them that the information is given in confidence. In order for this to happen workers are often asked by their employers to sign an agreement to this, which is called a confidentiality agreement.

Organisations must be held responsible for their decisions to pass on information. Good practice is to make sure that only the least amount of information that could identify an individual is passed on. Online services, particularly online banking and shopping, allow organisations to have access to private data such as names, addresses, phone numbers, financial records and so on. This information should not be passed from organisation to organisation without authorisation from the individual. Another action a company could take is to anonymise information. In its simplest form, **anonymised information** is simply not mentioning a person by name. However, we know that other information could enable the individual concerned to be identified, so this must be removed as well. Organisations should leave out as many personal details as possible. **Aggregated information** is another way of preventing individuals from being identified. This is where the personal details of a number of people are combined to provide information without individually identifying anybody. However, this may not always safeguard details adequately. An example could be a hospital which analyses data of its patients (without identifying individuals) who have a particular illness or disease. However, there may be only one patient with a particular disease and so it becomes obvious who that person is.

Individuals can keep their data confidential by not putting too much personal data on social media. Most employers do background checks on potential employees, which nowadays include looking at social media profiles. Personal information that might persuade a prospective employer not to employ an individual should not be posted. Insurance companies in many countries around the world often use the same approach and might charge much higher premiums (monthly payments) depending on the customer's lifestyle and in addition may use this as evidence when deciding not to pay out on a claim.

The photos and videos people take with their smartphone contain information, known as metadata, including the time and place they were taken (geotag). If these photos are intercepted, the individual's address or where they work can be discovered. However, smartphones do have software which can be used to either delete the geotag from images or disable them so they are not tagged.

5.1.2 Keeping personal data secure

There are a number of different precautions that can be taken in order to keep personal data secure. Some of these are described below.

Firewall

Firewalls are designed to prevent unauthorised network access. Organisations which store personal data tend to have several computers that form networks, many of which are connected to the internet. Without a firewall, these computers can be accessed by unauthorised users through the internet. Firewalls examine data coming into the network to see if it is allowable. It examines data packets and breaks them down into smaller pieces of information such as the IP address they came from. An IP (internet protocol) address is a combination of numbers that identifies each computer in a network. If it is an IP address that is not allowed, the firewall can block that traffic. It can prevent certain computers from gaining access to the network.

Firewalls do not always prevent hackers from accessing networks, however. Although a hacker's computer is prevented from accessing the network, the hacker could physically steal a computer that is permitted to access the network. More likely, they can use software which can change the IP address of their computer to one which is acceptable to the firewall.

There are two types of firewall, software and hardware. A software firewall is, as its name suggests, a piece of application software which is installed either on individual computers or on a network server. The installation of a software firewall is usually straightforward although the larger the organisation's network, the more complex it can become. It is a cheaper option than a hardware firewall. A hardware firewall is a physical device containing computer components which is connected between the router and the network server. Some hardware firewalls are also part of the router. Hardware firewalls can deal with more data packets per second than can be achieved with a software firewall. This is because they use purpose-built hardware whereas a software firewall has to use the resources of the server which is busy running all the other software needed to run a network. However, software firewalls can be upgraded without replacing the hardware making it a cheaper option as well as being easier to add new features.

Hardware firewalls are more able to resist malware because their operating system is different from typical network operating systems which hackers tend to be more familiar with. One consequence of this is that hardware firewalls require specialised IT skills to install and this usually means an IT expert has to be employed to monitor and manage the hardware firewall after installation. Hardware firewalls are mainly used by large organisations or those where data security is the top priority. Many home users or small business users will have a software firewall as it is much easier to install and configure than a hardware firewall. A hardware firewall, however, gives the network administrator control over how the network is used and it can protect other network devices that don't have built-in firewalls, such as printers and other smart devices.

Organisations dealing with healthcare or financial services process mainly sensitive data and often employ both types of firewalls. The same is true with companies processing credit and debit card payments as PCI DSS require this.

Penetration testing

A **penetration test**, sometimes referred to as a 'pen test', is when companies employ somebody to deliberately attack their computer network. They do this so that the authorised 'hacker' will identify the weaknesses in their system's security and the company can then take measures to improve it if necessary. Basically, it is a way to find out how easy it is to access a computer network and how well the measures being taken to protect the data are working and, if necessary, improve them. The purpose of doing this is to enable the company to secure personal data from illegal hackers who will attempt to gain unauthorised access to the system.

Authentication techniques

In order to prevent hackers accessing a computer network, users are required to log on. This means that they have to identify themselves to the system, so that it can be sure it is not a hacker trying to gain access. This is called 'authentication'. There are many ways in which a person can prove to a computer system that they are who they say they are:

» Typing in a **user ID** and password which only the user *knows*.
» Inserting or swiping a smart card which *belongs to* the user.
» Using **biometric data** which relates to a *unique physical characteristic* of the user.

Biometrics can involve the use of iris or fingerprint scanning, as these are both felt to be the best at providing unique data. Although quite expensive, iris scanners tend to be more effective as the scanner does not require as much cleaning after use and it also is more accurate since fingerprints can be affected by grease and dirt.

If only one of these methods is used, it would suggest that the system is not totally secure; at least two should be used when accessing personal data. For example, when somebody withdraws money from an ATM, they have to use something that belongs to them (their bank card) and something only they know (their PIN). Although many small transactions can be carried out with just a contactless card, for any transaction involving a lot of money a PIN has to be used as well. This is called twin- or two-factor authentication, sometimes referred to as multi-factor authentication, as it involves more than one method.

When using online banking, additional information such as the user's date of birth is often required. When a customer carries out certain transactions using a smartphone, some banks will send a one-time PIN or password in a text message for them to enter as part of the authentication process.

During the login process, keyboard presses can be detected by **spyware** and so drop-down options are often used for dates or PINs to be entered.

Levels of access

If hackers do gain access to a network, their ability to retrieve personal data can be limited by network settings created by a network manager. Different groups of users can be granted different levels of access to the data on the network. This is particularly the case with hospitals, for example, where doctors may be able to see the illnesses and diagnoses of their patients but administration staff may only be able to find out other, not health-related, information about patients. Often, the level of access granted to a user is related to their user ID, but some systems enable all users of the network to log on to the system. They then require the use of a particular smart card to access certain data.

Another example is the use of online shopping websites that require a login; customers will only see data that is relevant to them. However, if programmers employed by the company access the customer database, they will be able to view all the accounts. This is because they will have been given a higher level of access than the customers. With social networks, it is the owner of the data that can grant different levels of access. It is possible for individuals to amend settings so that only 'friends' are allowed to see their data, or they could allow both 'friends' and 'friends of friends' to see their data. On the other hand, if the setting is 'public', the data can be seen by everyone. Allocating different access levels to different groups limits the information that the different groups can see and the actions they are allowed to take.

Network policies

Network policies are sets of rules that allow companies to choose who is allowed to access their computer network and control their use of the network once they have gained access. Most companies now use the internet to carry out their business transactions, and as a consequence their computer networks have become vulnerable to attack. These attacks can allow competitors to gain knowledge of their operations; they can result in data being destroyed or provide access to any personal data that is stored. When workers join a company, they are normally required to sign an agreement, such as an acceptable use policy. This specifies what type of use is acceptable and what is not. They have to agree not to use the network for illegal, unethical or distracting non-work-related activities, such as downloading copyrighted material or spending time on social networking sites to communicate with friends. While not necessarily preventing hackers from outside the organisation attacking the network, it does help to limit what employees might be tempted to do with personal data.

Software updates

As well as being vital for updating a computer's operating system, software updates are often made available for different types of application software. Although these updates are useful in eliminating bugs and making the software easier to use, probably their most useful function is when they eliminate specific security weaknesses. If weaknesses are present in an operating system, hackers can take advantage of these in order to access the computer system. As soon as any major software company is made aware of vulnerabilities (security weaknesses), they produce updates which eliminate that risk. It is important for users to install updates as soon as possible in order to limit the amount of time hackers have to find and exploit these weaknesses. If a system or app is left without updating for a long time, more hackers may become aware of any vulnerabilities and use that information to gain access to personal information stored on the system or app. Operating systems and anti-virus software tend to be the main types of software that need regular updating. Certain types of application software will also need regular updating.

Other measures

There are other measures which can be taken to increase network security. Encryption has already been discussed in terms of the fact that data, even when illegally accessed, will not be understandable. The use of digital certificates was described in detail in Chapter 1. One other method relates to mobile networks and concerns the remote deletion of mobile (cell) phone or tablet data. If a device is lost or stolen, the owner can send a command to it using another phone that will completely remove any data, such as personal data, from it. Any of the individual's data that is not backed up on the cloud will be lost forever but it does mean that the person who has stolen the phone will be unable to see any of the owner's personal data. In order to receive the command to erase all the data, the device has to be turned on and connected to the internet. If a device is lost at an airport or a rail station, for example, the task of wiping the data may be straightforward. However, if whoever has stolen it wants to stop the data from being removed from the device, all they have to do is turn it off, remove the SIM card and switch it back on again and any personal data stored on the phone is now available. In order to send the command, the app has to be downloaded on to the phone being used to contact the stolen phone. The app allows the user to get the stolen phone to ring loudly for five minutes just in case it has just been mislaid rather than stolen.

> **Activity 5a**
> Write a sentence about each of **five** different methods of keeping personal data secure.

5.1.3 Preventing misuse of personal data

Before considering how to prevent the misuse of personal data, it is important to consider how personal data can be gathered by people who are not supposed to have access to it. Many students think that hacking into a computer in order to copy, delete or change data is very easy. It is not. With all the security measures outlined above, it is very difficult. However, there are other ways in which the personal data of individuals can be gathered by unauthorised persons. The most common methods, with the corresponding ways of preventing them, are outlined below.

Pharming

Computers connected to the internet all have a file called the **hosts file**. This is a basic **text file** which contains the name and the IP address of a number of URLs. When a user types a URL into a web browser, the computer first looks in the hosts file to find that URL. It then looks up the corresponding IP address and seeks to connect to the computer with that IP address. If it cannot find the URL in the hosts file it connects to the DNS server and looks for it there and uses the IP address to access the appropriate computer.

Pharming begins with a user downloading malicious software (malware) without realising they have done so. This software then corrupts the hosts file by adding URLs of banks, for example, to the hosts file and corresponding IP addresses which will take the user to a fake website when they enter that URL. The user then proceeds to type their bank details into this fake website, giving the fraudster all the information he or she needs to access that user's bank account by logging on to the real bank. Another type of pharming is when the fraudster hacks into the DNS server and corrupts the file on that computer, again causing the user to be redirected to a fake website. In both cases, the user will not realise that they have been tricked, since they will have typed in the correct URL and if they look in the browser URL window, they will see nothing out of the ordinary.

There are a number of ways that users can limit the chances of pharming, though none of these will completely prevent it. Using up-to-date anti-virus software is one way to prevent the downloading of software which changes the hosts file. Users need to make sure they install the latest software updates. An up-to-date browser can cause an alert to be raised that a fake website has been loaded. It is sensible to use a trusted, legitimate **internet service provider** (**ISP**). Digital certificates can be checked to make sure that the site is legitimate. Any site that requires the entry of personal data should begin with HTTPS. If it does not, it may well be a fake site, particularly if it has not got a coloured padlock icon next to it. It may be useful to check that the URL is indeed correct for that site. The actual fake website may have tell-tale signs such as poor grammar or spelling and this should alert a user that it may be a fake site.

Phishing

Phishing is when fraudsters try to obtain personal banking details such as usernames, passwords, and credit or debit card details using email. They pretend to be an official of the bank and leave a message which often directs users to enter personal information into a fake website which looks just like the legitimate site. It involves the use of an email in an attempt to get people to disclose their personal information. The email often includes a link to the fake website (a URL) inviting the receiver to go to that site. Some more primitive emails just ask the recipient to simply type in their bank details in a reply to the email. Others are more subtle than this, with the email containing something that instantly grabs the recipient's attention and requires them to take immediate action. It can be a message that tells them they have won the lottery but need to send personal information in order to claim the prize. The message usually contains a link to an email address where they must send this information. In addition to giving out their personal information (which could lead to identity theft), the recipient is then asked to send some money to cover the sender's fees and then they will receive their winnings. After they send the money to cover the fees, either they hear nothing from the sender ever again or they may even be asked to send extra sums of money!

Another type of phishing email informs the recipient that their account has been closed or blocked and they need to log on to unblock the account. An example is shown here of what happens after the website link is clicked on. The person has been told that their account has been frozen and they need to log on to change their settings so that the account can be unblocked.

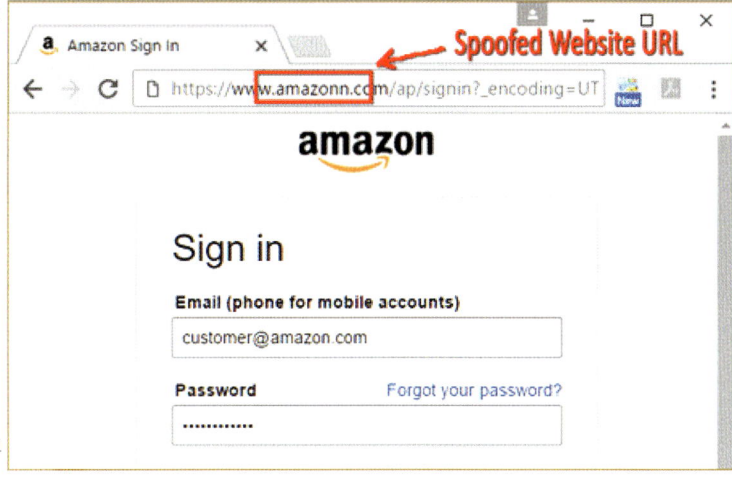

▲ **Figure 5.1** Fake website login

Here, the recipient has clicked on the link but for the purposes of this example, a fake email address has been typed in. Normally, an individual would be expected to type in their email address and password, allowing the fraudsters to gain access to the account. The URL is called a 'spoofed URL' as the hacker has given the site a name which is deliberately spelt in a way that is close to the name of an authentic site.

In order to avoid falling victim to these phishing scams, there are several things a computer user can do. It is important to use anti-phishing software on a computer connected to the internet. This identifies any content which could be interpreted as phishing contained in websites or emails. It can block the content and usually provides the user with a warning. It is often found within web browsers or email software. Not all web browsers provide this facility, however, so it is important to use one that does. It is a good idea to always have anti-virus and anti-spyware software running on a computer, and to update it at regular intervals. Phishing emails often contain grammatical and/or spelling mistakes, so it is important for users to look out for these. Users should never trust emails that come from people whose names they do not recognise. If an email looks suspicious, it is best practice to just delete it. Reputable companies or organisations will never ask for personal information, so that is usually a sign that it is a phishing email and, again, should not be trusted; the best action to take is probably deletion. If an email starts 'Dear customer' rather than using the receiver's name, it should also be treated with caution, as should emails asking the recipient to confirm their personal or financial information. Personal and financial information should never be sent in an email. If the email contains a message that the receiver has won a large amount of money or some other reason why they will benefit financially, it is likely to be a fake. Links placed within the email that are shorter than normal are used to hide the real URL and the best way of checking this is for the user to place the mouse cursor over the shortened link. This reveals the actual URL and the user can see straight away if it is suspicious. The best policy is never to click on such links.

Smishing

Smishing is a variation of phishing. The major difference is that it uses SMS (text messages) rather than email to send the message. The number of smishing attempts has increased since the introduction of smartphones, as it is so easy to activate a link within a text message. Just as with phishing, the main intention is to get the recipient to reveal their personal details.

There is a perception among most people that smartphones are more secure than laptops or PCs. However, this is not the case when it comes to smishing. In fact, the reason why there has been an increase in this type of scam is that people tend to be more vulnerable on their phones. They think there is less likelihood of being attacked on a phone than on a computer and so are more likely to respond to a smishing request. Some fraudsters are using text messages to get users to download an attachment which contains malware which, in turn, feeds personal data from the phone back to the fraudster. A smishing message is similar to a phishing message in that it often includes a link to the fake website, or it can just ask the recipient to simply type in their bank details in a reply to the text. The message usually contains a link to an email address where they must send this information. It can ask them to take immediate action. It can be a message that tells them they are entitled to a financial reward. Sometimes it contains a phone number asking the recipient to phone the bank or organisation using that number. When they phone, they are then asked for their personal details.

There are plenty of methods of prevention. Many digital security companies produce mobile protection software which users should have running in their smartphone. Users should also look out for all the same signs as in a phishing attempt, that is, spelling and grammatical errors, messages requiring immediate action or offering financial rewards, 'sign up now', or other pushy and too-good-to-be-true offers. They should never reply to such messages. Good practice is to open the sender's website itself rather than replying to a text with personal information included in it. A sensible action is to check the sender's phone number against the phone number of the company they claim to represent. Users should never type in personal or banking information, other than when using an organisation's official website. Receivers of a text should not click on links from senders they do not recognise. They should not click on any links in a text message since it is far safer to type the URL into a browser. They should also not phone any number contained in the text message.

Vishing

The word '**vishing**' is a combination of the words 'voice' and 'phishing'. As its name implies, it is the practice of making a phone call in order to get someone to divulge their personal or banking details. Vishing can take several forms; for example a fraudster, claiming to be from the bank, phones a customer telling them that their bank account has been accessed by a hacker and they need to change their password and that they will help them to do this. By giving the caller their account number and password, they have now allowed this fraudster access to their account. Even if the customer is not at home, the fraudster will leave a message requiring them to phone a particular number, which is actually the fraudster's. Often this phone number goes through to an answerphone message asking the customer to leave their account number and other personal details. An alternative is for the phone call to notify the receiver they have won some money or that they have won a prize. In both cases, the fraudsters charge a handling or redemption fee which they invite the person to pay by using their credit card number over the phone. Sometimes the fraudster will ask the customer to hang up and phone the bank to confirm, in an effort to convince the customer that the call really is from the bank. Meanwhile the fraudster has not disconnected the call so when the customer thinks they are phoning the bank they are actually still connected to the fraudster. They talk the customer through the process of logging on to the bank's website and the customer then enters their details in order to transfer money to their new account which has been set up by the bank. This account tends to be the fraudster's own bank account.

In order to avoid being a victim of vishing, it is good practice to use another phone to call the bank and ask to speak with the person who has just made the call. The main thing is not to give out login information over the phone, as a legitimate bank would never ask for it. The same goes for account information. The basic thing for customers to remember is to never give out any personal information over the phone. Banks will never ask for PINs or passwords. Customers should hang up, ignore them and block their number. These are all physical methods but smartphones have software included that can perform the blocking of numbers. There are also a variety of apps available which go beyond merely blocking the calls and keep a file containing the blocked numbers for all phone owners using the app. Users can browse through the file to see which numbers are blocked. One software solution is employed by large organisations whereby the software can filter numbers according to the likelihood that scams are being attempted.

5.1.4 Advantages and disadvantages of the different methods of preventing misuse of personal data

▼ Table 5.1 Advantages and disadvantages of methods of preventing misuse of personal data

Advantages	Disadvantages
Anti-virus software will also prevent viruses from deleting files on a computer as well as helping to prevent the user from downloading software which will corrupt the hosts file. (This is applicable to pharming and phishing.)	One of the disadvantages of the methods of trying to prevent pharming is that no particular method will completely remove the threat. For example, using anti-virus software to prevent the downloading of software which changes the hosts file will not necessarily work if the latest software updates are not installed. If your browser is not up to date you may not get an alert that a fake website has been loaded. Keeping the software up to date may prove expensive.
Deleting unauthorised emails will free up disk storage space and increase a computer's running speed. Never clicking on suspicious links within phishing emails will avoid the downloading of some viruses and therefore avoid such viruses deleting files or filling up the hard disk so it becomes unusable. (This is also applicable to pharming.)	One of the disadvantages of the methods of preventing phishing is that not all web browsers provide anti-phishing software. If anti-virus or anti-spyware software is not up to date it will not help prevent phishing. It is also possible that genuine emails may be deleted if all emails with minor grammatical or spelling mistakes or those sent from people whose names are not recognised or those that begin with 'Dear customer' are deleted. The use of anti-virus or anti-spyware software slows down the computer system. Often, anti-virus and anti-spyware providers have no customer support phone lines or if they do they are difficult and time-consuming to contact.
An advantage of using mobile protection software is that as well as alerting you to possible smishing messages it will protect against viruses and other malware. Deleting unauthorised texts (and text messages in general) will increase available memory and improve the performance of a mobile phone.	With smishing, the use of mobile protection software will slow down all the processes carried out by the phone. Just like you can delete emails to prevent phishing you can delete texts to prevent smishing, but you may delete texts even though they are genuine if you delete all texts with minor grammatical or spelling mistakes or if they are sent from people whose names are not recognised.
Using anti-vishing strategies, as well as possibly preventing fraud, will save the customer a lot of time which would ordinarily be spent on these phone calls.	With vishing customers may not be able to use another phone to call the bank to check as it is possible they only have one phone. Using a file containing all blocked numbers may lead to a genuine call being missed as the file may contain numbers that have been blocked for other reasons rather than vishing scams.

Activity 5b
1 Briefly describe the **two** different ways pharming attacks can be carried out.
2 Describe **four** methods of preventing smishing.

5.2 Malware

Malware is short for malicious software. It is the general term for computer programs which have been created with the deliberate intention of causing damage or disruption, or gaining access to a computer without the owner's permission.

5.2.1 Types of malware

The term malware covers all the different types of threats to computer security such as viruses, Trojan horses, worms, spyware, adware, **rootkits**, **malicious bots**, ransomware, as well as others.

Virus

A computer virus is a type of malware that is designed to spread from one computer to another, usually by means of the internet, causing changes in the way each computer operates as it spreads. They have the ability to replicate themselves, just like real viruses. Some types of virus delete the data on the disk or just corrupt or change the data. They insert themselves or attach themselves to another computer program. They often then lie dormant or inactive until a situation arises which causes the computer to execute its code. This situation, or event, can be a particular time or date. Symptoms that can indicate a virus is present are that popup windows suddenly start appearing frequently, the user's homepage is changed so that it is different to what it is normally, or their password is different, preventing the user from being able to log on. Viruses can cause large numbers of emails to be sent from the user's email account. The computer may frequently crash or its processing speed can noticeably slow down.

Trojan horse

Often shortened to 'Trojan', a Trojan horse is a malicious computer program which is used to hack into a computer. It enables the person who created it to take control of the computer it has infected. The name Trojan horse is a reference to the ancient Greek story, in which Greek soldiers laying siege to the city of Troy hid inside a wooden horse and deceived the Trojans into thinking that it was a peace offering. When the Trojans took the horse inside the city walls, the Greek soldiers let themselves out and attacked. Here it is the user who is deceived into thinking they have downloaded genuine software. Unlike computer viruses and worms, Trojans generally do not attempt to infect other files or replicate themselves. Instead, they are used for a number of purposes, the main one being simply to gain access to a computer so that the controller can discover the personal data of the owner. Another purpose is to delete files from the hard disk and Trojans can also be used simply to corrupt the data.

Worm

Worms are similar to viruses in that they replicate themselves. A worm will often exploit security holes in networks in order to spread throughout the network. Sometimes, their main purpose seems to be to continually replicate themselves and in so doing to occupy more and more disk space until the disk is full and can no longer function. They do not attach themselves to other programs or files and in that sense are said to be standalone programs. They are designed to spread by sending many copies to other computers in a network. This results in slowing down the traffic in a network. Some worms do not try to change the computers they are moving through but are designed simply to occupy more and more bandwidth in their attempt to slow down a network. When worms repeatedly replicate themselves, they start to use up the free space on a computer. One symptom on a computer with a worm is that the speed and performance are reduced. Another is that the amount of free storage space on the computer is noticeably reduced. Files going missing or new files suddenly appearing is another good indication that a worm is present.

Spyware

Spyware is malicious software that is designed to collect information about a computer user's activities without their knowledge. Data such as web browsing habits, email messages, usernames and passwords, and credit card information are passed to the hacker without the user having any idea what is happening. A keylogger is often used; this is a type of spyware which works by collecting a record of the user's keystrokes and results in the hacker receiving information about the user's credit card numbers and other sensitive or personal data. It does not replicate like a virus or worm but is just there to 'spy' on the computer. The only real indication that the user might notice that there is spyware on the computer is a reduction in processing power and bandwidth, since these are used by the spyware to communicate the information back to the hacker.

Adware

Malicious adware is software whose main purpose is to generate income for the originator or creator of the software. It is normally downloaded with free software, without the user's knowledge. It automatically generates advertisements. When the software is opened, advertisements may appear in the interface used by the software. It often keeps track of the internet sites the user visits and matches its advertisements to the types of goods or services that the user appears to be interested in. It often causes unrequested advertisements to be displayed in the browser when a user accesses the internet. Alternatively, the advertisements appear in popups. Generally, it is regarded as more of a nuisance than dangerous, but there has been a tendency in recent times for the adware to be linked to spyware. Symptoms can be unnoticeable, but some types of adware can slow down the performance of a computer and there tends to be a larger number than usual of popups.

Rootkit

Rootkit is a type of malicious software that is designed to install a set of tools in a computer which allows the attacker to have remote access to that computer continuously. It gives the attacker continuous privileged access to a computer and hides its presence deep within the operating system; the user is completely unaware that their computer has been infected. The different tools enable the attacker to discover the user's passwords and credit card details. The 'root' in rootkit is taken from the word used in Unix and Linux systems to signify the administrator account or somebody who has administrator privileges. It can be downloaded in a similar way to phishing by clicking on a link within an email, or a hacker could gain administrator privileges and install it remotely. The rootkit can change any security software such as an anti-virus to convince it that it is not there. It is also capable of removing the anti-virus software.

Malicious bots

Bot is short for internet robot and, as its name suggests, it performs tasks that are normally undertaken by a human. When used to gather information over the internet, bots are referred to as web crawlers. Without them, the smooth running of search engines, for example, would not be as efficient.

Unfortunately, there are many malicious bots. Like a worm, a malicious bot can replicate itself and is designed to feed back to a server; this is called a **botnet**, because it is in control of a network of infected computers. The botnet can then gather email addresses and from them generate spam to those and other addresses. They are capable of gathering information from different websites, such as date of birth from one site, health insurance details from another site, and address from another. They are extremely difficult to detect on a computer. Because of the increased use of the IoT, botnets are becoming a grave concern because they can control networks which consist of many different devices. This is largely because some of the devices in the IoT are relatively easy to hack into and susceptible to spreading bots.

Ransomware

Ransomware is a type of malware that blocks access to the user's data until a ransom is paid. An alternative approach is for the hacker to threaten to publish the user's data unless a ransom is paid. One of the most common ways of downloading ransomware is, again, a phishing email. The receiver clicks on what appears to be a link to a trusted file, but once it is opened the software, which can take the form of a worm or a Trojan horse, can take control of the user's computer. Among the many actions the software can perform once it has taken over the computer, the most common is to encrypt the user's files and send a message to the user demanding a ransom payment before it will decrypt them back.

Other types of malware

There are a number of other types of malware in addition to those described above. Two types are fileless malware and scareware.

Fileless malware is a type of malware that does not rely on files and leaves no evidence once it has been executed. It is very difficult for anti-malware software to detect and remove. It only resides in the main memory (RAM). Fileless malware does not perform any actions which affect the computer's hard drive. It ceases to work once the system is rebooted.

Scareware is a type of malware that tricks the computer user into thinking that their computer has been infected with a virus. It appears as a popup and seems to come from a genuine anti-virus provider. The user then pays the 'provider' to download the anti-virus and then discovers, too late, that it is a scam.

> **Activity 5c**
> 1 Describe the differences between ransomware and scareware.
> 2 Describe the differences between a Trojan horse and a worm.

5.2.2 How malware is used

There are a number of ways that malware can be used to carry out illegal activities. Three are described here.

Fraud

Computer fraud involves using a computer to take or alter electronic data, or to gain unlawful use of a computer or system to illegally benefit financially. Several different types of malware and general misuse of personal data have been described. For example, spyware collects a user's personal data, browsing habits and keystrokes. This can lead to credit card fraud as well as identity theft. Once fraudsters have gained a user's personal and financial data, they can either sell the information to other criminals or they can impersonate the user. They can use the user's financial data to ask the bank for a new PIN or even an extra card. They can buy goods via the internet using the credit card details they have obtained. They can also withdraw large sums of money from the user's bank account. Most credit card fraud victims are unaware of what has happened until it is too late.

Scareware, as we have seen, is used to obtain money under false pretences.

Phishing, vishing, smishing and pharming are intended to get the user to divulge their passwords, credit card numbers and bank account information so that the fraudster can access the user's account to withdraw money, make money transfers and also use the details to shop over the internet.

Ransomware, as we have seen, is used to blackmail users into paying large sums of money, usually in Bitcoin so that it cannot be traced.

Theft

As far as the uses of malware is concerned the word 'theft', in the context of 'data theft', is not literally the removing of information from the victim leaving them without it (as in, for example, theft of money). Data theft means that the hacker just makes a duplicate copy of the information for their own purposes. The purpose of data theft is often so that the hacker can make some form of monetary gain using the information. The perpetrator usually wants to sell the information or use it for identity theft. The information can be used to gain access to bank accounts so money can be withdrawn or transferred to the hacker's account. Credit cards can be created in the victim's name. As we saw in the section on fraud various methods can be used to steal the data including pharming, phishing, etc.

As we will see in the next section industrial spying is carried out to give groups or businesses an advantage over one another. Data theft is a major problem for many businesses because of potential lawsuits from customers whose information has been stolen, ransomware demands from hackers, the costs of restoring and repairing systems that have been hacked, the damage to a company's reputation and subsequent loss of customers as well as fines from regulatory bodies.

Industrial espionage

One dictionary definition of **industrial espionage** is 'spying directed towards discovering the secrets of a rival manufacturer or other industrial company'. It is usually the theft of business trade secrets. It used to be carried out by getting an employee to work for a rival company and spy from the inside; it is now more often carried out by hacking into databases or computer networks.

Malware has become a major tool in industrial cyber espionage, with the purpose of stealing information in the form of company secrets. For businesses, the internet has ceased to be a safe place. Malware is rapidly increasing in scale and can be found throughout the internet. When company employees go to a website or open an email, there is great risk of downloading malware in one form or another.

Malware can be used to access an employee's computer and the company information it holds. Regardless of what type of malware is being used, each one attempts to exploit weaknesses in software to gain access. There exists a form of malware that is designed to target a specific computer and thus lends itself to industrial espionage where a particular company is being spied on. Malware which is produced in order to attack any computer with vulnerabilities would serve no purpose to a hacker in this case, as they would get lots of information from unnecessary sources.

Hostile actors are people who organise themselves into teams of hackers with a collective aim. They include foreign states, criminals, groups of hackers with a common goal, as well as terrorists. Foreign states are usually best placed to conduct the most damaging cyber espionage and computer network attacks. Cyber espionage can be conducted in order to hack into specific business computer networks to steal large amounts of data without detection. This could include intellectual property, research and development projects, or a company's merger and acquisition plans. In the past, companies employed spies but now they are turning more and more to computer hackers to steal these secrets.

Intellectual property theft in the USA alone is estimated to cost companies hundreds of billions of dollars per year. Certain countries have been held responsible for these activities, but more private companies are now getting involved in this type of espionage. Groups of hackers are offering their services for hire for millions of dollars. Some are actually hacking into company secrets and offering the information they have gathered to the highest bidder. This activity has been made easier by the development of the Dark Web, where an auction-based marketplace exists. Most of the transactions are now taking place using Bitcoin. It is a bit disconcerting for companies to realise that their most confidential data may already be up for auction on the Dark Web. It is imperative for large corporations to have their own counter-espionage operatives and to make their IT systems secure. They need to investigate which of their secrets have already been offered at auction.

Sabotage

The term **computer sabotage** refers to making deliberate attacks which are intended to cause computers or networks to cease to function properly. The idea is that businesses, education establishments and other organisations are attacked in order that their normal operations are disrupted. It has been estimated that billions of dollars in the USA alone have been spent on legal fees so that damages could be paid out to victims of sabotage involving identity theft. A great deal of money has been spent on repairing computer systems in hospitals and banks.

Computer sabotage within organisations is often carried out by disgruntled employees intent on causing the organisation to lose money. Employees might make unauthorised attempts to view, disclose, retrieve, delete or change information by misusing the system privileges they have been granted. Some acts of sabotage are committed by former employees, perhaps unhappy with the way they lost their job. However, most attacks by employees or former employees are carried out remotely.

It is clear to see that not all sabotage is the result of sending malware, but it can consist of a virus being sent to a computer which prevents users from logging on, and it can take the form of distributing malware to allow hackers to illegally access an organisation's network.

Organisations need to guard against computer sabotage by taking measures to protect all their hardware and software. This will not only require a firewall and use of anti-virus software, but must include guidelines about the use of separate user IDs and passwords for each individual user of a computer, including advice to change passwords regularly.

5.2.3 Consequences for organisations and individuals

Malware poses a major threat to any organisation. It can ruin the organisation's security arrangement regarding its computer network and systems. As a result, it can disturb its business operations, leading to financial losses. Personal information can be accessed, leading to identity theft on a massive scale, as well as user IDs and passwords being compromised through the use of spyware. Some organisations, particularly banks, have had to pay out a great deal of money in the form of compensation when personal data has been stolen.

The three main implications of malware for an organisation are the loss of data and time and the costs it incurs. Keeping an organisation safe from viruses is often very expensive. Because there is so much malware being transmitted across the internet, it is important to plan ahead for any threats. An employee unwittingly clicking on a link in an email can release a virus that could delete all computer data stored on hard disk within an organisation, which would have serious consequences for the company. If a virus has infected one computer on a network, each computer has to be disconnected from the network and cleaned by using anti-virus software to remove viruses. Each computer must be clear of viruses before being reconnected to the network. Cleaning a computer while it is still connected to the network leaves it open to further infection. Disconnecting and cleaning each computer takes a lot of time and therefore cost. In addition, the organisation's IT department may not have the expertise to cope with such a massive undertaking of cleaning each computer, so outside experts may need to be hired to do the job, leading to further costs.

Looking at the impact on an individual and their personal computer or laptop, malware enables hackers to gain valuable information such as bank details, date of birth, email address and passwords. The hacker can then commit identity fraud. We have also seen how scareware can be used to fool users into paying for non-existent anti-virus software. The slowing down of their computer system can be inconvenient to say the least for an individual. Pranksters may simply want to see the havoc their programs can cause, whereas some hackers attack PCs with malware to make their reputations within the hacker world.

> **Activity 5d**
> Briefly describe **three** ways that individuals might be affected by the use of malware.

5.2.4 Prevention of malware

Use of prevention software

Anti-virus software must ensure all data is scanned for malicious code as it enters the company's network. (See Chapter 2 for more details on anti-virus software.) A firewall, which filters incoming traffic and prevents malicious software entering the system, should be used on all the external entry points to the network.

Anti-virus software must be kept up to date, running all the time, with scans scheduled to run at frequent intervals. Only one anti-virus program should be run at a time, as two different programs may conflict with each other. Anti-spyware software should be run as well if the anti-virus software does not incorporate an anti-spyware module, but it should not be in conflict with the anti-virus software. It is also a good idea to have a spam filter, particularly if the email software used does not have one.

Anti-malware software can be used. This may be different to anti-virus software, although these days the differences are becoming blurred. Anti-malware software can protect networked computers in two ways: it can guard against the entry of malware into a network and it can also remove malware that has managed to get into the computer network system.

Physical methods of prevention

Companies are advised to develop and implement anti-malware policies and ensure that they are consistently applied across the organisation. A record should be kept of any known malicious websites, which should then be blocked by the firewall from accessing the network. Dedicated stand-alone virus-checking computers which are equipped with anti-virus software should be provided. These must be capable of scanning any type of media. All employees should be educated so that they understand the risks from malware and are aware of the day-to-day procedures they can follow to help prevent malware infections. They should be encouraged to stop and think before clicking on links, but if they do, they need to inform the IT department as soon as possible. They should know that they must not connect any removable media or personally-owned device to the network. They need to be aware that they must report any strange or unexpected system behaviour to a technician or member of the IT department.

Good advice to both organisations and individuals is that operating systems and browser software should always be kept up to date. Software that is no longer used or old versions of application software should be removed. If software is no longer supported by the software company, then it will be open to malware attacks. Emails should be read with suspicion, since encouraging the user to download malware only works if the user is not paying sufficient attention to who sent the email and what it might be suggesting. In the phishing section, the things to look out for were described in detail: sender address, spelling/grammar errors, the true address of the link and so on. Strong passwords should be used, that is ones that are unique, changed often and do not relate to the owner's personal information.

Websites should have an appropriate padlock next to the URL. Users should always log off from their computers at the end of a session.

Practice questions

1 Describe **three** different items of personal data. [3]
2 Give **three** reasons why data should be kept confidential. [3]
3 Describe the different types of authentication techniques. [3]

The digital divide

In this chapter you will learn:
★ what the digital divide is
★ the causes and effects of the digital divide
★ the groups of people affected by the digital divide
★ about reducing the effects of the digital divide.

Before starting this chapter you should:
★ be familiar with the terms 'hardware', 'software', 'broadband', 'the internet' and 'wi-fi'.

We often take the use of the internet for granted. Most adults use the internet for everyday activities such as online banking and online shopping. Most young people use the internet for social media purposes. Many of us use the internet in one way or another and would be quite upset if we did not have fast access, such as good broadband. However, not everybody, for whatever reason, has access to, or perhaps wishes to use, the internet. There is said to be a divide between those that cannot, or will not, use the internet and those who can.

6.1 What is the digital divide?

The term **digital divide** relates to the gap between those people who are able to access modern technology and information and those with restricted or no access. It can be defined as the gap between people who have access to and use smartphones, television, personal computers, tablet computers, laptop computers and the internet, and those who do not.

When the phrase was first used, the digital divide just referred to the use of computers, but more recently it has been extended to include the use of telecommunications and broadband or internet.

The digital divide can be caused by the availability of hardware or software, or both. It can also be caused by individuals who do not have the access to use the technology, maybe those who live in rural areas, those with lower levels of education, or those living in poverty or in less industrially developed or technologically enabled countries.

A survey published by the Office for National Statistics (ONS) in the United Kingdom in 2018 showed that 5.3 million Britons have either never gone online or have not used the internet in the last three months. This number has reduced markedly since 2011, but this still amounts to about 10 per cent of the adult UK population. A graph demonstrating this trend is shown in Figure 6.1.

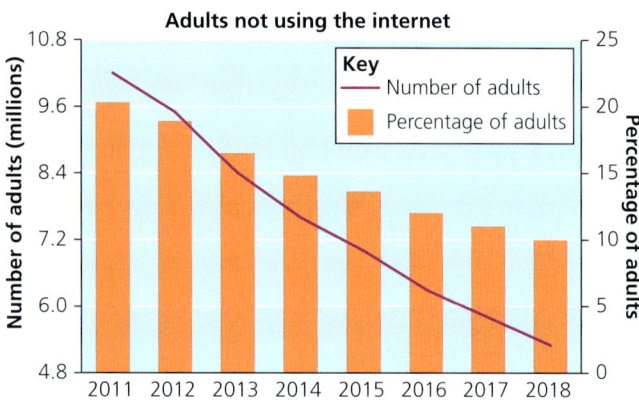

▲ Figure 6.1

The ONS report also referred to the Tech Partnership Basic Digital Skills framework, which describes five basic digital skills that can be used to measure digital inclusion and the activities someone should be able to carry out to demonstrate each skill. To be considered to have a digital skill, individuals need to be able to carry out one of the activities listed under it, as outlined below.

1 **Managing information**: using a search engine to look for information; finding a website visited before; downloading or saving a photo found online.
2 **Communicating**: sending a personal message via email or online messaging service; carefully making comments and sharing information online.
3 **Transacting**: buying items or services from a website; buying and installing apps on a device.
4 **Problem solving**: verifying sources of information online; solving a problem with a device or digital service using online help.
5 **Creating**: completing online application forms including personal details; creating something new from existing online images, music or video.

The report also referred to the Lloyds Bank UK Consumer Index 2018 which uses this framework to estimate the digital skills of the UK population. It estimated that the number of people in the UK lacking basic digital skills was declining. Despite this, in 2018, there was still a substantial proportion of people in the UK who were thought to have no basic digital skills (and were unable to do all or any of the activities described in the five basic digital skills).

6.2 Causes of the digital divide

At this point, it may be useful to consider the word *infrastructure*. We often hear phrases such as 'the infrastructure in rural areas is not as good as in urban areas'. What do we mean? The term infrastructure in an information technology (IT) context refers to the hardware, software and equipment used to provide IT services. This could include installation of electrical cables to enable the supply of electricity, installation of fibre optic cable for high-speed broadband, installation of mobile (cell) phone masts, installation of Fixed Wireless Access (FWA) Units and even launching satellites for the supply of broadband. In rural areas this infrastructure may be lacking. This is also likely to be the case in developing countries and may be due to the cost of such systems.

6.2.1 The availability of high and low performance computers

High performance computers is a general term referring to, in the main, personal computers which have fast data processing capability and a large amount of storage both in central memory and in backing storage. The term is not to be confused with High Performance Computing (HPC) which refers to the practice of combining the computing power of several computers. HPC enables organisations to process very large volumes of data at very high speeds using several computers and storage devices combined. Here we are concentrating on the individual use of high performance computers. Low performance computers, as the name implies, refers to those computers which, usually because of age, tend to be slow at processing data and also have far less storage than high performance computers.

A divide can also exist regarding the availability of high and low performance computers and wireless connections. In many parts of the world where people struggle to gain access to the use of computers, schools have been provided with low-cost refurbished computers from organisations based in more developed countries, particularly the USA. Unfortunately, these often tend to be lower specification models and these countries, though catching up, still tend to lag behind developed nations in their development of IT.

The differing levels of availability of high performance computers is a major contributory factor to the digital divide. If people only have access to low performance computers they will not be able to take full advantage of the benefits such as access to internet services. Slow performing computers together with lack of high speed broadband are major contributors to the digital divide for many of the groups described below.

6.2.2 Wireless connections

There are a variety of ways in which a wireless connection between IT devices can be created. The most common is wi-fi, whereby a user connects their computer or device wirelessly to a router. A second way is to use a smartphone to access a G4 or G5 mobile network. The connected smartphone can also be used for tethering. A device such as a laptop can connect to the mobile network via the the connected smartphone. This can prove expensive, however, as the laptop downloads full desktop web pages rather than mobile web pages which use up far less data. Many people who use smartphones to access the internet, for example, would benefit from the use of wi-fi hotspots in order to not have to pay high data charges. The absence of wi-fi hotspots can markedly reduce their opportunities for using the internet. Similarly, the absence of a G5 connection would also limit the smartphone user's access to the internet in terms of speed or being able to download videos. The lack of these opportunities creates a digital divide.

6.2.3 Geography

Geographical causes for the digital divide are visible both locally as well as worldwide. The IT infrastructure can vary between countries and between different areas within the same country.

Often, less economically developed countries (LEDCs) lack effective infrastructure when compared to more economically developed countries (MEDCs). This may be because MEDCs have more technology because of their greater wealth. An example is how the UK compares to a country such as Pakistan, which is an LEDC. In the UK an estimated 94.6 per cent of the population use the internet, compared to 35.0 per cent of the population of Pakistan.

Similar divides can sometimes be seen within countries, such as Italy for example. The people who live in the northern half are generally wealthier compared with those who live in the southern regions. As a result, they tend to have greater access to IT technologies.

Another geographical cause for the digital divide within individual countries relates not to lack of money, but location. In Russia, for example, there is a noticeable number of people who live in areas which do not have internet access. Northern parts of Russia are close to mountain ranges and it is difficult to provide the necessary IT requirements such as optical fibre to these areas, because of problems building such resources under mountains. Providers are much more interested in providing facilities for people living in the more densely populated areas, which are easier to supply and will provide more profit.

A similar divide can be seen in the UK, where living in certain areas means that you are less likely to have access to the required infrastructure. This is certainly true of the more rural parts of the country, where technologies such as 5G, or even 4G or fibre optic internet are not available. Again, this is often because suppliers of the technology tend to focus more on the more densely populated urban areas.

6.2.4 Fear of IT

Another cause of the digital divide is the fear of IT, caused either through lack of education in IT skills and knowledge, or because of age, since older individuals are simply less likely to be familiar with IT. IT is meant to make our lives easier, but sometimes it actually makes them more difficult. As IT devices become more sophisticated, they tend to become harder to use and learn about. Therefore, some people fear them. They believe that, when asked to use computers, telephone systems, or even ATMs, they might have to learn and understand sets of complicated instructions. Most people do find the instructions fairly easy to follow once they can be persuaded to try, but some do not. This can be particularly true of older individuals, who may find any type of change unsettling.

6.2.5 Economics

A third cause of the digital divide is economic factors. The wealth of the user will affect their ability to access IT technologies. As technological advances are made, every new innovation tends to be more expensive than its predecessor, at least in the short term. This means that the richer you are, the easier it is to have access to new IT technologies. It is in the interest of IT companies to create new innovations since they can sell these to people who are better paid, thereby making more money. There is little incentive to develop solutions that would be suitable for poorer people, because this would not result in the same level of profit. As a consequence, developments which are produced for the poorer members of society tend to be at a very basic level and are less likely to increase their IT skills, resulting in a widening rather than a narrowing of the divide. It is generally acknowledged that factors such as income affect whether people use IT in general and, more specifically, gain the skills needed to make effective use of the internet.

6.3 The effects of the digital divide

6.3.1 Inequality of access to all types of internet services and technology

The effects of the digital divide include inequality of access to all types of internet services. The word 'service' here is open to interpretation. It can mean the major services provided by the internet, such as email, the world wide web, video conferencing, instant messaging, VoIP and other resources which the internet makes it easier to access. For the purpose of this book, we will concentrate on the services which are provided by the world wide web, although, where appropriate, reference will be made to video conferencing and email.

Health

There is a vast amount of information on the world wide web relating to health. Many countries have health-based websites. These websites tend to contain details about medicines together with information about common illnesses and how to treat them. Some will simply recommend a visit to the doctor or hospital. If it is a minor illness, a website may advise individuals how to treat it themselves. Many doctors have their own websites through which patients can book an appointment, receive reminders by text on their smartphone and order prescriptions. Individuals are also able to buy medicines from private websites.

Unfortunately, however, these services can only be accessed by individuals with the necessary IT equipment, such as a PC, laptop or tablet with broadband, or a smartphone. Other reasons why such services may not be available to individuals include cost, remoteness (lack of signal) or lack of electricity supply.

It is not just the introduction of online services which has benefited people. There are more general uses of IT in medicine, with many hospitals now using online portable computer systems. X-rays can be accessed digitally and patients' health charts are now available on computers which are transported on trolleys to the patient's bedside. These computers are used by nurses to record the patient's pulse rate, temperature and blood pressure. Unfortunately, not all hospitals can afford the new technology. Some doctors working from home can access patient records on their own computers, but if they live in a rural area where broadband connections are poor or non-existent, such access may be limited. It is likely that patients in a more affluent urban society are likely to benefit more from such systems.

> **Activity 6a**
> Give **three** reasons why people may not be able to access health websites.

Education

Most schools these days have access to IT equipment and the digital divide in education is more likely to depend on the quality of broadband available. The divide between rich and poor schools has, in many countries, been reduced by governments providing money to the poorer schools in order to improve broadband access.

Students today are required to have a new academic literacy, different to that which was needed in the twentieth century, so that they can function effectively in the new digital age. In other words, academic literacy in the current digital educational environment is experiencing new challenges for students in their educational fields. There are a number of features of IT that students have to be familiar with. These involve accessing web-based course resources, using online resources such as spelling and grammar checkers efficiently, being able to participate in online discussions, and even straightforward tasks such as downloading lesson notes and presentations as well as being able to upload homework assignments. Students have to be able to perform these tasks as a matter of course. Students who are unable to perform these tasks will find their progress severely limited. Universities worldwide have taken advantage of recent advances in IT, and online submission of assignments and presentation of learning material is commonplace. This gives the more IT-literate students an unfair advantage over other less IT-literate ones.

The lack of education is a key factor as people educated to degree level are more likely to be able to use the internet to its full potential and computers in their daily lives compared to individuals who left education at the end of secondary school.

Although we live in an age where some countries still find it difficult to provide students with regular access to computers and the internet, discussions about the effect of the digital divide on education more often now are concentrated more on the development of technology-related skills as well as providing the training to make this happen. On the other hand, it is worth mentioning that the use of IT has led to the narrowing of the digital divide in certain situations, with students in certain remote areas benefiting from the use of IT to improve their education.

For adults, a lack of skill in using computers and IT equipment in general makes it difficult for them to access digital services. These services include aspects such as health, further education, online booking systems, online banking, online shopping, access to news, government websites and many more. Let us have a brief look at the effect of the digital divide on each of these.

Job opportunities

Most jobs in the modern world require IT skills. All businesses and manufacturing industries, and most supermarkets and shops, use IT in their everyday dealings. Obviously, job seekers must have these skills if they are to realise their hopes of a job. Job applicants who do not have these skills will tend to be unsuccessful in their attempts to find gainful employment.

Schools and colleges have courses to teach IT skills to students, but those in poor and rural areas in some developing countries often struggle to have enough computers to provide these courses. Richer schools do have sufficient computers and, consequently, can provide students with the necessary skills, thus contributing to the digital divide.

Another area affected by the digital divide is that of companies recruiting employees online. There are a number of recruitment services offered by job agencies. It is possible for an individual to create a curriculum vitae (CV – a summary of their qualifications, skills and talents) online.

The use of the internet for some time has been seen as an essential way for people both to look and apply for jobs. The question remains whether new technologies can even out the inequalities that exist among people trying to gain access to meaningful employment. A variety of obstacles remain, including access to the technology, ability to use it once accessed and levels of numeracy and literacy, which all play a part in limiting social equality for people from deprived backgrounds. Putting it simply, people without broadband are at a disadvantage when it comes to finding out about job opportunities or learning new career skills. The need to use a computer to make an online job application divides job hunters into two groups. These consist of those who have adequate IT knowledge and skills and those whose skills and knowledge are lacking. It is fair to say that many people who grew up with no access to the internet lack the appropriate technology and digital skills that modern jobs require.

Many organisations who pay good wages view computer skills as being essential if a person wishes to gain a well-paid position with them. Applicants who lack these skills may end up in low paid, less stimulating jobs. People without internet access or those lacking IT skills often find it difficult to gain the latest information concerning job opportunities. This is often the case for those who live in rural communities in developing countries. Often, they haven't got the money to buy a PC or laptop, and many do not have any computer skills at all. It may be that an applicant could be required to generate a CV online and then send it using email. If the applicant is a resident of a rural community in a developing country it may be that they are unable to access and use broadband. It could also be the case that they have no experience of typing documents and don't have an active email account. However, it is not only computer literacy that is crucial but also a greater awareness of the latest technological tools such as social media. As the technology improves and develops, job hunters can hugely increase their chances of finding their dream job by being familiar with and developing expertise in the technology.

As broadband becomes more evenly distributed, particularly in rural areas and developing nations, the gap will close but the development of IT skills among its users will become just as important. However, it is unfortunate but true that the impact of the digital divide will increase, giving a wider range of opportunities to those with good IT skills.

> ### Activity 6b
> Describe the ways in which people without access to the internet are affected when looking for employment.

Social interaction

The use of IT technologies has been fundamental in improving social interaction between people. Social networking websites have been very effective in making communication and staying in touch with friends and relatives so much easier. People can now keep in touch with friends or even make new friends using these sites. The use of the internet allows people to have these opportunities, but this can create a social divide between those who have developed these skills and those who lack them. Such division creates the possibility of social inconsistencies in communities where wealthy people can have computers and access the internet while poor people cannot.

Other uses for the internet

Online booking systems

Let us look at the tourist industry first. Generally, tourists use the internet to help them research and plan their holiday/vacation destinations. This has become more and more common and allows people to personalise their experience rather than having a travel agent plan it out for them. In developed areas, customers use PCs, laptops and smartphones to browse travel-related content and offers. In developing countries, internet access can often only be available to the small proportion of the population that has a relatively high income. Though this situation is improving, having access to the internet does not necessarily mean that people will use it wisely. In addition, there are people who have access to broadband and the internet but are not able to use it effectively because of lack of knowledge, literacy and language skills. Booking train tickets is mainly done online. Online booking systems are also used for planning train journeys. If you wish to travel by train and you do not have access to the internet or do not have the skills to use it, you will end up paying a lot more for your journey than if you booked in advance.

Online banking

The effect of the digital divide with regard to online banking has had less of an impact than on some online services as many people use smartphones to access their bank accounts. This has less of an effect on rural communities who do not have a good broadband connection. They simply use their phones instead. However, there remain rural communities who do not have a good phone signal and so still suffer, and of course people may not be able to afford to buy a phone or the data charges if they had one. With major banks closing many of their rural branches it is becoming increasingly difficult to carry out transactions of any sort.

Online shopping

There are two ways in which the digital divide affects online shopping. The first is between what are often regarded as the bricks-and-mortar retail outlets and those organisations that only have a digital footprint and no physical stores. Many main street or high street chainstores are finding it difficult to make a profit as more and more people shop online and the chains have often been slow to develop their online business. They are now finding it difficult to compete with those commercial organisations that are completely digital and have no physical presence. Many have ceased to trade.

The other way that online shopping has been affected by the digital divide is for the shoppers themselves. It is generally accepted that there are many advantages to being able to shop online: items tend to be cheaper; customers can compare products and prices at their leisure and can shop at a convenient time for them; food shops can remember the customer's shopping list and favourite brands and deliver at a time to suit them. If people are deprived of this access, particularly poorer people, they may end up paying more for their goods than richer people and the digital divide will remain in place.

Access to news

Those who suffer from the digital divide are at a distinct disadvantage when it comes to gaining access to news bulletins. Most newspapers are now also online, and it could be that in the not too distant future they cease to be printed and will go the way of utility bills and bank statements. Since January 2020, the BBC in the UK no longer has a teletext service on the television and has informed viewers that they will need to go online in future for news content outside its normal news bulletins. People who are disadvantaged as a result of the digital divide will find they have limited access to the latest news.

Government websites

People use government websites for a variety of reasons. These can be to:

» obtain information on statistics and government research
» download government forms or applications
» find information about government sponsorships and visas
» pay fines online or apply for government benefits.

If people do not have access to, or do not have the skills to use, the internet, then they are going to be at a distinct disadvantage. They will need to travel to the nearest big city with a government centre which deals with such matters. This obviously means they will have the expense of this travel and will have to spend a large amount of time travelling.

> **Activity 6c**
> 1 Describe the activities that students in school should have the IT skills to do.
> 2 Describe **three** activities that adults will not be able to carry out if they cannot access the internet.

6.3.2 Reducing the effects of the digital divide

One way to reduce the effects of the digital divide is to enable faster and cheaper access to the internet. This will reduce the divide between those who have access to the internet and those who do not. However, it does not get over the problem of those people who, even though they will have improved access, will not necessarily engage with the technology. People need to have the skills to use it as well as trust or a lack of fear when using it. Developing digital skills and digital literacy courses has been shown to lead towards positive effects on employability, social inclusion and general well-being. One major contribution that can be made is teaching the training and skills required to take advantage of such technology. People without these skills often welcome the opportunity to learn and develop their digital literacy skills. Training programmes should be developed in rural areas where familiarity with digital technology may be limited. It is important that these programmes cater for older adults to compensate for the natural decline in cognitive and physical capabilities that occur with age. Consideration should be given to how these programmes are delivered. Older adults tend to be slower to learn new information. Any attempt to hurry up the learning process can result in them worrying, becoming annoyed and, as a result, becoming unlikely to want to learn because of the fear of failure. It is advisable to deliver these digital literacy programmes in small, manageable pieces. Teachers should allow sufficient time after each stage in the process for participants to practise any new skills that have been acquired.

Reducing the digital divide requires a partnership between governments, the private sector, education providers and IT experts. Governments can provide incentives to private industries as well as investment for rural projects and introduce initiatives in state or local government education systems. The private sector can build up broadband and other communications infrastructure, but this cannot be done by one group alone; it needs collaboration between all these groups.

One initiative has been the use of power-line communication (PLC) which carries data through power cables. It is possible for data to travel down normal power cables and be distributed to households at the same time as electricity. This is often referred to as broadband over power lines (BPL).

In 2019, Comcast, one of the largest telecommunications companies in the USA, developed a series of initiatives in order to help people with disabilities who are on low incomes to gain access to the internet. Included was a grant to the American Association for People with Disabilities, intended to help with the delivery of digital literacy training programmes. They acknowledged that the internet is a wonderful thing provided people have the skills to use it.

The technology can be improved for individuals with health conditions or impairments. For example, aids that could be developed for apps and websites for those with visual impairments could include features such as text-to-speech conversion with speech and voice output. Braille displays could be included or on-screen text magnification. Voice recognition software could be embedded. The site or app could include an audio description of any graphics or videos. For those people with total or partial hearing loss there could be captions or subtitles for any videos. On smartphones, incoming voice calls could be converted to text/SMS displays as well as the use of vibrations or text alerts instead of audio alerts. Those with speech impairments would also benefit from the increased use of text-to-speech conversion. Individuals who have lost mobility or dexterity would benefit from an increased number of websites or apps using voice recognition systems, concept keyboards, joysticks and gesture-based interfaces.

Kenya is often referred to as Africa's digital heartland. As long ago as 2016, the Kenyan government produced a National ICT Policy. Its aim was to create sufficient incentives to the private sector to provide people with disabilities with access to a fast internet connection. Information providers, including the government, would also be required to produce content in suitable formats so that people with disabilities would be able to easily access such content.

Another section of the community which suffers from the digital divide is older adults. As we have already seen, this is often caused by fear of the new technology. To overcome this, there could be programmes within the community to teach older people how to use this advanced technology.

Children from poorer communities also suffer. One way to help them is to release cheap PCs or laptops to these school-age children. In the UK there are a number of companies that offer reduced prices for refurbished computers to children in families receiving state benefits. Unfortunately, the downside is that a broadband dongle costs almost as much as the PC itself.

There are a number of organisations that help governments in developing countries provide schools with computers. One such company is Computer Aid International. Computer Aid works with several non-profit-making organisations to provide PCs for use in schools and has done this in a number of African countries such as Rwanda, Burundi, Zambia and many others.

We have seen how people living in rural areas can have no or poor access to the internet. One solution is to create cyber or internet cafes in these areas. Many countries have rural communities whose sole access to the internet is through the use of internet cafes.

6.4 Groups affected by the digital divide

As a result of the causes above, the digital divide exists between the following groups of people:

- people in different age groups
- people in cities and people in rural areas
- people with differing levels of education
- people in different socioeconomic groups
- people with accessibility barriers due to learning difficulties or physical or sensory impairments
- people in more and less industrially developed nations, with different levels of technological awareness/infrastructure.

6.4.1 People in different age groups

Older generations may lack confidence when it comes to using IT, as they were not brought up being exposed to it constantly like younger generations. As a result, they may develop a 'fear' of technology making them more likely to avoid it. This leads to a digital divide being created between the old and the young.

This divide is increased because IT businesses wish to make large profits. This results in them concentrating on the larger groups of consumers. Consequently, new technologies tend to be aimed at younger people with very few modifications made to help older people use them more easily. One example is newer smartphones. Older people tend to be less dexterous with their hands and fingers which makes these devices more difficult to use than they are for younger people. In addition, young people seek new skills because they might become relevant later in life.

However, as people grow older, they tend to focus on what seems to matter today. They are more likely to concentrate on friendships and being emotionally involved with others rather than spending time on machines. As a consequence, many older people may not see the point of using computers for social networking.

German studies on the digital divide have shown that it may be too simplistic to maintain that the digital divide exists only between the younger and the older generations. Other factors, such as gender, education and socioeconomic status, come into play. They also suggest that internet usage among older adults is increasing but will continue to be less than that of younger users. Optimists could suggest that the increase in the number of older internet users would seem to indicate that the number of non-users will reduce and eventually reach zero. This, however, is doubtful as the type and number of obstacles against the increasing use among older users will still persist. It does not seem to be a priority for the developers of modern IT devices that older people find it very difficult to use tiny smartphones or that they have difficulty interpreting icons. A large number of older people find it difficult to read text on a screen or use a keyboard due to visual impairments or other physical disabilities.

Older adults tend to use well-established technologies and are slower at accepting new ones. It is generally accepted that the older generation are less inclined to use recent technologies compared with the younger generation.

Recent French studies showed that less than one in ten adults aged over 80 used the internet. As the age of the users reduced, the percentage of them using the internet gradually increased and of those aged under 30 almost 100 per cent were internet users. In the German studies it was seen that older people would only accept a new technological development after seeing that it had been successfully used by others and they could see clear benefits for themselves.

The ONS report mentioned earlier in this chapter showed that take-up of IT usage is slowest among older people and lower-income households. Its research showed more than half of over-50s had not used a computer in the past three months, compared with 13 per cent of people aged 16 to 30.

It may be a little simplistic to just divide people into young and old. This point was emphasised when the ONS in the UK published a document titled 'Exploring the UK's digital divide'. The table below shows how the lack of use of the internet increases with age but divides its findings into different age groups:

▼ Table 6.1 Age composition of internet non-users*, UK, 2011 to 2018

Age group	% of internet non-users						
	16 to 24	25 to 34	35 to 44	45 to 54	55 to 64	65 to 74	75+
2011	2	3	6	11	18	25	36
2012	1	3	5	11	17	26	37
2013	1	2	4	10	16	27	40
2014	1	2	3	9	15	27	43
2015	1	1	3	8	14	26	47
2016	1	1	2	7	14	26	49
2017	1	1	2	6	13	25	52
2018	1	1	2	5	12	24	55

* 'internet non-users' refers to those who have never used the internet or last used it more than 3 months ago

These statistics do not indicate the actual number of non-users within an age group but show what percentage of all the non-users exist within that age group. For example, in 2011, 2 per cent of all the non-users were aged between 16 and 24. In 2018, 55 per cent of all the non-users were at or over the age of 75. The study does not show whether there was a general decrease in the number of non-users but what it does show is that even if there was a decrease, the number of non-users aged 75 or over had not decreased by the same rate as those younger than 75. This would seem to suggest that while the number of non-users in the 75+ age group is probably declining, it is not declining at the same rate as other age groups.

> **Activity 6d**
> Describe the reasons why older people tend to be reluctant to use new technologies.

6.4.2 People in cities and people in rural areas

In its 2017 Digital Strategy, the UK government acknowledged that 'broadband and mobile must be treated as the fourth utility, with everyone benefiting from improved connectivity'. The other three utilities are gas, water and electricity. These tend to be available in many UK rural areas, yet broadband is not. Similarly, in 2018, the National Infrastructure Commission, the government's independent advisor on the UK's infrastructure needs, stated that digital connectivity was now 'an essential utility, as central to the UK's society and economy as electricity or water supply'. Poor digital connectivity in rural areas therefore has far reaching consequences for communities, economies and businesses.

In the USA, rural Americans have made large gains in adopting digital technology over the past decade, but they generally remain less likely than adults living in urban areas to have home broadband or own a smartphone. The percentage of Americans living in rural areas who have a broadband internet connection at home doubled between 2007 and 2018, according to the findings of a Pew Research Center survey carried out in early 2019 (the Pew Research Center is a nonpartisan US think tank based in Washington, DC; it provides information on social issues, public opinion, and demographic trends shaping the USA and the world). The proportion, however, was still only two-thirds of the population, a full 12 per cent fewer than other Americans. The graphs below demonstrate this.

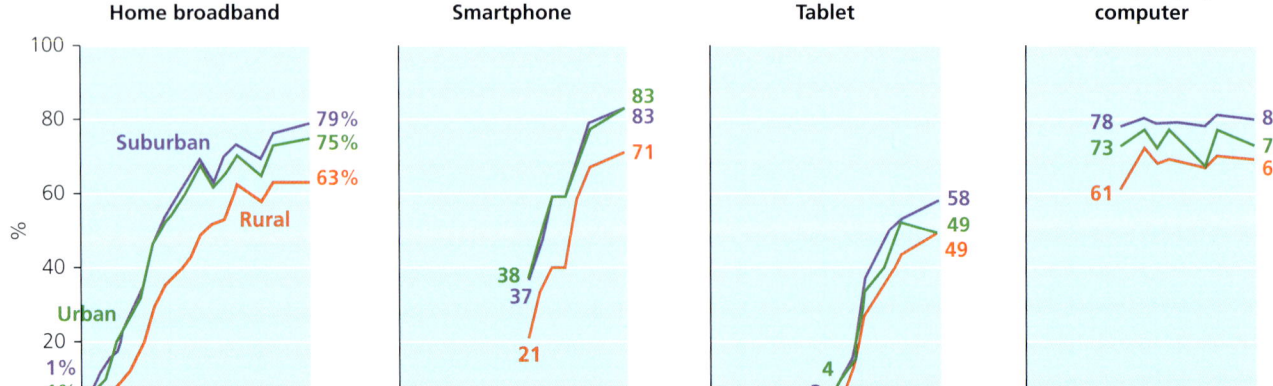

▲ **Figure 6.2** Technology ownership of US adults

In a recent BBC report, it was revealed that in 2019 India had more than 630 million internet subscribers, the second largest number of internet users in a single country. Indeed, the number of internet users in India is larger than the total populations of Brazil, Germany, Russia and Egypt put together. But for every Indian who had access to the internet, there was at least one who did not, and that person was most likely living in a rural area. In rural areas, 25 per cent of the population used the internet compared with 98 per cent in urban areas.

There are a number of factors making it difficult for people living in rural areas to take advantage of the latest technological developments. The three we will discuss here are the low populations of rural areas, the potentially challenging geographical features of rural areas, and the limited commercial potential for communications companies in installing the necessary infrastructure in these areas.

It is possible to get broadband anywhere there is a phone line. However, in order to obtain broadband that can reach fast speeds, a slightly more advanced infrastructure is required. This is fairly easy in urban areas; it is straightforward to dig up roads in order to install cables and a single fibre-connected cabinet in one street can connect several houses. However, in remote rural areas, because there are fewer houses and those houses tend to be further apart, people are further away from a single cabinet than would be the case in a more urban area. This means longer cables have to be used, which means slower speeds.

It can also be very difficult to lay fibre cables in rural areas as a result of geographical features. India, for example, has remote mountainous areas, thinly populated deserts and dense areas of forest, all of which make it difficult to install the necessary infrastructure. Any infrastructure that already exists tends to use a lower level of technology, making it difficult to install high-speed broadband. To improve the infrastructure costs a lot of money and broadband companies are reluctant to invest such sums for so few customers. For users who get broadband using the telephone network, the actual speed they achieve depends on how much of their connection uses copper phone lines. The further a signal has to travel over these lines, the weaker it becomes. So, where people live affects the speed.

Broadband providers will only install high-technology solutions where it is economically viable. So, while towns and cities might have a number of providers competing to persuade potential users to use their equipment, rural areas might only have one or two, and the most remote areas none. In urban areas, a given length of cable might pass dozens, or even hundreds, of homes and businesses. Rural broadband requires greater cable lengths as well as signal-boosting equipment and there are fewer potential customers from whom companies can retrieve the costs. All this makes companies less willing to invest in infrastructure. When they do, the extra costs incurred by providers are passed on to the customers.

To sum up, slow broadband speed is usually influenced by three main factors. These are how far the home is from the phone exchange, the level of technology used at the phone exchange, and the amount of competition there is among broadband providers.

What are the alternatives to broadband for people living in rural areas? There are a number of different options apart from cable, but they can be very expensive and many people who live in these areas might be in jobs which do not pay very large salaries. Wireless technologies are available, such as satellite broadband, short-distance radio links and mobile-phone broadband. Satellite broadband is technically available anywhere in a country. However, it tends to be slower and, again, more expensive than cabled broadband connections. The quality of connection can also be affected by bad weather. Radio connections vary a lot between areas. Fixed wireless requires customers to be within line of sight of a service tower. Speeds can, however, be very fast. Another type of wireless broadband is already widely available and that is through the use of smartphones. Many people have higher-speed connections at home and use mobile data on the move. However, this depends on the number of towers in the area and in rural areas there do not tend to be very many.

> ### Activity 6e
> 1. Give **three** reasons why people living in rural areas tend to have less access to broadband.
> 2. List **three** alternatives to cable-based broadband which people in rural areas might use.

6.4.3 People with differing levels of education

Where people have not gained many educational qualifications, this can restrict their chances of getting a well-paid job unless they are very lucky. Because of this, they have to work in lower-paying jobs, such as performing labour in construction or agriculture. More educated people earn more money throughout their lifetime leading to a higher quality of life. People who have had a lower level of education tend not to earn enough to provide for their families which pushes them deeper into poverty, making it less likely that they will invest in IT equipment or services.

An important factor that contributes to the digital divide is an individual's level of educational achievement and the related use of IT in schools and classrooms. Another factor is the divide that exists between educational establishments. In other words, inequalities in digital literacy can be considerable between schools that have a lot of IT equipment and fast broadband and use IT across the curriculum, and schools that do not.

It is generally thought that the digital divide among people still exists because people with a lower level of education do not understand the importance of the internet in their everyday lives. In a number of areas, particularly urban areas, the problem of providing cheap internet access has been overcome. Why is it, then, that a divide still exists within cities? Convincing less well-educated people that using the internet is beneficial is the real challenge for those who wish to narrow this divide. In the USA, the 2017 National Telecommunications and Information Administration report stated 'the percentage of respondents who say they don't have broadband at home because they have "no need" or "no interest" reached almost 60%, nearly double the percentage who consistently gave that response from 2003 to 2009'.

> ### Activity 6f
> Briefly describe **two** factors that contribute to the digital divide between well-educated people and less well-educated people.

6.4.4 People in different socioeconomic groups

Companies that produce hardware and software for computers used in the home have good reason to make further developments; selling new updated products leads to them making bigger profits. These products, however, can only be bought by those people who can afford them and this is obviously easier to do for people with large incomes. There is no need for companies to develop systems and products that would be suitable for people with low incomes because they would not make the same level of profit from them. As a result, those systems and products that are for lower income families tend to be less advanced, which means the people who buy these cheaper products and systems do not get the opportunity to increase their IT skills and knowledge. This state of affairs highlights two aspects of the digital divide: the gulf between people with high incomes and those with lower incomes and between those that have a good understanding and those that do not.

It is generally accepted that factors such as income affect whether people use IT in general and more specifically gain the skills to take advantage of using the internet. As household income increases, the level of basic digital skills also increases. The digital divide between people with low incomes and those with high incomes still persists, despite technological developments making devices more affordable and internet access seeming to be increasingly universal. Most research suggests that people with high incomes are still more likely than those with low incomes to have good access to digital resources. In a report published in May 2019, the Pew Research Center found that in the USA, roughly 29 per cent of people with household incomes of less than $30 000 a year did not own a smartphone, 44 per cent did not have broadband services, 46 per cent did not own a traditional computer, and the majority of low-income Americans were not tablet owners. This compares with all these technologies being present among most households earning $100 000 or more per year.

In a more recent survey by the Pew Research Center conducted between 25 January and 8 February 2021 in the USA, they discovered that generally, the use of the internet, installation of broadband and smartphone ownership had increased markedly. However there were still marked differences in this between different income levels. The graph below shows this:

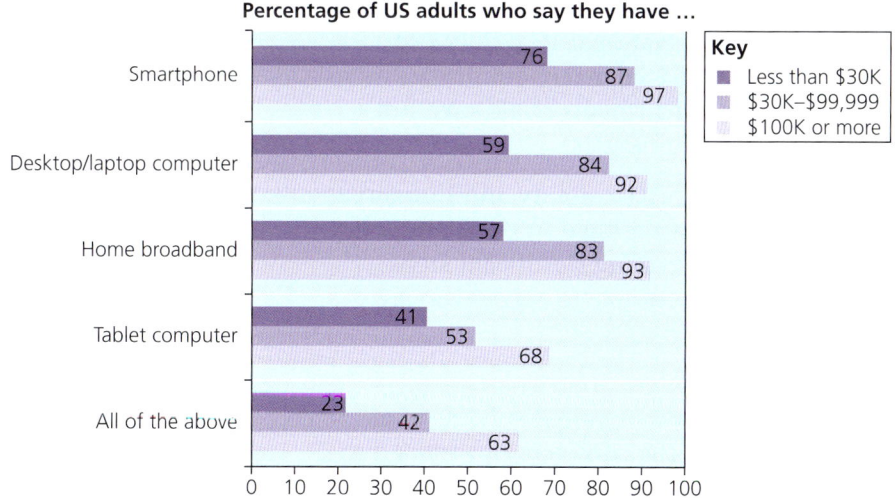

▲ Figure 6.3

From these statistics it can be seen that 24 per cent of adults with household incomes below $30 000 a year say they don't own a smartphone; 43 per cent do not have home broadband services and 41 per cent do not have a desktop or laptop computer. More than half of Americans with lower incomes do not own a tablet computer. Compare these findings with adults in households earning $100 000 or more a year where over 90 per cent have a smartphone, home broadband and a desktop or laptop computer. The percentage is still very high when you look at whether the higher income families have all of these devices and services compared with 23 per cent of lower income families.

Many people in poverty, often in less economically developed countries, simply cannot afford to own a computer or the cost of a broadband connection.

> **Activity 6g**
> Explain why IT companies may have contributed to the digital divide between high- and low-income earners.

6.4.5 People with accessibility barriers due to learning difficulties or physical or sensory impairments

Accessibility barriers in the context of the digital divide are conditions or obstacles that prevent individuals with health conditions or impairments from using or accessing digital resources as effectively as individuals without health conditions or impairments. People with health conditions or impairments have specific needs and have been disadvantaged for a long time regarding IT, despite it having the potential to improve their lives. People with physical impairments that affect movement of the hands, for example, are unable to use a touchscreen, mouse, or keyboard effectively; they need devices which are much easier to control. Unfortunately, many of these devices can be expensive to purchase. Another group of people who similarly experience a digital divide through the lack of access to personal devices and the internet are those with hearing or visual impairments.

Let us take Kenya as an example. Its constitution outlaws discrimination against people with health conditions or impairments. It requires that a person with an impairment or health condition is entitled to be treated with respect and dignity, is able to access educational institutions and facilities, has reasonable access to all places, public transport and information, and can access materials and devices, including for communications. Kenya's National ICT Policy of 2016 outlined strategies for creating an accessible ICT environment in the country so that people with health conditions or impairments could take full advantage of Information and Communication Technologies.

However, Judy Okite, founder of the Association for Accessibility and Equality, claims that many of the digital accessibility strategies outlined in the 2016 policy remained unfulfilled. While Kenya's government is making significant steps to move its services online, the platforms are not favourable to those who are visually impaired.

Research she was part of in 2018 showed that computers in some learning institutions had not been replaced for several years, the required software was not installed or was out of date, and staff managing the labs were not trained to teach users.

> **Activity 6h**
> Describe how people with impairments may have difficulties with new technology.

6.4.6 People in more and less industrially developed nations, with different levels of technological awareness/infrastructure

As well as the digital divide at the national level, there is a divide between countries. This is often called the **global digital divide**. There has always been a large gap between developing countries that perform poorly economically and the so-called developed countries that perform strongly. Unfortunately, the global digital divide can cause the economic divide to widen.

Some countries feel that they cannot afford to invest heavily in technology and the infrastructure required as they need to improve education, increase access to healthcare and increase job opportunities. However, investment by governments in IT and communications technologies can generate jobs and provide greater access to better health services and education.

A report published in late October 2018 by the United Nations Conference on Trade and Development said that by the end of the next decade (2029), growth, productivity gains, and human development will be determined by levels of integration into the digital economy. It also warned that to guard against new forms of inequality, the international community must do more to help developing countries close the connectivity gap.

Let us look at just one aspect of the use of IT in commerce: online shopping. It becomes apparent from the UN report that developing countries experience several advantages. It uses as an example Tuvalu, a tiny island in the South Pacific. It has fewer than ten streets that have names in the capital Funafuti and only about a hundred homes have a postal address. The report stated that even if everyone in Tuvalu had access to the internet, the delivery of goods which were purchased online would be difficult. In actual fact, the World Bank records that only 13 per cent of the country's population had broadband in 2016.

In other parts of the world, too, billions of people do not have bank accounts and credit cards, severely limiting their ability to buy goods online. In addition, it is often the case that in many developing countries, goods bought using the internet are not covered by consumer protection laws. Unlike developing countries, nations with the most developed economies tend to have efficient postal and parcel delivery systems, and most have laws which protect the consumer when buying goods online.

It is envisaged by the UN that by the end of the 2020s, economic growth will be governed to a large degree by information technology. In order for countries to flourish, people will have to acquire new skills and knowledge. It will be necessary and vital that the digital divide is closed, particularly where more than half the world's population has limited or no access to the internet.

Countries that have widespread access to the internet and, more specifically broadband, can trade with other countries far more easily than those who do not. Access to the internet, and IT in general, is an essential part of modern education. Consequently, nations where people only have limited access to these technologies, and the benefits they can bring to education, will be unable to compete effectively in the global economy. This follows on from Section 6.4.3 about the digital divide between people with differing levels of education.

Activity 6i
Using an example provided in the text, explain why opening up the online shopping market to people in developing countries might not be an attractive proposition to large supermarkets.

Practice questions

1. Describe **two** of the causes of the digital divide. [4]

2. Other than those living in rural areas and those living in urban areas, describe **two** groups of people who are affected by the digital divide, giving reasons why they are affected. [6]

3. The lack of a broadband connection affects people living in rural areas. Explain why these people might not have a broadband connection. [4]

4. Explain why having access to high-speed broadband does not necessarily solve the problem of the digital divide. [6]

7 Expert systems

In this chapter you will learn:
- ★ what an expert system is
- ★ what the components of an expert system are
- ★ how expert systems are used to produce possible solutions for different scenarios
- ★ what the concepts of forward and backward chaining involve

Before starting this chapter you should:
- ★ be familiar with the terms 'software', 'database', 'hypothesis' and 'information'.

7.1 What is an expert system?

Expert systems are computer-based systems used to solve problems, such as providing a diagnosis of something. They make use of the knowledge of a number of human experts within a specific field. They are unlike other problem-solving software in that they use reasoning to produce possible solutions. Data, often referred to as a **database of facts**, is stored within the computer and this is the knowledge part of a **knowledge base**. The system uses a set of rules, which represents the reasoning.

Expert systems can only be used in those areas where human experts already operate. Clearly defined rules and subject facts must already be in existence. Expert systems cannot be applied to problems in which there are missing facts, or in which rules can be interpreted in a variety of ways.

One way of approaching this topic is to think of the internet. The world wide web contains a huge amount of information or knowledge. This knowledge is used to solve a problem by you asking questions (making searches). You use the answers to these questions (searches), which we could argue are possible solutions. You then consider the possibilities before making a decision on the most appropriate solution.

Let us look at a use of an expert system in the field of medicine. It is used to help a doctor to make a diagnosis. The doctor types in the symptoms and the expert system produces a list of possible solutions, each usually with a percentage probability. The doctor then uses their experience to make a decision, choosing which of the possibilities produced by the system is likely to be correct.

Expert systems are not like most problem-solving programs. Problem-solving programs are produced to solve a specific problem. They can only solve that particular problem and it is difficult to adapt the program to alternative scenarios. The facts needed to solve the problem are an integral part of the programming. If the facts change, then the whole program may need to be rewritten. Compare that with expert systems, which are used to collect the knowledge of experts in order to help create a knowledge base.

There are alternative ways of looking at the make-up of an expert system, but for the purposes of our text we shall consider the system to be made up of a **shell** of the major components and a knowledge base which comprises the database of facts and the **rules base** (see Figure 7.1).

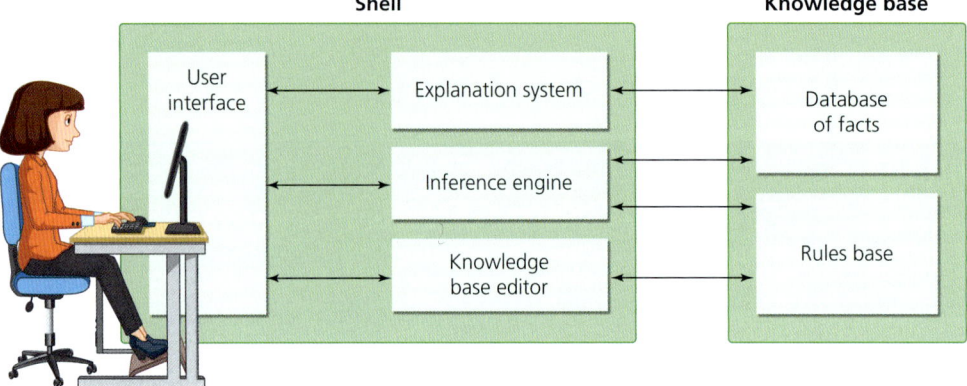

▲ Figure 7.1 A graphic representation of an expert system

As you can see, many of the components are housed within a shell. The shell often incorporates the **knowledge base editor**, which is basically software that allows the **knowledge engineer** to edit the rules and facts within the knowledge base to bring them up to date. It also contains the **explanation system**. This is the part of the system which explains the logical reasoning that the expert system has used to come to its conclusions.

The explanation system causes the expert system to explain to a user how a particular conclusion was reached and an alternative conclusion was not. It also must be able to explain why a specific fact was needed, or not, as the case may be. An expert system must be able to explain its reasoning and justify its recommendations.

The shell of the expert system is very important. A good expert system has a shell which can be used for solving different problems just by changing the knowledge base to one that matches the type of problem. The shell of an expert system, technically speaking, is an expert system with an empty knowledge base. It can be used for the development of other expert systems by adding a new knowledge base. It is possible that a diagnostic expert system could be used to diagnose car faults or medical illnesses just by changing the knowledge base.

Another important feature of an expert system is the user interface. This is the component that allows the user to interact with the computer which stores the expert system. It usually consists of a keyboard and mouse together with a monitor. The questions which need answering are displayed on the monitor. The user can then manoeuvre round the screen of the monitor using the mouse and can use the keyboard to type in their responses to the questions. Supplementary questions are often asked by the system and appear on the screen, which the user then answers using the keyboard. Validation checks are carried out to ensure the minimum number of invalid responses are entered by the user.

The **inference engine** is another important component of an expert system. It performs the reasoning of an expert system. It uses a series of IF…THEN statements within the rules base and interrogates the database of facts in the knowledge base. It uses inputs from the user to try and match them with the contents of the database of facts. It processes the rules to try and reach possible solutions. It can also generate questions via the user interface based on the inputs it has received so that it can use the answers to narrow down the number of possible solutions. In short, it produces a set of possible solutions comparing the inputs with the facts, using the rules contained in the knowledge base.

When an expert system is created, a knowledge engineer needs to be employed. Human experts or specialists have deep knowledge of their subject area and the engineer makes use of this by collecting information from them. This will consist of the experts' knowledge and any reasoning strategies that are used by these experts. The knowledge engineer must also find out from the experts the type of system they want. In addition, the engineer gathers data from databases that may exist for the subject area the expert system is being created for. The knowledge engineer must then design the knowledge base and the rules base. During the whole process the knowledge engineer and subject expert work together in order to define the problem the expert system will solve. It is the job of the engineer to integrate the knowledge into the knowledge base. Then the inference engine has to be created to provide the reasoning aspect.

> **Activity 7a**
> Describe the components of an expert system.

7.2 Different scenarios where expert systems are used

There are many different uses to which expert systems can be put. In the following scenarios reference is made to many of the components of these systems. You must also try to see how the components which are not mentioned would apply. The knowledge base editor software, for example, is used in much the same way regardless of the scenario. The same applies to explanation systems and knowledge engineers.

7.2.1 Mineral prospecting

Expert systems are used to help make decisions on where to drill for a specific mineral. They help geologists evaluate whether particular sites are likely to contain the mineral or ore being searched for. The system produces probabilities of minerals being found in particular locations. As well as producing a conclusion about the existence of a particular ore deposit, these types of system often provide explanations regarding their conclusions. The expert system's knowledge base will store models of successful drilling sites and their associated minerals together with a rules base. First of all, the geologist enters data about the types of rocks and geological structure as well as the minerals to be found at the site under consideration. The inference engine then compares this data with the models in the knowledge base using the IF…THEN rules in the rules base. The system may well ask the geologist further questions through the user interface and the resulting percentage probabilities of finding the mineral at this site are produced.

7.2.2 Investment analysis

Expert systems have been developed which act as assistants to financial advisors at many banks. The financial advisor still makes the final decision, but the expert system proves to be a very useful tool in helping the advisor make recommendations to would-be investors. When making a decision, it is important for the financial advisor to know four things about the investor:

- The amount the client wishes to invest.
- What level of risk the investor is willing to take.

» The desired rate of return.
» How long the investor is willing to make the investment for.

The user interface asks questions in the everyday language of the investor based on the above factors. Depending on the answers to the questions, the system may ask further questions through the user interface. The database of facts will consist of all the information the bank has gathered over the years about the performance and stability of different companies. The rules in the rules base will have been originally created by an expert investor. The expert system will, however, allow the financial advisor to add rules to the rules base as they see fit, based on their experience of personally dealing with the investor. The inference engine will search the database of facts using the rules from the rules base. The expert system will then produce recommendations for purchasing stocks or shares in a company, together with the reasons for recommending these particular stocks and shares.

7.2.3 Financial planning

Today, expert systems in personal financial planning are needed more than ever. Personal financial planning has become so complex that it is extremely difficult for one human expert to deal with it. The result is a need for computerised expert systems to assist in the process. An expert system is different to a normal computer program in that the problems it addresses tend not to have a unique solution. There are so many variables to deal with that only a few can be considered by one person in the time usually available. Expert systems can be used to create financial plans for individuals. Such plans include recommendations as to how individuals can manage their debts, reduce the amount of income tax they pay, and organise investments. These plans can also recommend the amount of insurance individuals should have and how they can plan for their retirement. They can also suggest short-term personal saving targets. Systems often involve the client completing a detailed questionnaire regarding their personal and economic situation, including income, taxes paid or due, investments, employee benefits, marital status, number of dependants, as well as any future obligations such as retirement plans or funding their children's university education. The results of the questionnaire are entered using the user interface. The inference engine analyses this information using the database of facts and a very large number of rules in the system. The database uses the results of books and articles which have been published relating to personal financial planning as well as information gathered from interviews with experts, such as professional financial advisors. A report can then be produced by the financial advisor as to the client's recommended financial plans. The report is based on a combination of a human expert's insights and the use of the results of the expert system. The report can serve as a financial action checklist for the client in order to achieve their financial plans.

7.2.4 Insurance planning

Insurance involves the use of underwriters. Underwriters are usually banks, insurance companies or investment companies. They guarantee to pay a certain amount of money to the person who has taken out the insurance if a particular event happens. Individuals or organisations can take out insurance for any number of circumstances. Life insurance is taken out against an individual's life. Should the person die then his or her relatives receive a fixed amount of money. Car insurance is when a person insures their car against damage or theft. Property insurance involves safeguarding against damage or destruction to property.

Insurance can involve the payment of one large sum of money or regular, usually monthly, payments. Some insurance policies are for a fixed amount of time. A car is usually insured for a year at a time. At the end of the year the driver has to renew their insurance. Obviously, if there has been no damage to the vehicle in that time the underwriter has made some money. The same applies to life insurance. A life insurance policy usually lasts for a fixed period of time, such as 10 or 20 years. If the insured person dies within that time the underwriter has to pay out to their relatives. If the person is still alive then the underwriter will be able to keep all the monthly payments as profit. There are many other types of insurance, for example health insurance. With any type of insurance, the underwriters have to assess the risk involved. How likely is it that they will have to pay out?

Expert systems are frequently used to assess the risk. In order to set up an expert system, insurance experts would be interviewed by the knowledge engineer and all the policies that the underwriter currently offers would be examined. The results would then be used to create the knowledge base, including the rules based on those used by the experts when arriving at their decision.

The process of calculating life or health insurance premiums (monthly payments) involves the consideration of a number of factors, such as the individual's age, gender, whether or not they smoke, and their general health. In general terms, the older a person is, the greater amount they have to pay for their insurance. Men tend to pay more than women as their life expectancy tends to be lower. Smokers pay a lot more than non-smokers, as again their life expectancy is shorter. If you are not in good health, then you are a bad risk to the underwriter and so will pay a lot more than a healthy person. An insurance expert system's user interface will ask questions of the user and based on their responses will, through the use of the inference engine, compare these inputs with the facts in the knowledge base. Using the rules which have been created, it will produce a recommendation for the cost of the individual's premium. It may, of course, recommend that the application be rejected. Many underwriters use human insurance consultants who will use the expert system's recommendations to make a final decision.

7.2.5 Car engine fault diagnosis

Like any type of mechanical equipment, motor vehicle engines can develop faults. When this happens, the driver has to seek the help of a mechanic (expert) who knows a lot about vehicle engines. Modern-day mechanics use expert systems to help them diagnose the fault. In order to create such an expert system, many expert mechanics have to be interviewed to gather together their knowledge and rules (systems) for diagnosing.

Expert systems are now an everyday aid to a car mechanic's ability to diagnose engine faults. The inclusion of complex electronic technologies into a typical car engine has led to the need for highly complex expert systems. A systematic approach to the diagnosis of faults is required, as the amount of knowledge required is so vast and there are many steps which need to be followed in order to identify the fault.

The use of expert systems is often based on the expert system's user interface asking questions regarding the problems with the car and the mechanic typing in the symptoms. The expert system would then supply a list of possible causes of the problem and the mechanic would decide which of the suggestions are most likely.

A second type of expert fault diagnosis system is a device which is plugged into the car engine's computer and questions it. We often think these are true expert systems, in that mechanics can read the results of testing from the expert system and identify the exact problem that has occurred.

What often happens, in reality, is that true expert systems inform the mechanic of a range of readings from sensors but do not detail the cause of the problems. The mechanic has to use their experience and expertise to actually diagnose the fault. However, new true expert systems have been developed in which the readings from the sensors are used by an inference engine, after comparison with those in the knowledge base, to produce a list of possible faults. Although, again, with this system, the mechanic makes the final decision.

7.2.6 Medical diagnosis

Since early computers were developed, it became apparent that they would be able to assist medical practitioners in their day-to-day work. Applications such as drug development and controlling the administration of drugs are commonplace. Computers are also used as aids when diagnosing illnesses. Because enormous amounts of data are required, it was originally thought that expert systems could only be used with large, expensive computers. However, with the development of more powerful, cheaper PCs and laptops, expert systems are now readily available to many medical practitioners.

The creation of a medical diagnosis expert system starts with the production of the shell followed by the knowledge base. The knowledge base is created by a knowledge engineer, who interviews a number of expert doctors and then shapes the knowledge into a form that can be easily accessed by the system. The components of an expert system are used to produce a list of possible diagnoses. The user interface allows the patient, or doctor, to enter symptoms and, at the end of the process, displays the output. The knowledge base provides the database of existing illnesses as well as the rules from the rules base, often IF… THEN rules. The IF part is the symptoms part and the THEN part is the illness or disease that is suggested. The inference engine searches the knowledge base for similar symptoms to those entered by comparing those typed in with those in the knowledge base. This is the reasoning part of the system. The knowledge base editor allows for the knowledge base to be increased by adding further rules and data. The explanation system clarifies how the expert system has produced its possible diagnoses.

Other types of medical diagnosis expert systems have been developed which allow the patient to interact with the system directly. However, these are only really suitable for minor complaints and such systems, after considering the symptoms, will often recommend the patient should visit a human doctor.

7.2.7 Route scheduling for delivery vehicles

For most manufacturing companies, the cost of transport is a major part of their outgoings. Therefore, it is essential that companies use their transport vehicles as efficiently as possible. It is important that they develop delivery schedules which prevent vehicles from travelling further than necessary. One of the most efficient ways of ensuring this is to use an expert system. Human experts have developed skills which enable them to solve such problems. It is important, therefore, to incorporate these schedulers' skills in an expert system. A major problem, however, is that there tends to be little agreement about the facts or rules required in such a knowledge base.

For example, a company which manufactures washing machines might have to deliver these washing machines to distribution points at the cheapest cost. If the company is situated in a country which occupies a large area, this makes the problem rather difficult. It could be that it has several distribution points in different parts of the country, each with several delivery vehicles, each of which may be a different size and may be available at different times. The problem then becomes, how do you ensure that the available vehicles provide coverage of all the distribution points but travel the least number of miles? The problem might also need to be broken down into the number of vehicles, size of vehicles, and loading of the vehicles. Vehicles have to be loaded so that each distribution point has quick access to its order. The vehicle has to be loaded so that when it arrives at a distribution point, machines do not have to be climbed over, or temporarily unloaded, to get to the order for that distribution point. In order to achieve this, the route for each vehicle has to be planned in advance.

An expert system can have the locations of each distribution point in its database of facts, together with the type and speed of the vehicle being used, the total available time and perhaps other constraints such as the topology or terrain of each road, whether it is mountainous or flat for example. This aspect of the problem is often referred to as the 'travelling salesman' problem. This refers to the times when a person was employed to travel to different places, to try to persuade people to buy their company's goods. The company would pay them commission, that is, they would receive a small percentage of each sale they made. It was, therefore, in their own interest to visit as many customers over a large area in as short a time as possible.

Although there are several linear programming algorithms available to solve such problems, they tend to take more time and more computer processing power to get to a solution. The expert system would examine the orders for the day, total the weights for the order for each location and the inference engine would match those against the available types of vehicle. The expert system would suggest an allocation of orders to each vehicle and suggest a list of machines in reverse order so that each vehicle had the first order loaded on to the vehicle last. The company's vehicle scheduler would take these suggestions from the expert system and decide on the number of vehicles needed and would know the total time to be taken.

7.2.8 Plant and animal identification

There are many hundreds of thousands of different species of plants and the number of different species of animals runs into millions. Identification of either plants or animals can prove difficult, particularly with rare species, but it has become extremely important with climate change affecting the existence of certain species. Experts are required to master a number of techniques and gather a great deal of knowledge in order to be able to identify plants and/or animals. They need additional tools to help them identify all the different species.

When considering plant or animal identification then, an expert system would be helpful. The inference engine would use a series of IF…THEN rules. Plants can be possibly divided into whether they are woody or herbaceous. IF they are woody the system, through the user interface, could ask another question, such as whether they have a single trunk, which would lead in turn to other questions based on the shape/type of leaf and so on. Eventually, the possibilities would be output through the user interface. The database of facts, which would be interrogated by the inference engine, would include the data regarding all knowledge about as many species as possible. This type of system would ask questions rather than inviting the expert to type in details about the plant. Animal classification or identification follows a similar pattern.

> **Activity 7b**
> 1. State the inputs to an investment analysis expert system.
> 2. State the outputs from an insurance planning expert system.
> 3. Describe the difficulties involved in producing an expert system for scheduling routes for delivery vehicles.

7.3 Chaining

When an inference engine interrogates the knowledge base, it imitates the way a human would use reasoning. It uses what is known as chaining. The reasoning that the inference engine uses consists of **backward chaining** or **forward chaining**. Often it uses a combination of both. Here we are going to see how both forms of reasoning work.

7.3.1 Backward chaining

Backward chaining is often referred to as goal-driven. This is because it starts with a goal or set of goals that basically establish which rules are to be followed. It works backwards from the goals. As we said earlier, the rules in an expert system are based on the IF...THEN construct. When an inference engine uses backward chaining, it explores the rules the system has been given until it finds one which has a THEN part that matches a required goal. If the IF part of that rule is known to be true, then it is added to the list of goals.

Let us look at an example of the use of backward chaining.

> **Example 7.1 Backward chaining**
>
> Suppose a zoo has just received a new animal from Africa. We shall call her Edwina and she has been delivered in an enclosed trailer. We cannot see inside the trailer, but we are given two facts about Edwina:
>
> Fact 1: Edwina has hooves.
>
> Fact 2: Edwina can run fast.
>
> We want to know whether Edwina has a striped coat. In terms of using an expert system, the goal is to decide whether Edwina has a striped coat. We will use a rules base containing the following four rules:
>
> Rule 1: IF X has hooves and can run fast, THEN X is a zebra.
>
> Rule 2: IF X cannot fly and X is a bird, THEN X is an ostrich.
>
> Rule 3: IF X is a zebra, THEN X has a striped coat.
>
> Rule 4: IF X is a giraffe, THEN X has irregular shaped spots.
>
> We call the part that begins with IF the premise (what we are supposing could be true). We will call the part that is based on the result of whether the premise is true, the THEN part, the conclusion.

Using backward chaining, an inference engine can decide whether Edwina has stripes in four steps. The goal that has to be proved is 'Edwina has a striped coat'.

Step 1: 'Edwina' replaces X in Rule 3 to see if its conclusion matches the goal, so Rule 3 becomes:

IF Edwina is a zebra, THEN Edwina has a striped coat.

Since the conclusion matches our goal 'Edwina has a striped coat', we now need to see if the premise – Edwina is a zebra – can be proved. So, the premise now becomes our new goal: 'Edwina is a zebra'.

Step 2: Again, replacing X with 'Edwina', in Rule 1, the rule becomes:

IF Edwina has hooves and can run fast, THEN Edwina is a zebra.

Since the conclusion matches the current goal 'Edwina is a zebra', the inference engine now needs to see if the premise – Edwina has hooves and can run fast – can be proved. The premise therefore becomes the new goal: 'Edwina has hooves **AND** can run fast'.

Step 3: Since this goal is a combination of two statements, the inference engine breaks it down into two sub-goals, both of which must be proved: 'Edwina has hooves' and 'Edwina can run fast'.

Step 4: To prove both of these sub-goals, the inference engine sees that both of these sub-goals were given as Fact 1 and Fact 2 at the start of this problem.

Therefore, the statement is true: 'Edwina has hooves AND can run fast'.

The premise of Rule 1 is true and the conclusion must be true: Edwina is a zebra.

It follows that the premise of Rule 3 is true and the conclusion must be true: Edwina has a striped coat.

Notice that Rules 2 and 4 were not used.

Because the selection and use of rules is based on the list of goals, this method is often called goal-driven.

7.3.2 Forward chaining

We are now going to compare the above method of reasoning with forward chaining. Forward chaining is often referred to as data-driven, because the data entered into the system determines which rules are selected and used. When an inference engine uses forward chaining, it explores the rules the system has been given until it finds one which has an IF part that is true. It takes the THEN part to add new facts. This process is repeated until a goal is achieved.

Let us take a different approach to our problem regarding Edwina the zebra.

> ### Example 7.2 Forward chaining
>
> Suppose that the goal is to decide the pattern of the coat of Edwina. Again, we can assume that Edwina has hooves and can run fast and that the rules base contains the following four rules:
>
> Rule 1: IF X has hooves and can run fast, THEN X is a zebra.
>
> Rule 2: IF X cannot fly and X is a bird, THEN X is an ostrich.
>
> Rule 3: IF X is a zebra, THEN X has a striped coat.
>
> Rule 4: IF X is a giraffe, THEN X has irregular shaped spots.
>
> Again, we have two known facts:

> Fact 1: Edwina has hooves.
>
> Fact 2: Edwina can run fast.
>
> With forward chaining, the inference engine can deduce that Edwina has a striped coat by using the following steps:
>
> **Step 1**: Since the facts we are given show that Edwina has hooves and Edwina can run fast, the premise of Rule 1 is matched by replacing X with 'Edwina' (IF Edwina has hooves and can run fast). The inference engine comes to the conclusion: Edwina is a zebra.
>
> **Step 2**: We can now use Rule 3 by replacing X with 'Edwina' (IF Edwina is a zebra). The inference engine comes to the conclusion: Edwina has a striped coat.

The difference in approaches is that backward chaining starts with us having a goal to achieve or an assertion we are trying to prove. We think Edwina has a striped coat, but we need to prove it. With forward chaining we do not know what type of coat Edwina has but we want to find out.

> ### Activity 7c
> Explain, by referring to the IF...THEN construct, the use of the words 'forward' and 'backward' when applied to methods of chaining.

7.3.3 Applications of forward and backward chaining

Generally, the choice of chaining strategy depends on the type of problem. Forward chaining, we know, is data-driven. The inference engine starts with the facts of the case and progresses to a goal or conclusion. This approach is thus determined by the facts available and by the premises that can be met. The inference engine attempts to match the condition (IF) part of each rule in the rules base with the facts which have been input. Forward-chaining systems are commonly used to solve more open-ended problems of a design or planning nature, such as establishing the configuration of a computer system. Backward chaining on the other hand is goal-driven. If there are very few possible goals or conclusions, this is a better approach to take. Backward chaining is best suited to problems where the possible conclusions are few and well defined.

Diagnoses

Expert systems have been associated with diagnosis applications for many years. Classification systems or diagnosis applications are best suited to backward chaining as there are few possible conclusions which need to be checked against the data. Backward chaining is the best choice if the goal can be sensibly guessed at the beginning of a consultation. Forward chaining is the best choice if all the facts are provided at the outset or there is not any sensible way to guess what the goal is at the beginning of a consultation. A backward-chaining system is more likely to cause questions to be asked which seem focused and in an order that appears logical to the user. In an expert system used for diagnosis, it is important that the system should seem to behave like a human expert. In this case, backward chaining, which is goal-driven, would usually be the most appropriate method. Diagnosis applications tend to have a fixed number of possible solutions which makes backward chaining more suitable.

To sum up, the choice of chaining method depends on the problem. If you have clear hypotheses, backward chaining is likely to be a better solution. It can be used for classification systems such as animals and plants. Forward chaining would be preferred if you have no clear hypothesis and want to see what can be concluded from the current state of affairs. Despite this, however, most expert system applications tend to use a mixture of forward and backward chaining. A number of diagnosis expert systems actually begin by using forward chaining which is data-driven, and this is why this approach was described in such detail in the medical diagnosis section on page 160. It is often the case that in a face-to-face consultation the doctor would be aware of all the facts of the case.

Gaming

How does an expert system work within computer gaming? Let us consider the most popular type of game, which involves the use of non-player characters (NPCs). The goal of the expert system in this case is to select which weapon is to be used during the course of the game by an NPC. Every game is a form of animation which consists of individual pictures called frames, in a sequence of moving images. The choice of weapon in a game will depend on a number of different variables that are changing in every frame of the game. The expert system is required to make a decision. The input variables may be the distance to the target, the type of target, and the range of the weapon. The system will need to output the type of weapon, why it was chosen, as well as the characteristics of the weapon. When setting up the system the knowledge engineer would have gathered from the game expert known facts such as individual weapon characteristics and their range, as well as which weapon is best for achieving the goal of eliminating the target. As the contents of the frame are constantly changing, the NPC will constantly have to re-evaluate the situation, much like a human player. Creation of an expert system will improve game play and provide a challenge, with the NPC making its own decisions on weapon selection.

The inference engine in such a system tends to use forward chaining. As it is data-driven, this method uses the data or facts which are input to the system and uses rules to determine a goal. Using the forward-chaining method, a series of IF statements are executed that sort through the data to THEN find a goal. The goal in this case is to select an appropriate weapon. These IF questions could, for example, be based on whether the target is a missile of some sort or another player. The THEN part will be the recommended type of weapon.

A good example of backward chaining in computer games is found in the game of chess. In chess, there are a fixed number of moves to reach a winning position. To compute all the possible ways of reaching a winning position, the system chains backwards and builds up a list of possible ways to get to the winning position. Backward chaining involves using all the possible premises which lead to getting a 'checkmate' conclusion. Backward chaining is used in gaming with games that have fewer rules.

Artificial intelligence

Artificial intelligence (AI) can be thought of as a piece of software that simulates or mimics the behaviour and judgement of a human being. It involves the system being able to learn, reason, and assess its own actions. In AI, forward chaining can be used to solve logic problems using rules and previous learning to find solutions to problems. An AI system might use forward chaining to look at the available information or data, answer a question or solve a problem. Forward chaining, as it is data-driven, works through the logic of a problem from beginning to end.

There are countless applications of AI: in the healthcare industry for dosing drugs (calculating the dose of drugs to be given to a patient) and suggesting different treatment for patients; in surgical procedures in operating theatres in hospitals; in computers that play chess and in self-driving cars. Each of these machines must be able to consider each action it takes, as each action will have an impact on the end result. In chess, its goal is to get its opponent in checkmate; in self-driving cars, its goal is to get from A to B without crashing. AI can also be used in the financial sector, where it can alert fraud departments to unusual spending patterns on a bank card or large amounts of money being deposited in bank accounts.

Another example of AI is in the field of robotics where basic expert systems are incorporated. These tend to use forward chaining where there is a clearly defined goal, such as moving from one spot to another, although the variables and obstacles may be numerous. Natural language processing (NLP) is another application of AI. The goal is for people to be able to converse with computers and computers being able to understand. Such systems tend to use a mixture of forward and backward chaining.

Artificial intelligence can also be used in manipulating social media. AI-driven fake bots can sense, think and act on social media platforms using human-like methods and simulating human behaviours. Unfortunately, these bots have a harmful side to them. They are able to provide wrong information to people (often referred to as fake news) as well as carrying out scams. They can even manipulate the stock market. AI systems can create fake media in the form of videos, voices, images and text. They can create realistic video portraits of an individual including control of facial expressions. They can replicate someone's voice, even creating a speech the speaker never made, making it easier to spread false information. They are able to produce text, which appears appropriate, based on a user's input. AI systems can create images that appear realistic – one actually won a prize in a world-renowned photography competition, fooling even the expert judges!

AI techniques such as machine learning and natural language processing are used to analyse data and learn what social media users like. Businesses can then supply more of that type of content and are able to analyse users' posts and the posts of other companies to identify new trends.

▼ Table 7.1 Advantages and disadvantages of expert systems

Advantages of expert systems	Disadvantages of expert systems
It reduces the time taken to produce a solution, for example, calculating an insurance premium or a doctor making a final diagnosis. Another example is it would be quicker for a car mechanic to find an engine fault as systems to diagnose faults in car engines can be directly connected to a car for immediate fault detection. This can save companies money as more cars can be fixed in a shorter time so either less money is spent on wages or more money comes in as more jobs can be processed in a shorter period of time.	Users such as mechanics, doctors, patients and insurance brokers will need training in how to use it, which takes a lot of time and comes with a financial cost.
In some areas such as car fault diagnosis, it can mean that a less skilled work force is required, therefore resulting in a lower wage bill.	Expert systems cost a lot to set up or buy and are fairly expensive to maintain.
A human expert can be tired and overworked and may miss a symptom, but the expert system is programmed to not forget to ask a question.	Expert systems lack common sense and when recommending certain drugs for a patient, an expert system may choose drugs that treat the illness but possibly without taking into consideration the side effects for that particular patient.

The solutions can be more accurate as the system uses the collective knowledge of several experts rather than relying on the knowledge of just one expert. This is particularly the case with detecting impending disasters such as avalanches, earthquakes, etc.	Expert systems can make mistakes if the knowledge base contains errors. It is not possible to program the expert system for every situation. This can result in incorrect decisions being made.
Some systems such as medical diagnosis, are always available and can be used 24 hours a day, 7 days a week, particularly when a doctor is not available.	The system will need to be updated from time to time leading to it being unavailable for a lengthy period at certain times.
The solutions of an expert system are consistent whereas a human expert can get tired or even emotionally involved which could cloud their judgement.	Experts may become dependent on the technology and may become deskilled if they always follow the expert system recommendations instead of using their own knowledge.
Systems can make more balanced decisions without being influenced emotionally.	
The expert system can be used worldwide, so can be used in parts of the world where experts or specialists are not available.	
Expert systems can produce several different recommendations for the human expert to choose from rather than a limited number within the single expert's experience.	

Practice questions

1 Doctors often use expert systems to help them diagnose illnesses. Explain how an expert system would work in this situation. [6]

2 An expert system consists of many components. Describe the role of a knowledge base in such a system. [4]

3 Expert systems often use forward chaining or backward chaining when producing suggestions. Explain what is meant by:
   a forward chaining [3]
   b backward chaining. [3]

8 Spreadsheets

In this chapter you will learn how to:
- create and edit the structure of a spreadsheet
- manipulate cells and their contents
- freeze panes and windows
- create formulas and use functions
- use validation rules
- format cells within a spreadsheet
- create a test plan and test the spreadsheet
- extract subsets of data within a spreadsheet
- sort the data within the spreadsheet
- create pivot tables and pivot charts to summarise data
- import and export data
- create macros to automate operations within a spreadsheet
- create a graph or chart appropriate to a specific purpose
- apply formatting to a graph or chart.

Before starting this chapter you should:
- have studied validation and verification (Section 1.4)
- understand the terms 'cell', 'row', 'column', 'sheet', 'tab', 'page', 'chart'
- be able to enter text, numeric data, formulas and simple functions into a spreadsheet
- be able to use editing functions such as cut, copy, paste and format painter
- be able to change the size of rows and columns within a spreadsheet
- be able to replicate formulas and functions in the spreadsheet
- understand the order of mathematical operations in formulas
- understand the types of mathematical averages: mean, median and mode
- be able to save a spreadsheet
- be able to print a spreadsheet displaying formulas or values
- understand the terms '**header**' and '**footer**' and why they are needed.

For this chapter you will need these source files:
- AlpHotels.csv
- Athletics.csv
- Bolton.csv
- Bolton2.csv
- BuildingProjects.csv
- BusAnalysis.csv
- BusTimes.csv
- Cardiff.csv
- CityTemp.csv
- Class.csv
- Client.csv
- Client1.csv
- Dundee.csv
- Employees.csv
- Employees2.csv
- Employees3.csv
- Flights.csv
- Flights1.csv
- Format.csv
- Gradebook.csv
- Hotels.csv
- JanSales.csv
- January.csv
- Motherboard.csv
- Operators.csv
- Orders.csv
- Projects.csv
- Quarter.csv
- Rain.csv
- Rates.csv
- Roles.csv
- RoomRates.csv
- Rooms.csv
- Rounding.csv
- Salary.csv
- SkiRace.csv
- Socket.csv
- Staff.csv
- Stock.csv
- Strings.csv
- Tasks.csv
- TBCWorkers.csv
- Test.csv
- TGCosts.csv
- TMC.csv
- TMCColour.csv
- TMCVehicles.csv
- TPAS.csv
- Traffic.csv
- TWTWorkers.csv
- WorkLocation.csv

8.1 Creating a spreadsheet

The spreadsheet structure will often be given to you by the organisation that you are creating it for. This may be given in diagram form, or as a specification sheet, or for larger organisations as two specification sheets, one for the design/layout and one for the organisation's house style. These specifications will often consist of font styles (including typeface and font sizes), colour schemes/themes, page styles (including page size and orientation), as well as other requirements like the use of gridlines and the visibility of row and column headings.

You will need to take these specific requirements into account as well as the intended user (audience) of the spreadsheet. Where specifications are not given, consider the font face, size and colour themes that would be appropriate for the user and pay attention to keeping the text easy to read, which can include considering the reading age and technical expertise of the intended audience.

Spreadsheet basics

A spreadsheet is a two-dimensional table split into rows and columns. It has a structure containing individual cells like this.

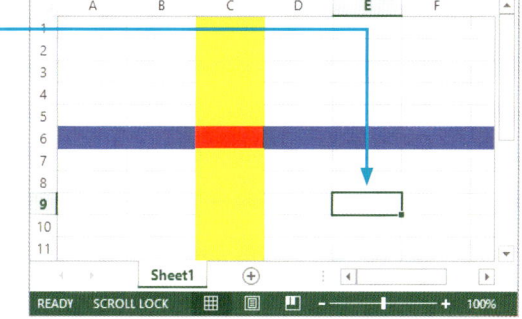

Each cell has an address. In this example, the cell with the cursor in it is called cell **E9** and the cell that has been coloured red is called cell **C6**. The red cell and all of the yellow cells are in column C, and the red cell and all of the blue cells are in row 6. A spreadsheet is sometimes called a sheet or even a worksheet. In *Microsoft Excel*®, many sheets/worksheets can be held within a single workbook.

Each cell in a spreadsheet can hold one of three things. It can contain:

» a number
» text, including hyperlinks, which is called a **label**
» a **formula**, which always starts with an = sign.

As a reminder, for this and all following chapters, the text following the Tasks demonstrates the techniques used to solve the Task and gives some example answers. These provide easy-to-follow step-by-step instructions, so that practical skills are developed alongside the knowledge and understanding.

Task 8a

Open the file **Staff.csv** in a new workbook. Place the file **Rates.csv** within the workbook as a new sheet called **Rates**. Add a new sheet called **Location** within the workbook by typing in the data so that it looks like this. Save the spreadsheet as **Task_8a**.

	A	B
1	Location code	Location name
2	H	Hypertown
3	J	Jamfrezzi
4	Y	Tawara

Importing data into Excel

Comma separated values (.csv) is a file format often used for source files. This file type can be imported into almost all spreadsheet packages on any platform, although the delimiters (usually commas) may be different for your regional settings. To attempt Task 8a, select the **File** tab, then **Open**. Locate the correct folder, but note that the .csv file will not be visible. Click on the drop-down menu for the file type to get a list of available files.

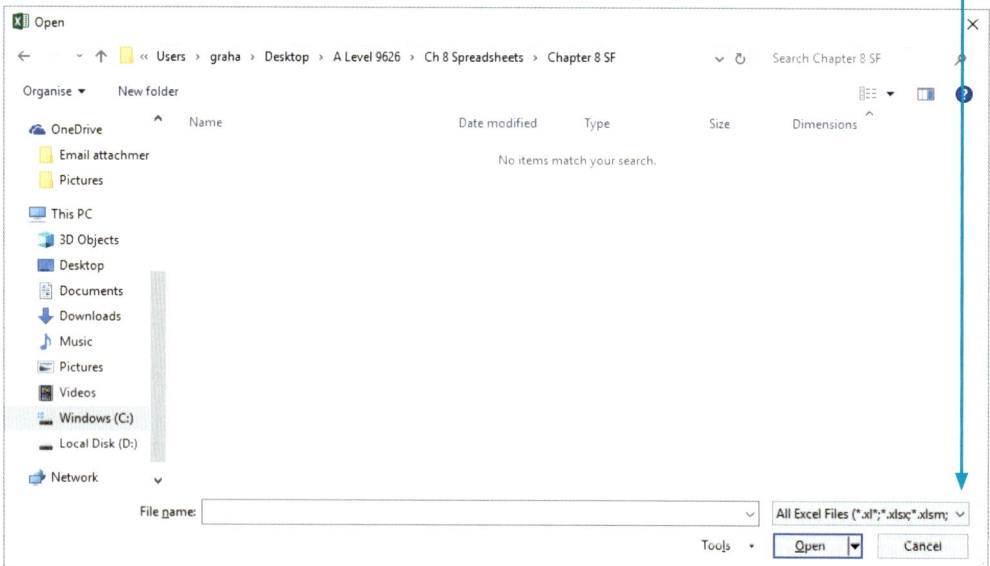

Select **Text Files** to see any **.csv** or **.txt** files. Double-click on the file **Staff.csv** to open it.

We need to import the file **Rates.csv** into a new worksheet within the current workbook.

Importing a .csv file into Excel

Data can be imported into a sheet using the **Data** tab.

In the **Get & Transform Data** group, select **From Text/CSV**

This opens the **Import Data** window.

Double-click on the file **Rates.csv** to open the text import window. Check that the delimiter is set to **Comma** before clicking on the [Load] button.

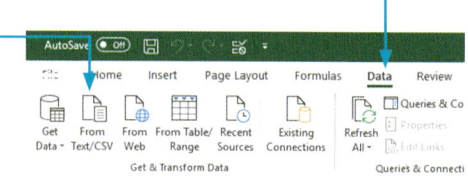

> **Advice**
>
> If you use regional settings with source files that do not use commas as delimiters, select the appropriate option for the regional settings of your source files.

The data has been imported into a new worksheet which is given a name by *Excel*, in this case 'Sheet1'.

The data may have been formatted by *Excel* with alternate green and white row shading. It can be reformatted from the **Table Design** tab, in the **Table Styles** section, by clicking the up arrow twice and selecting the plain white style which is on the left. Uncheck the tick box for the **Filter Button** to remove this.

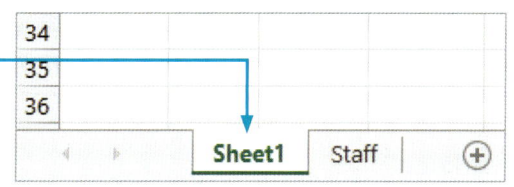

To rename this sheet, move the mouse over the sheet name 'Sheet1' and press the right mouse button to open the drop-down menu. Select **Rename** from this menu and the text 'Sheet 1' will be highlighted. Overtype the highlighted text with **Rates** and press **<Enter>** to make this change.

You now need to create a new worksheet called Location. To create a new worksheet, click on the **New Sheet** button.

This adds a new worksheet which can be renamed, and the new data entered in to it. Save your work with the filename **Task_8a** as a spreadsheet (not in .csv format).

Activity 8a

Open the file **TMC.csv** in a new workbook. Place the file **TMCColour.csv** within the workbook as a new sheet called **Colour**. Add a new sheet within the workbook called **Sites**. Type data into this sheet so that it looks like this. Save the spreadsheet as **Activity_8a**.

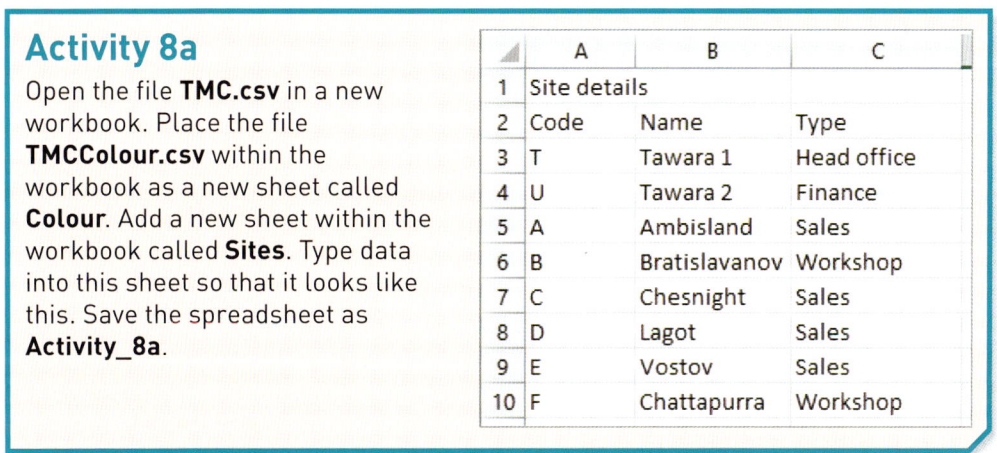

8.1.1 Create the structure

Page specifications

The page orientation, page size and margins all relate to printed versions of the spreadsheet. Although printed pages are not always required, you may be required to export your spreadsheet as a document (often as a **.pdf** file), so formatting your sheet for printing is still an important skill to learn.

To change the formatting, select the **Page Layout** tab. The **Page Setup** window (we will look at this later) can be opened using this arrow.

This is useful if you need to change more than one specification at once.

Page size is accessed using the drop-down arrow like this.

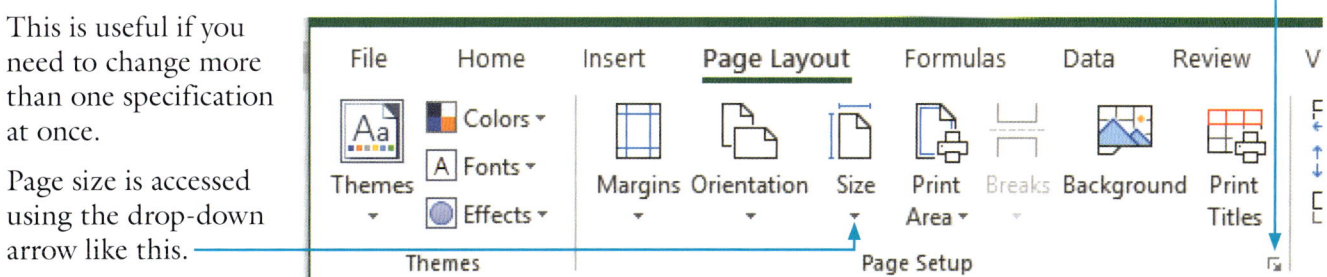

The drop-down menu offers a list of document sizes (pre-set by the software using your regional settings). Select the required page size from the list.

Advice
If you require a page size that is not shown, select **More Paper Sizes...** (at the bottom of the list) to open the **Page Setup** window. Use the drop-down **Paper size** list to select the required size.

To change the page orientation, use the drop-down arrow to display the options.

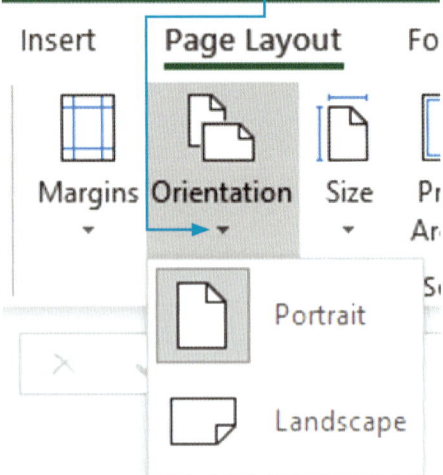
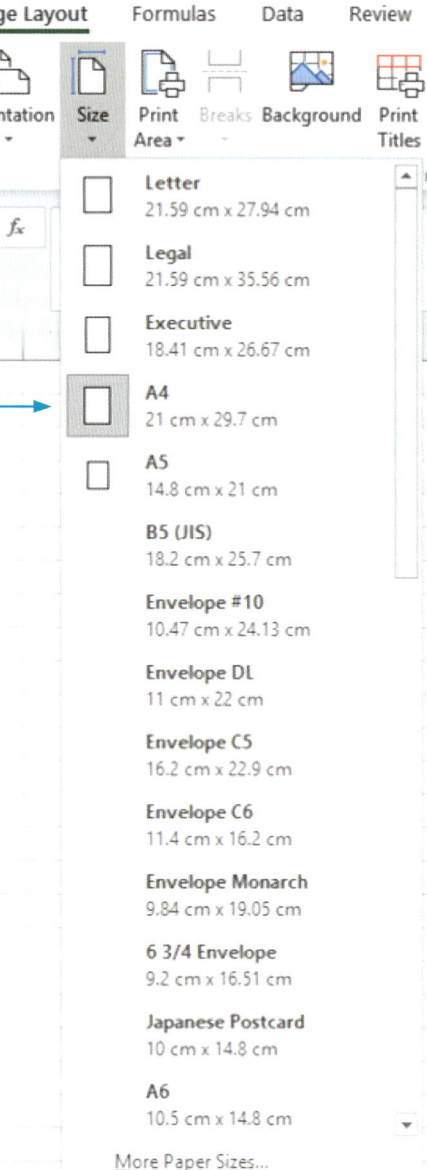

Select either **Portrait** or **Landscape** from the list.

Advice
Where specifications do not show portrait or landscape, select the most appropriate for the design of the printout. Make sure that all data, labels (and if appropriate formulas) are fully visible and in a font large enough to be read easily.

Page margins

Page margins contain the whitespace between the edge of the spreadsheet and the edge of the page. These can be changed to give documents a more professional look, for example placing data in the middle of a page rather than squashed into the top-left corner. They can also be changed to make more data fit on to a single page, without reducing its size and readability. Page margins can also be changed by selecting the **Page Layout** tab and using the drop-down arrow to display the options like this.

To open the **Page Setup** window in the margins tab, select **Custom Margins...**.

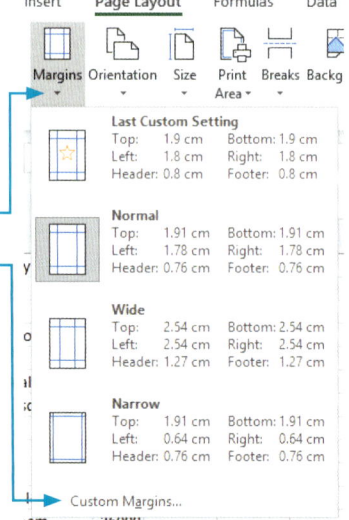

Remember this can also be selected directly from the Page Layout tab.

You can adjust the margin settings for Top, Bottom, Left and Right, then select the Print Preview button to see the effect of the changes.

Advice

If you are printing a document, make sure that the margins are within your printer's 'printable area'. These depend upon the make and model of your printer.

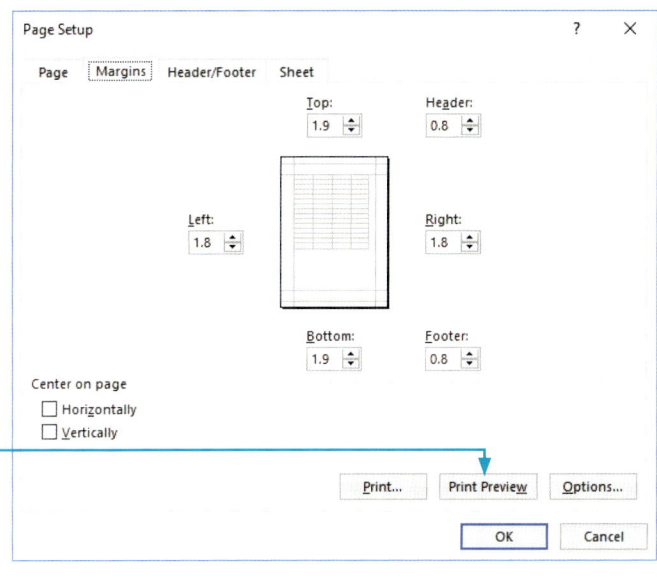

Headers and footers

Select the Page Setup window, then the Header/Footer tab like this.

To set up the header, select the Custom Header... button.

Make sure that the Align with page margins box is ticked. The header details can be entered into the left, centre or right sections by clicking in the chosen section. Text can be typed; the left button allows you to edit the text style, or automated fields can be included. These are selected using these buttons to insert the:

» page number

» number of pages

» date

» time

» filename with file path

» filename (no path)

» worksheet name.

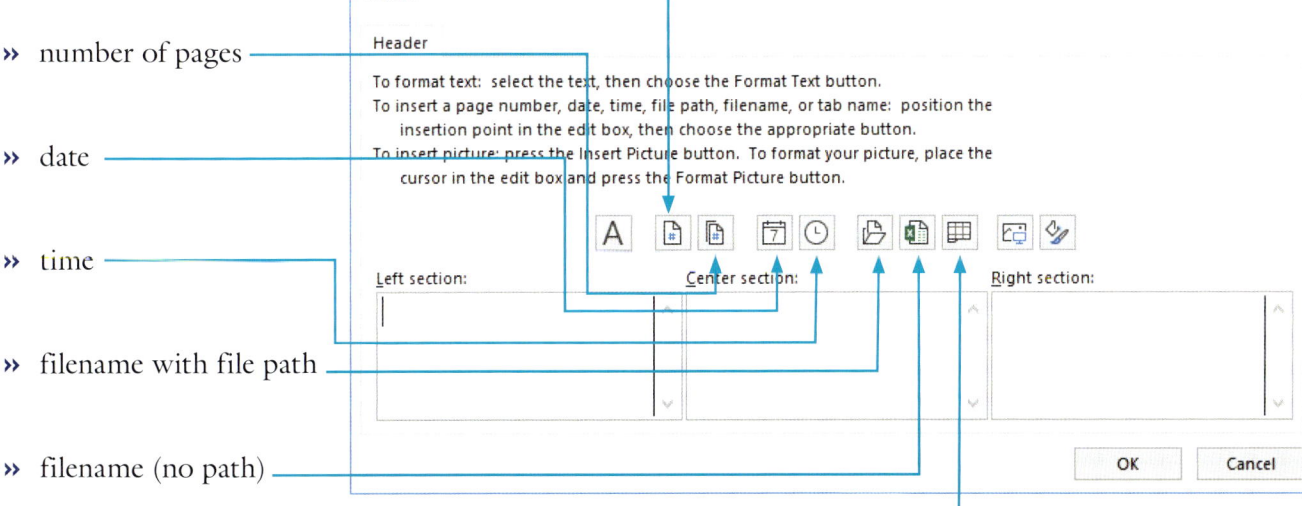

You may need to produce a document that fits on a single page (or on a specified number of pages). Before preparing your printout (or print preview if exporting into a document) make sure that all data is fully visible within each spreadsheet column and that there is no extra space in the column that can be removed. This is essential for both formulas and values printouts. To fit to a single page, open the **Page Setup** window and select the **Page** tab. The **Scaling** group is in the middle of this window. Click on the radio button for **Fit to:** like this.

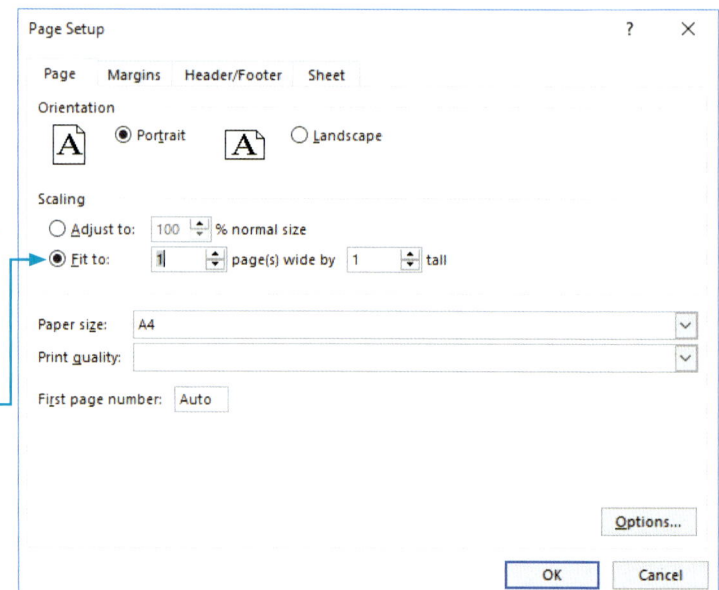

The number of pages can be changed if required. For a single page select 1 page wide by 1 tall, then click on **OK**.

Displaying gridlines, row and column headings

It is sometimes easier to work on your printed/exported spreadsheet with gridlines and row and column headings visible; it makes checking your formulas much easier.

Both of these options are in the **Page Layout** tab and can be found in the **Sheet Options** group. Click to place a tick in the relevant **Print** box.

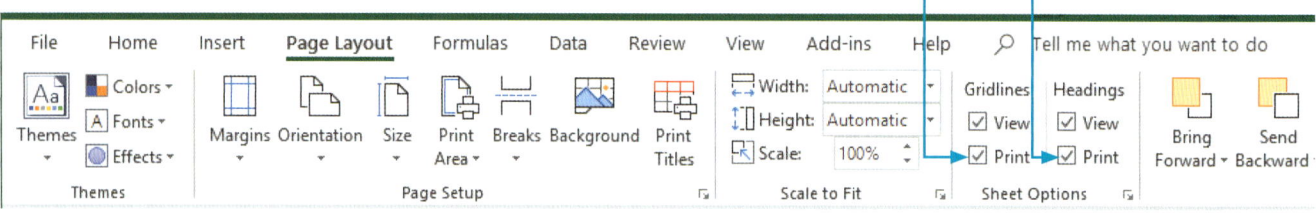

Task 8b

Open the file **Salary.csv** in your spreadsheet.

Using an A4 page size, place appropriate text in the header to display who created the spreadsheet and automated data to show when it was created. Place the filename and page numbering in the footer of the page. Place the contents of the spreadsheet on two pages with gridlines. Display all row and column headings. Save the spreadsheet as **Task_8b**.

Open the file **Salary.csv** in your spreadsheet. Before starting work, examine the data carefully. You will notice that not all data is visible. Select all the data using the <Ctrl> and <A> keys. An alternative method is to double-click here.

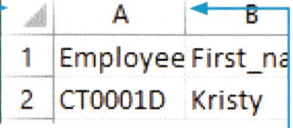

Double-click on the drag handle between the columns to resize all columns to match the data present.

All the data should now be visible.

> **Advice**
> Ensure you show all the contents of a cell (especially formulas).

Check that the paper size is set to A4. If not, edit the settings to make it the correct size. As all work should be of a professional standard, it is not enough to just type your name and select automated fields for the header. Consider what information is required, such as your name and when created. There are two required elements, so to improve layout we will use the left and right sections of the header to give the page balance. In the left section of the header enter text like 'Created by' followed by your name. On the right, enter the text 'Created on' followed by today's date, then the text 'at', followed by the time. Use the automated field buttons for this data.

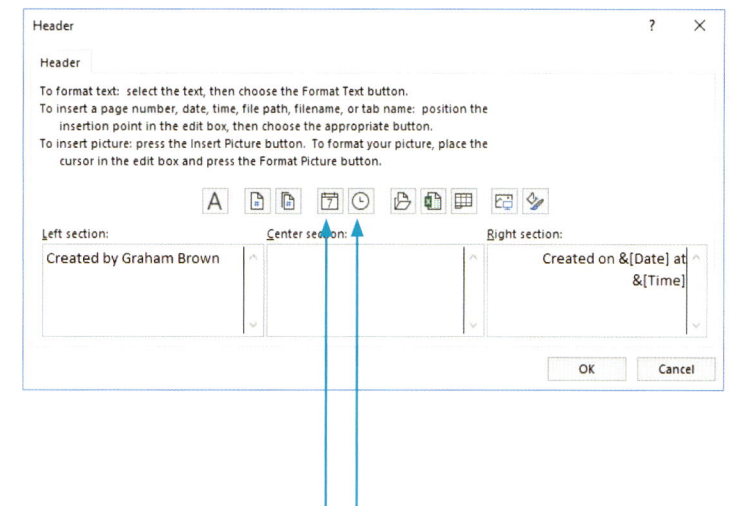

The design of the header in the spreadsheet may look like this.

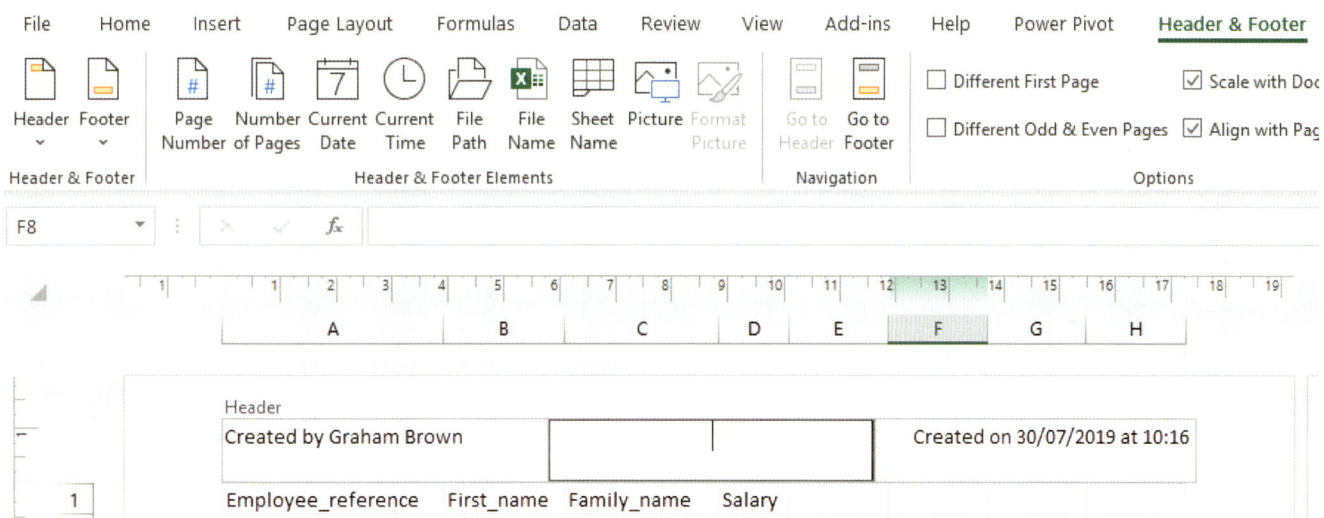

The print preview for the page will appear like this.

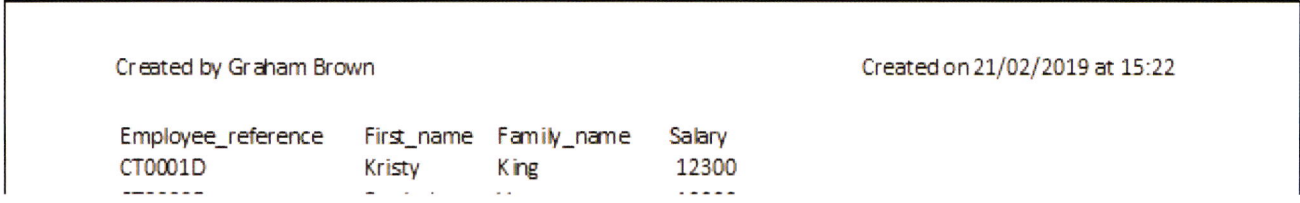

For the footer, again there are two required elements so use text and automated field buttons to add the filename and page numbering into the footer like this:

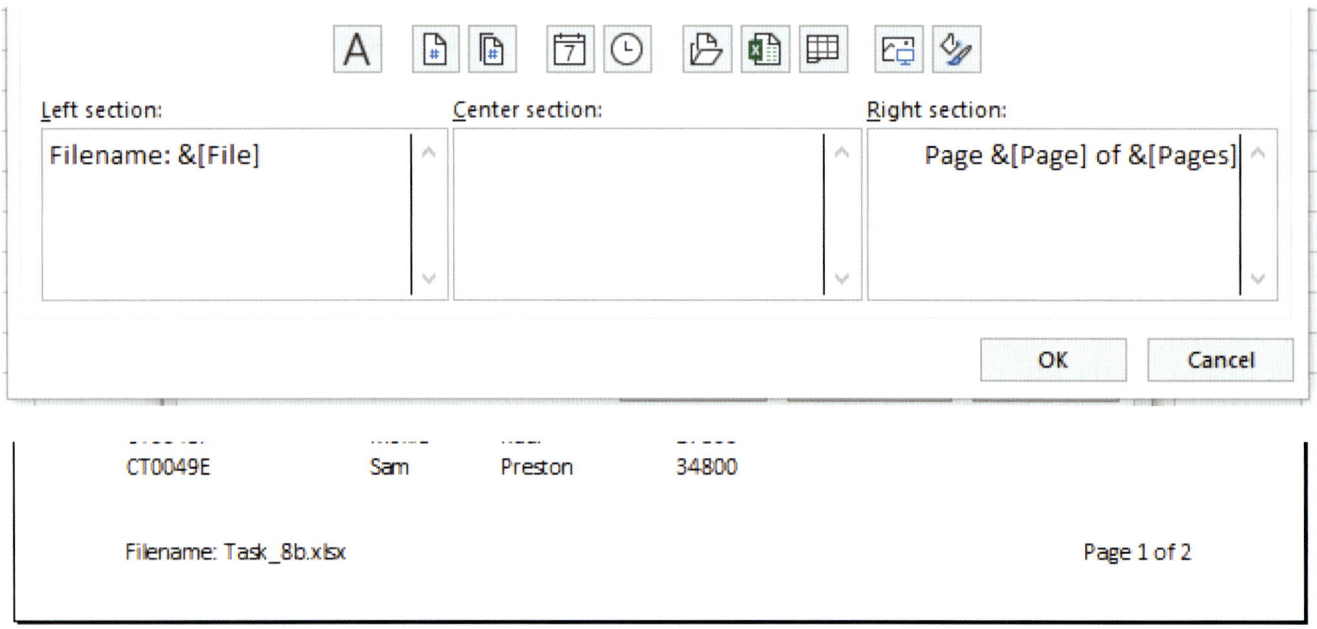

Tick both **Print** boxes to add gridlines and row and column headings to the sheet. The page now looks unbalanced, with the data to the left and lots of white space to the right.

Created by Graham Brown

Created on 23/02/2019 at 19:16

	A	B	C	D
1	Employee_reference	First_name	Family_name	Salary
2	CT0001D	Kristy	King	12300
3	CT0002E	Rupinder	Vas	10000
4	CT0003D	John	Mortlock	30000
5	CT0004C	Jamie	Smith	8900
6	CT0005A	Jack	Gonzalez	8200
7	CT0006B	Brian	O'Driscoll	25000
8	CT0007F	Jamal	Khan	8000
9	CT0008E	Laila	Aboli	25000
10	CT0009C	Laura	Smith	10000
11	CT0010B	Patrick	O'Malley	12500
12	CT0011B	Cameron	Garnham	36000
13	CT0012D	James	Smith	10000
14	CT0013D	Lauren	Garnham	6200
15	CT0014F	Linzi	Woodley	8000
16	CT0015B	Julia	Frobisher	16500
17	CT0016C	Dan	Dare	14000
18	CT0017G	Lea	Cabusbusan	28000
19	CT0018A	Jasmine	Patel	6400
20	CT0019F	Brett	Ling	8200
21	CT0020C	Wim	Van Hoffmann	17000
22	CT0021A	Bryan	Revell	10000
23	CT0022B	Thirumalar	Asokmani	10000
24	CT0023B	Brian	Guthrie	43000
25	CT0024C	Lilik	Kaznica	9000
26	CT0025F	Shelagh	O'Leary	9500
27	CT0026C	Isabel	Smith	9800
28	CT0027G	Leah	Temple	10000
29	CT0028E	Liam	Chi	8500
30	CT0029C	Bishen	Patel	10000
31	CT0030A	Henri	Ramos	10000
32	CT0031G	Jemma	Watson	8500
33	CT0032C	Joseph	McNevin	32000
34	CT0033F	Holly	Ling	27000
35	CT0034C	Greg	Mina	30000
36	CT0035A	Lee	Wan	13000
37	CT0036F	Sri	Paryanti	36000
38	CT0037E	Sean	O'Byrne	30000
39	CT0038C	Jake	Garnham	12500
40	CT0039F	Rupinder	Singh	11200
41	CT0040F	Sergio	Gonzalez	26000
42	CT0041C	Jovy	Singh	16000
43	CT0042E	Kiah	Wilson	23000
44	CT0043G	Hannah	Atkins	17500
45	CT0044G	Sebastian	Russell	21800
46	CT0045F	Adam	Newman	26200
47	CT0046C	Lara	Sanderson	34700
48	CT0047C	David	Tyler	28300
49	CT0048F	Mollie	Kaur	17800

Filename: Task_8b.xlsx

Page 1 of 2

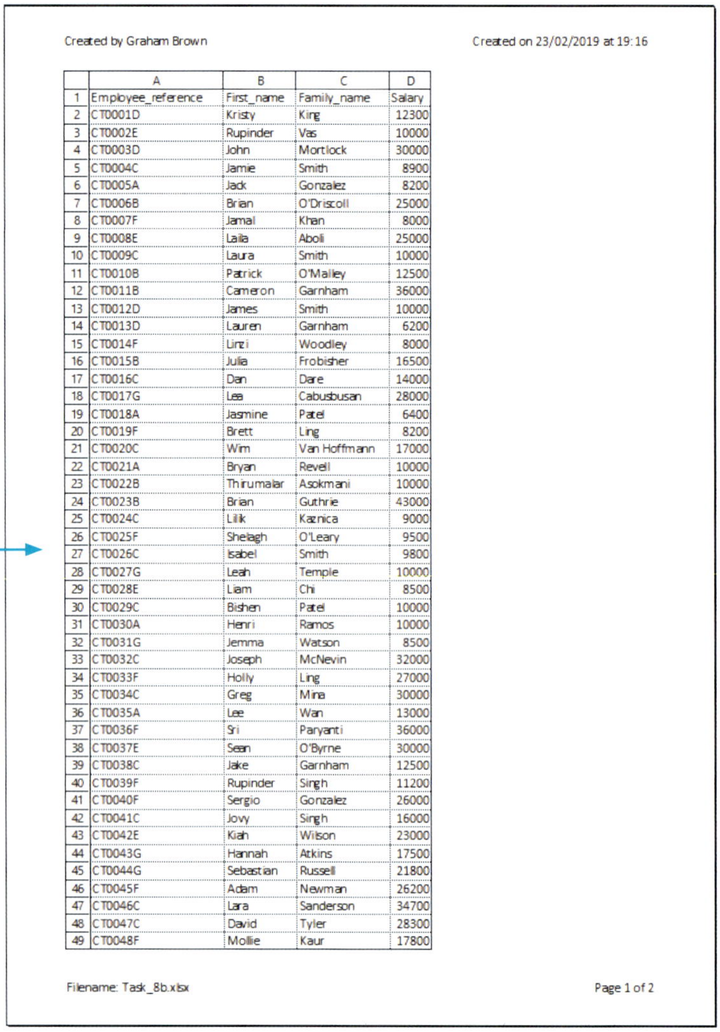

Increasing the left and right margin sizes will help but be careful to make sure that the left and right contents in the header and footer do not run together. In this case setting the margins to 5 cm works well. Because there is a large area of white space at the bottom of page 2, increasing the top and bottom margins will also improve the layout. Use trial and error to set these so that there is little extra white space but the spreadsheet fits on two pages. Setting the top and bottom to 3.6 cm works well. You can now change the header and footer position to 2 cm to place these vertically in the middle of the white space at the top and bottom. The page now looks like this.

Save your spreadsheet with the filename **Task_8b**.

> ### Activity 8b
> Open the file **Employees.csv** in your spreadsheet. Display the automated filename with its full file path in the header of the page. Place appropriate text in the footer to display who created the spreadsheet. Display the contents of the spreadsheet on a single A4 page with gridlines. Do not display row and column headings. Save the spreadsheet as **Activity_8b**.

Edit spreadsheet structures

Inserting cells, rows and columns

To insert a cell into a spreadsheet, click the right mouse button on the spreadsheet where you wish to insert the cell. From the drop-down menu select *Insert…* which opens the *Insert* window.

This will allow you to insert a cell (by selecting one of the top two options which creates the space for the cell by moving all other cells to the right or down, depending on your choice), or allow you to insert a row or a column.

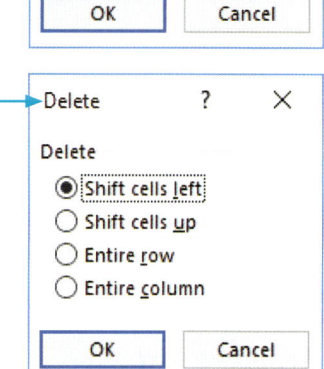

Deleting cells, rows and columns

To delete a cell from a spreadsheet, **right** click on the cell you wish to delete. From the drop-down menu select *Delete…* which opens the *Delete* window.

This will allow you to delete a cell (by selecting one of the top two options which removes the cell by moving all other cells to the left or up) or allow you to delete a row or a column.

Resizing rows and columns

You have already learnt how to resize all columns to fit the data, but sometimes rows and columns need manually adjusting. To manually adjust a column width (rows are the same process), click the left mouse button on the drag handle at the end of the column heading and hold it down like this.

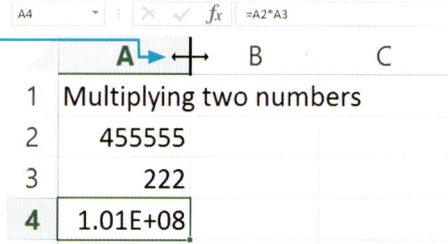

Drag the column to the required width. If you need to set a number of columns to the same width, click and drag the column headings to highlight all those that you wish to change. Click on one of the drag handles in this range and resize that one column. All other highlighted columns will match its width when you release the mouse button.

Hide and unhide rows and columns

Select the row/s or column/s that you wish to hide, using the methods described above, then click the right mouse button in the highlighted row/column heading to get a drop-down menu like this.

Click on the option to **Hide** this row/column. To unhide, select the rows/columns surrounding the hidden cells and use the **Unhide** option from this menu. Please note that you can also **Insert** or **Delete** row/s or column/s using this method.

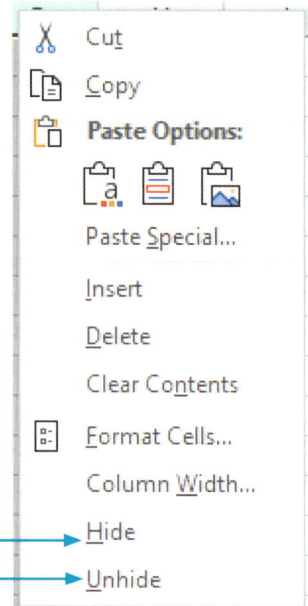

Task 8c

Use the file **TPAS.csv** to create a spreadsheet that looks like this:

Save the spreadsheet as **Task_8c**.

Advice

To repeat the last operation in *Microsoft Windows*, press **<Ctrl>** and **<Y>**. To undo the last operation, use **<Ctrl>** and **<Z>**.

Merging cells

For this task, you will need to merge cells A1 to C1 first and then cells A3 to C3. Highlight cells A1 to C1 and select the **Home** tab. In the **Alignment** group, click on the **Merge & Center** icon.

Repeat this process for cells A1 to A3. Highlight the spreadsheet and resize all columns to fit the displayed data.

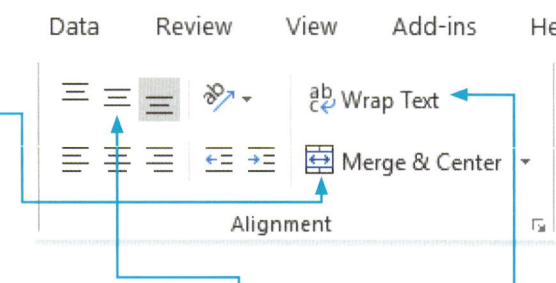

Wrapping text

Change the font size in the merged cell in row 1 to 24 point. The text may no longer fit the available space, if so (with this cell selected) click on the **Wrap Text** icon.

Use the drag handle to increase the row height, so that the text appears as shown in the question. Repeat this process to wrap the text in cell B5; you may need to adjust both row height and column width to make the cell appear like in the question. Highlight cells A5 to C5 and select the middle button for the vertical alignment like this.

Right align all cells between A6 and C17, then centre align all cells between B6 and B17. Select row 2 and reduce the height of this row to half the height of row 3. Repeat this for row 4. Check that your finished spreadsheet matches the diagram in Task 8b. Save your spreadsheet with the filename **Task_8c**.

> ### Activity 8c
> Create a spreadsheet that looks like this:
>
>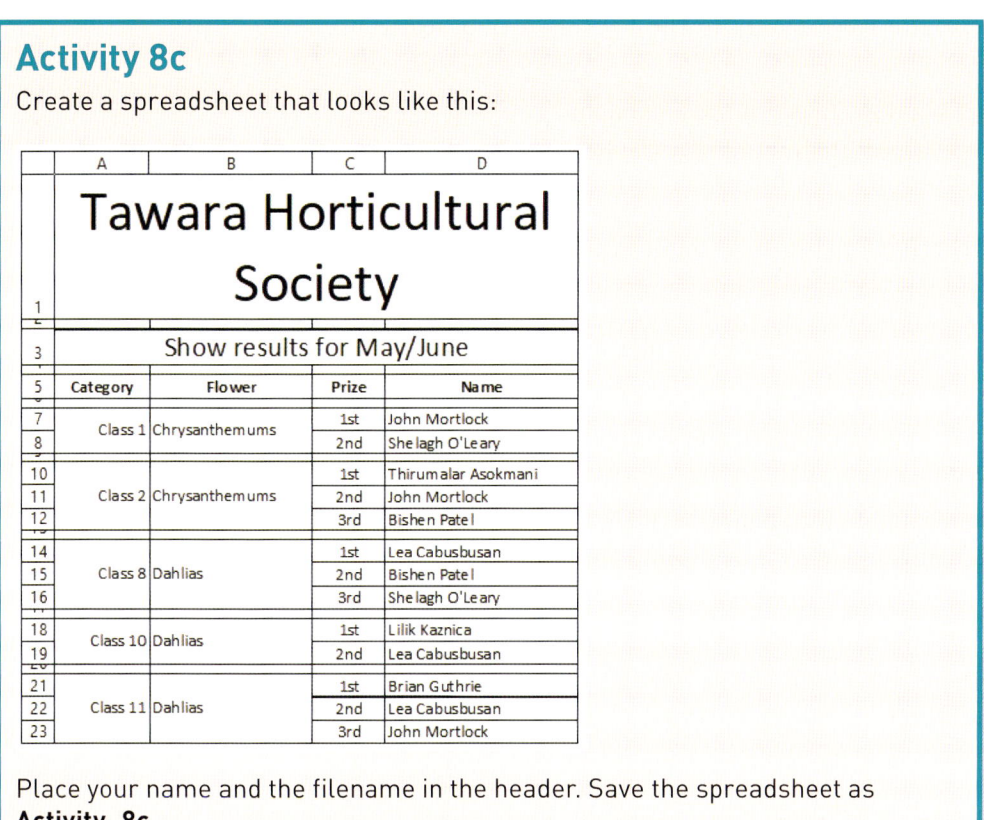
>
> Place your name and the filename in the header. Save the spreadsheet as **Activity_8c**.

Protecting cells and their contents

You can protect parts of the spreadsheet to stop anyone changing, deleting or moving important data within the sheet. This can involve protecting cells within a worksheet, a worksheet within a multi-layered workbook, or the entire workbook.

> ### Advice
> Make sure that you have read Section 1.4 Checking the accuracy of data before attempting this Task.

> ### Task 8d
> Open the file that you saved as **Task_8c**. Protect all cells between A1 and C5 and between A6 and A17 inclusive. Use the password **Hodder** to protect your sheet. Save the spreadsheet as **Task_8d**.

Identify and highlight the individual cell or cells that you will protect, in this case the first range is between A1 and C5. Select the **Home** tab and the drop-down menu for the **Alignment** group.

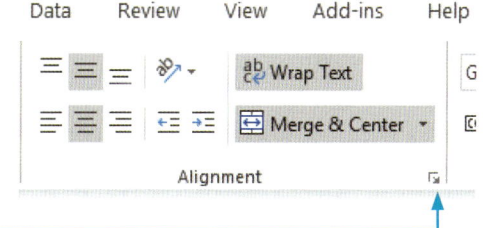

This opens the **Format Cells** window. Select the **Protection** tab. Make sure that the tick box for **Locked** contains a tick, before clicking on [OK]. Repeat this process for cells A6 to A17. Make sure that the other cells B6 to C17 do not have the tick box for **Locked** displayed.

To activate protection, move to the **Review** tab. In the **Protect** group, select **Protect Sheet**.

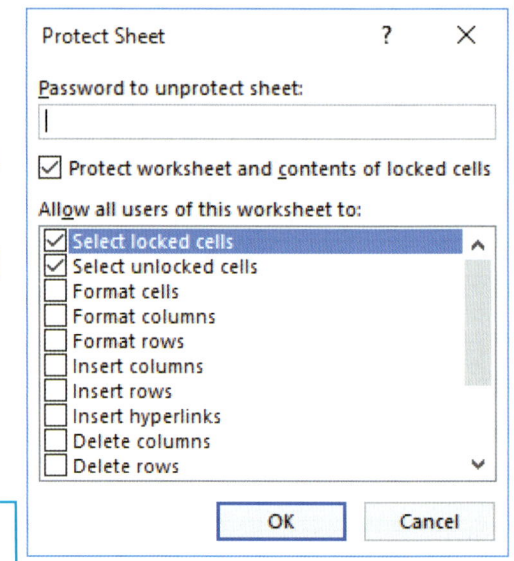

This opens the **Protect Sheet** window. Because the locked and unlocked ranges of cells have been defined, *Excel* has selected the first two tick boxes. Any tick boxes that are not selected mean that those functions are removed when the sheet is protected, for example the formatting of any cells in the sheet will not be allowed (all the options to format will be 'greyed out' from the menu). Type the password, in this case 'Hodder' into the box and press [OK]. You will need to re-enter this password and click [OK] for **double-entry verification** by *Excel*.

The same method can be used to protect whole rows or columns, by selecting the rows or columns rather than a range of cells.

Protecting the workbook protects only the structure of the workbook, to prevent sheets being added to it or deleted from it. It is accessed using the **Protect Workbook** icon.

To unprotect the sheet and edit these cells, use the **Review** tab, and in the **Protect** group select **Unprotect Sheet**. You must enter the password followed by [OK].

Activity 8d

Open the file that you saved as **Activity_8a**. Protect all cells in the Sites sheet and cells A2 to C18 in the Colour sheet so that they cannot be edited. Allow all cells in the TMC sheet to be edited. Use your name as the password to protect your workbook. Save the spreadsheet as **Activity_8d**.

Advice

Passwords can be removed by selecting the **File** tab, the **Info** section, then using the **Unprotect** hyperlinks for each sheet, or for the workbook, by double-clicking on the drop-down menu for **Protect Workbook**.

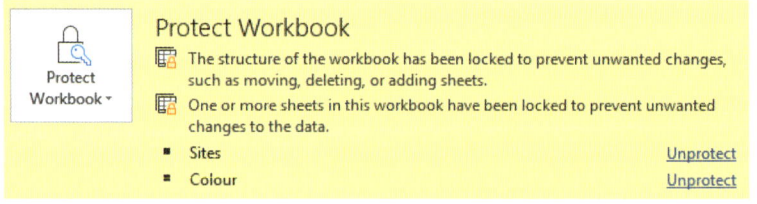

Split window and freeze panes

It is sometimes easier to work on a large spreadsheet by being able to see two different parts of the same sheet on the screen at the same time. Two methods that can be used for this are:

» split the window
» freeze panes within the window.

Split windows

Windows can be split horizontally, vertically or both. To split a window horizontally, select the row directly below the position you require the split. Select from the **View** tab, in the **Window** group, the **Split** button.

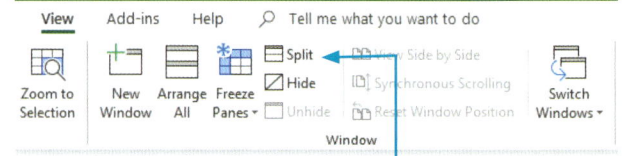

To remove the split window, click the *Split* button again.

To split the window vertically, select the column after the position you require the split, then click the *Split* button.

To split the window both horizontally and vertically, select the cell below and to the right of the required split position, then click the *Split* button.

Task 8e

Open and examine the file **Stock.csv**. Split this so that both types of stock can be viewed together. Save the spreadsheet as **Task_8e**.

'Open and examine the file.' Before starting work you need to understand the contents and structure of the files given. In this case the Stock file is split into two different parts; the top part for motherboards and the lower part for processors. The split must appear below row 120 and above row 121. Highlight row 121. Select from the *View* tab, in the *Window* group, the *Split* button. You can now scroll independently in both windows. Scroll up in the top window so the window looks like this.

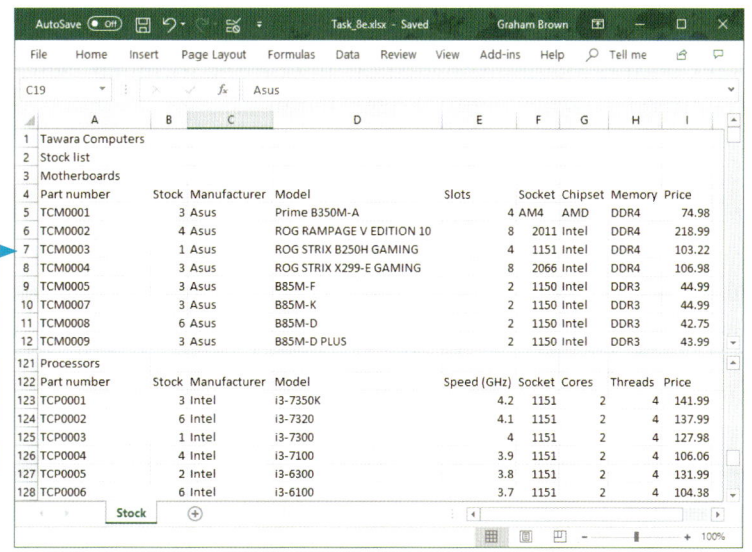

Save the spreadsheet as **Task_8e**.

Task 8f

Open and examine the file **Motherboard.csv**. Freeze the panes so that the part number and stock columns and rows 1 to 4 are always visible. Save the spreadsheet as **Task_8f**.

Freeze and unfreeze panes

Freezing a pane works in the same way as a split window, but it will stop part of the sheet from scrolling. Open and examine the file **Motherboard.csv**. Move the cursor to cell C5, which is the first cell that will be allowed to scroll. Select from the *View* tab, in the *Window* group, the drop-down menu from the *Freeze Panes* button.

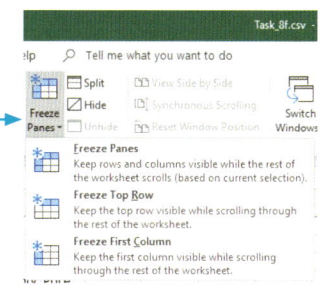

Click on the *Freeze Panes* button; now rows 1 to 4 and columns A and B will be frozen (fixed) within the window, so the spreadsheet looks like this:

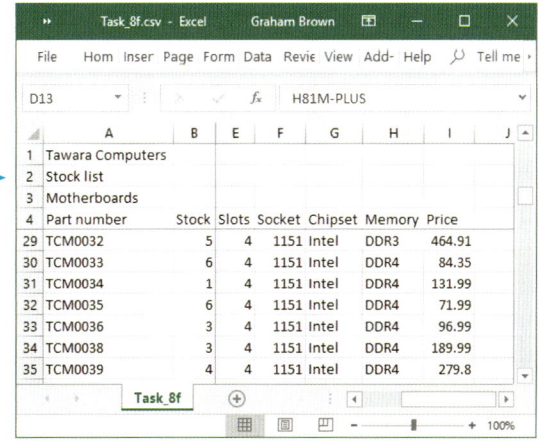

To unfreeze the panes, select the drop-down menu from the **Freeze Panes** button, then the **Unfreeze Panes** option.

8.1.2 Create formulas and use functions

Display formulas

Formulas are always preceded by an = sign when entered into a spreadsheet. The formula held in the current cell is displayed in the **Formula** bar, along with the address of that cell like this.

To display all formulas in your spreadsheet, select the **Formulas** tab, in the **Formula Auditing** group, select the *fx* Show Formulas icon.

This button toggles to switch on and off the formulas view, which may be required to give evidence of your methods.

Create formulas

Mathematical operators

Simple mathematical operators are placed in a formula and can be used to add, subtract, multiply, divide and calculate indices (powers) of a number. For addition, use the + symbol; for subtraction, use the - symbol; for multiplication, use the * symbol; and for division, use the / symbol. Indices are calculated using the ^ symbol, so the contents of cell A2 squared (x^2) would be typed as =A2^2

> ### Task 8g
> Open and examine the file **Operators.csv**. Place two numbers of your choice in cells B2 and B3. Calculate in cell:
> » B5, the sum of the two numbers
> » B6, the difference between the two numbers
> » B7, the product of the two numbers
> » B8, the contents of cell B2 divided by the contents of cell B3
> » B9, the contents of cell B2 to the power of the contents of cell B3.
>
> Check that the formulas have worked. Save the spreadsheet as **Task_8g**.

Open the file **Operators.csv**. Enter the number 4 in cell B2 and 2 into cell B3. The values were carefully selected to make the maths easy when you check your calculations. The more difficult calculations are likely to be the division and indices. These numbers were selected so that the 4 divided by 2 gives an easy result and the 4 to the power of 2 is reasonably easy (4 × 4).

It is wise to perform all calculations by hand before entering the formulas, to make sure that you understand the formulas that you are using and to find the results of the calculation before the computer has shown you its results. These calculations may look like this.

Number X	4
Number Y	2
X+Y	4+2=6
X-Y	4-2=2
X*Y	4*2=8
X/Y	4/2=2
X^Y	4^2=16

» **Addition:** Enter the formula **=B2+B3** in cell B5.
» **Subtraction:** Enter the formula **=B2-B3** in cell B6.
» **Multiplication:** Enter the formula **=B2*B3** in cell B7.
» **Division:** Enter the formula **=B2/B3** in cell B8.
» **Indices:** Enter the formula **=B2^B3** in cell B9.

Advice
While other functions like SUM, PRODUCT or POWER could also be used to perform these tasks, using mathematical operators provides the most efficient method.

	A	B
1	Mathematical operators	
2	First number - X	4
3	Second number - Y	2
4		
5	Sum of X and Y	6
6	Difference between X and Y	2
7	Product of X and Y	8
8	X divided by Y	2
9	X to the power Y	16

To check that the formulas have worked correctly, compare your original paper-based calculations above with the values in the spreadsheet.

Save the spreadsheet as **Task_8g**.

Indices

Task 8h
Create a spreadsheet to calculate x^n when you are given the value of x and n. Use this spreadsheet to calculate 3^4, π^2, 16^{-2} and $16^{0.5}$. Save the spreadsheet as **Task_8h**.

Indices are sometimes called powers. They are commonly used in mathematical and scientific formulas. To calculate values like 3^4 or x^2, a spreadsheet provides you with the ideal tool. You could calculate 3^4 by multiplying (=3*3*3*3) or with the formula =3^4. However, this will not allow these figures to be changed without changing the formula. A better solution is to create a small spreadsheet model like this.

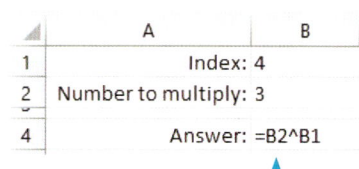

By changing the values in B1 and B2, we can get different answers that will be shown in cell B4. To calculate 3^4, place 4 in B1 and 3 in B2 which will display 81 in cell B4. To calculate π^2, place 2 in B1 and 3.14 (the value of **Pi** rounded to two decimal places) in cell B2, which will display 9.8596 in cell B4.

Advice
When we learn to use functions later in this chapter, we can replace 3.14 with a Pi (π) function.

To calculate 16^{-2}, place -2 in B1 and 16 in cell B2, which will display 0.00390625 in cell B4. To calculate $16^{0.5}$, place 0.5 in B1 and 16 in cell B2, which will display 4 in cell B4. You can see that $16^{0.5}$ is the same as $\sqrt{16}$. Save your spreadsheet as **Task_8h**.

Activity 8e
Create a spreadsheet to calculate x^n for any given values. Use this spreadsheet to calculate 4^4, 24^7, 0.7^{-3}, 5^3, x^4 where $x=3.7$. Save the spreadsheet as **Activity_8e**.

Absolute and relative cell referencing

An **absolute reference** is a cell reference that will not change when a formula is replicated. All or part of a cell reference can be made absolute by adding a $ sign in front of the part to 'fix' in place. A **relative reference** is a cell reference that will change when a formula is replicated. This spreadsheet shows some examples where formulas in cells A2, F2, A10 and F10 have been replicated in both directions:

	A	B	C	D	E	F	G	H	I
1		Absolute referencing					Relative referencing		
2	=$A$30	=$A$30	=$A$30	=$A$30		=F30	=G30	=H30	=I30
3	=$A$30					=F10			
4	=$A$30					=F11			
5	=$A$30					=F12			
6	=$A$30					=F13			
7	=$A$30					=F14			
8									
9	Absolute referencing - column fixed					Absolute referencing - row fixed			
10	=$A30	=$A30	=$A30	=$A30		=A$30	=B$30	=C$30	=D$30
11	=$A31					=A$30			
12	=$A32					=A$30			
13	=$A33					=A$30			
14	=$A34					=A$30			
15	=$A35					=A$30			

Named ranges (met later in this chapter) are another example of absolute referencing.

Using absolute and relative cell references

> **Task 8i**
> Create a spreadsheet to calculate the times tables. Cell A1 should contain the number of the table and the table should be calculated between 1 and 10. Save the spreadsheet as **Task_8i**.

Create a new spreadsheet to look like this (in formula view).

Highlight cells A3 and A4 as shown, then use the drag handle in cell A4 to drag the numbers down from A4 to cell A12.

	A	B
1	2	Times Table
2		
3	1	=A3*$A$1
4	2	

In cell B3 the formula is **=A3*$A$1** which contains both absolute and relative referencing. The reference to cell A1 (with the $ symbols) is an absolute reference and the reference to cell A3 is a relative reference.

Select B3 and use the drag handle to replicate (copy) this formula down into cells B4 to B12. You can see from this view that the reference in cell B3 to cell A3 has been changed as the cell has been replicated because it uses relative referencing. However, the reference to cell A1 has not been changed during the replication because absolute referencing has been used.

	A	B
1	2	Times Table
2		
3	1	=A3*$A$1
4	2	=A4*$A$1
5	3	=A5*$A$1
6	4	=A6*$A$1
7	5	=A7*$A$1
8	6	=A8*$A$1
9	7	=A9*$A$1
10	8	=A10*$A$1
11	9	=A11*$A$1
12	10	=A12*$A$1

To return to the values view of the spreadsheet, select the **Formulas** tab, in the **Formula Auditing** group, then select the *fx* Show Formulas icon. Save your spreadsheet as **Task_8i**.

Named cells and ranges

It is difficult to remember which cell holds which data, especially in a large spreadsheet. When an individual cell or an area of a spreadsheet is going to be referred to many times within spreadsheet formulas, it is often a good idea to give it a name. This name should be short and meaningful. In the case of a large spreadsheet, it is easier to remember the name of a cell, for example VAT or AveMiles, rather than trying to remember the cell reference, for example AC456 or X232. Once a cell or a range of cells has been named, you use this name in all your formulas within that workbook.

> **Task 8j**
> Open the file **TBCWorkers.csv**. Format the spreadsheet as shown.
>
> Name cell B2 **Pay**, cell E2 **Tax** and cells B6 to C11 **Hours**. Save the spreadsheet as **Task_8j**.
>
> | | A | B | C | D | E | F | G | H |
> |---|---|---|---|---|---|---|---|---|
> | 1 | | | Tawara Book Company | | | | | |
> | 2 | Rate of pay | | £10.20 | | Tax Rate | | 40% | |
> | 3 | | | | | | | | |
> | 4 | | | Hours worked | | | | | |
> | 5 | Name | | Week 1 | Week 2 | Total | Gross | Tax | Net | % Time |
> | 6 | Brian Sargent | | 14 | 10 | | | | |
> | 7 | Graham Brown | | 9 | 12 | | | | |
> | 8 | Karla Brown | | 6 | 5 | | | | |
> | 9 | Nigel Sinclair-Banks | | 8 | 8 | | | | |
> | 10 | Stuart Morris | | 5 | 6 | | | | |
> | 11 | Zovita Moon | | 2 | 2 | | | | |
> | 12 | Total | | | | | | | |

Open the file and find cell B2. You must name this cell 'Pay'. Right click the mouse in this cell to get the drop-down menu and select **Define Name...** to open the **New Name** window. In the **Name:** box, overtype the suggested name with the word **Pay**. Add suitable text in the **Comment:** box so that the window looks like this.

Repeat this to name cell E2 **Tax**.

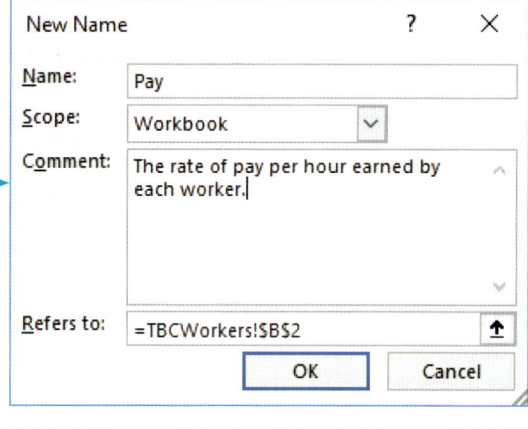

To create the named range called 'Hours', you must highlight the cells between B6 and C11. Click the right mouse button within the highlighted range to get the drop-down menu. Enter the text **Hours** into the **Name:** box and add an appropriate **Comment:**. Check that the **New Name** window looks like this before clicking on OK.

The name of the range is only visible in the **Name** box if all the cells in the range are highlighted.

This named cell and range will be used in later tasks.

Save this spreadsheet as **Task_8j**.

If you have already created a named cell or range and wish to view it or edit it, you can open the **Formulas** tab, in the **Defined Names** group select the **Name Manager**.

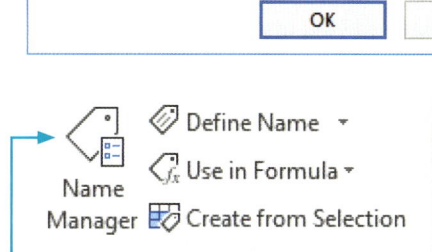

The **Name Manager** looks like this.

Double-click on the name that you wish to view or edit.

Use functions

Function names

A function has a predefined name like **SUM** or **AVERAGE** that performs a particular calculation. It is an operation built into the spreadsheet. There are lots of these

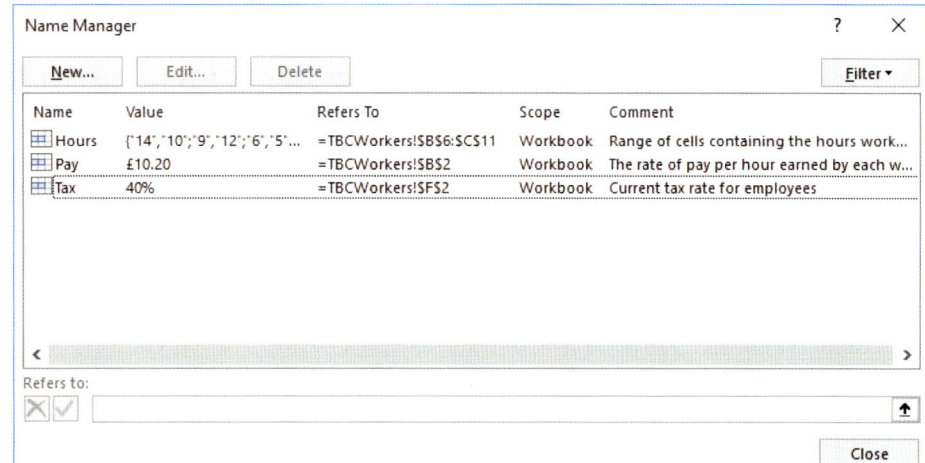

functions in *Excel*, many of which are beyond the scope of this book, but each has a reserved function name. If a question asks you to choose your own name for a cell or range, try to avoid using these function names.

SUM function

The SUM function adds two or more numbers together. In Task 8g, you used the mathematical + operator and the formula =B2+B3 to add the contents of two cells together. With only two cells to be added, this was the most efficient way of doing this. If there had been more figures to add, particularly if they were grouped together in the spreadsheet, using the SUM function would have been more efficient.

> ### Task 8k
>
> Open the spreadsheet you saved as **Task_8j**. Use functions to calculate in the appropriate cells the:
> - total number of hours worked each week
> - total hours worked by each person
> - total hours worked.
>
> Calculate in cells B13 and C13 the average number of hours worked by an employee that week. Calculate in cell B14 the largest and in cell B15 the least number of hours worked each week. Add appropriate labels to this data.
>
> Save the spreadsheet as **Task_8k**.

Advice

When reading the question stem, you will notice that it states: 'Use functions to calculate ...'. This means that, apart from its inefficiency, you cannot use a formula like =B6+B7+B8+B9+B10+B11 as this does not contain a function. If a question stem states 'Place a formula ...' then it is fine to use a function within the formula.

Advice

An alternative method is to enter **=SUM(** then drag the cursor to highlight cells B6 to B11, enter **)** and press the **<Enter>** key.

Open the spreadsheet. In cell B12 enter the formula =SUM(B6:B11). This should give the value 44. Replicate this formula (using the drag handle into cell C12). This should place the formula =SUM(C6:C11) in this cell and give the value 43. In cell D6 enter the formula =SUM(B6:C6) and replicate this formula down to cell D12. In this case we have used the SUM function rather than the formula =B6+C6 because it is likely that extra weeks may be added into the spreadsheet, in which case the SUM function would be more efficient. The value in cell D12 should be 87.

We will now try two alternative SUM functions in cell D12 that should also return the value 87. These are: =SUM(B12:C12) and =SUM(Hours). This works because in Task 8j we gave the name 'Hours' to all the cells in the range B6:C11.

Advice

An alternative way to use the SUM function without typing is for you to use **AutoSum**. Move the cursor into the required cell, select the **Home** tab and find the **Editing** group. Click on the **AutoSum** icon.

This will place the SUM function into the current cell and *Excel* will attempt to work out which cells to use; if they are not correct, drag the cursor to highlight the required cells.

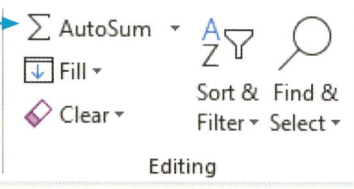

There are many ways of using the SUM function, some of which are shown in Table 8.1.

▼ Table 8.1 Ways of using the SUM function

Function	Equivalent formula	What it does
=SUM(B4:B8)	=B4+B5+B6+B7+B8	Adds up the contents of all the cells in the range B4 to B8
=SUM(D3,D8,D12)	=D3+D8+D12	Adds up the contents of the cells D3, D8 and D12
=SUM(D5:D8,F2)	=D5+D6+D7+D8+F2	Adds up the contents of the cells in the range D5 to D8 and the contents of cell F2
=SUM(MyRange)	None	Adds up the contents of all the cells within a named range, in this case called MyRange

AVERAGE function

To find the average number of hours worked, click the cursor into cell B13 and use the AVERAGE function to calculate the mean (average) of a list of numbers. Enter the formula =AVERAGE(B6:B11). This should give the value 7.333333. Replicate this function into cell C13 and add an appropriate label (like Average hours) in cell A13.

There are many ways of using the AVERAGE function, some of which are shown in Table 8.2.

▼ Table 8.2 Ways of using the AVERAGE function

Function	Equivalent formula	What it does
=AVERAGE(B4:B8)	=(B4+B5+B6+B7+B8)/5	Calculates the mean of the cells in the range B4 to B8
=AVERAGE(D3,D8,D12)	=(D3+D8+D12)/3	Calculates the mean of the cells D3, D8 and D12
=AVERAGE(D5:D8,F2)	=(D5+D6+D7+D8+F2)/5	Calculates the mean of the cells in the range D5 to D8 and cell F2
=AVERAGE(MyRange)	None	Calculates the mean of the cells in a named range called MyRange

MAX function

MAX is a function used to select the largest (maximum) value within a list of numbers. To calculate the largest number of hours worked over the two-week period, click in cell B14 and enter =MAX(Hours). This uses the named range 'Hours' and returns the value 14.

MAXA function

MAXA is a function used to select the largest (maximum) value within a list that may contain numbers, text and logical values (TRUE or FALSE). This function has the same syntax as the MAX function, for example =MAXA(Hours) will return the largest numeric value from the named range called 'Hours' even if that range were to contain text or logical values.

MIN function

MIN is a function used to select the smallest (minimum) value from a list. To calculate the least hours worked, select cell B15 and enter =MIN(Hours). This uses the named range 'Hours' and returns the value 2. Save this spreadsheet as Task_8k.

Activity 8f

Open the file **January.csv** as a spreadsheet. Place your name and filename in the header.

Insert new rows between rows 2 and 3 and between rows 1 and 2. Format the top of the spreadsheet to look like this.

	A	B
1	Tawara Motor Company	
2		
3	Number of days worked in January	
4		
5		Days
6	Pauline	26

Enter functions in cells B16 to B19 to calculate, using a named range, the total number of days worked by all employees, the average number of days worked, and the maximum and minimum values.

Save the spreadsheet as **Activity_8f**.

Whole numbers and rounding

INT function

In mathematics, the word '**integer**' is used to describe a whole number (with no decimals or fractions). In *Excel*, the **INT** function takes the whole number part of a number and ignores all digits after the decimal point, also known as truncating. Open a new spreadsheet and enter 84.56789 into cell A1. Enter in cell A2 the formula **=INT(A1)** which will return the value 84. Change the value in cell A1 to see the changing results in cell A2.

Advice

Setting a cell as an integer value will truncate the contents of a cell to remove the decimal/fraction part of the number. This is *not* the same as formatting a cell to 0 decimal places, which stops the decimal/fraction part from being displayed but not from being used in a calculation.

Rounding

Often a spreadsheet produces an extremely accurate answer, with more digits than we really need. Rounding involves taking a number and making it shorter, although it is still approximately equal to the original. This new value stored may be slightly more or slightly less than the original. For example: if a cell holds the value $23.486532, we may only need it to contain the number of dollars and cents. It is therefore a good idea to shorten and store this value to two decimal places to make it more meaningful. This is called rounding.

ROUND function

Task 8l

Open the file you saved as **Task_8k**. Insert a new row between rows 13 and 14. Copy the contents of cell A13 into cell A14. In cells B14 and C14, round the corresponding values in row 13 to one decimal place. Hide row 13. Save the spreadsheet as **Task_8l**.

Open **Task_8k**. Insert the new row between rows 13 and 14. Move the cursor into cell A14 and enter the forumla **=A13**. This takes the contents of A13 and copies it into A14. This function would allow the contents of A13 to change and A14 will change with it rather than the repeated use of copy and paste when cells are edited.

In cell B14, enter the formula **=ROUND(B13,1)**. This formula uses the **ROUND** function, which takes the content of cell B13 and rounds the number to one decimal place. Replicate this function into cell C14 using the drag handle. If the next digit is five or more, the number will be increased by one. For example in cell B13 the value is 7.3333333, so the content of B14 is 7.3 but in cell C13 the value is 7.166667, so the content of C14 is 7.2 as it has rounded up the first decimal place because the second was a 6. Hide (do not delete) row 13. Save the spreadsheet as **Task_8l**.

Rounding to zero digits is the same as using the INT function. Rounding with a positive number of digits rounds to that number of decimal places. Rounding with a negative number of digits rounds to the nearest 10, 100 and so on, as shown in Table 8.3.

▼ **Table 8.3** Effects of using the ROUND function

Number to be rounded	10928.375
=ROUND($B$1,3)	10928.375
=ROUND($B$1,2)	10928.38
=ROUND($B$1,1)	10928.4
=ROUND($B$1,0)	10928
=ROUND($B$1,-1)	10930
=ROUND($B$1,-2)	10900
=ROUND($B$1,-3)	11000

Advice
Rounding a cell will remove part of the number. This is not the same as formatting a cell (which will not remove part of the number). Rounding will affect any calculations based upon this cell.

ROUNDUP function

The **ROUNDUP** function behaves like ROUND, except that it always rounds a number up.

Task 8m
Create a new spreadsheet like this:

Use this spreadsheet to calculate the number of packs of bricks required to build a wall requiring 1100 bricks. Bricks can only be bought in whole packs of 390. Save the spreadsheet as **Task_8m**.

	A	B
1	Brick buying calculator	
2	Number of bricks required:	
3	Number of bricks in a pack:	390
4	Number of packs required:	

Open a new spreadsheet and create the spreadsheet as shown. Enter the value 1100 in cell B2. Enter in cell B4 the formula **=B2/B3** which returns the value 2.820513. In order to calculate the number of whole packs of bricks (as we cannot buy 0.820513 of a pack), we can use the ROUNDUP function. Edit the formula in B4 so that it becomes **=ROUNDUP(B2/B3,0)** which rounds up to 0 decimal places (a whole number) and now returns a value of 3. Save the spreadsheet as **Task_8m**.

ROUNDDOWN function

The **ROUNDDOWN** function behaves like ROUND and ROUNDUP, except that it always rounds a number down (this is also called truncating a number).

Task 8n

Create a new spreadsheet like this:

Use this spreadsheet to calculate the value of Pi (π) truncated to a number of decimal places.

Save the spreadsheet as **Task_8n**.

	A	B
1	ROUNDDOWN Pi	
2	Value of Pi:	
3	Number of places to round to:	
4	Truncated value:	

Open a new spreadsheet and create the spreadsheet as shown. Enter **=PI( )** in cell B2, which is the function to display the value of Pi (π). Enter the value 3 in cell B3. Enter in cell B4 the formula **=ROUNDDOWN(B2,B3)** which returns the value 3.141. Change the value in cell B3 to truncate Pi to a different number of decimal places. Save the spreadsheet as **Task_8n**.

Activity 8g

Open the file **Rounding.csv** in your spreadsheet. Edit it to look like this.

Enter a formula in cell B3 to calculate the whole number part of cell B2.

Enter formulas in cells B8 to B10 to round, round up and round down the contents of cell B6 to the number of decimal places specified in cell B7.

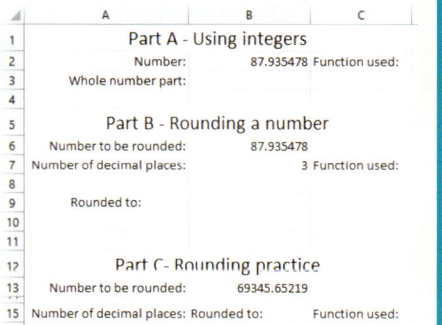

Enter formulas in cells B16 to B26 to round the contents of cell B13 to the number of decimal places specified in the corresponding cell in column A.

For each function entered in column B, place the name of the function in the corresponding cell in column C. Try different values in this spreadsheet to see the effects of the different functions. Save the spreadsheet as **Activity_8g**.

Counting

There are two main functions used to count the number of data items in a spreadsheet and two conditional counting functions.

Task 8o

Open the file **Hotels.csv** and format each day to look like this.

Place, in the appropriate cells, calculations for the number of hotels and, for each room type, the number of:
» rooms available
» hotels with available rooms
» hotels without available rooms.

Save the spreadsheet as **Task_8oText**.

Edit the data in cells in the range B5:D18 that contain the word None, so that these cells become blank cells. Edit the formula you placed in cell B22 so that it works for the new data.

Save the spreadsheet as **Task_8oBlanks**.

Open the spreadsheet and format it as shown in the Task. To calculate the number of rooms of each type available, place in cell B20 the formula =SUM(B5:B18) which will return the value 47. Use the drag handle to replicate this into cells C20 and D20.

COUNT function

COUNT looks at the cells within a given range and counts the number of these cells containing numeric data. To calculate the number of hotels with rooms of each type available, you must count the number of numeric cells. Place the cursor in cell B21 and enter the formula =COUNT(B5:B18) which will return the value 10. Use the drag handle to replicate this into cells C21 and D21. Note that the COUNT function ignores any cells containing labels or blank cells.

COUNTA function

The COUNTA function works in a similar way to the COUNT function. Rather than counting just the number of numeric values, this function counts the number of cells that display either a numeric value or text. It will not count any blank cells within the range. To calculate the number of hotels, move the cursor into cell B23. Enter the formula =COUNTA(B5:B18) which will return the value 14. It is not necessary to replicate this function as this data will not change for different room types.

In *Excel*, there is not a count function for just text values, so the COUNTA and COUNT functions will both be used to calculate the number of hotels with no room of this type available. Enter in cell B22 the formula =COUNTA(B5:B18)-COUNT(B5:B18) which will return the value 4. Use the drag handle to replicate this into cells C22 and D22. Save the spreadsheet as **Task_80Text**.

Highlight the cells in the range B5:D18 and from the **Home** tab, in the **Editing** section, select the drop-down menu using the **Find and Select** option.

Use the **Replace** option from the list and replace the text 'None' with a blank cell like this.

Click on the **Replace All** button to change the data.

COUNTBLANK function

The COUNTBLANK function works in a similar way to the COUNT and COUNTA functions. This function counts the number of blank cells within a given range. To re-calculate the number of hotels with no rooms of a particular type available, move the cursor into cell B22. Enter the formula =COUNTBLANK(B5:B18) which will return the value 9. Save the spreadsheet as **Task_80Blanks**.

Activity 8h

Open the file **Class.csv**. This spreadsheet lists all the students in a year, with their class and the grades scored in Maths and English. Format the top of the spreadsheet and each person's Grades box to look like this.

Format the bottom of the spreadsheet to look like this.

Place a formula in cell C102 to count the number of students in the class. Place a formula in cell C103 to count the number of grades gained by the whole year in both Maths and English. Place a formula in cell C104 to count the number of blank values in both Maths and English.

Fit the spreadsheet so that it would print on three A4 pages.

Save the spreadsheet as **Activity_8h**.

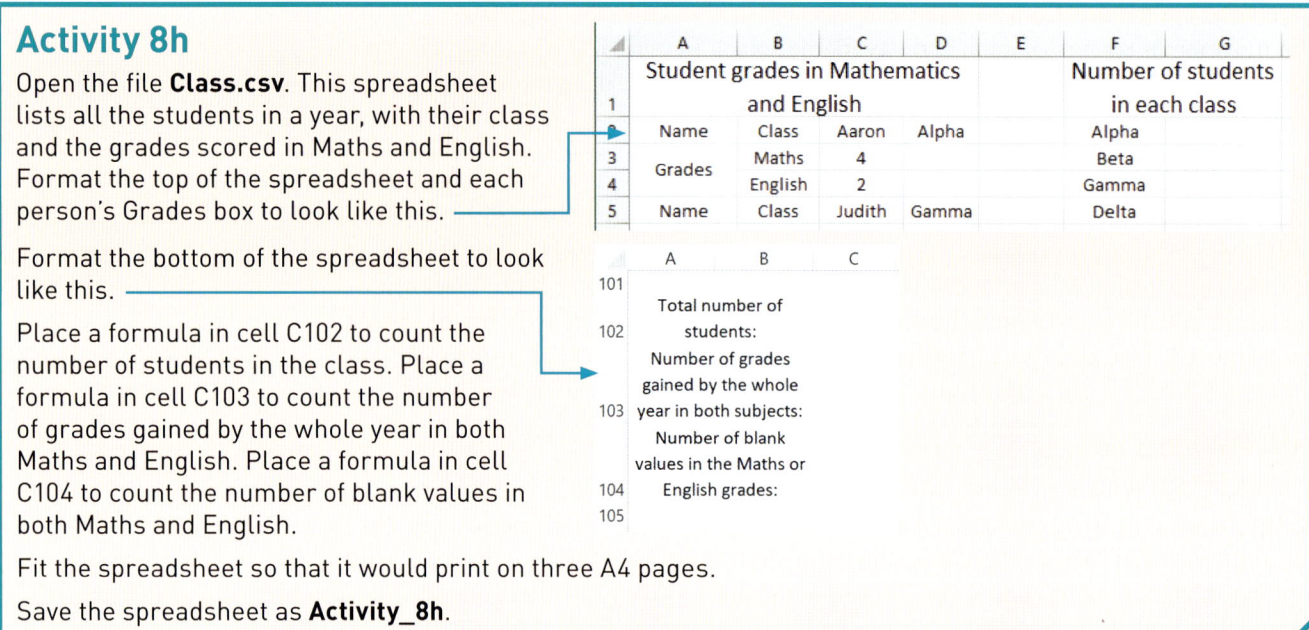

Nested functions

Nested functions involve placing one function inside another one. Sometimes a formula can contain several other functions nested within each other.

Task 8p

Open the file you saved as **Task_8k**.

Edit the formulas in row 13 to calculate the average number of hours worked each week, rounded to one decimal place.

Save the spreadsheet as **Task_8p**.

Study your solution to Task 8l, where you inserted a row, rounded the contents of the average calculation, then hid the row containing the unrounded data.

Open your solution to Task 8k. Edit the formula in cell B13 so that the AVERAGE function is nested within the ROUND function like this =ROUND(AVERAGE(B6:B11),1).

This is a much more efficient solution than Task 8l and calculates the average value first, then rounds this to one decimal place to give the value 7.3. Replicate this function into cell C13. Save the spreadsheet as **Task_8p**.

Activity 8i

Open the file you saved as **Activity_8f**.

Edit the formula in cell B17 to round the calculated average to two decimal places. Save the spreadsheet as **Activity_8i**.

Conditional functions

Conditional functions all test a condition and then return different answers or perform different operations depending on the value of that condition.

IF function

The **IF** function is the most commonly used conditional function. It looks at a given condition and performs an operation if the condition is met, or a different operation if the condition is not met. The operation could be to place a value or label in the cell, it could involve a reference to another cell, or it could involve a more complex calculation.

An IF function contains a pair of brackets and within the brackets three parts, each separated by a comma. An example of an IF function is =IF(A6=4,B6*0.04,"No discount offered"). The first part is a condition; in this example, it is testing to see if cell A6 contains the number 4. The other two parts are what to do if the condition is met and what to do if it is not met. In this example, if the condition is met, the contents of cell B6 will be multiplied by 0.04 and the answer will be displayed. If the condition is not met, 'No discount offered' will be displayed.

> ## Task 8q
>
> Open the file saved as **Task_8oText**.
>
> Add a new label **Family availability** in cell E3. Place formulas in cells E5 to E18 to display **Family rooms available** or **Not available** depending upon the contents of column D.
>
> Save the spreadsheet as **Task_8q**.

Open the file and place the cursor in cell E3. Enter the label 'Family availability'. Place the cursor in cell E5 and enter the formula =IF(D5="None","Not available","Family rooms available").

Do not use absolute referencing in this formula as the reference to cell D5 needs to change when you replicate the formula. Replicate this formula so it is copied into cells E5 to E18. Save your spreadsheet as **Task_8q**.

There are many ways of using the IF function, some of which are shown in Table 8.4.

In a spreadsheet we wish to enter formulas into cell G1.

▼ **Table 8.4** Ways of using the IF function

Function	What it does
=IF(B6="C","Car","Not a car")	Checks the contents of cell B6. If the contents match the letter "C" with no additional spaces, etc., then cell G1 displays "Car". If cell B6 contains anything other than the letter "C", then cell G1 displays "Not a car".
=IF(B7=3,"March","Not March")	Checks the contents of cell B7. If the contents are the value 3, then cell G1 displays "March". If cell B7 contains anything other than the value 3, then cell G1 displays "Not March".
=IF(B7<4,"1st Quarter","")	Checks the contents of cell B7. If the contents are less than the value 4, then cell G1 displays "1st Quarter". If cell B7 contains a value of 4 or more, then cell G1 displays a null string (no contents are shown).
=IF(B7<=3,"1st Quarter","")	Checks the contents of cell B7. If the contents are less than or equal to 3, then cell G1 displays "1st Quarter"; if not, cell G1 displays a null string.
=IF(OR(B7=4,B7=5,B7=6),"2nd Quarter","")	Checks the contents of cell B7. If the contents are any of the values 4, 5 or 6, cell G1 displays "2nd Quarter"; if not, cell G1 displays a null string.
=IF(AND(B7>3,B7<7),"2nd Quarter","")	Checks the contents of cell B7. If the contents are greater than 3 AND less than 7, cell G1 displays "2nd Quarter"; if not, cell G1 displays a null string.
=IF(NOT(OR(B7<4,B7>6)),"2nd Quarter","")	Checks the contents of cell B7. If the contents are NOT less than 4 OR NOT greater than 6, cell G1 displays "2nd Quarter"; if not, cell G1 displays a null string.

IFS function

The **IFS** function looks at a number of given conditions and performs a different operation if one of these conditions is met. If none of the conditions are met it returns an error message, so careful use of error trapping may be needed with this function (we will look at error trapping later in this Task). This is just a more efficient method of nesting IF statements.

> ### Task 8r
> Open the file **WorkLocation.csv** and for each employee, in column C, identify where they are working, given that project Alpha is in London, project Beta is in Milan and project Gamma is in Tunis.
>
> Save the spreadsheet as **Task_8r**.

Open the file and place the cursor in cell C2. Enter the formula:
=IFS(B2="Alpha","London",B2="Beta","Milan",B2="Gamma","Tunis")

Do not use absolute referencing in this formula as the references to cell B2 need to change when you replicate the formula. Replicate this formula so it is copied into cells C3 to C11.

Error trapping

ISERROR function

The ISERROR function is used to detect if an error will occur when a formula is used, and returns True if there is an error or False otherwise. Although this can be used to detect errors, there is a better alternative that detects the error and allows us to trap the error so that it does not appear as *Excel* would generate it. The IFERROR function can be used to indicate the error or in some cases correct the error.

IFERROR function

When this formula is replicated an error occurs in cells C10 and C11 as the three conditions in the IFS function are not met. Rather than allow Excel to display this error, in this case as #N/A, we can add error trapping to the function. For this we will include an IFERROR function in each of the replicated cells. Move the cursor into cell C2 and edit the formula so that it is: =IFERROR(IFS(B2="Alpha","London",B2="Beta","Milan",B2="Gamma","Tunis"),"Project location unknown")

This has used the original IFS function, highlighted in yellow, and placed this within an error trapping routine. Replicate this formula so it is copied into cells C3 to C11. Rather than the generated #N/A error message, cells C10 and C11 now display 'Project location unknown'.

Save your spreadsheet as **Task_8r**.

Conditional functions

There are many conditional functions that can be used in *Excel*. Many involve the use of the IF (or IFS) function together with a mathematical function like COUNT, AVERAGE, MAX, MIN. We will study some of these but will not cover all of these functions. If you understand the structure of the IF and IFS functions and the mathematical functions, then other functions like COUNTIFS, MAXIF, MINIF, MINIFS and so on can be used in the same way as the examples shown.

Conditional counting

COUNTIF function

If conditional counting is required with a single condition, then **COUNTIF** is the most efficient function to use.

> ### Task 8s
> Open the file saved as **Task_8q**. In cells B22 to D22, enter single more efficient functions to calculate the number of hotels with no room of that type available.
>
> Save the spreadsheet as **Task_8s**.

Open the file saved as **Task_8q**. The current method counted the number of cells containing any data or label, then subtracted the number of cells that contain a number. It is more efficient to use a single COUNTIF function which will count the number of times a condition is met. Place the cursor in cell B22 and enter **=COUNTIF(B5:B18,"None")**. This function will check each cell in the range B5 to B18 and will only count those that contain the text/string "None". Replicate this function into cells C22 and D22 using the drag handle. Save your spreadsheet as **Task_8s**. The formula/method used in Task 8q would be valid but at AS level you would need to have used COUNTIF as the use of the most efficient formula/method is essential.

> ### Activity 8j
> Open the file that you saved as **Activity_8h**. This spreadsheet lists all the students in a class. Next to each student's name is the name of the class that they are in.
>
> Place formulas in cells G2 to G5 that use both absolute and relative referencing to count the number of students in each class.
>
> Set the print area of the spreadsheet to show cells F1 to G11 only.
>
> Save the spreadsheet as **Activity_8j**.

COUNTIFS function

If conditional counting is required with two or more conditions, then the **COUNTIFS** function offers us the most efficient solution.

> ### Task 8t
> Open the file saved as **Task_8s**. Enter in cell A24 the text **Number of hotels with no available rooms**.
>
> Apply appropriate formatting to this text. Enter in cell B2 a function to perform this calculation. Edit the data as five double rooms have now been booked at the Kaiserhof Hotel.
>
> Save the spreadsheet as **Task_8t**.

Open the file saved as **Task_8s**. Enter into cell A24 the label 'Number of hotels with no available rooms'. Use the format painter tool to format this text to match cells A5 to A23. Enter into cell B24 the formula **=COUNTIFS(B5:B18,"None",C5:C18,"None",D5:D18,"None")**, which will check the first cell in the range B5 to B18 and if it contains the string "None", will move on to check the cell in the same row within the range C5 to C18. If this cell also contains the string "None", it will then check in the same row within the range D5 to D18 and will only count if it also contains the string "None", before moving on to test the next row. It does this until it reaches cell D18. This function should return the value 0 to start with, as no hotel is full. Follow the instruction to edit the data because five double rooms have now been booked at the Kaiserhof Hotel by entering in cell C15 the string "None". This will now change the resulting value in cell B24 to 1.

Activity 8k

Open the file that you saved as **Activity_8j**.
Enter the following text into the spreadsheet and format it like this.

Place formulas in cells C105 to C109 to display the relevant values.

Set the print area of the spreadsheet to show cells A105 to C109 only.

Save the spreadsheet as **Activity_8k**.

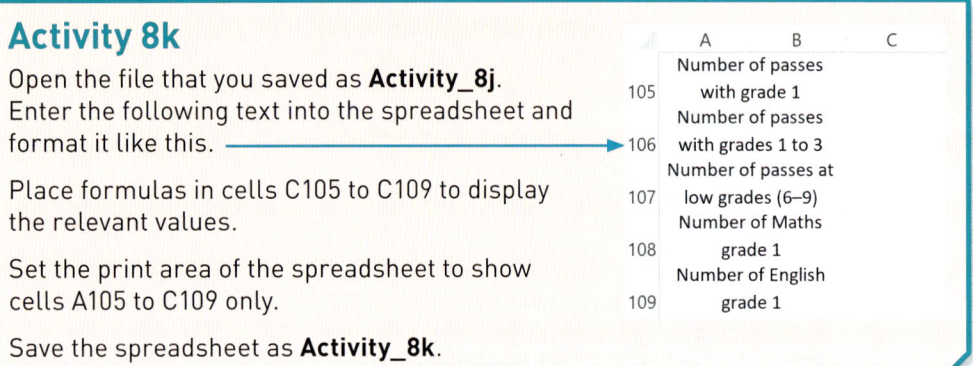

Task 8u

Open the file **TWTWorkers.csv** in a spreadsheet. Edit and format the spreadsheet to look like this.

Enter functions into each green cell to calculate the required values, rounded to the nearest whole number.

Save the spreadsheet as **Task_8u**.

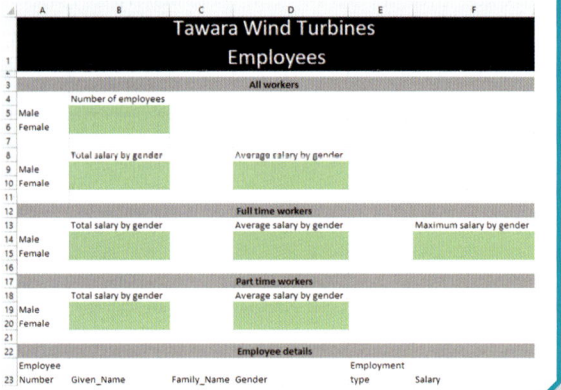

Open the file and format it as shown in the question using the skills learnt earlier in the chapter. Enter in cell B5 the formula =COUNTIF($D$24:$D$52,A5), which counts any cells in the range D24 to D52 that contain the same as the contents of cell A5, in this case 'Male'. Because we have used the correct absolute and relative cell references we can use the drag handle to replicate this down into cell B6.

Advice

In *Excel*, the <f4> key toggles between absolute and relative referencing. To use it, highlight the cell or range to be changed (for example in the formula in cell B5, highlight the range $D$24:$D$52) and press the <f4> key until the correct referencing is found. This will toggle between $D$24:$D$52, D$24:D$52, $D24:$D52 and D24:D52.

Conditional sum

SUMIF function

If a conditional total is required using a single condition, then the **SUMIF** function offers the most efficient solution to a task.

Enter in cell B9 the formula =SUMIF($D$24:$D$52,A9,$F$24:$F$52), which compares the contents of each cell in the range D24 to D52 with the contents of cell A9, in this case 'Male'. If these values or strings match, then the value contained in the corresponding cell in the row within the range F24 to F52 is added to the total. Use the drag handle to replicate this into cell B10.

SUMIFS function

If a conditional total is required with two or more conditions, then the **SUMIFS** function offers the most efficient solution to a task.

The formulas needed for cells B14 and B15 both require two conditions to be met. For B14 the employee has to be male and work full time. The syntax used for SUMIFS, compared to that used for SUMIF, is very different. With SUMIFS, the first range given will be the range containing the cells to be added together. The next part of the function will be a range of cells to be tested and also the first condition to be tested. Then will come another range of cells to be tested followed by the second test condition. This is repeated for all the conditions to be tested.

Our Task only requires two conditions to be met, so will consist of the range to be summed and two further ranges and their associated criteria. Enter into cell B14 the formula
=SUMIFS($F$24:$F$52,$D$24:$D$52,A14,$E$24:$E$52,"Full").

This adds the value of each row held within the range F24 to F52 to the total, only if both these criteria are met:

» the contents of the cell (in this row) in the range D24 to D52 contain the same as cell A14
» the contents of the cell (in this row) in the range E24 to E52 contain the label "Full".

Use the drag handle to replicate this into cell B15.

Similar formulas are required in cells B19 and B20 to calculate the total salary of part-time male and part-time female workers. Copy the formula from cell B14 and paste it into cell B19. Edit this formula so it becomes
=SUMIFS($F$24:$F$52,$D$24:$D$52,A14,$E$24:$E$52,"Part").
Use the drag handle to replicate this into cell B20.

Conditional average
AVERAGEIF function

If a conditional mean (average) is required using a single condition, then the **AVERAGEIF** function offers the most efficient solution to a task.

Enter into cell D9 the formula
=AVERAGEIF($D$24:$D$52,A9,$F$24: $F$52), which works in a similar way to SUMIF, except the calculated part gives the mean average rather than the total of all cells where the condition is met. An efficient method of doing this is to copy and paste the formula from B9, then edit the function name.

The resulting value in D9 is 18636.36364. This needs to be rounded to no decimal places, so amend the formula to become:
=ROUND(AVERAGEIF($D$24:$D$52,A9,$F$24:$F$52),0).

Use the drag handle to replicate this into cell D10.

AVERAGEIFS function

If a conditional mean (average) is required with two or more conditions, then the **AVERAGEIFS** function offers the most efficient solution to a task.

Copy the function from B14, paste it into D14 and edit it to become
=ROUND(AVERAGEIFS($F$24:$F$52,$D$24:$D$52,A14,
$E$24:$E$52,"Full"),0). Use the drag handle to replicate this into cell D15.

Similar formulas are required in cells D19 and D20 to calculate the average salary of part-time male and part-time female workers. Copy the formula from cell D14 and paste it into cell D19. Edit this formula so it becomes
=ROUND(AVERAGEIFS($F$24:$F$52,$D$24:$D$52,
A19,$E$24:$E$52,"Part"),0). Use the drag handle to replicate this into cell D20.

Conditional maximum
MAXIFS function

If a conditional maximum is required with two or more conditions, then the MAXIFS function offers the most efficient solution to a task.

Copy the function from B14, paste it into F14 and edit it to become:
=ROUND(MAXIFS($F$24:$F$52,$D$24:$D$52,A14,
$E$24:$E$52,"Full"),0). Use the drag handle to replicate this into cell F15. Save the spreadsheet as **Task_8u**.

Activity 8l

Open the file that you saved as **Activity_8k**.

Add appropriate functions in cells G2 to G5 to count the number of students in each class. Set the print area of the spreadsheet to show cells F1 to G5 only. Save the spreadsheet as **Activity_8l**.

Advice

If you want to wrap text within a cell so that it wraps at a particular point, click the cursor at that point in the text, then hold down <Alt> and press <Enter>.

Activity 8m

Open the file **Gradebook.csv**.

Format the top of the spreadsheet to look like this.

	B	C	D	E	F	G	H
1		Average grade attained by each class in each subject					
2		English	Maths	Science	History	Geography	IT
3	Alpha						
4	Beta						
5	Gamma						
6	Delta						

Add appropriate functions to cells C3 to H6 for the average grades for each subject for each class, rounded as whole numbers. Set the print area of the spreadsheet to show cells B1 to H6 only.

Save the spreadsheet as **Activity_8m**.

Mathematical averages

We have already used the AVERAGE function to calculate the mean value and calculated conditional average calculations using the mean. There are also functions used to calculate the values for the median and mode.

Task 8v

Open the file **Test.csv** in a spreadsheet. Format the spreadsheet to look like this.

Fourteen students have yet to sit the test. Enter functions into cells C2 to C4 to calculate each type of average score. Round the mean score to one decimal place.

Save the spreadsheet as **Task_8v**.

	A	B	C
1	Average test scores		
2		Mean	
3		Median	
4		Mode	
5			
6	Test scores		
7	First_Name	Family_Name	Score
8	Caitlin	Akhtar	1

Open the file and format it as shown in the question using the skills learnt earlier in the chapter. Enter into cell C2 the formula =ROUND(AVERAGE($C$8:$C$80),1), which will return the rounded value 5.5. The range has been extended from C66 to C80 as 14 students have yet to sit the test.

MEDIAN function

The **MEDIAN** function is used to find the middle value when all data items are listed in ascending order. Enter in cell C3 the formula =MEDIAN($C$8:$C$80), which will return the value 6.

MODE function

The **MODE** function is used to find the piece of data (in this case test score) that occurs most frequently. Enter in cell C4 the formula =MODE($C$8:$C$80), which will return the value 7 as more students scored 7 on the test than any other score.

Nested IF functions

Sometimes nested IF functions are required to display one of several different possible results, which are dependent on different conditions that have been tested. It is important for you to work in a logical order. You may have to use three or more conditions involving numerical values, so work through the smallest to the largest numerical values, or vice versa. For example, exam results may be graded fail (less than 40%), pass (40% or more), merit (60% or more), or distinction (80% or more). A sensible approach would be to deal with 40 first, 60 next and finally 80. It would be extremely difficult to start with 60. Do *not* start with middle values; this will give incorrect results.

Task 8w

Open the file you saved as **Task_8v**. Place formulas in cells D8 to D66 that display:
» **Excellent** if the test score was 8 or more
» **See your tutor** if the test score was less than 4
» **More revision needed** for any other test score.

Save the spreadsheet as **Task_8w**.

Before starting this task on the spreadsheet, reorder the list so that the conditions follow a logical pattern, for example lowest to highest like this:

» Score <4 then **See your tutor**
» Score <8 then **More revision needed**
» Score 8 or more then **Excellent**.

Open the file and enter in cell D8 the formula =IF(C8<4,"See your tutor",IF(C8<8,"More revision needed","Excellent")).

If the condition is TRUE then the first result is produced ("See your tutor" is displayed in the cell) but if it is FALSE it moves on to the second result which is in the form of another IF condition. Be very careful to get the brackets correct; each condition has one open and one close bracket. When you work through this formula, it checks whether the value is less than 4 first; if so, it displays the matching text. Then if it is not true, it checks if the value is less than 8 next; if so, it displays the matching text. As the only other possible result is that the value must be greater than or equal to 8, the remaining piece of text ("Excellent") is displayed. There is, therefore, no need to include another condition (IF C8>=8, "Excellent"). Replicate this formula into cells D9 to D66. An alternative method for this would be to use the IFS function studied earlier in the chapter. Save the spreadsheet as **Task_8w**.

Activity 8n

Open the file you saved as **Activity_8i**.

Edit and format the top of the spreadsheet to look like this.

	A	B	C
1	Tawara Motor Company		
3	January work patterns		
5		Days	Notes
6	Pauline	26	
7	Gwynn	5	

Place formulas in cells C6 to C14 that display:
» **Too many days** if they worked 23 days or more
» **See manager** if they worked less than 19 days
» **Within tolerance** for any other number of days.

Save the spreadsheet as **Activity_8n**.

SUBTOTAL function

SUBTOTAL is a function that allows a number of functions to be used in one cell. The first variable is a number between 1 and 10, which tells it which function to use. The remaining part of the function contains the range of cells this function is applied to. For example, if the function =SUBTOTAL(1,A10:A14) is used, it will calculate the AVERAGE of the range A10 to A14. If this function was changed to =SUBTOTAL(9,A10:A14), it would calculate the SUM of this range. There are many ways of using the SUBTOTAL function, some of which are shown in Table 8.5.

▼ **Table 8.5** SUBTOTAL function codes and their use

Code	Function
1	AVERAGE
2	COUNT
3	COUNTA
4	MAX
5	MIN
6	PRODUCT
7	STDEV
8	STDEVP
9	SUM
10	VAR

Simple lookup functions

The term 'lookup' means to look up from a list. There are four functions that can be used for this. These are: **LOOKUP, HLOOKUP, VLOOKUP** and **XLOOKUP**. Data may be 'looked up' (referenced) from the same worksheet or workbook or from an external data file. Using external data files is good practice in industry, but it is important that you do not edit the external data files (unless instructed to).

LOOKUP function

The list of data that is looked up is contained in a range, often consisting of two columns or two rows. LOOKUP is used to look up information using data in the first row or the first column of the range of cells and returns the value contained in the cell corresponding to the matched value in the first row or column. This is probably the least useful of the three formulas, especially as the VLOOKUP, HLOOKUP and XLOOKUP functions offer the user more flexibility.

HLOOKUP function

HLOOKUP is a function that performs a horizontal lookup of data. This should be used when the values that you wish to compare your data with are stored in one row. The values to be looked up are stored in corresponding cells in the rows below it.

> ### Task 8x
> Open and examine the file **TMCVehicles.csv**. Place formulas in the 'Type' column to return the name of the type of each vehicle.
>
> Save the spreadsheet as **Task_8x**.

Open the file. If a question asks you to examine the file, please do so carefully; in this question it is very important to recognise that the codes in row 1 are not in alphabetical order.

Enter into cell C6 the formula =HLOOKUP(B6,$B$1: $H$3,3,0). This formula looks up and compares the contents of cell B6 with the contents of each cell from left to right in the top (horizontal) row of the range B1 to H3. Note that we have not included cells A1 to A3 and A5 in this range, as these are row headings and not part of the data to be looked up. When it finds a match, the ',3' instructs *Excel* to take the label (or value) stored in the third row, which is directly under the matched cell. The final attribute, in this example, the '0' (or 'False') instructs *Excel* to find an exact match in the top row. This is critical in data which is not sorted into order. Replicate your formula down to C13. Save the spreadsheet as **Task_8x**.

	A	B	C	D	E	F	G	H
1	Code	A	B	C	D	T	E	F
2	Wheels	4	2	2	4	6	4	4
3	Name of type	Car	Motorbike	Scooter	Van	Truck	ATV	Tractor
4								
5	Make	Tcode	Type					
6	T-Swift 3	A						
7	T-Jinmee	A						
8	T-Racer	B						
9	T-Bee	E						
10	T-Haul	T						
11	T-Plough	F						
12	T-Court	A						
13	T-Bone	C						

> ### Advice
> The alternative for this final attribute is a '1' (or 'True') which does not give correct results in unsorted data, but gives an approximate match rather than an exact match. This is the default value if the final '0' (or 'False') is missing. Try changing this attribute in all formulas from B6 to B13 in **Task_8w** and study the results. You will notice that the result for row 10 shows Tractor rather than Truck.

VLOOKUP function

VLOOKUP is a function that performs a vertical lookup of data. This should be used when the values that you wish to compare your data with are stored in a single column. The values to be looked up are stored in the columns to the right of these cells. The lookup data can be stored either in the same file or in a different file.

Task 8y

Open and examine the file **Orders.csv**. Place formulas in the 'Customer name' column to return the name of the customer for each order, using the file **Client.csv**.

Save the spreadsheet as **Task_8y**.

Open and examine the file **Client.csv**. You will notice that client codes are vertically listed but are not sorted into order. The file should look like this.

	A	B
1	ClientCode	Client
2	R	Rootrainer
3	Q	Quattichem
4	H	Hothouse Design
5	A	Avricom
6	V	Binnaccount
7	L	LGY
8	S	Rock ICT
9	T	Tawara Motor Company
10	U	Tawara Academy

Leave this file open. Open the file **Orders.csv**, format the column width to look as shown below and place the cursor in cell C3. Because the data in the client file is stored in vertical columns, a VLOOKUP is the most appropriate formula. Enter the formula =VLOOKUP(B3,Client.csv!$A$2:$B$10,2,FALSE) into this cell. This formula will look up and compare the contents of cell B3 with the contents of each cell in the left (vertical) column of the range A2 to B10 within the file Client.csv. When entering this formula, you can add the yellow highlighted section of the formula by moving the cursor into this file and dragging it to highlight all of the cells in both columns, so it includes the lookup value and the result. The '2' in the formula tells *Excel* to look in the second column of this range. The 'False' (or '0') condition forces an exact match. When it finds a match, it will take the value or label stored in the second column which is to the right of the matched cell. Replicate this formula into cells C4 to C22. Save the spreadsheet as **Task_8y**. The first few results should look similar to this.

	A	B	C	D
1			Orders	
2	Order number	Code	Customer name	Amount
3	700001	U	Tawara Academy	632.87
4	700002	V	Binnaccount	969.3
5	700003	S	Rock ICT	706.72

Activity 8o

Using suitable spreadsheet software, open and examine the data in the files **AlpHotels.csv** and **Rooms.csv**.

Open the file **RoomRates.csv** and save this as a spreadsheet. Row 1 contains the room type code and column A contains the hotel code. Enter formulas in appropriate cells in row 2 to display the room type for each code.

Enter formulas in the appropriate cells to display, in column:
» B the name of each hotel
» C the address of each hotel
» D the zip code for each hotel.

Merge cells B2 to D2. Hide column A and row 1. Save the spreadsheet as **Activity_8o**.

XLOOKUP function

XLOOKUP is a new function in Excel that can be used to perform either a horizontal or a vertical lookup of data. This is similar to HLOOKUP and VLOOKUP but is more powerful and flexible than either of these. It will also reference data stored in rows/columns before the lookup value. It therefore allows backward referencing within an array. The values to be looked up can be stored to either the right or left or above or below the lookup array. The lookup data can be stored either in the same file or in a different file.

Advice
XLOOKUP and the following Task can only be attempted with the latest versions of *Excel*. If you have older software, skip this section and move on to Advanced lookup functions.

Task 8z
Open the file **Tasks.csv**. Insert formulas in the CurrentTask column to look up the client, using the TaskCode for the lookup value and the file **Client1.csv**. Make sure that you use both absolute and relative referencing within your function.

	A	B
1	Client	TaskCode
2	Rootrainer	1
3	Quattichem	2
4	Hothouse Design	3
5	Avricom	4
6	Binnaccount	5
7	LGY	6
8	Rock ICT	7

Open and examine the file **Tasks.csv** and click the left mouse button to place the cursor in cell C3. The Task instructs you to use the file **Client1.csv** for the lookup. Open this file in a new spreadsheet. Examine the layout of this file to decide which type of lookup formula to use. **Client1.csv** looks like this.

At first glance, because it is stored with the lookup data in vertical columns, a VLOOKUP may appear the correct function, but the TaskCode column is to the right of the Client column and VLOOKUP cannot do backwards referencing. Therefore, the most appropriate function to use for this task would be XLOOKUP. Enter the formula =**XLOOKUP(B3,Client1.csv!$B$2:$B$8, Client1.csv!$A$2:$A$8,"Not found",0,1)** into this cell. This formula will look up and compare the contents of cell B3 with the contents of each cell in the range B2 to B8 within the file **Client1.csv**. When entering this formula, you can add the yellow highlighted section of the formula by moving the cursor into this file and dragging it to highlight all the cells in this range, so it includes only the TaskCodes (the lookup array). You can also add the green highlighted section of the formula by moving the cursor into this file and dragging it to highlight all the cells in this range, so it includes only the names of the clients (the return array). The text "Not found" is displayed if the TaskCode is not found during the look up. The number 0 is the match mode, the 0 tells the function that we want an exact match only. The final number 1 instructs the function to search through the items first to last as the TaskCodes are in ascending order in the source file. When it finds a match, it will take the value or label in the return array which is the range A2:A8. Replicate this formula into cells C4 to C24. Save the spreadsheet as **Task_8z**. The first few results should look similar to this.

	A	B	C
1	Current client list		
2	Name	TaskCode	CurrentTask
3	Laila Aboli	6	LGY
4	Greg Mina	4	Avricom
5	Sri Paryanti	6	LGY
6	Bishen Patel	6	LGY
7	Rupinder Singh	3	Hothouse Design
8	Sergio Gonzalez	5	Binnaccount
9	Rupinder Vas	1	Rootrainer
10	Bryan Revell	1	Rootrainer
11	Henri Ramos	7	Rock ICT

Advanced lookup functions

Simple lookup functions enable many solutions but have limitations. You have to work out whether vertical or horizontal is required and you can only look down a list (in VLOOKUP) or to the right (in HLOOKUP) or either way using XLOOKUP. Other more advanced lookup solutions such as **INDEX** and **MATCH** can work with both horizontal and vertical ranges of data. INDEX and MATCH together become a nested function, which makes this type of solution more complex, although it offers greater flexibility. We will first look at each function individually, then how to combine them for a two-way lookup.

INDEX function

The INDEX function returns values from a given location in a table, where the user specifies the row and column position of the item in the table.

Task 8aa

Tawara's top U17 athletes have been timed for the 100 m sprint. The results have been stored in **Athletics.csv**. Open and examine this file, save it as a spreadsheet and format it like this.

The sporting directors are choosing the athletes for the 100 m relay teams. They have decided to use the fastest runner in position 4, then the next fastest in position 1, then the next fastest in position 2 and the fourth fastest in position 3. The fifth and sixth fastest runners will be selected as reserves. Enter formulas in the appropriate cells so that the teams (in running order) and reserves are displayed.

Save the spreadsheet as **Task_8aa**.

Open and examine the file **Athletics.csv**; you will notice that lists of athletes are split into male and female and ordered with the fastest athletes at the top. Format the spreadsheet as shown.

Enter in cell B6 the formula **=INDEX($A$16:$C$40,2,2)**, which looks in the range of cells between A16 and C40, finds the second row in that range, then returns the value held in the second column, which in this case is the name of Jade Naylor. Jade was the second fastest runner, so will run first in the relay race. In cell B7, a similar formula **=INDEX($A$16:$C$40,3,2)** finds the third row in that range and returns the second cell within that row, which returns 'Paige Rice'. Copy, paste and edit these formulas for all 12 athletes. The formulas and results should look like this.

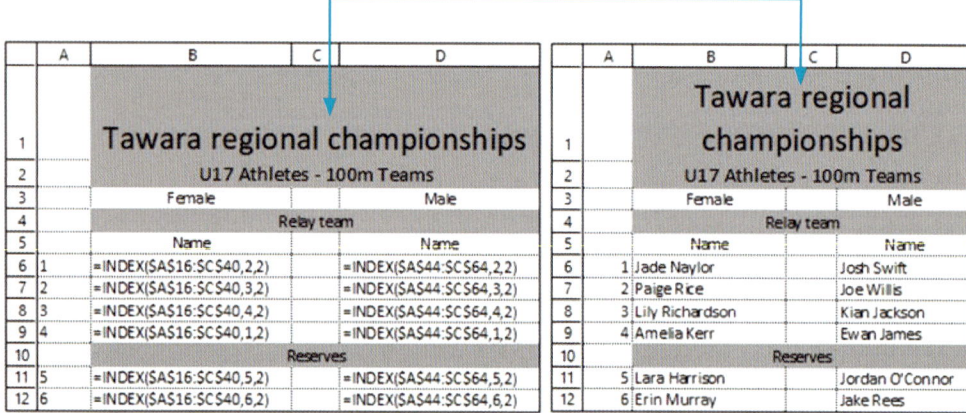

Save the spreadsheet as **Task_8aa**.

MATCH function

The MATCH function searches for a specified item in a range of cells and returns the relative position of that item in the range.

Task 8ab

Open and examine the file **SkiRace.csv**. Skiers finish the course as fast as possible but go down the course one at a time.

Open and examine this file, save it as a spreadsheet and format it like this.

Without changing the data, display the data for the skier who had:
» a starting position of number 64
» finished first.

Save the spreadsheet as **Task_8ab**.

	A	B	C	D	E
1		Locate a skier using:			
2	Starting order:			Finishing position:	
3	List reference:			List reference:	
4	Entry number:			Starting position:	
5	First_Name:			Entry number:	
6	Family_Name:			First_Name:	
7	Finishing position:			Family_Name:	
8					
9	Entry number	Starting order	First_name	Family_Name	Finishing position
10	AT00001	30	Henry	Gale	111

Open and examine the data in the file. The reference position in the list of skiers needs to be located, as the order of the list will not change. Enter the test data **30** into cell B2, as this is the first item in the list to be referenced. Enter in cell B3 the formula **=MATCH($B$2,$B$10:$B$165,0)**, which matches the contents of cell B2 (in this case the test data 30) with the first 30 it finds within the range B10 to B165. The final ',0' instructs *Excel* to find an exact match. This test data gives a list reference of 1 as this is the first position in the list. Change cell B2 to 77, which returns a list reference of 2.

Place in cell E3 the formula **=MATCH($E$2,$E$10:$E$165,0)** and **111** as test data in cell E2, which will return a value of 1 (the first item in the list).

Advice

The alternatives for this final match-type attribute is a 1 for **Less than**, if an exact match cannot be made, or -1 for **Greater than**. Try changing the test data in B2 to 30.6 where an exact match cannot be found, then change this attribute to 1, then -1 and study the results. You will notice that if the data is not in ascending order with the 1 attribute it gives errors, and if the data is not in descending order with the -1 attribute it gives errors.

INDEX and MATCH together

When INDEX and MATCH are used together, they allow us to look up a value in a table using both rows and columns. We will study that later, but first let us get used to using the functions together.

In a copy of Task 8ab, enter in cell B4 the formula **=INDEX($A$10:$E$165, MATCH($B$2,$B$10:$B$165,0),1)**, which nests a MATCH function within the INDEX function. The formula looks within the range A10 to E165, using the row position (list reference derived using the MATCH function, highlighted here in yellow) and the column position set to the value 1. The yellow portion is the function placed in cell B3 to work out the row position within the data.

Advice

For this Task we could have placed the formula **=INDEX($A$10:$E$165,B3,1)** in B4, which would give the same result; however this is often less efficient as it uses two cells to perform the task rather than one.

Place similar formulas in cells B5 to B7 and in E4 to E7 so that they look like this:

	A	B		D	E
3	List reference:	=MATCH($B$2,$B$10:$B$165,0)		List reference:	=MATCH($E$2,$E$10:$E$165,0)
4	Entry number:	=INDEX($A$10:$E$165,MATCH($B$2,$B$10:$B$165,0),1)		Starting position:	=INDEX($A$10:$E$165,MATCH($E$2,$E$10:$E$165,0),2)
5	First_Name:	=INDEX($A$10:$E$165,MATCH($B$2,$B$10:$B$165,0),3)		Entry number:	=INDEX($A$10:$E$165,MATCH($E$2,$E$10:$E$165,0),1)
6	Family_Name:	=INDEX($A$10:$E$165,MATCH($B$2,$B$10:$B$165,0),4)		First_Name:	=INDEX($A$10:$E$165,MATCH($E$2,$E$10:$E$165,0),3)
7	Finishing position:	=INDEX($A$10:$E$165,MATCH($B$2,$B$10:$B$165,0),5)		Family_Name:	=INDEX($A$10:$E$165,MATCH($E$2,$E$10:$E$165,0),4)

Now change B2 to contain **64** and E2 to contain **1**. Save the spreadsheet as **Task_8ab**. The spreadsheet should look like this.

	A	B	C	D	E
1		Locate a skier using:			
2	Starting order:	64		Finishing position:	1
3	List reference:	43		List reference:	19
4	Entry number:	AT00085		Starting position:	39
5	First_Name:	Steve		Entry number:	AT00034
6	Family_Name:	Walsom		First_Name:	Ieuan
7	Finishing position:	4		Family_Name:	Williams

Task 8ac

Open and examine the file **CityTemp.csv**. This shows the average temperature in some American towns and cities for each month.

Use this to create a spreadsheet so that the name of the city and the month are entered into cells B3 and B4 and the average temperature for that time and place are returned in cell B5.

Use this to display the temperature in Denver in February.

Save the spreadsheet as **Task_8ac**.

Open and examine the data in the file. Place test data 'Akron' (which is a short name so is quick to type and near the top) in cell B3 and 'Jan' in B4. Enter in cell B5 the formula =INDEX(B8:M253,MATCH(B3,A8:A253,0), MATCH(B4,B7:M7,0)), which nests two MATCH functions within the INDEX function. It looks within the range B8 to M253, using the row position (list reference) highlighted in yellow and the column position highlighted in green. The row position (yellow) matches the contents of cell B3 with the names of the cities in the range A8 to A253 and the ',0' forces an exact match. The column position (green) matches the contents of cell B4 with the abbreviated names of each month in the range B7 to M7; again the ',0' forces an exact match. The value returned for the test data should be -3.8 degrees. It is important to get the order of this correct with the row positions before the column positions. Change the data so that B3 contains 'Denver' and B4 contains 'Feb'. Save the spreadsheet as **Task_8ac**.

Activity 8p

Open the file **TGCosts.csv** and save this as a spreadsheet. Format the spreadsheet to look like this.

Two lookup tables have been provided, so make sure you select the correct table for this activity. The length and width of the required granite are entered into cells B4 and B5 by the user. The price will be displayed in cell B6. If a customer wants granite that is smaller than one of the stock sizes, it will be cut from the nearest size and the price for that size will be charged.

Use the spreadsheet to calculate the price of granite with a length of 262 and a width of 148. Save the spreadsheet as **Activity_8p**.

Task 8ad

Open the file **Projects.csv** and transpose the data into the cells below and to the right of cell A15 so that the Tasks are listed along the top.

Save the spreadsheet as **Task_8ad**.

Transposing data

TRANSPOSE function

The **TRANSPOSE** function changes a range of cells from being arranged horizontally to vertically and vice versa. It is the same as taking data in columns and making that data appear in rows, or from rows into columns. Open and examine the file **Projects.csv.** Enter into cell A15 the formula **=TRANSPOSE(A1:G10)** which will transpose the data so that a copy of the original data looks like this.

	A	B	C	D	E	F	G	H	I	J
1	Project completion	Outline plans	Planning	Ground work	Building	Fixings	Total number of days			
2	Task 1	20	37	34	40	13	144			
3	Task 2	9	22	33	35	13	112			
4	Task 3	17	38	28	40	12	135			
5	Task 4	15	41	21	30	20	127			
6	Task 5	27	31	18	36	20	132			
7	Task 6	5	44	51	35	15	150			
8	Task 7	27	17	49	37	16	146			
9	Task 8	24	14	16	32	13	99			
10	Task 9	19	34	16	31	20	120			
15	Project completion	Task 1	Task 2	Task 3	Task 4	Task 5	Task 6	Task 7	Task 8	Task 9
16	Outline plans	20	9	17	15	27	5	27	24	19
17	Planning	37	22	38	41	31	44	17	14	34
18	Ground work	34	33	28	21	18	51	49	16	16
19	Building	40	35	40	30	36	35	37	32	31
20	Fixings	13	13	12	20	20	15	16	13	20
21	Total number of days	144	112	135	127	132	150	146	99	120

Save the spreadsheet as **Task_8ad**.

Working with strings

Concatenate strings

A string is a name given to a list of text characters, but it can include numbers. The word concatenate means to join (link together) and there is both a function and an operator in *Excel* to perform this.

Task 8ae

Open and examine the file **Employees.csv**.

Place a formula in cell:
» H2 to display the employee's full name
» I2 to display the employee's name in the format 'Family_name: Given_name'.

Place a function in cell:
» J2 to display the employee's full name
» K2 to display the employee's name in the format 'Family_name: Given_name'
» L2 to display the entire address with the postcode, with each part separated with a comma then a space.

Replicate these for all employees. Save the spreadsheet as **Task_8ae**

The easiest method of joining two string values is to use the & (ampersand) operator. Open and examine the file **Employees.csv**. Enter in cell H2 the formula =B2&" "&C2, which takes the contents of cell B2, adds a space, then the contents of cell C2, so that it displays 'Kristy King'.

Enter in cell I2 the formula =C2&": "&B2, which takes the contents of cell C2, adds the text in the speech marks (a colon and a space), then the contents of cell B2 so that it displays 'King: Kristy'.

This method cannot be used to answer the final three questions as the question explicitly asks us to 'Place a function in …' and the ampersand operator is not a function, because functions have names.

CONCATENATE function

The **CONCATENATE** function joins up to 30 string items together and displays the result as text. Enter in cell J2 the formula **=CONCATENATE(B2," ",C2)**, which takes the contents of cell B2, adds a space, then the contents of cell C2 so that it displays 'Kristy King'. Compare this function and its result with that placed in cell H2.

Enter in cell K2 the formula **=CONCATENATE(C2,": ",B2)**, which takes the contents of cell C2, adds the text in the speech marks (a colon and a space), then the contents of cell B2 so that it displays 'King: Kristy'. Compare this function and its result with that placed in cell I2.

Enter in cell L2 the formula **=CONCATENATE(D2,", ",E2,", ",F2)**, which returns the address for this employee stored as a single string, with comma delimiters so that the data can be extracted from it again if required.

Replicate these formulas for all employees. Save the spreadsheet as **Task_8ae**.

Extracting data from strings

Data can be extracted from strings in many different ways. There are three functions that are most often used for this process. They extract characters from the left, right or middle of the string. As well as extracting data, you can identify the length of the string, and there are other functions that allow you to find the position of a character (or substring) in a string.

> ### Task 8af
>
> Open and examine the file **Employees2.csv**.
>
> Place formulas to extract from the employee reference number, in cell:
> » D2 to display the first three characters
> » E2 to display the last five characters
> » F2 to display the code for the workplace which is stored as the third character
> » G2 to display the third to the seventh characters (inclusive).
>
> Place a formula in cell:
> » H2 to display the length of the employee's name
> » I2 to find the position of the colon in the name
> » J2 to display the employee's family name
> » K2 to display the employee's given name
> » M2 to display the name of the town or city where the employee lives.
>
> Replicate these for all employees. Save the spreadsheet as **Task_8af**.

LEFT function

The **LEFT** function extracts a number of characters from the left-hand side of a string.

Open and examine the file **Employees2.csv**. Enter in cell D2 the formula **=LEFT(A2,3)**, which displays the first three characters from the contents of cell A2, so that it displays 'CTD'.

RIGHT function

The **RIGHT** function extracts a number of characters from the right-hand side of a string.

Enter in cell E2 the formula **=RIGHT(A2,5)**, which displays the last five characters from the contents of cell A2, so that it displays '0001D'.

MID function

The **MID** function extracts a number of characters from the middle of a string. To use the MID function, you must specify the string to be extracted from, then the start position within the string, then the number of characters to be extracted.

Enter in cell F2 the formula **=MID(A2,3,1)**, which extracts from the contents of cell A2, starts at the third character, and extracts just a single character, so that it displays the letter 'D'.

Enter in cell G2 the formula **=MID(A2,3,5)**, which performs a similar extract but takes five characters rather than one (as the third to seventh character requires five), so that it displays the string 'D0001'.

LEN function

The **LEN** function counts the number of characters within a string. This function is very easy to use and is useful when included in nested functions to solve more complex tasks.

Enter in cell H2 the formula **=LEN(B2)**, which counts all the characters held in B2 including the colon and the space and returns the value 12.

Searching within a string

There are two functions in *Excel* that allow you to search within a string and both work in a similar way. The **FIND** function allows case-sensitive searching but cannot be used with wildcard characters. The **SEARCH** function is case-insensitive but allows searching with a * (wildcard symbol). Both would provide similar solutions to these tasks but for these exercises we will only study the FIND.

FIND function

The FIND function returns a numeric value that represents the position of a character or substring within a string.

Enter in cell I2 the formula **=FIND(":",B2)**, which looks character by character for the string ":", in this case a single colon. When it finds the first colon in the string held in B2, it returns the numeric position of that character, which in this case is 5.

This will be needed to allow us to split the name into two parts, as most names are not the same length. The family name is stored as the first part of the name in cell B2 so we will use the LEFT function for the extraction. Enter in cell J2 the formula **=LEFT(B2,FIND(":",B2)-1)**, which takes the left characters from B2. To find the number of characters to be extracted, *Excel* has located the position of the colon in the string (as we did in cell I2) and has extracted all characters up to that point. The -1 removes the colon itself from the extract. This returns the string 'King'.

The given name is stored as the second part of the name in cell B2, so we will use the RIGHT function for the extraction. This is slightly more complex, because we have to have to work out the length of the overall string and subtract the number of characters up to the colon, as well as the space, to get the number of characters in the given name. Enter in cell K2 the formula **=RIGHT(B2,(LEN(B2)-FIND(":",B2)-1))**, which takes the right characters from B2. To find the number of characters to be extracted, *Excel* has calculated the length of the string, then subtracted the length of the family name and the -1 removes the space. This returns the string 'Kristy'.

The extraction of the name of the town or city from the address requires two stages. As FIND (and SEARCH, had we chosen to use that) finds the position of the first character (or substring) searching from left to right, we must find the first comma in the address, extract the text after it (not including the space) and then remove the text including and after the second comma. It is not necessary to use two cells for this but it will be easier for you to understand the method.

Enter in cell L2 the formula =MID(C2,FIND(",",C2)+2,50), which although it uses a MID function, extracts the right characters from C2 without having to calculate its length. The length of each address, each town and each postcode may be different. We cannot work out the exact number of characters to extract as there are two commas in the address string. A figure that is too large has been chosen for the number of characters, in this case 50, to cater for any long addresses that may be added in the future. This formula returns 'Dundee, DU9 3UH'.

Enter in cell M2 the formula =LEFT(L2,FIND(",",L2)-1), which extracts the left characters up to (but not including because of the -1) the comma and returns 'Dundee'.

These two functions could have been entered as a single function, but the efficiency of this is debatable as the content of the calculation in L2 is used twice within M2. Enter into cell N2 the formula =LEFT(MID(C2,FIND(",",C2)+2,50),FIND(",",MID(C2,FIND(",",C2)+2,50))-1), which is the combined result. Which do you think is the most efficient method?

Replicate these formulas for all employees. Save the spreadsheet as **Task_8af**.

Testing cell contents

There are three *Excel* functions that can be used to test whether a cell contains text, numbers or non-text items. Each of these functions returns either 'True' or 'False' and can be used in string manipulation. For these three examples we will check the contents of cell A5 in a new spreadsheet. After entering the formula, change the contents of cell A5 to see the results.

ISTEXT function

This function, for example =ISTEXT(A5), checks to see if a cell contains text (including special characters).

ISNUMBER function

This function, for example =ISNUMBER(A5), checks to see if a cell contains numeric data.

ISNONTEXT function

This function, for example =ISNONTEXT(A5), checks to see if a cell is blank (has no contents).

> ### Activity 8q
>
> Open the file **Employees3.csv**, examine it carefully and save this as a spreadsheet.
>
> Enter formulas in columns D, E and F to display each employee's address separated into its three parts.
>
> Save the spreadsheet as **Activity_8q**.

Changing case

Strings can be set into upper case (capitals), lower case or a mixture of the two. We are now going to create a mixture of upper-case and lower-case text by forcing text to be upper and lower case and then concatenating the parts.

> ### Task 8ag
>
> Open and examine the file **Strings.csv**.
>
> Enter a function in cell:
> - B5 to display the original string in capitals
> - B6 to display the original string in lower-case text
> - B7 and B8 to compare string A and string B
> - B9 and B10 to return the ASCII values of strings A and B
> - B12 to return the character from the given ASCII value
> - B13 to return the length of the original string
> - B16 and B17 to return the values in binary and hexadecimal
> - B20 and B21 to return the values in decimal and hexadecimal
> - B24 and B25 to return the values in binary and decimal.
>
> Save the spreadsheet as **Task_8ag**.

UPPER function

The **UPPER** function returns the contents of a string as upper-case characters.

Open the file and place the cursor in cell B5. Enter the formula **=UPPER(B1)**, which will return the contents of cell B1 set into upper case.

LOWER function

The **LOWER** function returns the contents of a string as lower-case characters.

Open the file and place the cursor in cell B6. Enter the formula **=LOWER(B1)**, which will return the contents of cell B1 set into lower case.

Comparing strings

Excel can compare strings in many ways. Most of these methods ignore case. For example, enter in cell B7 the formula **IF(B2=B3,"Y","N")**, which will compare the two cells and return 'Y' if they are the same and 'N' if they are not the same. In this case the result is 'Y' which tells us that *Excel* does not compare case with an IF function. It also ignores case with other functions including VLOOKUP, HLOOKUP and INDEX and MATCH.

EXACT function

The **EXACT** function compares two strings to check that they are exactly the same (including case).

Enter in cell B8 the formula **=IF(EXACT(B2,B3),"Y","N")**, which returns the **expected result** of 'N', as the two strings are not identical.

Convert strings to ASCII values

It may be necessary to convert a string into an ASCII value, which is the code used by a computer to identify each letter or symbol on the keyboard.

CODE function

The **CODE** function returns the ASCII (numeric) code for a given character. If more than one character is present in a string, this function returns the code of the first character.

Enter in cell B9 the formula =**CODE(B2)**, which returns the value 65. Enter in cell B10 the formula =**CODE(B3)**, which returns the value 97. We can see that the ASCII codes for these two strings are very different.

CHAR function

The **CHAR** function returns the ASCII character for a given numeric code.

Enter in cell B12 the formula =**CHAR(B4)**, which returns the ASCII character for this numeric code, in this case the character 'q'.

Enter in cell B13 the formula =**LEN(B1)**, which returns the value 11, as there are five letters + one space + five numbers.

Working with number bases

We are familiar with using numeric values in our spreadsheet; these are all displayed in base 10 (decimal), but sometimes we need to convert numbers between different number bases, often to/from base 2 (binary) or base 16 (hexadecimal). In *Excel* there are some functions that convert these values.

DEC2BIN function

The DEC2BIN function converts a decimal (base 10) number into its binary (base 2) equivalent.

Enter in cell B16 the formula =**DEC2BIN(B15)**, which returns, for the decimal number 78, the value 1001110.

DEC2HEX function

The DEC2HEX function converts a decimal (base 10) number into its hexadecimal (base 16) equivalent.

Enter in cell B17 the formula =**DEC2HEX(B15)**, which returns, for the decimal number 78, the value 4E.

BIN2DEC function

The BIN2DEC function converts a binary (base 2) number into its decimal (base 10) equivalent.

Enter in cell B20 the formula =**BIN2DEC(B19)**, which returns, for the binary number 11111111, the value 255.

BIN2HEX function

The BIN2HEX function converts a binary (base 2) number into its hexadecimal (base 16) equivalent.

Enter in cell B21 the formula =**BIN2HEX(B19)**, which returns, for the binary number 11111111, the value FF.

HEX2BIN function

The HEX2BIN function converts a hexadecimal (base 16) number into its binary (base 2) equivalent.

Enter in cell B24 the formula =**HEX2BIN(B23)**, which returns, for the hexadecimal number FE, the value 11111110.

HEX2DEC function

The HEX2DEC function converts a hexadecimal (base 16) number into its decimal (base 10) equivalent.

Enter in cell B25 the formula **=HEX2DEC(B23)**, which returns, for the hexadecimal number FE, the value 254. Save the spreadsheet as **Task_8ag**.

Date and time functions

Although date and time values in *Excel* are displayed as dates/times, each one is actually stored as a number. The integer part of the number stores the date and the decimal part of the number stores the time. The number 1 represents 1 January 1990, the number 2 represents 2 January 1990 and so on. Times are stored as parts of the day, 0.5 is 12 noon, so a value of 2.5 would be 12 noon on 2 January 1900. 1 January 2022 at 6 a.m. would be stored as 44 562.25 because this is 44 562 days after 1 January 1900 and 6 a.m. is exactly a quarter of the way through the day.

> **Advice**
>
> Dates shown in this chapter are in day, month, year (UK) format. If you use month, day, year (US) format you will need to adjust your formulas accordingly.

DATE function

The **DATE** function allows you to enter the year, month and day parameters of a date and turn these into a number that *Excel* will recognise as a date. For example, if we enter into a spreadsheet cell the formula **=DATE(2024,5,16)**, *Excel* will display the date 16/05/2024 (if the cell is formatted as Date). If you change the format of the cell from Date to General the cell will display 45428 which is the number that *Excel* uses to store this date.

TIME function

The **TIME** function allows you to enter the hour, minute and second parameters of a time in the 24-hour clock and turn these into a number that *Excel* will recognise as a time. For example, if we enter into a spreadsheet cell the formula **=TIME(17,25,23)**, *Excel* will display the time 5:25 PM (although this format may change if a different custom format for the time has been selected). If you change the format of the cell from Time to General the cell will display 0.72596065 which is the number that *Excel* uses to store this time.

WEEKDAY function

WEEKDAY is used to return a number between 1 and 7 from a given date. If the day is a Sunday, 1 is returned, Monday is 2, and so on.

> **Task 8ah**
>
> Open and examine the file **JanSales.csv**. For each order, display in column F the day of the week that the order was placed. Save the spreadsheet as **Task_8ah**.

Open and examine the data in the file. Enter in cell E2 the formula **=WEEKDAY(C2)**, which returns the code for the day, in this case '2'. Enter in cell F2 the formula **=VLOOKUP(E2,$H$2:$I$8,2,0)**, which looks up the code and displays the text from the adjacent cell, in this case 'Monday'.

Replicate both formulas for all orders. Save the spreadsheet as **Task_8ah**.

Task 8ai

Create a new spreadsheet like this.

Place functions in cells B2, B3 and B4 to display the day, month and year parts of the date in cell B1.

Save the spreadsheet as **Task_8ai**.

	A	B
1	Date:	20/05/2022
2	Day:	
3	Month:	
4	Year:	

DAY function

DAY is a function that is used to return the day part of a given date, as a number between 1 and 31.

Create a new spreadsheet as shown and enter in cell B2 the formula **=DAY(B1)**, which will return the value 20.

MONTH function

MONTH is a function that is used to return the month part of a given date, as a number between 1 and 12.

Enter in cell B3 the formula **=MONTH(B1)**, which will return the value 5.

YEAR function

YEAR is a function that is used to return the year part of a given date.

Enter in cell B4 the formula **=YEAR(B1)**, which will return the value 2022.

Save the spreadsheet as **Task_8ai**.

Task 8aj

Create a new spreadsheet like this.

Place functions in cells B5, B6 and B7 to display the number of days, months and years, respectively, between the first date and the second date.

Save the spreadsheet as **Task_8aj**.

	A	B
1	First date:	20/05/2022
2	Second date:	23/11/2023
4	Difference between dates in:	
5	Days:	
6	Months:	
7	Years:	
9	Using DATEDIF	
11	Difference between dates in:	
12	Days:	
13	Months:	
14	Years:	

Calculate the number of days between two dates – Method 1

Create a new spreadsheet as shown. To calculate the number of days between two dates, do not use any date functions, just subtract the first date from the second by entering in cell B5 the formula **=B2-B1**, which will return the result of the calculation in date format as 05/07/1901. To display the number of days, you need to format this cell as a number. Select the **Home** tab, in the **Number** group click on the drop-down menu for formatting like this.

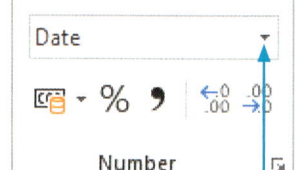

Select the top option for **General** formatting and the value shown in the cell will change to 552.

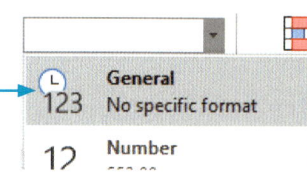

Try swapping the two dates entered so that the first date is 23/11/2023 and the second date is 20/05/2022. This will result in a negative value in cell B5. To avoid negative values, you can use the ABS function.

Calculate the number of years between two dates – Method 1

At first glance it would appear that to calculate the number of years between two dates, you need to subtract the YEAR function of the first date from the YEAR function for the second date. This does find the difference in years between two dates, but if the two dates were 31/12/2022 and 1/1/2023 then this would return a one-year difference even though there is only really a one-day difference. It is more accurate to use the number of days and divide this by 365.25 (which is the approximate number of days per year).

Enter in cell B7 the formula =INT((B2-B1)/365.25), which will return the value 1.

DATEDIF function

Although undocumented in *Excel*, the DATEDIF function will calculate the date difference between two dates and display the number of days, months or years between the two dates. The syntax for this function is =DATEDIF(start date, end date, return code).

The start date and end date MUST be in the correct order and the different return codes can be seen in Table 8.6.

▼ Table 8.6 Ways of using the DATEDIF codes

Code	What it does
"D"	Returns the difference in days between the start date and end date.
"M"	Returns the difference in months between the start date and end date.
"Y"	Returns the difference in years between the start date and end date.
"MD"	Returns the difference in days, ignoring the months and years, between the start date and end date.
"YM"	Returns the difference in months, ignoring the days and years, between the start date and end date.

Calculate the number of days between two dates – Method 2

Enter in cell B12 the formula =DATEDIF(B$1,B$2,"D"), which will return the value 552.

Calculate the number of months between two dates

Enter in cell B13 the formula =DATEDIF(B$1,B$2,"M"), which will return the value 18.

Calculate the number of years between two dates – Method 2

Enter in cell B14 the formula =DATEDIF(B$1,B$2,"Y"), which will return the value 1.

Save the spreadsheet as **Task_8aj**.

Task 8ak

Create a new spreadsheet like this.

Place functions in cells B3, B4 and B5 to display the hour, minute and second parts of the time in cell B1.

	A	B
1	Time:	18:07:03
3	Hours	
4	Minutes	
5	Seconds	

Save the spreadsheet as **Task_8ak**.

HOUR function

The **HOUR** function is used to return a number between 0 and 23 from a given time.

Create a new spreadsheet as shown and enter in cell B3 the formula **=HOUR(B1)**, which will return the value 18, because *Excel* stores the time using the 24-hour clock.

MINUTE function

MINUTE is a function that is used to return a number between 0 and 59 from a given time.

Enter in cell B4 the formula **=MINUTE(B1)**, which will return the value 7.

SECOND function

SECOND is a function that is used to return a number between 0 and 59 from a given time.

Enter in cell B5 the formula **=SECOND(B1)**, which will return the value 3. Save the spreadsheet as **Task_8ak**.

Task 8al

Create a new spreadsheet like this.

Place functions in cells B5, B6 and B7 to display the difference between the two times in terms of the number of:

» hours
» hours and minutes
» hours, minutes and seconds

between the first time and the second time.

	A	B
1	First time:	20:53
2	Second time:	23:46
4	Difference between times in:	
5	hours	
6	hours and minutes	
7	hours, minutes and seconds	

Save the spreadsheet as **Task_8al**.

The same calculation will be performed for all three of these different results, by subtracting one time from the other (and if required, using the ABS function to ensure a positive result). The format of the resulting answer can be specified using the **TEXT** function, depending whether the result is required in hours, hours and minutes, or hours, minutes and seconds.

Enter into cell B5 the formula **=TEXT(ABS($B$2-$B$1),"h")**, which will find the difference between the two cells, make sure that it is a positive difference and then format the cell as text displaying only the hour portion. This displays a 2 in this cell.

Use the drag handle to replicate this formula into cells B6 and B7. Edit the formula in cell B6 to **=TEXT(ABS($B$2-$B$1),"hh:mm")**, which formats the result to hours and minutes. The "hh" forces the hours to display as two digits like this '02:53'.

Edit the formula in cell B7 to **=TEXT(ABS($B$2-$B$1),"hh:mm:ss")**, which formats the result as hours, minutes and seconds and displays '02:53:00'. Save the spreadsheet as **Task_8al**.

> **Task 8am**
>
> Open and examine the file **BusTimes.csv**.
>
> For each journey, if the bus was late, calculate the number of minutes it was late. Calculate the total number of minutes that the buses were late.
>
> Save the spreadsheet as **Task_8am**.

This problem is not as simple as it appears. One solution would be to check if the bus was late, then to calculate how late it was. As the data is formatted and displayed as text, it needs converting into a numeric value so that the total minutes late can be calculated. This solution assumes that no bus is going to be more than a few hours late.

Open and examine the file. Enter in cell E2 the formula =IF(D2>C2,TEXT(D2-C2,"hh:mm"),"00:00"), which displays '00:00' if the bus was on time or early and displays the time a bus was late as a text string in "hh:mm" format if the bus was late.

Enter in cell F2 the formula =(LEFT(E2,2)*60+RIGHT(E2,2)), which multiplies the left two characters of E2 (the hour part of the late column) by 60 to turn them into minutes. This is added to the two right characters of E2 (the minutes part) to give the total minutes late stored as an integer value. This only works on the data stored in cell E2 because it is stored as text. It will not work on the data in cells such as C2 and D2, which are not stored as text.

Replicate these formulas for all journeys. Enter in cell F119 the formula =SUM(F2:F118), which returns a value of 467. Save the spreadsheet as **Task_8am**.

> **Task 8an**
>
> Create a new spreadsheet like this.
>
> Make sure that all cells are appropriately formatted as date and time. Calculate the flight duration in hours and minutes.
>
> Save the spreadsheet as **Task_8an**.
>
> | | A | B |
> |---|---|---|
> | 1 | Flight departure: | |
> | 2 | Date | 03 July 2023 |
> | 3 | Time | 15:35:00 |
> | 4 | Flight arrival: | |
> | 5 | Date | 04 July 2023 |
> | 6 | Time | 06:30:00 |
> | 8 | Flight duration: | |

As long as cells B2 and B5 are formatted as date and cells B3 and B6 are formatted as time, the following solution works.

Enter in cell B8 the formula =TEXT((B5+B6)-(B2+B3),"hh:mm"), which displays 14:55 in the correct format.

Save the spreadsheet as **Task_8an**.

> **Activity 8r**
>
> Open the file **Flights.csv** and save this as a spreadsheet. For each flight, calculate its duration and display this in hours and minutes. Calculate the average duration for all flights.
>
> Save the spreadsheet as **Activity_8r**.

Visually verify the accuracy of your data entry

The most common reason for making mistakes is the lack of accurate data entry. Although it appears obvious, each label and item of data entered should be carefully checked to make sure that there are no spelling errors or errors in the spacing or case (letters with upper or lower case) of the labels entered. You should visually verify this by comparing the label or data with that provided in the task.

8.1.3 Validation rules

Data entry errors can also be reduced by adding validation rules to your spreadsheet.

> **Task 8ao**
>
> Open and examine the spreadsheet you saved as **Task_8v**.
>
> Fourteen students have yet to sit the test. All test scores must be whole numbers. The highest score is 10. Apply appropriate rules to ensure only acceptable data can be entered.
>
> Save the spreadsheet as **Task_8ao**.

Open and examine the spreadsheet. The task requires you to validate the data entry, although the word is deliberately not used. Fourteen students have yet to sit the test, so highlight all cells in the range C8 to C80, which is where the data has been and will be entered. Select the **Data** tab, in the **Data Tools** group, and click on the **Data Validation** icon.

This opens the **Data Validation** window. In the **Allow** section, click on the top drop-down menu arrow.

Select **Whole number** from the list. The validation window changes and the **Data** group becomes available. Click on the drop-down list and select **between** (if it is not already selected). Place 0 in the **Minimum** value box, which is the lowest possible test score, and in the **Maximum** value box, place 10.

Although we have now set the validation rule, we must also enter appropriate text that will be displayed if a user enters data that does not meet the rule. Select the **Error Alert** tab.

Enter in the **Title** box the text **Data entry error!** to inform the user that an error has occurred. Enter in the **Error message** box the text 'Please enter a whole number between 0 and 10 (inclusive)'. These messages clearly inform the user that they have made a mistake and more importantly what they should do to rectify this error.

It is also useful to display an input message before the user enters data, telling them what data is acceptable. To do this select the **Input Message** tab.

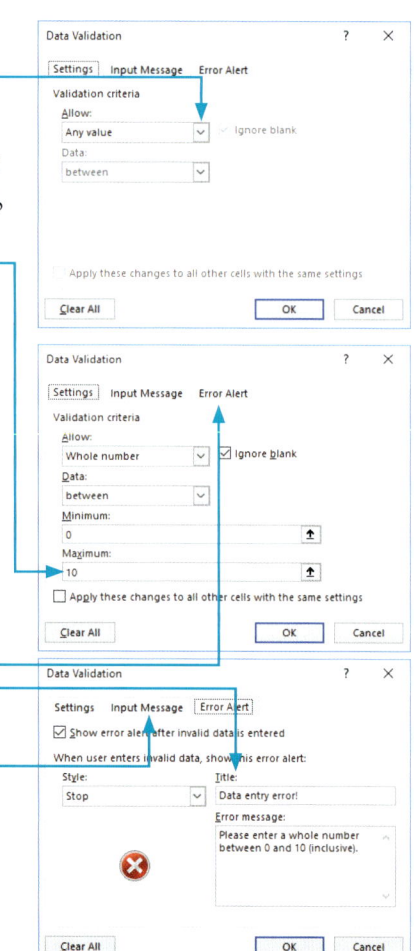

Enter an appropriate Title which informs the user what data is being entered. Add an appropriate Input message that informs the user what data entry is accepted before they make a data entry error.

Click on the OK button to activate the validation rule. We will test this rule in Section 8.2. Save the spreadsheet as **Task_8ao**.

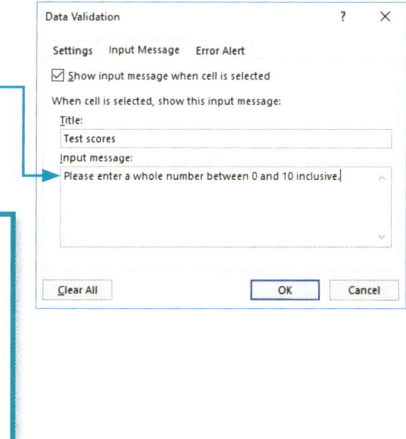

Task 8ap

Open and examine the spreadsheet you saved as **Task_8f**.

New products need to be added to the list. The only chipset manufacturers used by this company are AMD and Intel. A list of supported socket types can be found in the file **Socket.csv**. Apply appropriate rules to ensure only acceptable data can be entered.

Save the spreadsheet as **Task_8ap**.

Open and examine the spreadsheet. The task requires you to validate the data entry for the chipset manufacturers in column G, although not the top four cells. Highlight a range of cells from G5 to (at least) G130. Open the Data Validation window and select the Settings tab. In the Allow group, select List from the drop-down menu. In the Source box type 'AMD,Intel' like this.

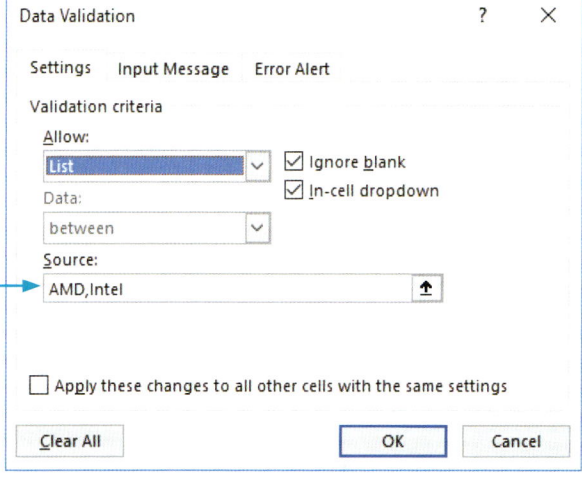

Add appropriate input and error messages before clicking on OK. For the supported socket types, open and examine the file **Socket.csv**. It contains a list of the supported socket types with a header row. Copy cells A1 to A12 and paste these as cells into a new worksheet in your Task_8f workbook. Rename this worksheet **Socket**. Select the Task_8f worksheet and highlight a range of cells from F5 to (at least) F130. Open the Data Validation window and select the Settings tab. In the Allow group, select List from the drop-down menu. Click the left mouse button in the Source box and drag the cursor in the worksheet Socket to highlight cells in the range A2 to A12 like this.

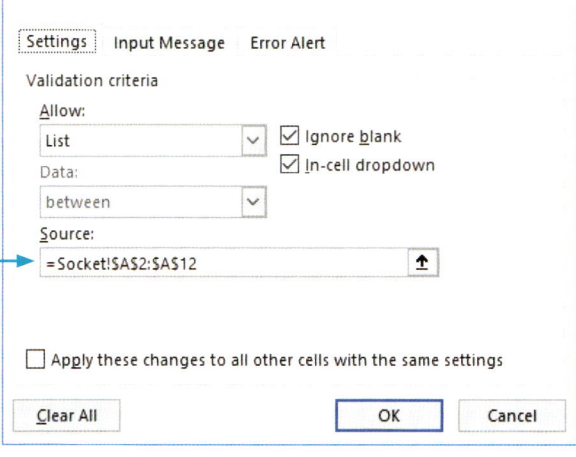

Add appropriate input and error messages before clicking on OK. We will test this rule in Section 8.2. Save the spreadsheet as **Task_8ap**.

Advice
You cannot use external references with data validation in *Excel*. Copy the data from the external source into your current worksheet or workbook.

Task 8aq

Open and examine the spreadsheet you saved as **Task_8ac**.

Ensure that only valid data can be added in cells B3 and B4.

Save the spreadsheet as **Task_8aq**.

Open and examine the spreadsheet. The formula placed in cell B5 will generate a #N/A error if invalid data is entered into B3 or B4. Place the cursor in cell B3. Open the **Data Validation** window and select the **Settings** tab. In the **Allow** group, select **List** from the drop-down menu. Click the left mouse button in the **Source** box and highlight cells A8 to A253 like this.

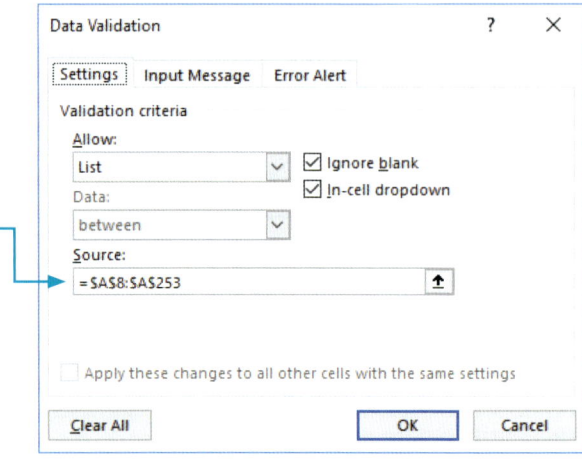

Perform similar actions to set validation in cell B4 using the cell range B7 to M7 as the data source.

As you move the cursor into each of these cells, a drop-down list of valid cities/months appears to the right of the cell. This will assist the user as they can now click on the item instead of typing, which will reduce typographical errors.

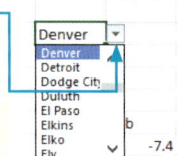

Save the spreadsheet as **Task_8aq**.

Advice

Selecting from a list with data validation in *Excel* offers the user a drop-down list of valid options to click on.

Activity 8s

Open the file you saved as **Activity_8m**.

Apply appropriate validation to ensure that all grades entered are within the range 1 to 9 inclusive.

Save the spreadsheet as **Activity_8s**.

8.1.4 Formatting

Format cells

As we have already seen when using date and time functions, the format *Excel* stores data in is not the same as the format used to display that data.

Task 8ar

Open and examine the file **Format.csv**.

Format cells in the range B3 to C12 to match the display style described in column A.

Save the spreadsheet as **Task_8ar**.

The quickest way to format cells is to highlight cells B3 to D3, select the **Home** tab and in the **Number** group, use the drop-down menu to select the cells as **General** number format.

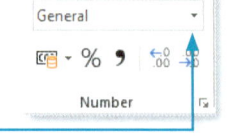

	A	B	C	D
1	Original	42	0.25	2.5
3	General	42	0.25	2.5
4	Number	42.00	0.25	2.50
5	Currency	£42.00	£0.25	£2.50
6	Short date	11/02/1900	00/01/1900	02/01/1900
7	Long date	11 February 1900	00 January 1900	02 January 1900
8	Time	00:00:00	06:00:00	12:00:00
9	Percentage	4200.00%	25.00%	250.00%
10	Fraction	42	1/4	2 1/2
11	Scientific	4.20E+01	2.50E-01	2.50E+00
12	Text	42	0.25	2.5

Repeat this for each display format. Your results may differ from this, because the regional settings and installation settings of *Excel* are likely to be different. Although this is the quickest method, it does not give you the flexibility to change the settings like the number of decimal places displayed.

Format numbers

Highlight cells in the range B4 to D4. Select the *Home* tab, and in the *Number* group click on the dialogue box launcher.

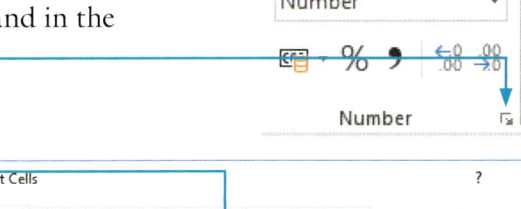

This opens the *Format Cells* window. Use this to change the number of decimal places of this number from 2 to 1.

You will notice that the numbers displayed have changed:

From	42.00	0.25	2.5
To	42.0	0.3	2.5

The numbers stored have not changed; for example if you enter in cell F4 the formula *=2*C4*, it will return the value 0.5 rather than 0.6.

As you can see, the figure displayed in C3 is not the one used for the calculation. It is important for you to use functions like ROUND and INT, rather than just formatting where appropriate.

Format currency

Examine the formatting of cells B5 to D5. The computer has assumed that because my machine is configured for the UK, the settings required are £ sterling. Change these three cells into Australian dollars. To do this select these three cells, then open the *Format Cells* window. Use the drop-down menu in the *Symbol* box to select the required currency, in this case $ English (Australia).

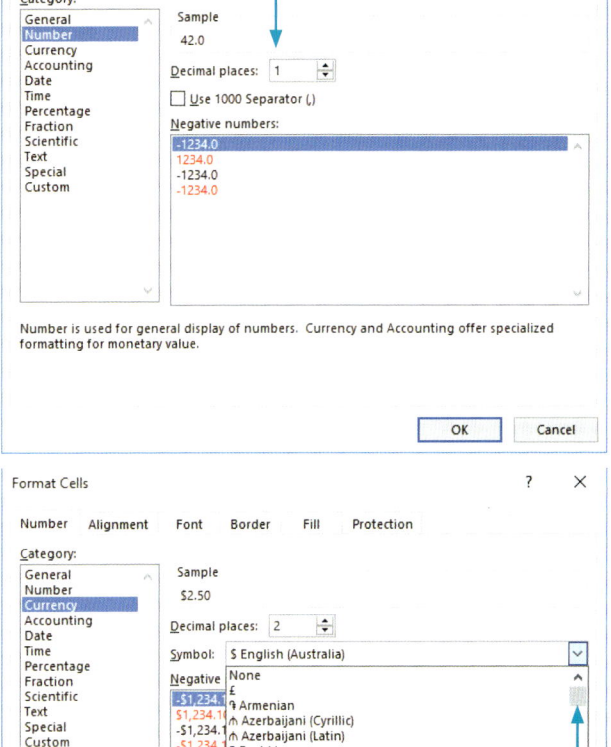

Some currencies like Japanese yen have no decimal places and so will need to be formatted to zero decimal places. If the currency symbol that you are looking for (for example ¥, the symbol for the Japanese yen) does not appear in the drop-down list, use the ISO code for the country, in this case JPY. These are found at the bottom of the menu list.

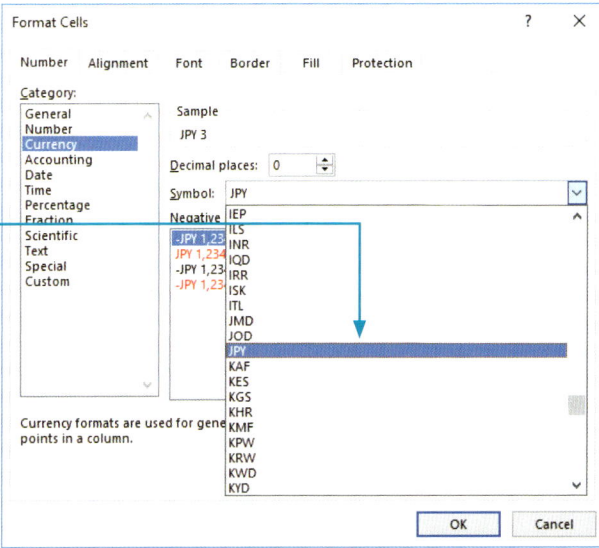

Format dates

Excel holds a number of preformatted styles for dates. These are selected by highlighting and opening the Format Cells window. Use the drop-down menu in the Type box to select the required date format.

You may find that your list is different to this. These differences are due to the settings applied to your software, both in *Windows* and in *Excel*. Different regional settings will offer different format options, for example in the United Kingdom the date for 25 March 2023 may be formatted as 25/03/2023, yet on a machine in the USA it may be formatted as 03/25/2023. The regional variations of these settings can be changed using the drop-down list for Locale (location):.

This allows dates in long-date format (those with a text component), like 25 March 2023, to be formatted into different languages, such as 25 de marzo de 2023 in Spanish. You need to know how to produce different versions of a spreadsheet, formatted for audiences in different parts of the world.

Remember earlier in the chapter, we also applied some specialised date formats to cells using the text function. This can be used to specify formats that are not available in these drop-down lists. Enter in cell F6 the formula =TEXT(B6,"dd mm yyyy"), in G6 =TEXT(B6,"dd mmm yyyy"), in H6 =TEXT(B6,"dd mmmm yyyy"). What do you notice about the results?

Format times

Time values can be formatted in a similar way to the date. Different format types can be selected from a list of Type: options. Again, you can restrict the types to different regional variations using the Locale (location): list.

Format percentages

This is again set in the the Format Cells window. Highlight cells B9 to D9 and examine the formatting. These cells have automatically been formatted to two decimal places, which is not appropriate for the figures displayed. Change it to display integers by setting the Decimal Places box to 0.

It is important to remember that if you want to display 25% in a cell, you must enter 0.25 before formatting the cell as a percentage value. Before working with percentages, make sure that you understand the maths needed to change decimals to percentages.

> **Advice**
>
> A shortcut to format cells as percentage values is to use the **%** icon in the Number group of the Home tab.

Format fractions

Examine the cells B10 to D10. You can see that they have been formatted as fractions. This feature can display simple fractions but is limited. You can change the number of digits for the parts of the fraction and the type of fraction required, including equivalent fractions when using simple fractions like quarters and eighths. Open the Format Cells window and change the formatting in C9 to display the fraction in eighths like this.

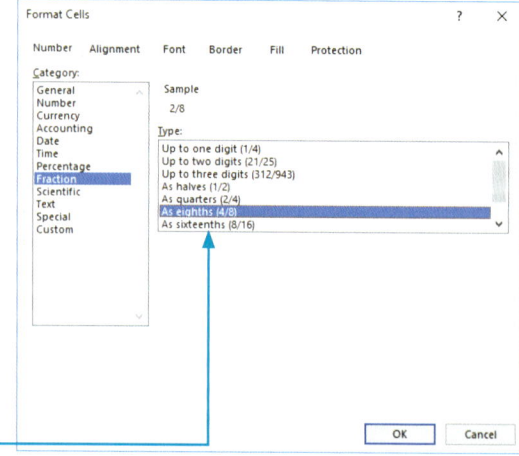

Save the spreadsheet as **Task_8ar**.

Cell alignment

In the **Home** tab, locate the **Alignment** group. This group contains icons to format the alignment of labels and data within a cell. The first three icons along the top row are to set the vertical alignment to the top, middle or bottom:

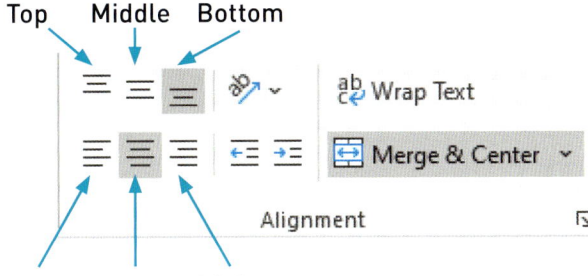

The fourth icon on the top row changes the text direction and we will study that in the next section of this chapter. The first three icons along the bottom row are to set the horizontal alignment to the left, centre or right.

Text orientation

It is sometimes difficult to get spreadsheets to appear as we want, especially when trying to fit them to a single page or screen. Sometimes it is because the labels along the top of columns are too long; at other times you want a merged label to the left of the data, but this does not fit well on the page or screen. One solution is to rotate the text in those cells by ninety degrees so that it is vertical within the cell rather than horizontal.

Task 8as

Open and examine the spreadsheet you saved as **Task_8o**.

Insert a new column to the left of A and format it to look like this.

Save the spreadsheet as **Task_8as**.

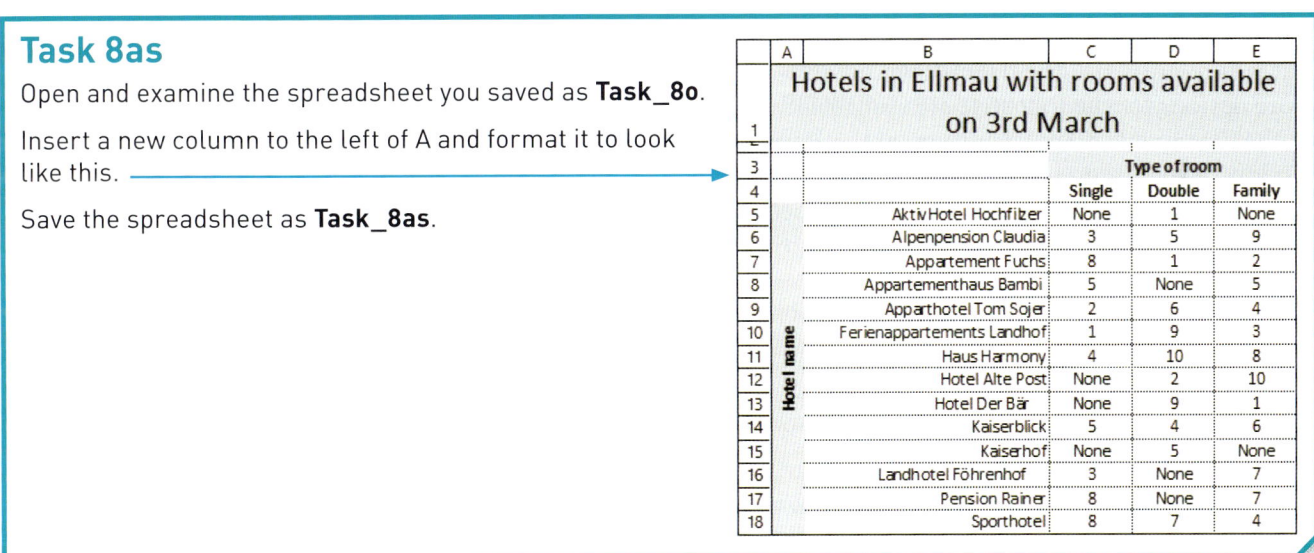

Open the spreadsheet and insert a new column A. Select cells A1 to E1. Locate the **Home** tab, and in the **Alignment** group click on the **Merge & Center** button. This will remove the existing merge and a second click will merge the cells to include the new column. Insert a new row 3, making this the same height as row 4. Move the label **Type of room** into C3, then merge cells C3 to E3. Move the label **Hotel name** from B5 to A6 and delete row 5. Merge cells A5 to A18. Place the cursor in this cell, click on the **Home** tab and in the **Number** group, click on the dialogue box launcher. This opens the **Format Cells** window in the **Alignment** tab.

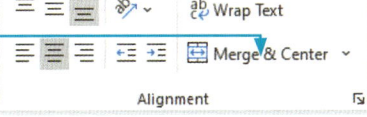

Locate the **Orientation** box, grab the red drag handle and rotate it to a vertical position like this.

Format this cell so that it is centre-aligned horizontally and vertically. We do this by staying in the **Format Cells** window and setting the **Horizontal** and **Vertical** boxes to centre aligned.

An alternative method would be to use the alignment icons in the **Home** tab, **Alignment** group on the toolbar.

Save the spreadsheet as **Task_8as**.

Format cell emphasis

Font size

When you want to change the font size within a spreadsheet, you must select the cell or cells to be changed. The font size of the text within a cell is changed from the **Home** tab, within the **Font** group. You can either use the drop-down menu to select the font size or type a new value into the box.

Font style

When you want to change the font style within a spreadsheet, you must select the cell or cells to be changed. The font style (typeface) of the text within a cell is changed from the **Home** tab, within the **Font** group. To select the font style that you require, use the drop-down menu.
More options for the font style are available if we click on the dialogue box launcher to open the **Format Cells** window in the **Font** tab. Enhancements such as bold and italic are available from the **Font style** menu, single or double underline is available from the **Underline** drop-down menu, and other effects such as superscript and subscript are also available here.

Font colour

When you want to change the font colour within a spreadsheet, you must select the cell or cells to be changed. The font colour within a cell is changed from the **Home** tab, within the **Font** group. To select the font colour that you require, use the drop-down palette.

Cell colour and shading

To fill a cell with a solid background colour, select the cell or cells to be coloured. From the **Home** tab, within the **Font** group, select the background colour that you require using the drop-down palette next to the fill icon.

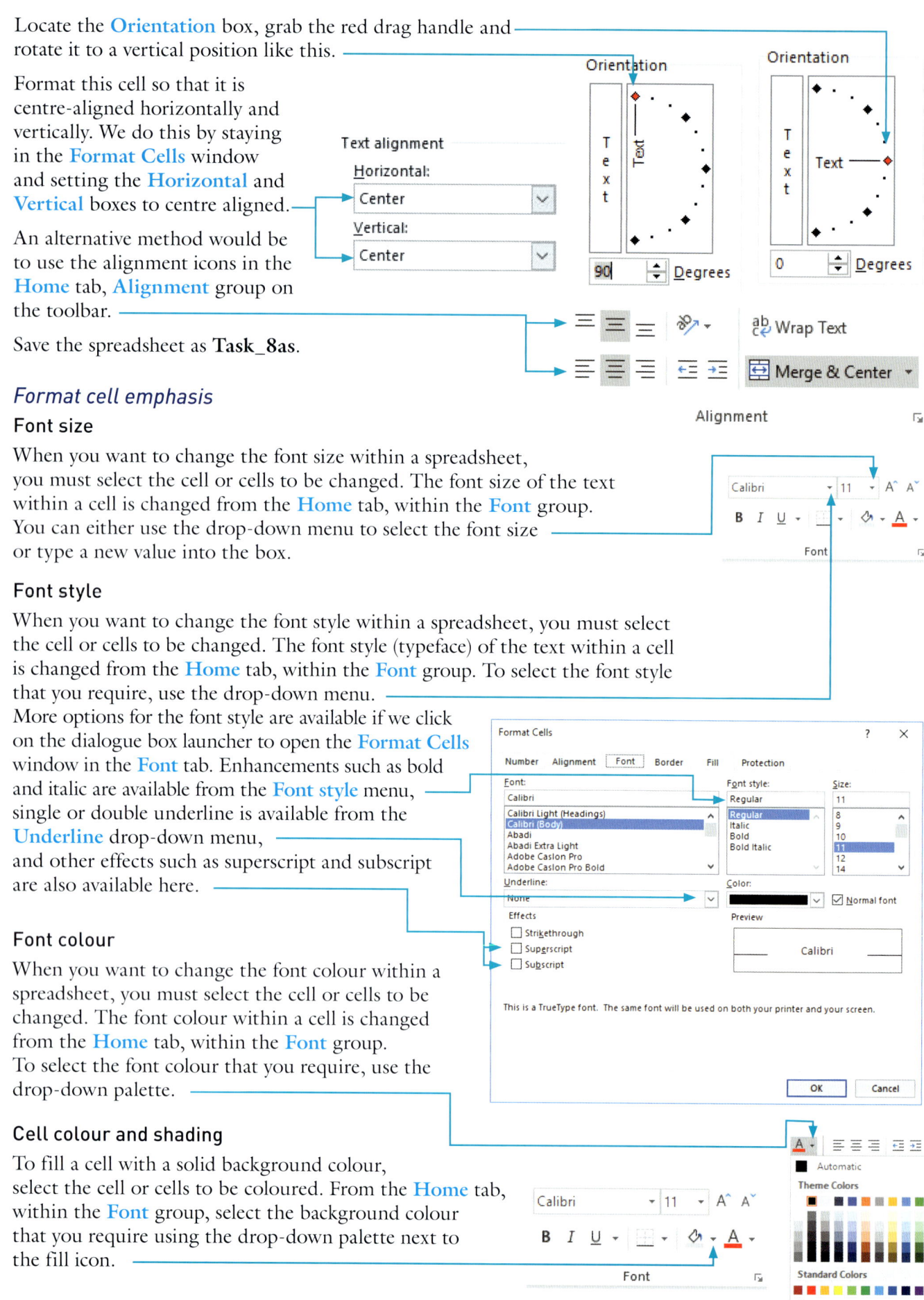

Other fill effects like gradient fill effects using a combination of two colours and different types of shading are also available using the **Format Cells** window and the **Fill** tab.

Pattern fills are created by selecting the **Pattern Color** using the drop-down palette, and the style of shading using the **Pattern Style** drop-down menu.

Each time you select these options a sample fill is shown in the **Sample** section.

Gradient fills can be located by selecting the ![Fill Effects...] button.

This opens the **Fill Effects** window, where different options are available to give a range of one and two colour gradient fills.

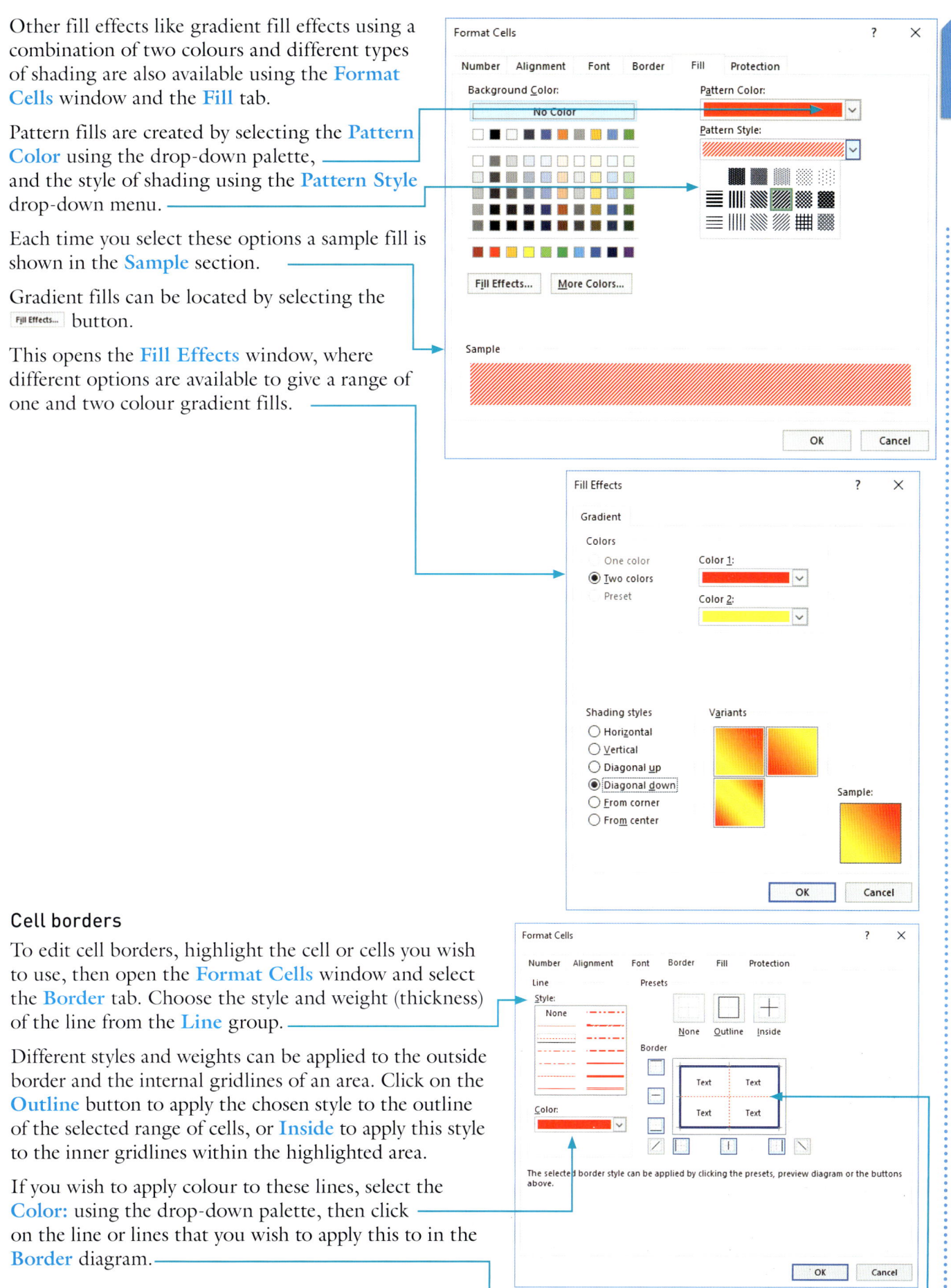

Cell borders

To edit cell borders, highlight the cell or cells you wish to use, then open the **Format Cells** window and select the **Border** tab. Choose the style and weight (thickness) of the line from the **Line** group.

Different styles and weights can be applied to the outside border and the internal gridlines of an area. Click on the **Outline** button to apply the chosen style to the outline of the selected range of cells, or **Inside** to apply this style to the inner gridlines within the highlighted area.

If you wish to apply colour to these lines, select the **Color:** using the drop-down palette, then click on the line or lines that you wish to apply this to in the **Border** diagram.

In this example, the outside of the highlighted area will be set to a thick dark blue line and all internal gridlines will be set to a dotted red line. To do this click on all four internal lines in the Border section, then click on OK.

Adding comments

Comments or notes can be added to cells to add extra information about the contents of that cell, without changing the contents or affecting any calculations. These are added (almost like sticky notes) to a cell by clicking on the cell with the right mouse button and selecting New Comment from the drop-down menu. Type your comment or note in the box and, if it is a comment, it will be attributed to your user name as the start of a comment thread. If another user edits the comment, their reply will be added to your comment in the same thread, along with the date and time.

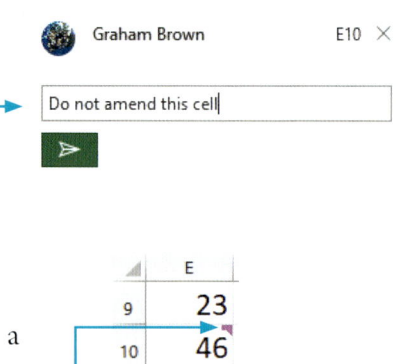

Use the ▶ button to save your comment. The cell will now appear with a small comment shape in the top-right corner when seen on screen but the comment will not show when printed.

> **Advice**
>
> In previous versions of *Excel*, these Comments were called Notes and were inserted by clicking the right mouse button and selecting New Note. Notes did not allow others to reply in a thread.

> ### Task 8at
>
> Open the file saved as **Task_8j**.
>
> For each employee, calculate the total hours worked, the gross pay, tax due and net pay. Calculate the totals for columns B to G.
>
> Calculate the percentage contribution in terms of time to the company. For each cell displaying the percentage time, set the background colour of the cell to:
> » orange if an employee contributed between 12% and 18% of the total hours
> » red if an employee contributed less than 12% of the total hours, and format the text to be white
> » green if an employee contributed more than 18% of the total hours.
>
> Save the spreadsheet as **Task_8at**.

Place in cell B12 the formula **=SUM(B6:B11)** and replicate this to cell G12. Place in cell D6 the formula **=B6+C6**, which returns the value 24. Place in cell E6 the formula **=D6*$B$2**, which returns the value 244.80. Place in cell F6 the formula **=E6*$E$2**, which returns the value 97.92. Place in cell G6 the formula **=E6-F6**, which returns the value 146.88. Place in cell G7 the formula **=D6/$D$12**. Format this cell as a percentage to one decimal place, so that it returns the value 27.6%. Replicate these formulas (and the formatting) for all employees. Place in cell B12 the formula **=SUM(B6:B11)** and replicate this for columns C to G inclusive.

Conditional formatting

Conditional formatting is the most efficient method used to change the display format (usually the font or background colour within a cell), depending upon the contents of the cell.

Highlight cells H6 to H11 inclusive as conditional formatting will be applied to all of them. Select the **Home** tab, in the **Styles** group, and use the drop-down menu for the **Conditional Formatting** icon.

Select **New Rule** to open the **New Formatting Rule** window. Select the second option for **Format only cells that contain**, which changes the layout of the window. Enter data into the **Format only cells with:** section so that it looks like this.

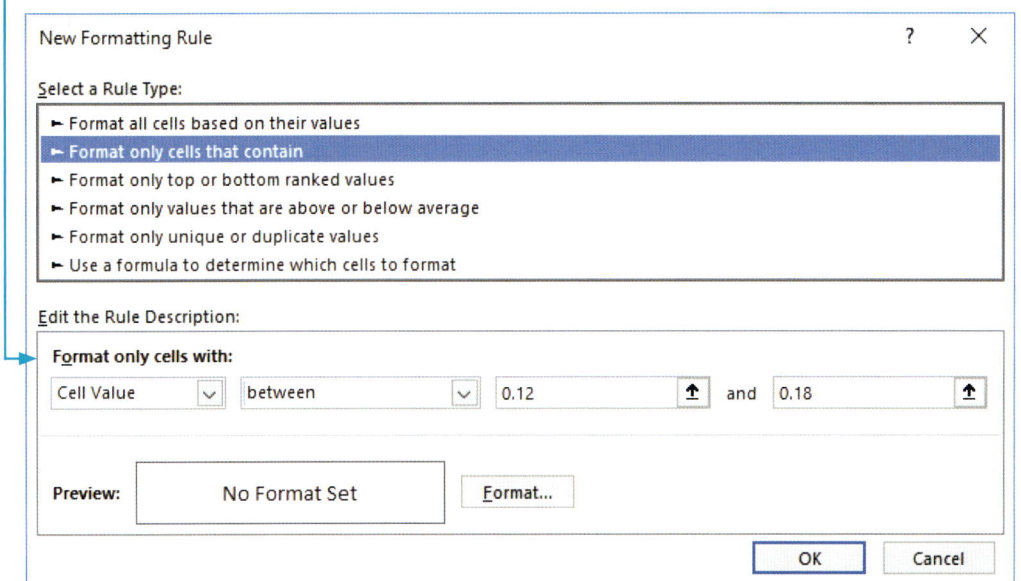

Note that the 12% and 18% have been entered as their decimal equivalents. When the rules have been entered, click on the **Format...** button to open the **Format Cells** window. Select the **Fill** tab, then double-click on an orange colour from the palette, then click **OK**. Click again and the first conditional formatting rule has been set. To add the second rule, repeat the process. Use the drop-down menu that contains **between** and change it to **less than**. Change the value to 0.12 and the format to white text on a red background so that the **Format only cells with:** window looks like this.

Click **OK** to set this rule and repeat the process to add a third rule, changing the rule to **Format only cells with:** greater than 0.18 as green. The formatting rule will look like this.

The completed spreadsheet looks like this:

To edit conditional formatting rules after they have been created, highlight the range of cells, in the **Home** tab, **Styles** group select **Conditional Formatting**, then choose the last option from the list to **Manage the rules**. All three rules created are visible in the **Conditional Formatting Rules Manager** window.

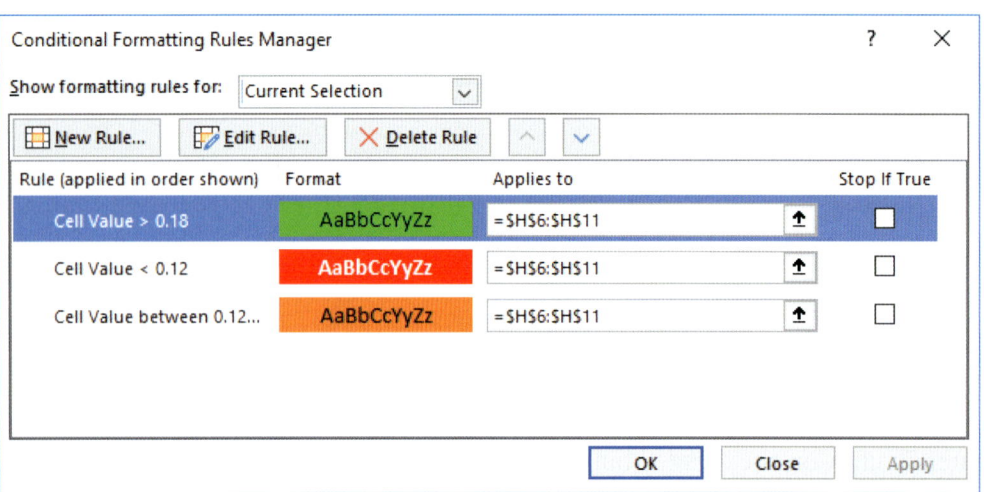

To edit a rule, click on the rule to select it and then click on the **Edit Rule...** button. This opens the **Edit Formatting Rule** window, which is identical to the **New Formatting Rule** window that we have used before. Save the spreadsheet as **Task_8at**.

> **Advice**
>
> You don't always need to use conditional formatting that involves numeric cell values; sometimes comparing cell contents with text strings or finding duplicate values is required.

> **Activity 8t**
>
> Open the file you saved as **Activity_8c**.
>
> Apply coloured backgrounds to cells in the range C7 to C23 to display:
> » gold if the cell contains 1st
> » silver if the cell contains 2nd
> » bronze if the cell contains 3rd.
>
> Save the spreadsheet as **Activity_8t**.

8.1.5 Factors to be appreciated when creating a spreadsheet

Requirements

The first thing to consider when creating a spreadsheet is to understand what it is required to do. These requirements are usually provided to you by the person who has asked for the spreadsheet to be created and must be addressed by you. It may be a request for it to perform certain calculations which you must ensure the spreadsheet will be able to perform. Its appearance may well be specified and you must try to make sure those wishes are carried out. Some requirements such as it should be easy to use and understand may not be specifically mentioned by the person requesting the spreadsheet but must, nevertheless, form part of your considerations.

Purpose

The purpose of the spreadsheet is the reason for its need. Is it just going to present data nicely or will the user need access to formulas within the spreadsheet? Are the users going to be beginners? Do certain cells need to be protected to prevent users accidentally changing the contents of these cells? In other words, what is it designed to do?

Audience

When creating a spreadsheet you need to appreciate the type of audience who will see/use it. This includes the age of your audience and the specialisms, if any, of your audience. By specialism we mean are they scientists, administrators or artists for example? The appearance of the spreadsheet, its font size, colours used and layout must all be considered. Choice of font and font size are discussed in more detail in Chapter 11 regarding video production, but the same rules apply here. A younger audience will need a larger font size and a font type that will appeal to such an audience. A spreadsheet for an adult audience should be more formal in appearance. A scientist would probably need access to the formulas and more scientific functions and would need a user guide which is appropriate for this use. A business or government administrator would not necessarily need access to formulas but would need access to the use of graphs, for example, as well as alternative means of presenting the data. An artist, however, might only be interested in keeping financial records of their paintings and so require a more simplistic type of spreadsheet.

All the above factors need to be considered when creating a spreadsheet in order that the spreadsheet produces the results required in a format that satisfies the intended audience. It is therefore important that these considerations are addressed in the creation of the spreadsheet.

8.2 Testing a spreadsheet

Designing a test plan and choosing your test data are the most important parts of testing any spreadsheet. All formulas and all validation rules need to be tested to ensure that they work as you expected them to. If you test every formula in a spreadsheet thoroughly, the number of possible errors is reduced when you use the spreadsheet with real data. For formulas containing conditions (for example IF, COUNTIF, INDEX, MATCH and so on) you must choose data that will test every part of the condition. If you are testing calculations, use simple numbers that make it easier for you to check them. Be careful to test each part of the spreadsheet with **normal data** from which you would expect your formulas to produce appropriate results, with **extreme data** to test the boundaries, and with **abnormal data** which you would not expect to be accepted. Carefully check that all formulas work as you expect them to by using simple test data.

> ### Example 8.1 Test the validation rules set in Task_8ao
>
> As there are 14 students yet to sit the test, we will need to ensure the validation rule works for cells C8 to C80. We should really create test plans for each of these cells. As all test scores must be whole numbers we will need to test with integer as well as decimal values, test with two normal data items and two abnormal data items (decimals). As the highest score is 10 and the lowest must be 0, we need to test both extreme items **0** and **10** and some other selected normal data items. We also need to test with two abnormal data items that are outside the acceptable range.
>
> You need to create the test plan *before* entering any data into the spreadsheet. Test plans should be created for each cell in the range C8 to C80. A test plan for these validation rules for cell C8 would be similar to that shown in Table 8.7.
>
> ▼ **Table 8.7** Sample test plan for Task_8ao: cell C8
>
> | Test type | Validation | Cell | C8 | |
> |---|---|---|---|---|
> | Data entry | Data | Expected result | Actual result | Remedial action |
> | 0 | Extreme/Normal | Data accepted | | |
> | 1 | Normal | Data accepted | | |
> | 4 | Normal | Data accepted | | |
> | 8 | Normal | Data accepted | | |
> | 10 | Extreme/Normal | Data accepted | | |
> | -14 | Abnormal | Validation error message | | |
> | 1.3 | Abnormal | Validation error message | | |
> | 5.6 | Abnormal | Validation error message | | |
> | 1004 | Abnormal | Validation error message | | |

Example 8.2 Test the validation rules set in Task_8ap

As new products need to be added to the list, we will need to ensure the validation rule works for cells G5 to G130 (or more) and for F5 to F130 (or more, depending on the range of the entered rule). One example test plan for each validation rule should be enough. As the only chipset manufacturers used by this company are AMD and Intel, we will need to test with at least two normal data items and two abnormal data items (mis-spelled data). As the list of supported socket types can be found in the file **Socket.csv**, we need to test with three normal data items including those at the start and end of the list and two abnormal data items (mis-spelled data).

Two test plans are needed for these validation rules: one for cell G5, which would be similar to that shown in Table 8.8, and one for cell F5, which would be similar to that shown in Table 8.9.

▼ Table 8.8 Sample test plan for Task_8ap: cell G5

Test type	Validation	Cell	G5	
Data entry	Data	Expected result	Actual result	Remedial action
AMD	Normal	Data accepted		
Intel	Normal	Data accepted		
AME	Abnormal	Validation error message		
Intem	Abnormal	Validation error message		

▼ Table 8.9 Sample test plan for Task_8ap: cell F5

Test type	Validation	Cell	F5	
Data entry	Data	Expected result	Actual result	Remedial action
775	Normal	Data accepted		
AM3	Normal	Data accepted		
FT1	Normal	Data accepted		
776	Abnormal	Validation error message		
1FT	Abnormal	Validation error message		

Advice
There are no examples of extreme data in either of these tests as text data held in a list cannot have extremes.

For each test that you perform, write down each test data item and the expected results before trying each number in the cell. Check that the **actual result** matches the expected result for every entry. If not, change the validation rule before starting the whole test process again.

Check you have used the right ranges within formulas by testing that everything works before using real data in your model. If you find an error during testing, correct it, and then perform all of the tests again, as one change to a spreadsheet can affect lots of different cells. It is very important to test extreme data (where it exists) as it can reduce errors such as greater than being used rather than greater than or equal to.

Task 8au

Create this spreadsheet to calculate the surface area of a closed box.

Test the spreadsheet. Save your test plan as **TP_8au.rtf**.

Use the finished spreadsheet to calculate the surface area of a box 3.4 cm x 4 cm x 2.7 cm.

Save the spreadsheet as **Task_8au**.

	A	B
1	Dimensions in cm	
2	Length	
3	Width	
4	Height	
5		
6	Calculation	=B2*B3+2*(B2*B4+B3*B4)

Create and save the spreadsheet to appear as shown. Highlight cells B2 to B4 and add validation like this.

This rule excludes sides with no length or negative values, as a box cannot have sides with a negative length.

Four test plans are needed, three for the validation rules, one for each cell, although only one example has been shown here in Table 8.10 and one test plan for the formula in cell B6 which can be seen in Table 8.11.

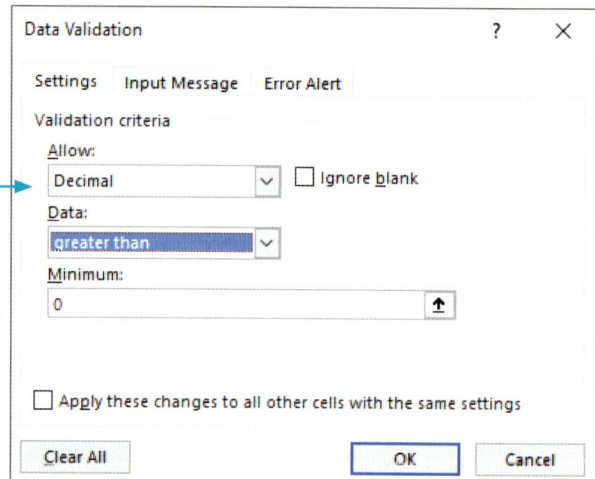

▼ **Table 8.10** Sample test plan for Task_8au: cell B2

Test type	Validation	Cell	B2	
Data entry	Data type	Expected result	Actual result	Remedial action
0.00001	Normal	Data accepted		
42.6	Normal	Data accepted		
1000	Normal	Data accepted		
0	Abnormal	Validation error message		
-0.001	Abnormal	Validation error message		
One	Abnormal	Validation error message		
-60	Abnormal	Validation error message		

Now we need to plan the test for the formula. The first sizes chosen are for a box 1 cm square. As a closed box has six sides and each is 1 cm², the total surface area should be 6 cm². The second items of test data will be 2 cm sides, again six sides, each with a surface area of 2 × 2 = 4 cm², so the total surface area will be 6 × 4 = 24 cm². The third items of test data will be for a box that is not a cube with square sides, with a length of 2, width of 2 and height of 3. This means that the base and top should each be 2 × 2 = 4 cm² and the four sides should be 2 × 3 = 6 cm², so the total surface area will be 4 + 4 + 6 + 6 + 6 + 6 = 32 cm². The fourth items of test data will have a length of 10, width of 5 and height of 2. This means that the base and top should each be 10 × 5 = 50 cm², two sides should be 10 × 2 = 20 cm², two sides should be 5 × 2 = 10 cm², so the total surface area will be 50 + 50 + 20 + 20 + 10 + 10 = 160 cm². We should also include decimal values to test it works for those, so the fifth items of test data will have a length of 10, width of 1.5 and height of 2. This means that the base and top should each be 10 × 1.5 = 15 cm², two sides should be 10 × 2 = 20 cm²,

two sides should be 1.5 × 2 = 3 cm², so the total surface area will be
15 + 15 + 20 + 20 + 3 + 3 = 76 cm². The initial plan would look like this.

▼ Table 8.11 Sample test plan for Task_8au: cell B6

Test type			Formula	Cell	B6
Data entry in cell:			Expected result	Actual result	Remedial action
B2	B3	B4			
1	1	1	6		
2	2	2	24		
2	2	3	32		
10	5	2	160		
10	1.5	2	76		

Use the test data from Table 8.10 to test the validation rules. Do they work as expected? Complete your table as you try each test.

Use the test data from Table 8.11 to test the formula. Does it work as expected? Complete your table as you try each test. The completed table with the results of each test is shown in Table 8.12.

▼ Table 8.12 Completed test plan for Task_8au: cell B6 – Version 1

Test type			Formula	Cell	B6
Data entry in cell:			Expected result	Actual result	Remedial action
B2	B3	B4			
1	1	1	6	5	Yes – change formula?
2	2	2	24	20	Yes – change formula?
2	2	3	32	28	Yes – change formula?
10	5	2	160	110	Yes – change formula?
10	1.5	2	76	61	Yes – change formula?

We must now identify where the error has occurred and change the formula. The first piece of test data is probably the most useful here as it shows that only five sides have been calculated. Correct the formula in cell B6 so it becomes **=2*(B2*B3+B2*B4+B3*B4)**. Attempt all the tests for cell B6 again. The resulting test plan should look like Table 8.13. Save your test plans as **TP_8au.rtf**.

▼ Table 8.13 Completed test plan for Task_8au: cell B6 – Version 2

Test type			Formula	Cell	B6
Data entry in cell:			Expected result	Actual result	Remedial action
B2	B3	B4			
1	1	1	6	6	No
2	2	2	24	24	No
2	2	3	32	32	No
10	5	2	160	160	No
10	1.5	2	76	76	No

> **Activity 8u**
>
> Open the file you saved as **Activity_8s**.
>
> Create a test plan to test the validation placed in cell C9. Use this to perform the tests on this cell, complete the test plan and save the test plan as **Activity_8u.rtf**.
>
> Save the spreadsheet as **Activity_8u**.

8.3 Using a spreadsheet

This section includes searching for data extracts, sorting data, summarising data, and exporting data. We do not need to cover importing data, as we have already imported data many times in this chapter.

8.3.1 Extract data

Search using text filters

This means getting *Excel* to search through data held in a spreadsheet to extract only rows where the data matches your search criteria.

> **Task 8av**
>
> Open the file you saved as **Task_8ap**.
>
> A customer requires a motherboard manufactured by MSI, that uses an AMD chipset and DDR3 or DDR4 memory. Extract from all the data only the motherboards that meet these requirements.
>
> Save the spreadsheet as **Task_8av**.

Open the file that you saved as **Task_8ap** and highlight cells A4 to I20, which is the range of data to be searched, and the headings for each column. Select the **Data** tab and find the **Sort & Filter** group. Click on the **Filter** icon to display an arrow in the top-right-hand corner of each column.

To start this task, you need to use this drop-down arrow in the Manufacturer column, to select those boards manufactured by MSI.

When you click on the **Manufacturer** arrow, a small drop-down menu appears like this.

In the lower part of the drop-down menu, click on the tick box for (Select All) to remove all of the ticks from every box. Then tick only the MSI box, then [OK] and the spreadsheet will not display any other manufacturer.

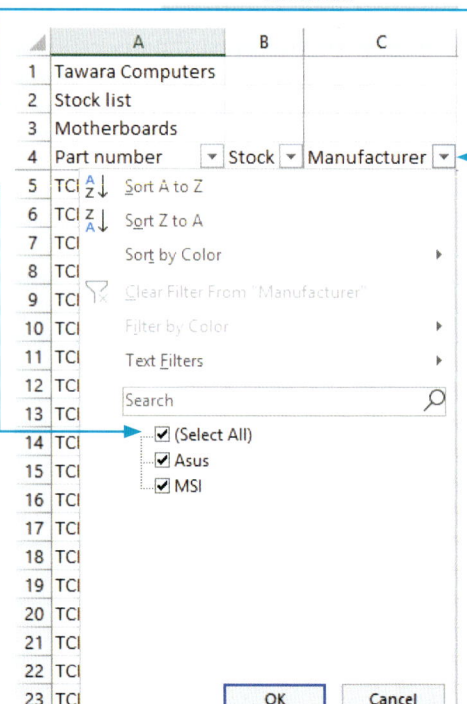

Use the drop-down menu in the Chipset column and select only those motherboards with an AMD chipset. What we have done so far is equivalent to an **AND** search using two columns. We have selected only the motherboards made by MSI AND that have an AMD chipset. For the next stage we are going to search in a third column, this time the Memory column, and the values that are required are DDR3 **OR** DDR4 so we need to tick both the DDR3 and DDR4 boxes.

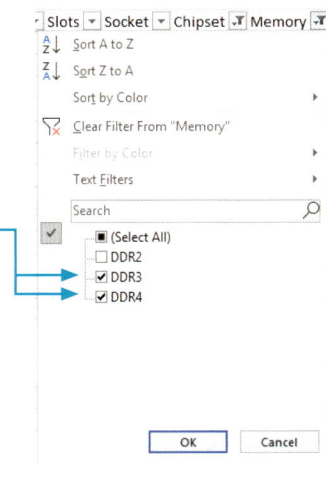

The result of this complex search displays 27 possible motherboards that meet the customer's requirements.

Save the spreadsheet as **Task_8av**.

Search using numeric data

Examine Table 8.14 carefully to relate how comparison operators can be interchanged with text in AutoFilter searches.

▼ Table 8.14 Comparison operators and their equivalent text

Operator	Text	Example:	
=	Equal to / Like	X=Y	X is equal to Y
<>	Not equal to	X<>Y	X is not equal to Y
>	Greater than	X>Y	X is greater than Y
<	Less than	X<Y	X is less than Y
>=	Greater than or equal to	X>=Y	X is greater than or equal to Y
<=	Less than or equal to	X<=Y	X is less than or equal to Y

Task 8aw

Open the file you saved as **Task_8av**.

This list meets the customer's needs but there are too many items to select from. Extract from all the data only the motherboards to meet these requirements. The customer wants to spend less than $130 but does not want to risk using a cheaper motherboard costing less than $80. Extract from the results of **Task_8av** only the motherboards to meet these requirements.

Save the spreadsheet as **Task_8aw**.

Open the file that you saved as **Task_8av** and select the drop-down menu for the Price column. For this task it is possible to tick or untick all the boxes, depending on whether each price listed meets both criteria, but this is inefficient. Instead, click on **Number Filters**.

From the drop-down list, select **Between** to open a **Custom AutoFilter** window. Enter the two values as shown. Make sure, using the drop-down lists in the **Price** group, to change 'is less than or equal to' to 'is less than' to match the wording of the task, like this.

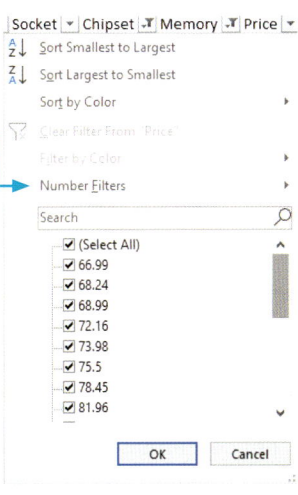

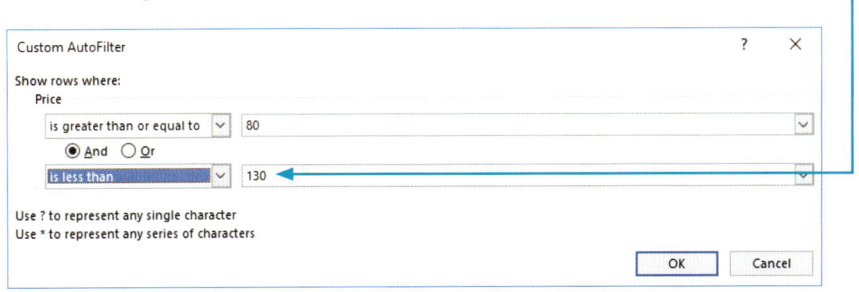

This search is again a form of AND search as the price must be greater than or equal to $80 *and* less than $130. The list of products for the customer will look like this:

	A	B	C	D	E	F	G	H	I
1	Tawara Computers								
2	Stock list								
3	Motherboards								
4	Part number	Stock	Manufacturer	Model	Slots	Socket	Chipset	Memory	Price
99	TCM0113	2	MSI	78-G41 PC Mate	2	FM2	AMD	DDR3	82.89
108	TCM0122	5	MSI	370 Gaming Pro Carbon	4	AM4	AMD	DDR4	117.84
109	TCM0123	5	MSI	B350 Gaming Plus	4	AM4	AMD	DDR4	81.96
110	TCM0124	2	MSI	350 GAMING PLUS	4	AM4	AMD	DDR4	122.14
111	TCM0128	4	MSI	350 TOMAHAWK	4	AM4	AMD	DDR4	94.35
113	TCM0130	4	MSI	S-7778 (2AEO)	4	FM2	AMD	DDR3	129.96
115	TCM0132	1	MSI	88X-G43	4	FM2	AMD	DDR3	94.35

Save the spreadsheet as **Task_8aw**.

Task 8ax

Open the file you saved as **Task_8aw**.

This list meets the customer's needs, but the customer has been advised that Model S-7778 (2AEO) is not suitable for their needs. Refine the extract from the results of Task_8av so that this model does not appear.

Save the spreadsheet as **Task_8ax**.

Open the file that you saved as **Task_8aw** and select the drop-down menu for the Model column. For this task it is possible to untick the box for this model, but we will remove it using a **NOT** operator. Select the drop-down list for model and click on **Text Filters**. From the drop-down list to the right, select **Does Not Contain** to open a **Custom AutoFilter** window. Use the right-hand drop-down and scroll down the list of models to highlight S-7778 (2AEO), like this.

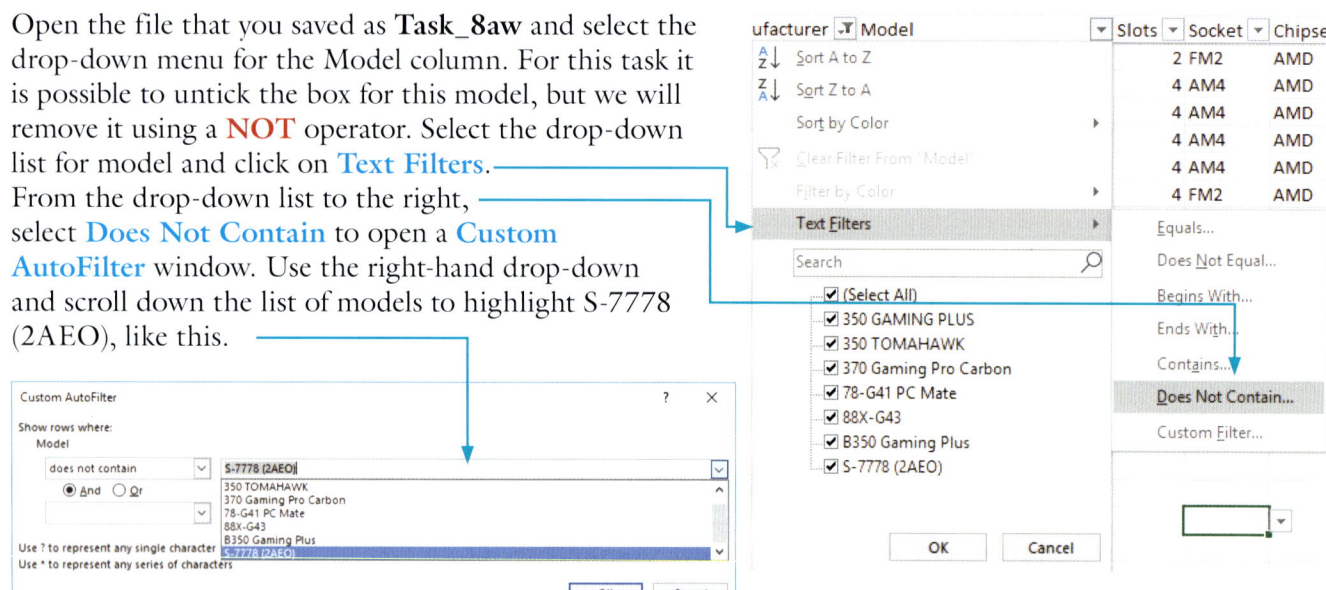

Click OK to refine the list. Save the spreadsheet as **Task_8ax**.

Search using date and time

Task 8ay

Open the file **Flights.csv**.

A passenger travelled on an early morning flight in January 2022. They could not remember the flight number but knew that they travelled on one of three days. Extract from this data all flights that departed before 9 a.m. on 16, 17 or 18 January 2022.

Save the spreadsheet as **Task_8ay**.

Open the file, save it as a spreadsheet and highlight cells A1 to E115. Select the **Data** tab, find the **Sort & Filter** group and click on the **Filter** icon. Use the drop-down arrow in the Depart_Date column to select the **Date Filters** drop-down menu to open the **Custom AutoFilter** window.

> **Advice**
>
> There is not scope in this book to cover all the different filters available here, or in the Text and Number Filter menus. Spend time experimenting with them to find out how they all work.

Use the drop-down menus and scroll down to select the start and end dates for departure like this.

Accuracy in the exact dates entered is crucial in order to produce the correct results. Click **OK** to create the extract.

Use the drop-down arrow in the Depart_Time column to select the **Number Filters** drop-down menu and the option for **Less Than...** to open the **Custom AutoFilter** window. To reduce typing errors, use the drop-down list on the right to select the time value.

Click **OK** to create the list. This will return these three flights. Save the spreadsheet as **Task_8ay**.

	A	B	C	D	E
1	Flight_code	Depart_Date	Depart_Time	Arrive_Date	Arrive_Time
44	TA2043	18/01/2022	08:10	18/01/2022	16:15
51	TA2050	16/01/2022	08:30	16/01/2022	10:50
64	TA2063	17/01/2022	06:30	17/01/2022	18:30

> ### Task 8az
>
> Open the file **Flights1.csv**.
>
> Identify all flights with a duration of less than three hours that arrived in February.
>
> Save the spreadsheet as **Task_8az**.

Open and examine the file **Flights1.csv**. This file has the dates and times together in a single column. Select the **Home** tab. In the **Number** section of the toolbar, format the contents of the cells in column D as time. Place in cell D2 the formula **=C2-B2** to calculate the duration for this flight. Replicate this formula for all flights. Highlight all the cells. Select the **Data** tab, find the **Sort & Filter** group and click on the **Filter** icon. Use the drop-down arrow in the Duration column to select the **Time Filters** drop-down menu to open the **Custom AutoFilter** window. Set the duration and value like this.

8.3 Using a spreadsheet

237

Click [OK] then use the drop-down menu for the Arrival column to select **Date Filters**. Near the bottom of the drop-down list is an option for **All Dates in the Period**; select this and then February from the sub-menu. Click [OK]. The results will be:

	A	B	C	D
1	Flight_code	Departure	Arrival	Duration
87	TA2086	03/02/2022 17:50	03/02/2022 18:50	01:00:00
88	TA2087	04/02/2022 04:10	04/02/2022 05:10	01:00:00

Save the spreadsheet as **Task_8az**.

Search using substrings

> **Task 8ba**
>
> Open and examine the file you saved as **Task_8ap**.
>
> A customer requires a motherboard made for gaming. The customer can remember that the model name she requires starts with 170. Extract from all the data only the motherboards to meet these requirements.
>
> Save the spreadsheet as **Task_8ba**.

Open and examine the file. You will notice that the only reference to the word 'Gaming' is in the Model column. Highlight cells A4 to I120 and select the **Data** tab, in the **Sort & Filter** group, click on the **Filter** icon. Select the drop-down arrow for the Model column and the option for **Text Filters**. From the sub-menu select **Contains...**.

Enter 'Gaming' into the box like this.

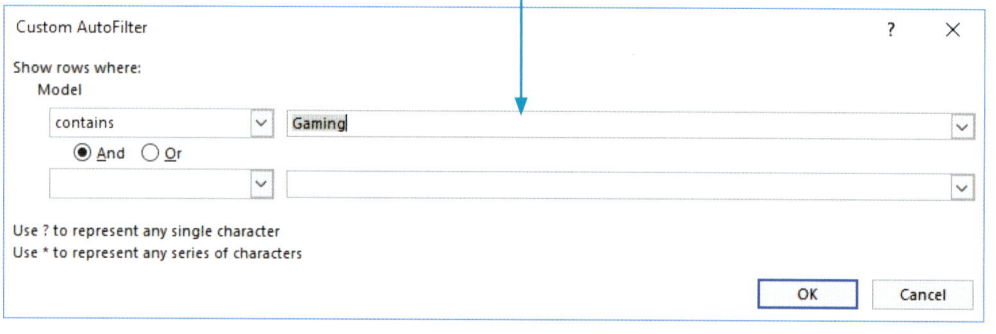

As this **Custom AutoFilter** window will also be used to find the model name beginning with 170, we add a second criteria to this filter. Make sure that the **AND** radio button is selected rather than OR and select **begins with** as the second filter type. Enter the text '170' in the final box like this.

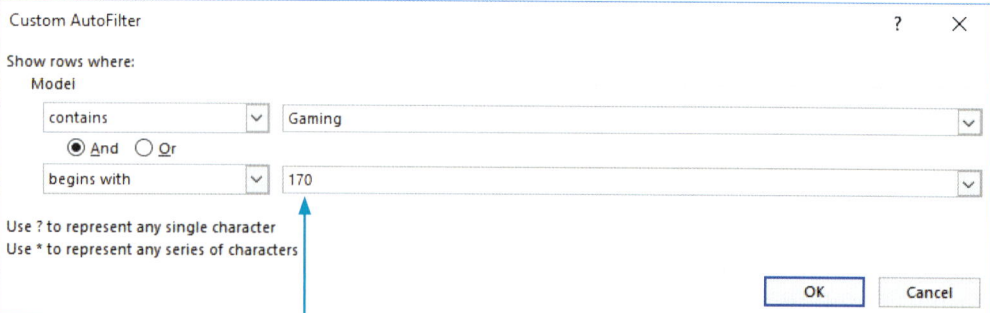

Click [OK]. The results will be:

	A	B	C	D	E	F	G
1	Tawara Computers						
2	Stock list						
3	Motherboards						
4	Part number	Stock	Manufacturer	Model	Slots	Socket	Chipset
82	TCM0095	4	MSI	170A Gaming PRO	4	1151	Intel
85	TCM0098	4	MSI	170A GAMING M3	4	1151	Intel
86	TCM0099	2	MSI	170A GAMING M5	4	1151	Intel
87	TCM0100	3	MSI	170A KRAIT GAMING	4	1151	Intel

To select a criterion that ends with some text, use the same method and select **ends with** as the filter type. Save the spreadsheet as **Task_8ba**.

Activity 8v

Open the spreadsheet you saved as **Activity_8s**.

Extract a list of students who attained both Maths and English at grade 1.

Save the spreadsheet as **Activity_8v**.

Activity 8w

Open the file you saved as **Activity_8r**.

Extract a list of flights that departed after 8 a.m. on either 7 January 2022 or 21 January 2022.

Save the spreadsheet as **Activity_8w**.

Activity 8x

Open the file you saved as **Activity_8q**.

Extract all the employees with names beginning with the letter S, who live in the centre of Bolton or Dundee.

Save the spreadsheet as **Activity_8x**.

Sort data

Before you sort data, make sure that you select all the data for each item to be sorted. A common error is to select and sort on a single column and therefore lose data integrity. Table 8.15 shows examples of correct and incorrect sorting on students' names for a spreadsheet containing their test results in Maths and English. Note how the results for each person have been changed when sorting without highlighting all the data.

▼ Table 8.15 Sorting data with correct and incorrect data selected

Original data			Sorted correctly with all data selected			Sorted with only the name column selected		
Name	Maths	English	Name	Maths	English	Name	Maths	English
Sheila	72	75	Karla	52	75	Karla	72	75
Marcos	64	34	Marcos	64	34	Marcos	64	34
Vikram	61	44	Sheila	72	75	Sheila	61	44
Karla	52	75	Vikram	61	44	Vikram	52	75

Task 8bb

Open and examine the file saved as **Task_8af**.

Hide columns L, N and columns B to I inclusive. Sort the data into ascending order of town/city, then ascending order of family name, then into ascending order of given name.

Save the spreadsheet as **Task_8bb**.

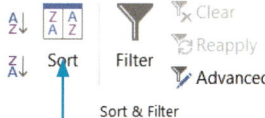

Open and examine the file. Hide the specified columns so that only columns A, J, K and M are visible. Highlight all visible cells. Select the **Data** tab, find the **Sort & Filter** group and click on the **Sort** icon.

This opens the **Sort** window. As all the data was highlighted, including the column headings in row 1, tick the check box for **My data has headers**.

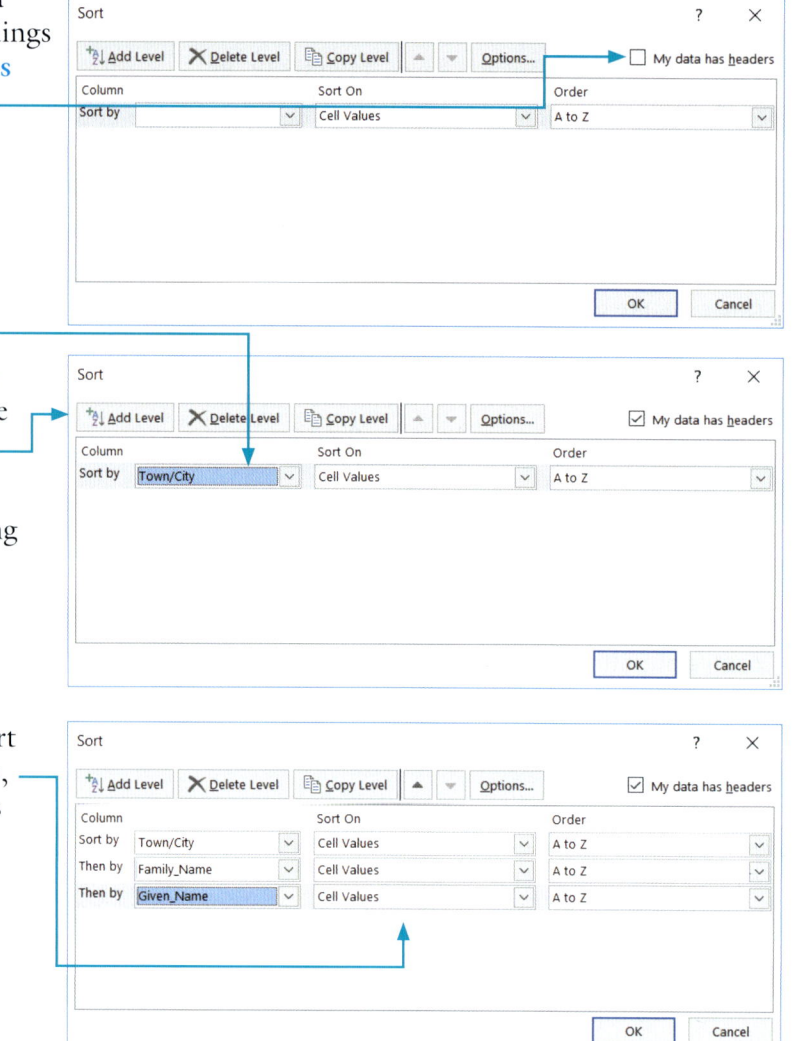

Use the drop-down arrow in the **Sort by** section to select the **Town/City** column as the primary sort.

Check that the right two text boxes contain 'Cell Values' and 'A to Z' (ascending) for the sort order. Click on the **Add Level** button.

Use similar methods to select the Family_name column as the secondary sort, checking that the right two text boxes contain 'Cell Values' and 'A to Z' for the sort order. Use similar methods to select the Given_name column as the tertiary (third-level) sort, checking that the right two text boxes also contain 'Cell Values' and 'A to Z' for the sort order. When the sort window looks like this, click OK to complete the sort. Save this as **Task_8bb**.

Advice

Data can be sorted into descending order rather than ascending order by selecting **Z to A** rather than **A to Z** in the **Order** box.

Activity 8y

Open the file you saved as **Activity_8b**.

Sort the data into ascending order of family name, then into ascending order of first name.

Save the spreadsheet as **Activity_8y**.

Pivot tables

A pivot table can analyse and summarise data into a two-dimensional table. It is used to see patterns and trends in data.

Task 8bc

Open and examine the file **BusAnalysis.csv**.

Analyse this data to count the number of buses between each destination. The 'buses from' should provide the row headings and 'buses to' should be the column headings.

Repeat this analysis, with the same display layout, to count the number of buses that were five or more minutes late between each destination.

Use these analyses to display the percentage of buses that were five or more minutes late on each route.

Show the information from the first analysis in graphical form.

Save the spreadsheet as **Task_8bc**.

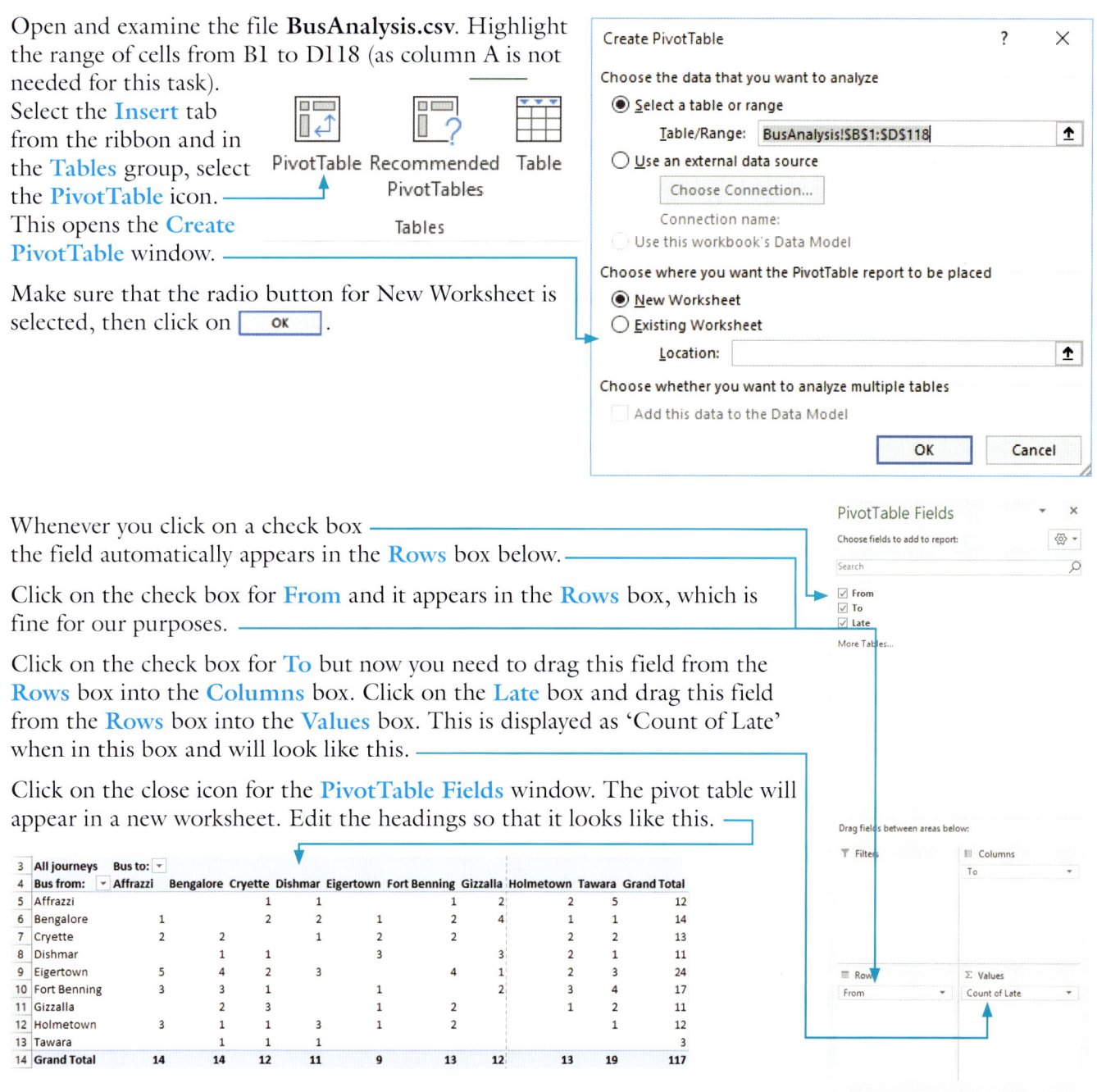

Open and examine the file **BusAnalysis.csv**. Highlight the range of cells from B1 to D118 (as column A is not needed for this task). Select the **Insert** tab from the ribbon and in the **Tables** group, select the **PivotTable** icon. This opens the **Create PivotTable** window.

Make sure that the radio button for New Worksheet is selected, then click on OK.

Whenever you click on a check box the field automatically appears in the **Rows** box below.

Click on the check box for **From** and it appears in the **Rows** box, which is fine for our purposes.

Click on the check box for **To** but now you need to drag this field from the **Rows** box into the **Columns** box. Click on the **Late** box and drag this field from the **Rows** box into the **Values** box. This is displayed as 'Count of Late' when in this box and will look like this.

Click on the close icon for the **PivotTable Fields** window. The pivot table will appear in a new worksheet. Edit the headings so that it looks like this.

All journeys	Bus to:								
Bus from:	Affrazzi	Bengalore	Cryette	Dishmar	Eigertown	Fort Benning	Gizzalla	Holmetown	Tawara	Grand Total
Affrazzi			1	1		1	2	2	5	12
Bengalore	1		2	2	1	2	4	1	1	14
Cryette	2	2		1	2	2		2	2	13
Dishmar		1	1		3		3	2	1	11
Eigertown	5	4	2	3		4	1	2	3	24
Fort Benning	3	3	1		1		2	3	4	17
Gizzalla		2	3	1		2		1	2	11
Holmetown	3	1	1	3	1	2			1	12
Tawara			1	1	1					3
Grand Total	14	14	12	11	9	13	12	13	19	117

Excel has counted the number of journeys made from one location to another during this period of time. You will notice that in cell A3, it has attempted to give a heading for the pivot table, but this is incorrect, as the count includes times where the bus was not late. Enter in A3 a new label 'All journeys'. To make it easier to understand the pivot table, change the label in A4 to 'Bus from:' and change the label in B3 to 'Bus to:'.

For the second pivot table, filter the original data in the BusAnalysis sheet so that only buses that were five or more minutes late are visible. Copy and paste all columns into a new worksheet in the workbook (otherwise all other data will appear in your pivot table). Highlight all the data in this new worksheet apart from column A. Create a new pivot table as described above but place the pivot table in cell A17 of the same worksheet that contains the first pivot table. It should look like this.

	A	B	C	D	E	F	G	H	I	J	K
17	5 minutes late	Bus to:									
18	Bus from:	Affrazzi	Bengalore	Cryette	Dishmar	Eigertown	Fort Benning	Gizzalla	Holmetown	Tawara	Grand Total
19	Affrazzi				1				2		3
20	Bengalore				1	1		2	1		5
21	Cryette	1	1		1				1		4
22	Dishmar				1		1				2
23	Eigertown	2	1	1	1		1		1	2	9
24	Fort Benning	1	1	1		1				1	5
25	Gizzalla								1		1
26	Holmetown							2			2
27	Tawara		1	1							2
28	Grand Total	4	4	3	4	3	3	3	4	5	33

To display the percentage of buses that were five or more minutes late on each route, we are going to copy the contents of each cell in the lower pivot table. To do this, enter into cell A31 the formula =A17 and replicate this to the right and down to fill all cells in the range A31 to K41. Do not copy the Grand Totals. To calculate the first percentage, type in cell B33 the formula =IF(AND(B19>0,B5>0),B19/B5,""). (Type this formula; do not click in cells B19 and B5 or the formula will be much more complex.) This will only calculate the percentage if both B19 and B5 contain a number greater than zero. It will remove division by zero errors or cells displaying 0%. Format this cell as a percentage and replicate this to the right and down to fill all cells in the range B33 to K41. The completed analysis should look like this.

	A	B	C	D	E	F	G	H
31	5 minutes late	Bus to:						
32	Bus from:	Affrazzi	Bengalore	Cryette	Dishmar	Eigertown	Fort Benning	Gizzalla
33	Affrazzi				100%			
34	Bengalore				50%	100%		50%
35	Cryette	50%	50%		100%			
36	Dishmar					33%		33%
37	Eigertown	40%	25%	50%	33%		25%	
38	Fort Benning	33%	33%	100%		100%		
39	Gizzalla							
40	Holmetown						100%	
41	Tawara		100%	100%				

Pivot charts

The final part of this task requires us to show the information from the original data in the BusAnalysis sheet in graphical form. Click on the filter button to remove the filter which we used when creating the pivot table. Highlight the range of cells from B1 to D118 (as we did for the first part of Task 8bb). Select the **Insert** tab from the ribbon and in the **Charts** group, select the **PivotChart** icon.

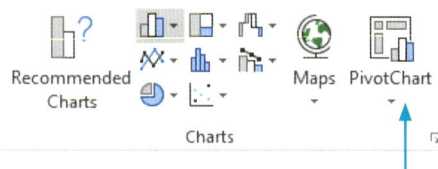

This opens the **Create PivotChart** window. Make sure the radio button for a **New Worksheet** has been selected then click on OK . When the **PivotChart Fields** pane opens, click on the check box for each of **From**, **To** and **Late**. The fields will automatically appear in the **Axis (Categories)** box. Leave the **From** field where it is but drag the **To** field into the **Legend (Series)** box and drag the **Late** field into the **Values** box. The resultant chart should look like this.

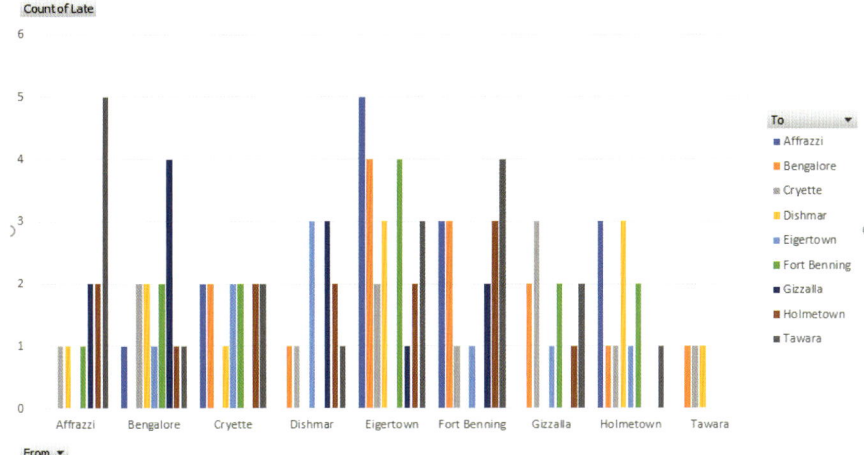

It is easy to see that buses leaving Tawara and Cryette are rarely late compared to those leaving other towns. Save the workbook as **Task_8bc**.

> ### Activity 8z
> Open and examine the file **Roles.csv**.
>
> Analyse this data to count the number of employees working in each department at each site. Use the sites for the column headings and the departments for the row headings.
>
> Display the same information in graphical form.
>
> Save the spreadsheet as **Activity_8z**.

Import and export data

We learnt how to import data at the start of the chapter. The only input formats that you are likely to experience are comma separated values (.csv) and text (.txt) files, and both have similar properties. It is unlikely that your spreadsheets will be exported into either of these formats, as they do not preserve the formulas that you have used. However, should you be required to export a spreadsheet into either of these formats, from the File tab, use the Save As option and select the appropriate file type from the drop-down menu.

Export in portable document format

It is more likely that you will have to export the values view of your spreadsheet into portable document format (.pdf).

> ### Task 8bd
> Open and examine the file you saved as **Task_8at**.
>
> Export the values view of this spreadsheet as a single page in portable document format.
>
> Save the spreadsheet as **Task_8bd**.

Open and examine the file. Ensure that the contents of all columns are fully visible. Select the File tab, then Export. Click on the Create PDF/XPS button.

Enter the filename **Task_8bd**. Set the format as .pdf, then click on Publish.

8.4 Graphs and charts

8.4.1 Chart types

Selecting the most appropriate chart type is often difficult to work out, so it is helpful to consider the options of a pie chart, a bar chart, and a line graph.

Pie charts

If you are asked to compare percentage values, a pie chart is often the most appropriate type because pie charts *compare parts of a whole*. An example would be comparing the percentage of males and females in a class.

Bar charts

Bar charts *show the difference* between different items. A bar chart is traditionally a graph with vertical bars, but this is called a 'column graph' in *Excel*. To create a vertical bar chart you would need to use the 'column chart' and for a horizontal bar chart (with the bars going across the page) you would need to use the 'bar chart'. An example would be showing the prices of five different items for sale.

> **Advice**
> Do not use stacked column charts or stacked bar charts instead of bar charts, as these display percentages or parts of a whole.

Line graphs

Line graphs are used to *plot trends* between two variables. An example would be plotting the distance travelled on a journey against time. You could then find any point in time on the graph and be able to read the corresponding distance, even if the exact distance had not been measured at that time.

> **Advice**
> Keep your charts simple, do not add features that are not needed. A simple chart is more efficient and effective.

8.4.2 Create a chart

Selecting the data

To create a chart, you have to highlight the data that you wish to use. This is easy if it is all **contiguous data** (where the data is placed together and can be selected in a single range). If you need to highlight **non-contiguous** data, hold down the <Ctrl> key while selecting your ranges of data (if you do not, the previously selected data is no longer highlighted).

> **Task 8be**
> Open and examine the file **Bolton2.csv**. Calculate the number of employees and the average salary for each department. Produce an appropriate chart to display the number of employees for each department.
>
> Save the spreadsheet as **Task_8be**.

Open the file. Enter in cell B20 the formula
=COUNTIF($B$2:$B$17,A20) and in cell C20 the formula
=AVERAGEIF($B$2:$B$17,A20,$C$2:$C$17). Replicate these

formulas for each department. Highlight only cells A18 to B24. The highlighted data will be the cells used to produce the graph and should look like this.

The cells containing the column headings (A19 and B19) have been included as they will be used as the labels in the chart, although they will be edited later to make them more appropriate.

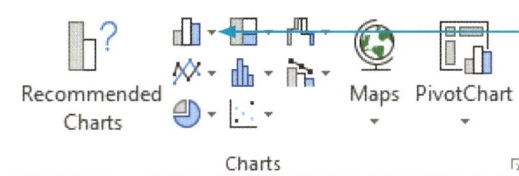

Selecting the chart type

Decide what type of chart you will need for this task. Look at the data and decide if it compares parts of a whole, shows trends between two variables or shows a difference. In this task the data shows the different numbers of employees in each job type, so a bar chart is the most appropriate chart type, and in this case we can use a vertical bar chart.

Create a bar chart

Select the **Insert** tab and find the **Charts** group. Select the **Vertical bar chart** icon (labelled **Column** in *Excel*) to access the drop-down menu to select the chart type like this.

> **Advice**
> The Recommended Charts icon will often select the most appropriate chart type, but not always.

From the **2-D Column** section, choose the first option.

The bar chart will be generated and look similar to this:

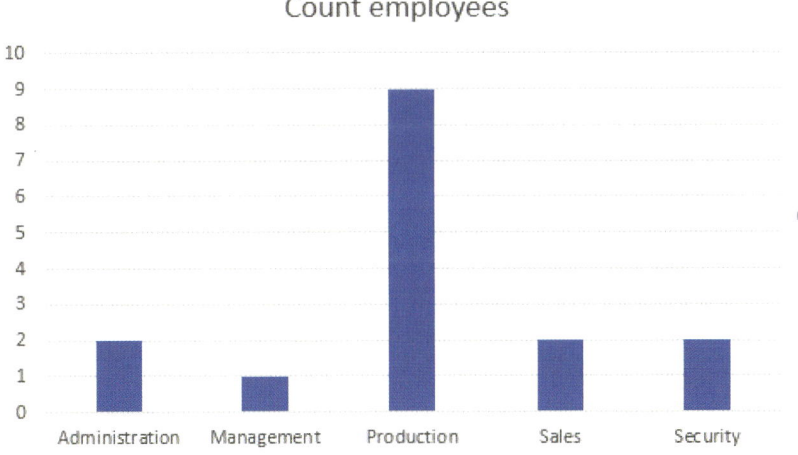

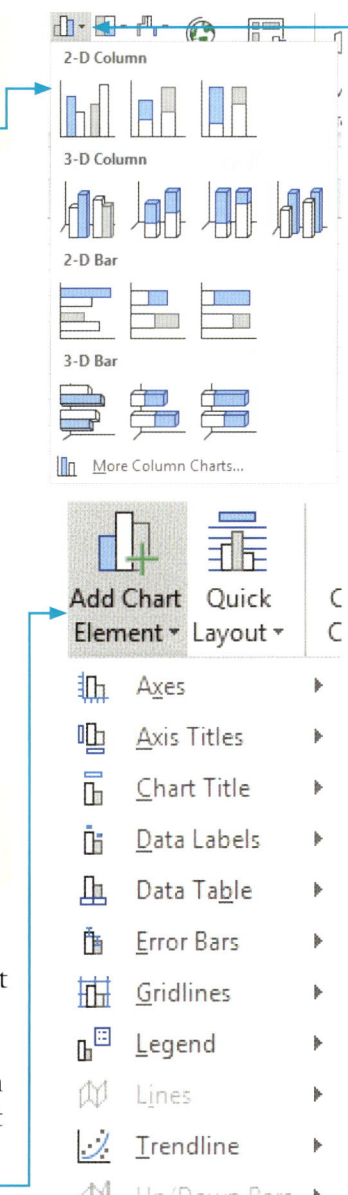

> **Advice**
> If you select the wrong chart type, you can always change this later.

Formatting the chart

Even though this chart appears to be finished, it does not tell a user clearly what it represents. To change the chart title, click within the text 'Count employees'. Edit this text to give a clear description of the chart, like 'The number of employees in each department at the Bolton site'. Each axis should also be given an axis title; first click on the chart to select it. Axis titles (and many other chart features) are found in the **Design** tab in the **Chart Layouts** group. Select the icon for **Add Chart Element** to get a drop-down menu like this.

8 Select **Axis Titles** from this menu. Select the option for **Primary Horizontal** and type in 'Department' to add this. Repeat this operation for **Primary Vertical** and type in 'Number of employees' to add the second title. Click the left mouse button in each title and edit them like this.

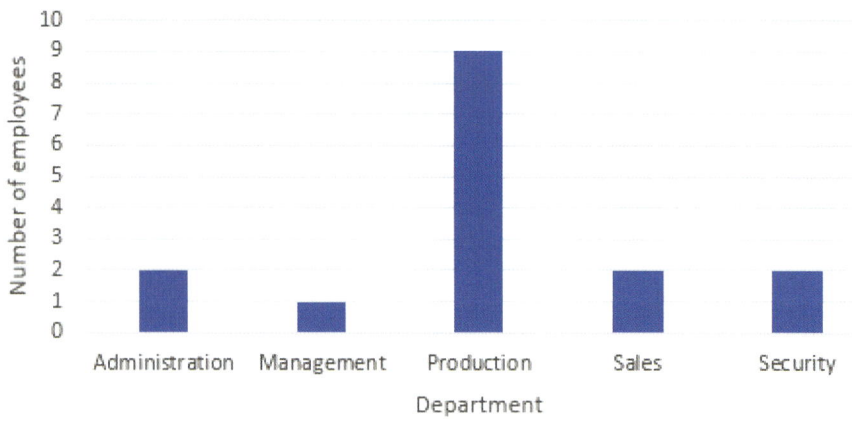

Save the spreadsheet as **Task_8be**.

> ### Task 8bf
> Open the file you saved as **Task_8be**. Produce an appropriate chart to compare the relative salaries for each department. Emphasise the department with the highest average salary.
>
> Save the spreadsheet as **Task_8bf**.

Create a pie chart

Open the file. We have already calculated the average salaries for each department. This needs displaying as parts of the whole, suggesting a pie chart. Highlight the non-contiguous data holding the <Ctrl> key while selecting the cells like this.

	A	B	C
19	Department	Count employees	Average salary
20	Administration	2	22500
21	Management	1	62000
22	Production	9	33111.11111
23	Sales	2	35000
24	Security	2	21000

Select the **Insert** tab and find the **Charts** group. Select the **Pie Chart** icon to access the drop-down menu to select the chart type. From the **2-D Pie** section, choose the first option.

The pie chart will be generated and look similar to this.

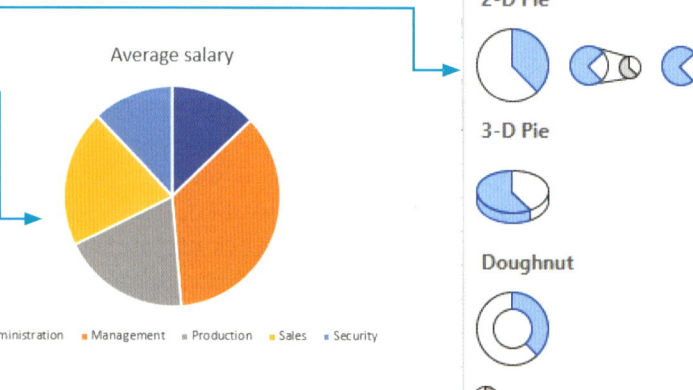

246

Change the title of the chart to a more appropriate one like 'Comparative salaries for each department at the Bolton site'. The question also requires you to emphasise the department with the highest average salary. As this is a pie chart, the best emphasis will be to withdraw the largest segment from the pie chart. Click on the segment for Management, continue holding the left mouse button down and gently drag the segment to the right so that it is extracted from the rest of the chart, like this.

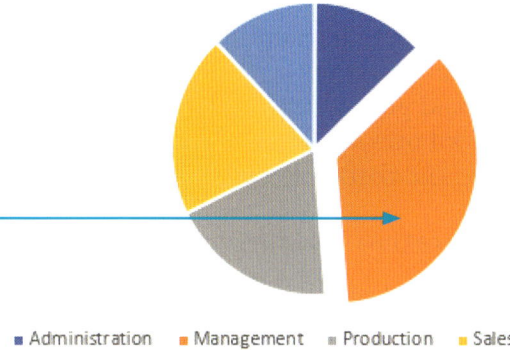

In this chart, the legend looks a little unusual at the bottom of the pie chart. To move it, click on the chart to select it, then select the **Design** tab in the **Chart Layouts** group, then select the **Add Chart Element** icon. Select **Legend** from the drop-down menu and, for this example, select the **Right** position.

The legend moves like this.

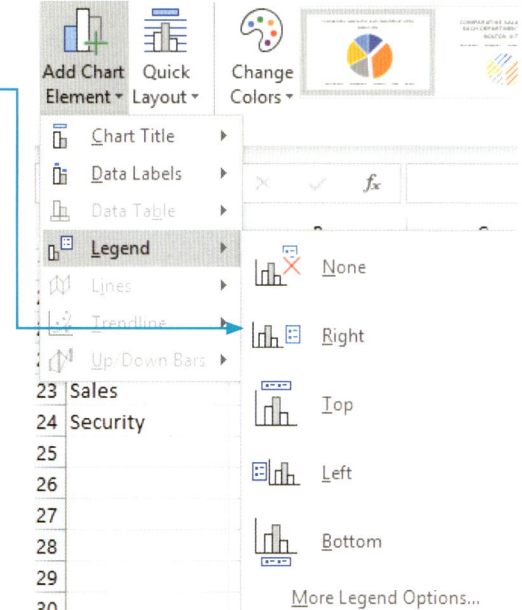

To add data to the segments of the pie chart, use the **Design** tab in the **Chart Layouts** group, and select the **Add Chart Element** icon. Select **Data Labels** from the drop-down menu, then select the **More Data Label Options…** from the bottom of the sub-menu.

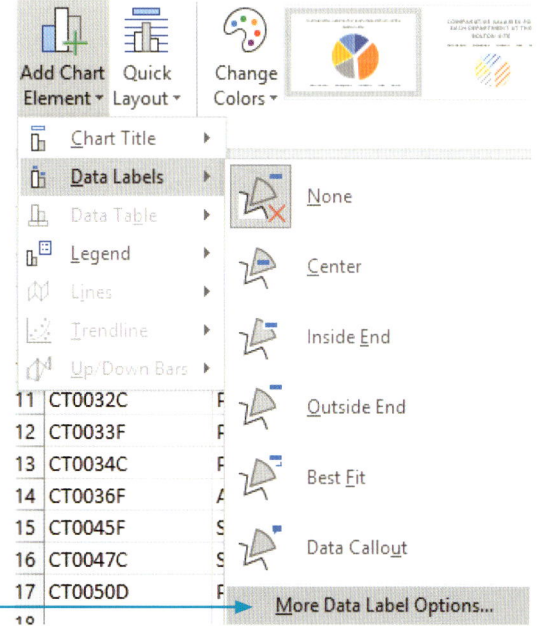

This opens the **Format Data Labels** pane. Click on the **Bar Chart** icon (shown here in green) to open the **Label Options** tab.

This tab can be used to display the values like this.

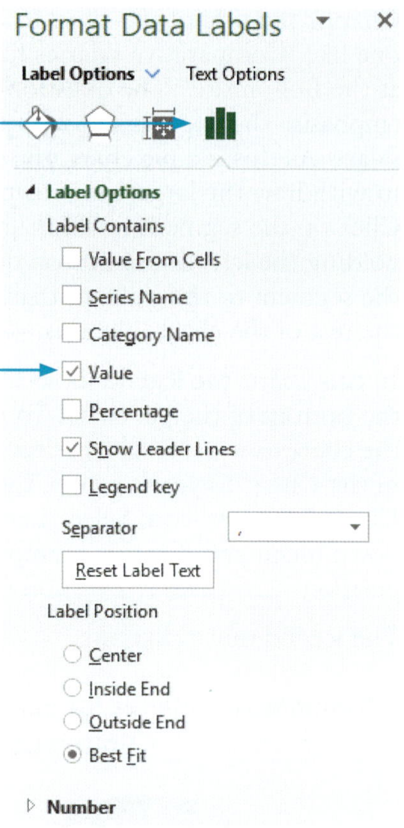

For this chart it would be more appropriate to display the percentage for each segment. Remove the tick from the Value check box by clicking on it and place a tick in the **Percentage** check box. As you change the tick boxes, the chart changes, so you can constantly check your changes. The chart will look like this.

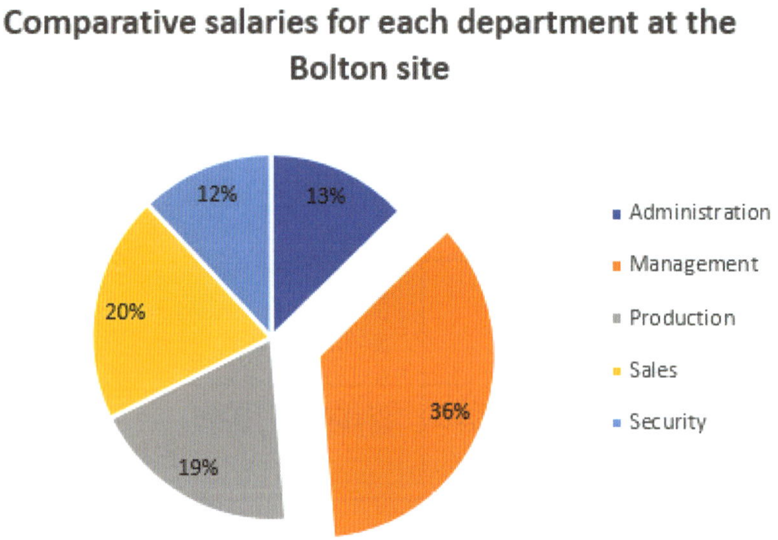

Comparative salaries for each department at the Bolton site

- Administration 13%
- Management 36%
- Production 19%
- Sales 20%
- Security 12%

Save the spreadsheet as **Task_8bf**.

Task 8bg

Open and examine the file **Rain.csv**. Create an appropriate graph or chart to compare the cumulative monthly rainfall for the towns of Tawara and Affrazzi.

Save the spreadsheet as **Task_8bg**.

Create a line graph

Open the file and highlight cells A1 to C15. Decide what type of chart you will need for this task. Again, look at the data and decide if it compares parts of a whole, shows trends between two variables or shows a difference. In this task, the data shows trends between the date and the total amount of rainfall that had fallen by that date. Because specific dates are used and the rainfall is cumulative, a line graph is the most appropriate chart type. As there are two towns shown in the data, you will make a comparative line graph using both data sets. Select the **Insert** tab and, in the **Charts** section, select a **Line** graph and the top-left icon in the **2-D Line** section.

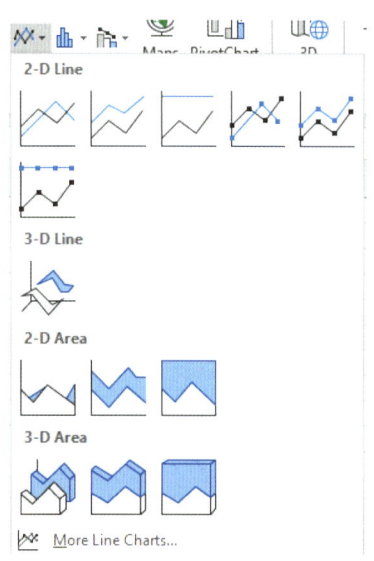

The line graph will be generated and look similar to this:

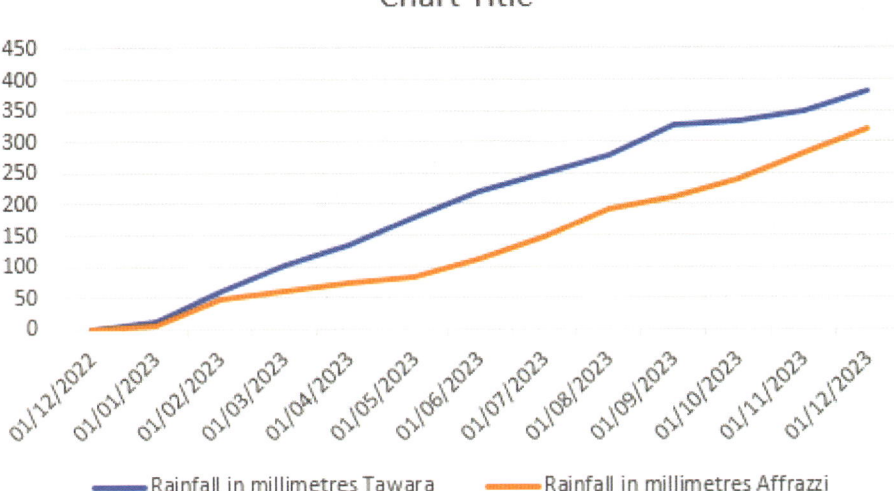

The chart will need appropriate labels, so the text 'Rainfall in millimetres' needs entering as the value axis title and an appropriate chart title like 'Comparing cumulative rainfall between the towns of Tawara and Affrazzi' needs adding. Add a category axis title of 'Date'. The finished line graph will look like this:

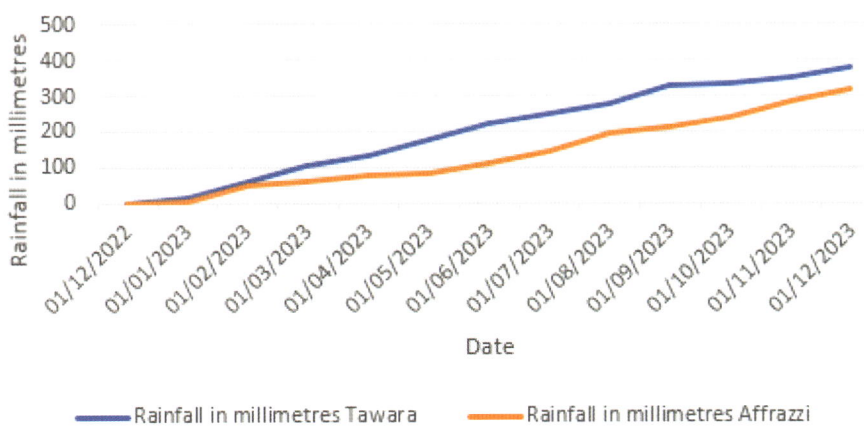

Save the spreadsheet as **Task_8bg**.

> ### Task 8bh
> Open and examine the file **Rain.csv**. Create an appropriate graph to show a comparison of the rainfall and average temperatures for each month in Tawara. Add a second value axis to the chart for the temperature. Label and scale these axes appropriately. Export the chart in portable document format.
>
> Save the spreadsheet as **Task_8bh**.

Create a combo chart

Open the file and highlight the dates and data for Tawara; this is in cells A1 to B15 and D1 to D15. Select the **Insert** tab then, in the **Charts** section, select the **Insert Combo Chart** icon.

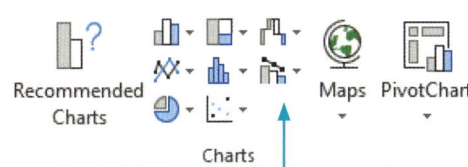

8 Use the bottom option to **Create Custom Combo Chart**, which allows you to compare two values using bar charts and/or line graphs and opens the **Insert Chart** window set to Combo (*Excel* attempts to predict the correct chart type), like this.

Both data series show trends between two variables (with both rainfall and temperature plotted against the date), so using line graphs for both series would be the most appropriate chart type. To make this happen, choose the **Chart Type** as **Line** for both series using the drop-down menus, selecting the first diagram of a line graph from the options.

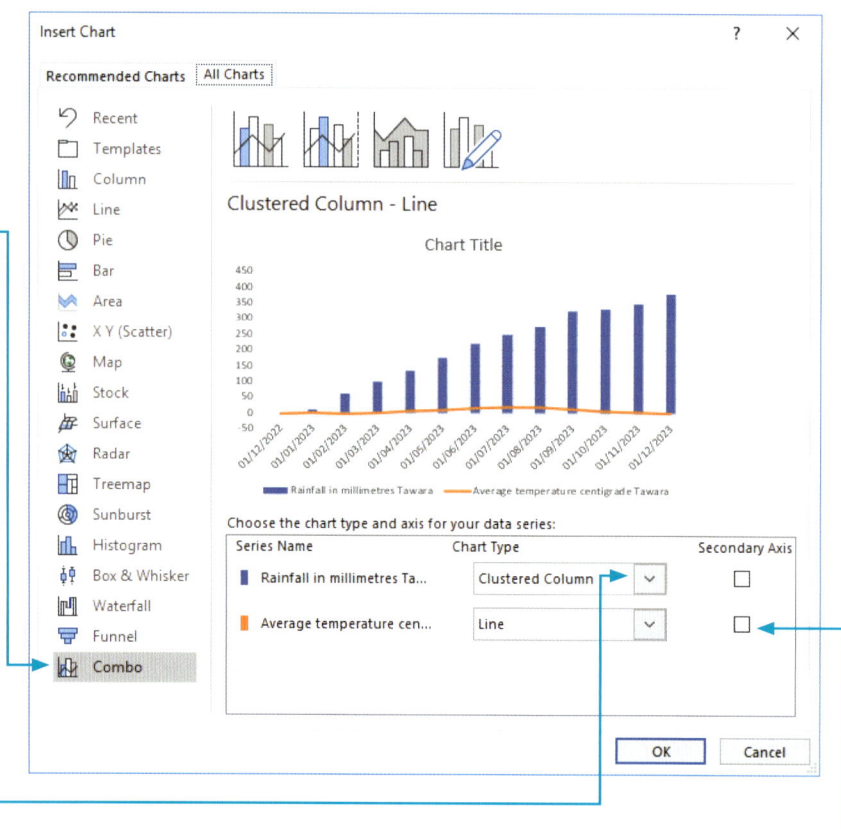

Use secondary axes

It is difficult to read the values for the temperature, so adding and scaling a second value axis will make it easier to read the graph. Click the left mouse button to place a tick in the check box for **Secondary Axis** for the temperature data series (the one shown in orange).

Your graph will now look similar to this.

Click on OK to create the chart.

Excel has attempted to scale these axes, but you are now going to adjust them further. You will change the primary axis so that it is set between 0 and 400 and the secondary axis so that it is set between −4 and 22. These values have been extracted from the original data: the total cumulative rainfall is 381 mm (therefore we will choose 400, so that the scale can go up in steps of 50); the temperature changes between −3 and 21 degrees (therefore we will use −4 and 22 so the scale can go up in steps of two). For this axis it would be acceptable to use the values −5 to 25 suggested by *Excel*.

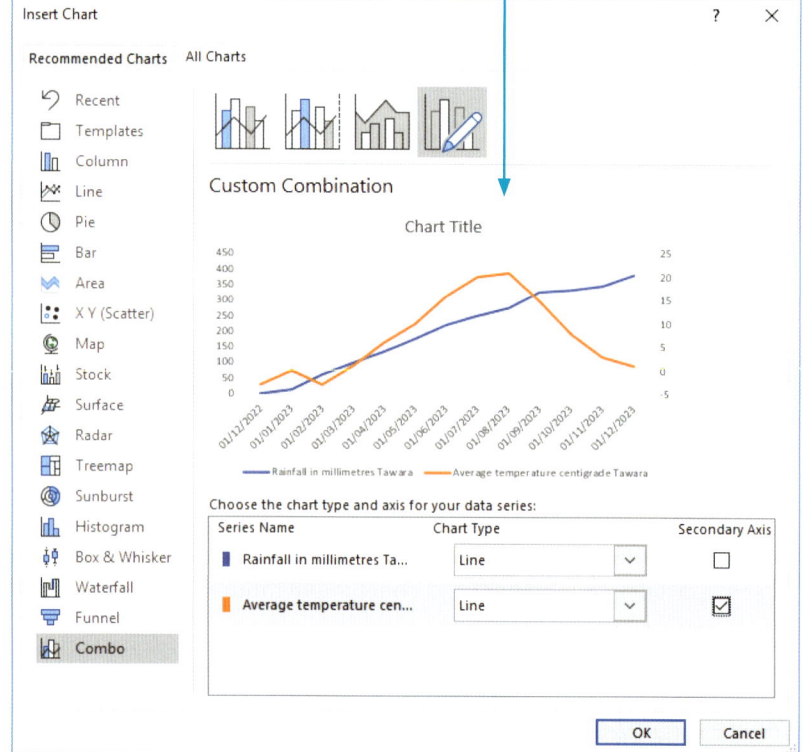

To change the primary axis values, click the right mouse button on the axis labels and from the drop-down menu select **Format Axis**.

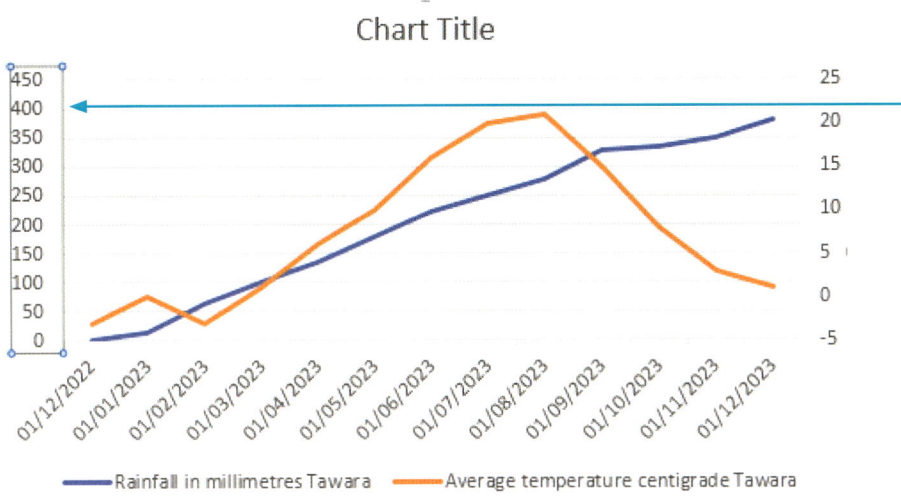

This opens the **Format Axis** pane on the right-hand side of the window. In the **Axis Options** section, the axis **Bounds** are set to 0 and 450. We want the bounds set to 0 and 400, so edit the **Maximum** boundary to 400.

Press the **<Enter>** key while hovering over the chart, or click the left mouse button on the chart to refresh it.

Click the left mouse button on the secondary axis. In the **Format Axis** pane change the axis settings to a **Minimum** value of –4, a **Maximum** value of 22 and, in the **Units** section, set the **Major** unit to 2. The **Minor** unit will change automatically.

It is important to label these axes appropriately. Label the primary axis (the left one) 'Cumulative rainfall in millimetres' and the secondary axis 'Average daily temperature'. Label the category axis 'Date'. Label the chart with a meaningful title, such as 'Comparison of rainfall and temperature in Tawara by <your name>'. These changes should leave the chart looking like this:

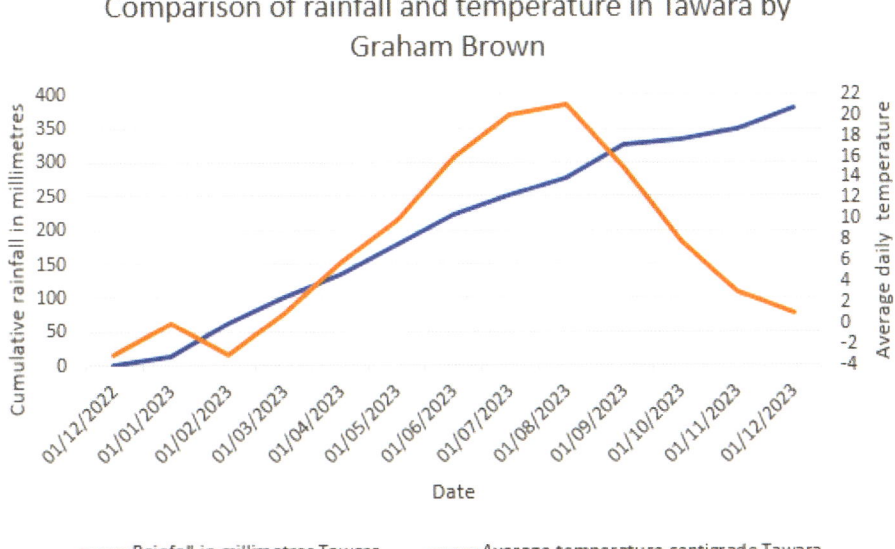

Save the file as **Task_8bh**.

Export graphs and charts

The final step of this Task was to export the chart into portable document format (a .pdf file). Click on the chart to select it, click on the **File** tab, select **Export** from the menu on the left, then double-click on the button for **Create PDF/XPS**. Enter the **File name:** as **Task_8bg.pdf**, then click on Publish.

> ### Activity 8aa
> Open and examine the file **Traffic.csv**.
>
> Create an appropriate graph to show a comparison of the maximum and minimum daily temperatures and the number of cars travelling on a road each day during the month of January. Add a second value axis to the chart for the two temperatures. Label and scale these axes appropriately. Export the completed chart in portable document format.
>
> Save the spreadsheet as **Activity_8aa**.

Practice questions

Tawara Construction is developing a spreadsheet to manage and analyse data about building projects.

1. Open the file **BuildingProjects.csv**.
   Insert a new row 12 and in cell A12 enter the text **Project 10**. [1]

2. Replace the word Task with the word Project wherever it appears in the spreadsheet. [1]

3. Enter the following data for Project 10:
Outline plans	21
Planning	30
Ground work	43
Building	37
Fixings	11
   [1]

4. Enter the following dates into these cells: [1]

Cell	Date
B3	28 August 2018
B4	16 October 2018
B5	2 November 2018
B6	30 November 2018
B7	1 December 2018
B8	2 December 2018
B9	30 December 2018
B10	5 February 2019
B11	2 March 2019
B12	1 April 2019

5. Delete row 2 from the spreadsheet. [1]

6. Centre align and embolden the contents of all cells in row 1 and all cells in column A. [2]

7. Place functions in cells C2, D2 and E2 to extract the day, month and year as numeric values from the start date of this project. [3]

8. In cell K2, calculate the end date for the project using the start date and the number of days it will take for outline planning, planning, ground works, building, and fixings. [2]

9. Format all the date cells in columns B and K into the format dd/mm/yyyy. [1]

10. Column L must display the end date of the project followed by the quarter. For example, the date 21/04/2019 should be shown as **21 April 2019 [Quarter 1]**.

    Insert a formula in cell L2 to display the date in K2 in the required format using the data from the file **Quarter.csv**.

    Replicate this formula for all events. [14]

11. Replicate the formulas used in steps 7, 8 and 10 for all events. [1]

12. In cell B14, use a function to count the number of events that end in the first financial quarter of 2018. You may include formulas in the working column to help you.

    Replicate this function for all events. [7]

13. Create a header which says **Project Completion** and a footer which contains your name and automated fields for today's date and the time. [4]

14. Sort the event data so that the end date of the most recent event is at the top.

    Save the spreadsheet with the filename **EQ1**. [2]

After analysing the spreadsheet your manager has noticed that some of the figures are incorrect. The company counts any business in January, February and March as taking place in the previous year. He knows that there were three projects that were completed in the tax year 2018.

For example: in the 'Full end date' column, 5 February 2019 is shown as 5 February 2019 [Quarter 4], but this must be shown in the 'Completion dates during' column as 2018 [Quarter 4].

15. Correct your formulas to show the correct counts for the *Completion dates during* section of the spreadsheet. Do NOT change the data displayed in the *Full end date* column. [18]

16. Change the header to **Project completion – corrected**.

    Save the spreadsheet with the filename **EQ2**. [1]

    [Total 60 marks]

9 Modelling

In this chapter you will learn:
- about what-if analysis
- about the characteristics of modelling software
- about the need for computer models
- about the effectiveness of spreadsheet models
- about the use of a model to create and run simulations
- about a number of situations where modelling is used
- how to use what-if analysis
- how to test a spreadsheet model.

Before starting this chapter you should:
- have studied Spreadsheets (Chapter 8).

For this chapter you will need these source files:
- TuckShop.csv
- Widget.csv

9.1 Modelling

Modelling means many things to many people. Many students confuse computer modelling with physical modelling, such as creating prototypes. A prototype can be a physical model which has been built to a smaller scale for use in trials and testing. Prototypes can be very costly since they can be damaged during testing and may have to be rebuilt, which means that extra raw materials have to be bought. In more recent times it has become possible to use a computer model to test new cars, for instance, at the design stage without having to build and then damage actual cars.

For our purposes, modelling means the use of computers to represent real-life situations. These models are usually financial, scientific or mathematical. It is the creation of a computer-based simulation to see what would happen in certain circumstances. It is not the creation of a working model, like the prototype of a new car or a model of a new building, which are also known as 3-D (three-dimensional) models.

Computer models can be complex and often require expensive purpose-built software and a mainframe computer to operate. Spreadsheet software can also be used to create a computer model, but often businesses require a more sophisticated method or solution.

9.1.1 What-if analysis

Later in this chapter you will see how to use what-if analysis, but before you do it is worth considering what it is. It is basically asking a question such as *what* will happen *if* I perform a certain action. For example, if you have $5, *what* will happen *if* you spend $4? The answer is quite easy to predict; you will have $1 left. Unfortunately, not all situations are so straightforward.

Consider a company which is making steel ingots. It employs a number of workers in its foundry to produce the ingots, and a number of office staff to deal with company orders and organise the payroll. It also has an office manager and a foundry manager.

Look at this spreadsheet which displays data for one week. It shows the employees and how much they earn. It also shows how many ingots are produced and how much money they are sold for. In order to work out the company's weekly profit, the running costs have to be taken into account. Fortunately, in our model, this figure does not change from week to week.

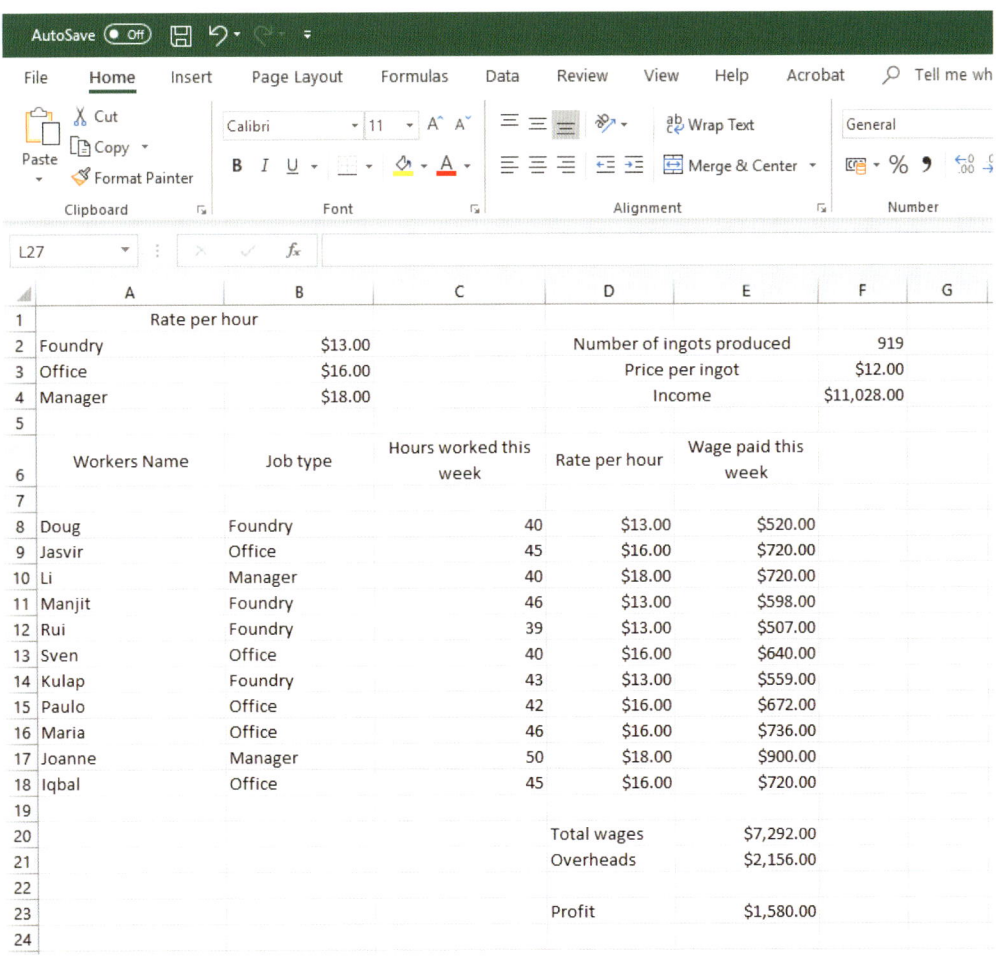

What-if analysis can be used on this spreadsheet. The foundry workers have been complaining for some time that they are not paid enough. The owner of the company, Morag, could use the spreadsheet to perform some what-if analysis. She could ask the question 'What would happen to my profits if I increased their pay from $13 to $14?' This would be easy to do just by changing the value in cell B2 to 14. The value in cell E23 would change, revealing the new profit. In this case profits would reduce from $1580 to $1412. She could then ask the question 'If I now increase the price of the ingots to $12.50, would that make up the difference?' The answer is that it would make more money. She could experiment with this model to see the effect on her profits of increasing the wages of any, or all, of the workers. She could also investigate the effect on her profits of increasing the price she sells the ingots for, or indeed of the company producing more ingots.

Another use of this spreadsheet could be to use the **Goal Seek** feature. You will learn how to use this later in this chapter. For now, we will concentrate on what it can let you do. It enables you to be more precise when you are using what-if analysis. Suppose you know the value you want a formula to produce; however, you want to know what values to enter into the formula. This is where Goal Seek helps. The Goal Seek feature, available in most spreadsheet software, is a what-if analysis tool that enables you to find the input values needed to achieve a goal or objective. To use Goal Seek, you select the cell containing the formula that will return the result you are seeking. You then input the value you want the formula to return and enter the cell reference of the input value that the spreadsheet can change to reach the target. In our example spreadsheet above, Morag could change the value in B2 to 14, as before, but then use the Goal Seek feature. She could enter the required profit as $1580. She could then ask it to tell us how many ingots (cell F2) the company would have to produce to keep the profit at $1580. Alternatively, she could ask it to tell us what price she would need to charge for each ingot (cell F3), if the number of ingots produced was left at 919. Goal Seek will be studied in more detail later in the chapter.

9.1.2 The characteristics of modelling software

As in most systems, modelling software has inputs often using the keyboard or prepared data files, a set of embedded rules used to manipulate the data, and outputs. There is a significant difference between a computer model and a calculation. Sometimes both models and calculations process large quantities of numeric data. However, the rules governing the model can be changed, whereas they cannot be changed in a calculation.

Spreadsheet software is often used to create computer models. The characteristics of spreadsheets which make them so suitable for this purpose include the ability to replicate formulas throughout a model but also influence this replication through the use of absolute and relative cell referencing. This makes sure you only increment the parts of a formula you need to by using relative cell referencing. The use of absolute cell referencing ensures that those parts of the formula you wish to remain constant do so.

Another characteristic is the use of cell protection. This ensures that the cells which are crucial to your model are not changed accidentally by someone who is using the spreadsheet.

There are also user interface forms. These make it easier to input values into the model by showing clearly where data is to be entered.

With spreadsheets you can create macros. Macros are a way of storing the keystrokes you make on your keyboard. They make it easier to create more complex formulas or functions.

One spreadsheet characteristic that makes it easy to use models is automatic recalculation. This means that when you alter data in a spreadsheet any formula which makes use of that data is automatically recalculated. It is not necessary to evaluate a formula every time you change data. Conditional formatting in spreadsheets allows you to highlight certain values that match specific criteria.

This means that you can see when positive values produced by a formula change to negative if a formula is recalculated. You can see the change instantly.

Spreadsheets also make use of graphs or charts to reveal trends over a given period of time.

9.1.3 The need for computer models

There are a number of reasons why computer models are useful and necessary.

Cost

As has already been implied, one of the major reasons we need models is because of cost. We always have to pay for our mistakes. If engineers are designing structures such as bridges and high-rise buildings, or designing cars and planes, it is very expensive to correct mistakes. If a computer model is used, then the costs of any mistakes practically disappear.

Testing

Another reason is the ability to test the design. It would be impossible to build several different types of building just to see which is most likely to survive heavy flooding. However, it can be done relatively easily with a computer model.

Prediction

It is impossible to travel through time, so in order to predict events over a large time span, such as climate change or population growth, computer models are used as they are the best way of making predictions. Another use of computer models is to accurately predict when storms will occur and the severity of them, thus enabling early warnings to the public about how to protect themselves and their property.

Time-efficiency

Modelling also saves time as well as money. Car manufacturers can perform more tests in a shorter space of time with a computer model than by arranging tests of prototypes with crash-test dummies inside them over and over again. Computer models make it possible to run many more tests.

Planning

Simulations are a specific type of computer model and are very useful in planning for rare or unexpected events. For example, they can be used to help predict and plan for the after-effects of natural disasters, such as volcanic eruptions, earthquakes, and tsunamis.

Safety

Some situations are dangerous, such as experimenting with a nuclear reactor. A computer model is a much safer way to do this.

9.1.4 The effectiveness of spreadsheet models

You learnt in Chapter 8 how to create, edit, and use a spreadsheet. Here we are going to learn about their benefits and drawbacks.

Benefits of spreadsheets

One benefit you will already have noticed is that calculations can be performed a lot more quickly than if you were writing them down or even using a calculator.

One of the reasons for this is that formulas are recalculated automatically. Imagine you have a cell, E2, with a formula in it such as B2*C2*D2, like this.

	A	B	C	D	E
1					
2		5	4	6	120
3					

E2 — =B2*C2*D2

Every time you type a new value in cell B2, the formula will automatically recalculate and produce a new value in E2. Here you could type 7 into cell B2 and the answer 168 would automatically appear in E2. If you were doing it manually, even on a calculator, it would be slower as you would have to type in 7*4*6. Obviously, the more complex the formula, the more time you save.

It is also possible to automatically import data into a spreadsheet from a database. Suppose you want to create a spreadsheet and you already have a database which contains the data you are going to use. Most spreadsheet software allows you to select the database as your source and then load the data directly from that database. This saves you the trouble of having to manually input data into your spreadsheet.

Spreadsheet models are very useful for observing patterns which take place over a long period of time. They are particularly useful when studying animal population growth. To observe population growth in real time would take many years. If an accurate computer model is created, the results will be known quite quickly. Also, if certain variables changed within the animal's environment, the effect would be noticed much sooner. Otherwise, waiting for the results could take years to happen in real life.

Another useful property of spreadsheet models is the automatic updating of graphs. In a business spreadsheet, for example, graphs might be produced which show the profits, output, costs, and income, among other variables. It is usual to show the trends over a period of time. These are useful when making predictions about the future prospects of a company. Often, what-if scenarios are carried out, with some values within the model changing in order to show how this would affect future trends. Any graphs which show these trends are automatically updated as values are changed within the model. New graphs are produced automatically.

Another feature of spreadsheet models is that they can remove human fallibility. This means that the likelihood of decisions being made based on personal feelings is no longer a problem. It is still the case that, for example, individual representatives of banks or credit unions are faced with making a decision regarding offering a mortgage or loan to someone so the person can buy a house. Sometimes the decision is a personal one and could even depend on how the representative was feeling that day! More and more, spreadsheet models are being used to make these decisions. The obvious benefit of this is that the decisions made by spreadsheet models are consistent for the same inputs, which is not always the case with different humans.

Another benefit of spreadsheet models is that templates exist for regularly used spreadsheets. Users can save time setting up the spreadsheet by using a template which might just need to be slightly amended to suit the particular situation. Users do not always have to be experts because the templates already exist which means they do not need to have great computing expertise.

Validation rules can also be built into spreadsheet models. This means that data input errors are reduced as the validation will ensure only data which is reasonable is allowed to be entered. This would have saved the ticket organisers of the London 2012 Olympics™ a lot of embarrassment. There were often issues with people not getting the tickets they asked for. Some of this was due to the high demand for tickets. However, the Olympic Committee sold more tickets than the number available for some swimming events. It was reported that the error occurred when a member of staff typed in '2' instead of '1' and entered '20 000' into a spreadsheet rather than the correct figure of 10 000 remaining tickets. The error appeared when organisers compared the number of tickets sold with the actual seating arrangements at the venues. They had to get in touch with the ticket holders and try to persuade them to accept seats at other venues. Of course, if proper validation procedures had been in place, this would never have happened.

Drawbacks of spreadsheets

However, there are some properties of spreadsheet models which reduce their effectiveness. One such problem occurs because it is very difficult to include every possible variable in a spreadsheet model. In population growth models, it is very difficult to accurately predict patterns. There are numerous variables that affect the study of a particular species; factors such as the number of predators, birth rate, illness and so on, all need to be considered. In addition, the variables involved tend to be considered only at the present time. It may be difficult to then predict what the variables may become over the length of time the model is covering. It becomes easy to leave out some of these variables.

In financial spreadsheet models, there may be a number of variables which are fairly straightforward to identify. However, there may be problems in predicting future patterns of spending, borrowing and investing, as it is difficult to accurately model human behaviour; it is just not possible to quantify or turn it into a number.

Another aspect of modelling which makes it difficult to predict trends is the effect that sudden emergencies will have on real-life behaviour. The banking crisis of 2008 forced governments to make decisions. Some governments reacted by putting more money in to their economy. Other governments cut spending. There was no way of predicting how individual governments would react. As a result, economic spreadsheet modelling became problematic.

Some situations will need software to be specially designed and/or computer experts to create a model, and neither of these is inexpensive to purchase or invest in. Spreadsheets are often used for implementing a new financial recording and reporting system. Aspects such as invoicing can be complex if different types of invoices and methods of producing these are required for different companies. In this case, it is often easier to bring in expertise from outside. Another aspect that spreadsheet users sometimes fail to take into account is the fact that it may take a very long time to produce a complex model. Various phases are used when producing a model. The problem has to be identified and broken down into tasks which make it easier to solve. The inputs and outputs of the model have to be identified. The user requirements have to be taken into consideration. The spreadsheet model has to be created and then it needs to be tested. Documentation for the user has to be produced. Finally, the model has to be improved and updated as necessary. All this contributes towards the effectiveness of the model, but it can take time.

When an expert has been through all these stages, the staff at the organisation which is going to use the spreadsheet need to be trained to use this model. This, of course, creates an extra expense to the organisation on top of paying the creator(s) of the spreadsheet model.

One other factor which affects the effectiveness of a spreadsheet model is human errors introduced to the model by users. This can cause even the most sophisticated model to give out inaccurate results when incorrect data is input. Remember the mistakes made at the 2012 Olympics when a member of staff typed in '2' instead of '1' and entered '20 000' into a spreadsheet rather than the correct figure of 10 000 remaining tickets? Another problem can occur due to the complexity of some spreadsheets. We have seen one of the key phases of producing a model is the testing of it to make sure it does what we want it to. Having data spread over different workbooks and worksheets can make them very difficult to test thoroughly.

9.1.5 Uses of computer models

There are many types of computer model. Here, we will look at some examples of the uses of modelling. Remember, you will be required to apply and adapt all the knowledge you have acquired in the preceding sections: what-if analysis, the characteristics of modelling software, the need for computer models, and the effectiveness of spreadsheet models for each of the following uses.

Financial forecasting

Earlier in this chapter we looked at Morag's company and we saw a simple spreadsheet model which could be used for financial forecasting. Companies need to be able to make a profit. It is important to them that they can model possible future events. Our model was fairly simplistic but demonstrated how a model can be used to predict profits. It involved calculating the costs involved in producing ingots and subtracting these from the income obtained when selling the ingots. This type of model tends to make use of what we call off-the-shelf spreadsheet software. This is software which has been produced for general use and does not take into account the needs of specific users.

Details of income and outgoing costs were entered into the spreadsheet and a formula linking the two was used to calculate the profit. We can see, using the model, the effect on the amount of profit a company makes of changing the variables. This is making use of the what-if feature of spreadsheets. It is also making use of the automatic recalculation feature of spreadsheets which means you do not have to ask the computer to run the calculation every time you change individual prices. The fact that we can change the variables is the most important property of spreadsheets when using them as models. User interface forms make it easier to input quantities/costs into the model.

Conditional formatting is another spreadsheet feature that is beneficial in this context. It allows you to highlight particular values that match particular criteria. For example, you can set cells which are showing a profit to a particular colour. This enables you to see immediately which items are making a profit.

However, most companies require more intricate models than this. Companies need to be able to predict the effects of changing the rate of selling a product. Sales of certain products vary depending on the time of year. There is no point spending money on producing a certain number of a type of product if it is not going to sell anywhere near that amount. The company would end up making a loss.

In our model, we assumed that the company's overheads were fixed. This is not always the case. We might want to observe the effects of changing rates of taxation, energy prices, rentals or mortgage payments on a company's property.

This would obviously require a more complex model. Complex models take into account the current economic climate and its effect on customers' willingness to spend money. They also take into account the effect of purchase taxes when predicting the number of products which will be sold over time. This is often represented in the form of a graph showing selling trends.

Graphs are also used to illustrate forecasts, such as which goods are likely to make profits over a period of time, and allow you to compare the profit levels and sales of different goods; you can see the profit made by each product very clearly. The model can then be used to predict profits based on the rate of selling.

As we have seen, one of the most useful features of spreadsheets is the use of the Goal Seek facility. This can be used to produce results showing the number of products that must be sold in a given period for the company to make a specific profit. Alternatively, you can use it to see what price you need to charge for certain goods in order to make that profit.

Another feature of a spreadsheet is the use of absolute and relative cell referencing. This ensures you only increment the parts of a formula you need to, such as the prices or costs of individual items to see their effect.

Complex models used by financial organisations can be used by several employees. Some of these employees may be inexperienced in the use of spreadsheets. One of the most useful features of spreadsheets is the use of cell protection. This prevents these inexperienced users changing formulas or data by accident. This is particularly useful when protecting cells containing fixed costs such as overheads, making sure that only senior staff can amend these as changes occur.

Macros are another useful feature of spreadsheets; if a company wanted to compare the effect of different costs at the same time, a macro can be used which makes it simpler to create more complex formulas or functions.

More sophisticated financial models are often used by governments to predict the future of a nation's economy. Most governments collect money in the form of taxes as well as import and export duties. They use this money to spend on services such as the police, army, education, roads, and public transport. Most countries have some type of purchase tax, so that when goods are bought in shops some of the price paid is sent by the store keeper to the government. Most countries also have some form of income tax. People who work have to give some of their income to the government. Governments can use computer models to see how much extra money would be raised if they increased taxes. They can do this for a variety of scenarios. Alternatively, they could change the amount they spend and see what effect this would have on the need to raise or decrease the tax rate. There are other variables, such as interest rates, that could be changed to observe their effect on the model.

Population growth

Population growth is the study of how populations of any type of organism, such as animals, bacteria, fungi or plants, increase in size over time. Population growth tends to fall in to one of three classifications. The first is termed 'exponential growth'. This means that the number of animals increases in an unlimited way. An example of this is the explosion of the rabbit population after their introduction to Australia. Although prior to 1859 there had been some rabbits introduced to Australia with no noticeable consequence, it was in this year that a farmer called Thomas Austin imported 24 wild English rabbits and set them free on his land. It is claimed that the population had increased within six years to 22 million. This may be an exaggeration, but it is, however, documented that rabbits had become so numerous that within ten years of their introduction in 1859, two million could be shot or trapped annually without having any noticeable effect on the population. By 1950, the population was estimated to be 600 million. It shows that given enough food, water and other resources required for life, populations can increase exponentially without limit. The exponential growth rises rapidly when animals such as rabbits have several young instead of just two. As long as exponential growth is in effect, the areas that experience it become more densely populated regardless of the number already included in the population.

However, it is unusual for populations to grow in an unlimited way because there are natural limiting factors which stop the population increase. Limiting factors include the lack of food, water or even shelter. If organisms cannot find enough food and water, they will have fewer or no young and the rate of population growth goes down. If there are predators or disease, population growth is also reduced. Limiting factors have the greatest effect on large populations that have grown quickly. This was the case in Australia where the rabbit population reduced to 100 million. Although there were no real predators, the deliberate introduction of myxomatosis into Australia in the 1950s, a killer disease for rabbits, caused a reduction in the rabbit population. Exponential growth with limiting factors, such as disease or lack of food, is often referred to as a 'logistic model'.

Another type of model is the 'chaotic model', whereby rabbits breed exponentially but eventually the amount of grass and water is insufficient to sustain the population and so the rabbits die out. Some will survive and the grass will grow back, so the cycle repeats itself. Because the numbers might be low and the time taken to reestablish the population varies, the growth is often referred to as being chaotic.

A population-growth computer model tries to predict the population of an organism. It assumes that the organism reproduces according to a set of rules. There are a number of variables, such as how often the organism reproduces, how many new organisms it produces each time, limiting factors (such as limited resources like food and water), natural death rates and types of predators.

Models provide a way of understanding how populations of organisms change over time or when compared to each other. Many patterns can be identified by using computer modelling. Other studies can be made such as the age distribution within a population and how this changes. Computer models are used to calculate the maximum harvest for farmers and also to help with environmental conservation. Computer models are also used to understand the spread of parasites, viruses, and disease. They are also useful when modelling the future of particular endangered species.

Weather systems

Weather forecasting used to involve several weather forecasters and days of calculations. The arrival of satellites and computer models has really helped forecasters to make more reliable predictions. Weather presenters on television will sometimes refer to the uncertainty of their predictions because the computer models are not in agreement. The technical term given to producing weather forecasts using a computer is 'numerical weather prediction' (NWP). The computer needs to know the current state of the atmosphere, which is provided to it through observations from weather stations and weather balloons as well as satellites. These readings are fed into models that analyse the data and combine it with the most recent forecasts to produce a 3-D model of the Earth's atmosphere. It uses mathematical equations that represent horizontal and vertical air motions, temperature changes and moisture processes among other variables to calculate what the atmosphere might look like. Computer models have to include hundreds of mathematical equations to produce a weather forecast. A lot of computing power is needed to run a computer model. It is usual for supercomputers to be used in such activities. As we have seen, these supercomputers are able to perform quadrillions of calculations every second!

The UK Met Office uses the Unified Model (UM) approach. This is a numerical model of the atmosphere used for both weather and climate applications. It is a model which is being continuously developed by the Met Office and its partners. The Met Office uses a seamless modelling approach, whereby a single group of related models can be used to make predictions over a number of timescales involving short-term weather forecasts as well as long-range climate forecasts.

The models used to take variables like atmospheric pressure, humidity, rainfall, temperature, wind speed, and wind direction, which are recorded all over the Earth's surface, into account. The model is suitable for NWP, seasonal forecasting and climate modelling, with forecast times ranging from a few days to hundreds of years. Furthermore, the UM approach can be used both as a global and a regional model.

For the weather forecasting aspect, atmospheric models can be used to compare the current conditions with previous days at similar times of the year with the same or similar conditions. This enables a reasonably accurate forecast to be made. Maps are produced showing isobars, temperatures, and wind speed. Predictions can be made because similar weather conditions will have been observed over very long periods of time.

Climate change

Climate can be measured using a number of variables (quantities that change) such as rainfall, hours of sunlight, wind speed and temperature, among others. Temperatures in most regions of our planet vary according to the time of year. Summer in the southern hemisphere tends to be December to February, while in the northern hemisphere it is June to August. The temperature differences between different regions of the world cause differences in atmospheric pressure which, in turn, lead to winds, storms and, in some cases, hurricanes. Many scientists believe that the increasing levels of carbon dioxide in the atmosphere, which is leading to hotter temperatures for our planet, will give rise to major changes in our climate. Climate change is already beginning to happen in the Arctic and Antarctic. This does not necessarily mean that all regions on the planet will become hotter, but some will. The result will be more changeable weather conditions, such as extremely high levels of temperature, and rain- and snowfall becoming far more frequent than they are now. Forecasting the possible changes to the Earth's climate, in other words the long-term patterns of weather for the entire planet, is obviously going to be more complex than relatively short-term weather forecasting for specific regions of the planet.

For a long time now, the Earth's climate has been monitored and records have been kept of the actual values of the variables involved over a number of years. This data can be input into a computer model, which is a large and very complex program that runs on a supercomputer. It is made up of a collection of equations representing different parts of the climate. Each equation contains variables like temperature, rainfall, the amount of carbon dioxide in the atmosphere, and sea level. It shows how each of these affects the other variables. The variables can be changed to see what effect they will have on the climate. These equations together give an approximation of the Earth's climate. The model can be used to predict trends and examine the effect of increasing and decreasing each variable. When the model is created, it is tested. The weather for any particular year in the past is known. The data for several years prior to that year are entered into the model. The model's predictions can be compared with the real data for that year and so its accuracy can be judged. The model is then refined until the results accurately match the known climate. If it makes accurate predictions, it can be used to see what will happen in the future. The accuracy, however, is limited by the number of years into the future the model extends.

In order to create a computer model, climatologists have to make several assumptions about how the climate works. Problems arise because climate is affected by a large number of variables and relating all these variables to each other is a very difficult task. When you consider that even the largest supercomputer has its limitations, there are a number of people who are very sceptical about the accuracy of computer models in this scenario. The sceptics think that these models are too approximate and may not represent how climate functions in real life. Many experts are not convinced about the accuracy of global models. Models take into account the effect of greenhouse gas emissions such as water vapour, carbon dioxide, methane, nitrous oxide and ozone, including car exhaust gases. Since it is not possible to predict future government policies, developments in technology or economic forces relating to these factors, it is difficult to predict with any accuracy what their effect on the climate will be in the future. Some countries are discouraging the use of internal-combustion-engine-powered vehicles and positively advocating the use of electric cars. How fast this change will take place, and what its effect will be, is difficult to calculate. However, looking into the future, as climate scientists have more and more data to work on and more powerful computers are developed, the models will become better able to predict climate change.

Queue management

Queue management, sometimes referred to as 'waiting line management', is crucial to many aspects of the service industry. The aim is to reduce the time customers wait to be served and therefore improve the quality of the service provided. Queue management deals with cases where customer arrival is unpredictable and varies a great deal for different times of the day. A service business can reduce its costs leading to increased profitability by managing queues effectively. The length of time customers have to queue incurs a cost to the business, because there is a cost associated with putting extra employees at checkouts to reduce service time. Queue management is used to weigh up one cost against the other and provide solutions to these problems for the management of the business. If customers are walking away unhappy because of the lack of supermarket checkout operators, for example, the business should weigh up the cost of hiring more staff against the value of increased income and maintaining customer loyalty.

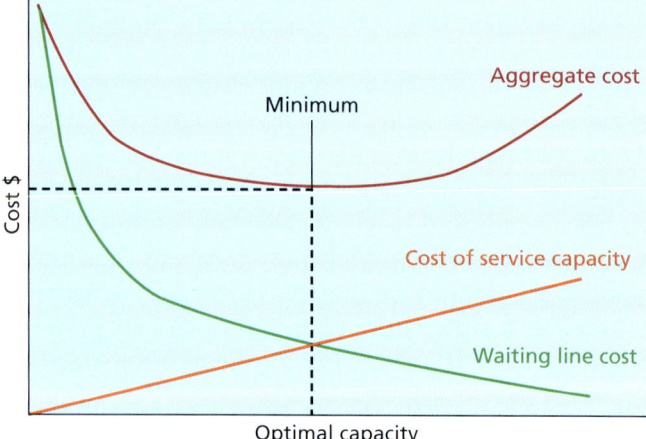

▲ Figure 9.1 Service capacity versus cost

This is illustrated by the graph in Figure 9.1, which shows the relationship between service capacity (for example the number of operating checkouts) and queuing cost (the cost the supermarket must bear when there are queues).

To begin with, the cost of waiting in line is at a maximum when the business is at its lowest service capacity. As service capacity increases, there is a reduction in the number of customers in the queue and in their waiting times, which in turn decreases the cost of queuing. The ideal total cost is found where the service capacity and waiting line curves meet. This is fine for the queuing at supermarkets but what about other queuing systems?

In order to be able to solve the problem of queue management, it is important to understand the characteristics of a queue. Queuing, or waiting in line, is a normal feature of daily life, for example customers queue in bank branches in order to be served by a bank teller, cars queue up for petrol or diesel, and people queue for service at a checkout in a supermarket. It follows that business managers will need to employ people who will work on mathematical formulas which will reduce the waiting time and satisfy their customers without incurring additional costs. There are many other forms of queuing, such as data packets travelling across the internet or even programs running on your computer.

There are a number of factors that need to be taken into account when studying queue management. Customers can arrive one at a time, as a group or even en masse. They can arrive evenly over a period of time. The number of customers can be finite, such as the boarding of an airplane when it is known exactly how many passengers will be queuing to board the flight, or it can be what is referred to as infinite, which is a queue management term for when the number is unpredictable such as the number of customers in a supermarket during the course of a day.

Queue management theory also has to take into account the type of queue set-up in particular situations. Some set-ups involve just one queue and one outcome, such as queuing at a drive-through car wash. You have to wait for the car in front of you to be washed before you can drive your car through. Some have one queue but several outcomes, such as at a fast-food drive-through where there may be one queue of cars but you could be required to first place your order at one window, drive on and pay for your food at a second window, then finally collect your food at a third window. Some have more than one queue but only one outcome, for example at an airport when boarding a flight there may be two queues – for business and economy passengers respectively – but only one outcome, passing through the same gate to board the plane. The business passenger queue boards first, then afterwards the economy passengers board. The final type is several queues and several outcomes. This is best illustrated by the example of a launderette where there may be a number of queues for each washing machine, then once customers have used the washing machine, they proceed to form queues for each dryer.

The equations needed to cover all possible eventualities in queue management are extremely complex and can involve the use of an advanced form of mathematical formulas called differential equations, as well as Petri nets, a sophisticated mathematical modelling technique.

Let us consider a supermarket which has an automatic queue measurement system using computers. A number of the large supermarket chains use such systems. Automatic queue measurement systems can use people-counting sensors at entrances to the supermarket and above checkouts. They accurately detect the number and behaviour of people in the queue. Built-in predictive algorithms can provide information on how many checkouts will be needed to move customers through quickly. This is done as the queues are forming and the information can be displayed on computer monitors or tablet devices. A range of information is provided, such as the length of the queue at any given time and how long the waiting time for customers is. Management teams can be automatically alerted before the waiting times get too long. Outputs produced by such systems are:

» the number of people entering the supermarket
» queue length

- average wait time
- amount of time each checkout operator is idle
- total wait time.

Other software is available for more complex queuing systems.

Traffic flow

Traffic flow is the study of the flow of vehicles on the road, and their interactions with other vehicles, pedestrians and methods of traffic control present on the road. Free movement of traffic is affected by many factors like speed, the number of heavy vehicles, the number of lanes and intersections present along the road. Mathematicians and civil engineers, particularly, carry out research into this subject. Their aim is to attempt to understand and develop a road network which causes traffic to flow smoothly with few hold-ups.

Traffic flow on a road is calculated as being the number of vehicles using a particular road in one hour. This is done by counting the traffic at a specific point on a road for a given length of time. What time of day it is will affect the traffic count. Particular attention is paid to the traffic flow during peak hours. Traffic-flow measurement is used on the road to identify places where congestion occurs. This often leads to decisions being made as to whether traffic lights or roundabouts will be needed or, indeed, the need for variable speed control. Variable speed control is the use of speed limits to slow down traffic on busy motorways or highways; the normal speed limit for that road is temporarily suspended. Traffic flow depends on two variables: the speed of the traffic and the density of vehicles on the road (the number of vehicles per given length of road). Traffic flow can be measured using sensors, (induction) loop detectors, video cameras, and by manually counting the number of vehicles.

Since as long ago as the 1920s, attempts have been made to produce a mathematical theory of traffic flow. However, even with the use of sophisticated computer processing power, it has still proved difficult to produce a general theory that can be applied consistently to modern traffic flow as it exists in reality. Current traffic models are based on observed conditions as well as theoretical techniques. These models are then developed so they can forecast the amount of traffic, taking into account factors such as increased vehicle use as well as changes in the methods of transport people use, such as changing from taking the bus to taking the train or car.

Traffic-flow computer models use a series of complex algorithms. Some models treat each mode of transport (such as vehicle, bus, train, tram, cyclist or pedestrian) as a unique object, since each mode has its own behaviour and targets. Other models concentrate on traffic-flow characteristics such as density, flow, and average speed of the traffic.

Construction

Models are frequently used in the construction industry. The most common type of software used is computer-aided design (CAD). It is used by architects and engineers to create precise drawings or technical illustrations. CAD software can be used to create two-dimensional (2-D) drawings or 3-D models. CAD software is used to design structures such as bridges and buildings, and can also be used to generate animations and other presentational material. The software often allows for the inclusion of additional information, such as dimensions and descriptions of components.

CAD packages also incorporate libraries. A library, in this sense of the word, is a collection of pre-drawn parts and features that you can insert into drawings.

◀ **Figure 9.2** Laptop showing a 3-D drawing using CAD

If a building is being designed, a library could be used which would have parts such as doors, windows, staircases and roofs available. All that is needed is to select the required parts, changing the dimensions to suit. CAD packages produce 3-D views and different viewing angles. Most packages can create 3-D walkthroughs so that the viewer can actually see inside buildings or other structures. Printers are used to produce hard copies of designs; however, it is more usual for plotters to be used to produce blueprints. More recently 3-D models can be produced using 3-D printers.

Advantages and disadvantages of CAD

CAD can allow a user to design a part much faster than traditional drawing. There is a library of shapes and designs so only minor modifications need to be made. However, work can be lost because of the sudden breakdown of computers.

Using computer-aided design software, it will be much easier to make any changes because you can fix the errors and modify the drawings more easily. It is also more accurate. The number of errors that used to occur with manual designs is significantly reduced. However, it takes time for users to learn how to use the software and with every new release of the CAD software the user has to update their skills which can take time. It is expensive to regularly update CAD software or operating systems.

The different parts of the design can be reused over and over again and CAD tools make it easier to save files of the drawings and store them in a way that they can be used time and again.

Open source CAD software is available for free, though it tends to be limited in scope, but the better forms of CAD software can be expensive to buy, and it costs a lot of money to train workers to use the software.

Building information modelling

Building information modelling (BIM) is an extension of CAD. It utilises many of the aspects of CAD but has additional benefits. CAD refers to the use of computer systems to aid in creating designs using high quality drawings. BIM uses software that applies CAD concepts to designing buildings. It is used to create models that include not just the physical aspects, such as walls, roofs and windows, but also heating, ventilation, and air conditioning (HVAC) and the electrical aspects of a building.

The US National BIM Standard Project Committee uses this definition:

> Building information modelling (BIM) is a digital representation of physical and functional characteristics of a facility. A BIM is a shared knowledge resource for information about a facility forming a reliable basis for decisions during its life-cycle; defined as existing from earliest conception to demolition.

BIM is a new process and methodology used by teams of architects, engineers and contractors who work together to design and build a commercial building using the same database and computer model. This allows the team to analyse and visualise design decisions long before a project starts. BIM allows a team to bring all its designs, including CAD models, into a single database. Using the Cloud, BIM provides access to this database to all project members. The use of BIM extends beyond the planning and design phase of a project and covers the whole of the building life-cycle.

9.2 Simulations

A computer simulation is when a computer is used to imitate an event as it happens in the real world. A simulation is created usually from a mathematical model which represents the real system. It consists of computer programs, which are written to include the main features of this real-world system. The model is the representation of the system, whereas the simulation is how the system performs over time.

Computer simulations are often used to represent environments that may be difficult or even dangerous to replicate in real life. Natural-disaster planning and nuclear-science research come under these categories.

To begin with we will look at the use of computer simulators in training pilots and in learning to drive a car.

Advantages and disadvantages of simulations

With a simulator you do not have to worry about the cost of repairing and replacing damaged cars or airplanes, but buying a simulator with a fast processor and large amounts of memory can be very expensive, and the technical support for such systems can prove to be more expensive than regular servicing of airplanes and vehicles.

You may need fewer driving or flying instructors resulting in lower costs but, on the other hand, a driving school using a simulator from scratch will need to retrain driving instructors, costing money and taking time.

Simulators can provide results that are generally difficult to measure such as reaction times, but the unpredictability of human behaviour means that it may be impossible to create simulations that cover all eventualities.

With a simulator you can see how a system might work before you design or modify it, resulting in fewer mistakes in the system, but inaccurate output from the simulation can still occur as the formulas and functions used may not accurately represent all possible scenarios.

A simulator prevents injury or even death in the event of a car or plane crash, but it may make drivers or pilots overconfident and too casual regarding their abilities.

Some events, such as extreme weather conditions, can be recreated fairly easily without having to wait for them to happen in real life. However, there is sometimes not sufficient data to produce a mathematical model.

9.2.1 Pilot training

As the number of people who travel by air increases, the number of pilots will need to increase too. Similarly, as new aircraft are developed, the need to train existing pilots to fly these new aircraft increases. This, obviously, has to be done as cheaply and safely as possible. It is for this reason that airlines use flight simulators. A simulator is designed to react to a pilot's instructions just like an actual aircraft would. Many students think that the flight simulator games available for a PC are authentic simulators. In actual fact, a flight simulator used by airlines is a purpose-built cabin as shown in Figure 9.3. Inside the simulator it looks just like a real cabin in an aircraft with all the controls of a real aircraft, as you can see in Figure 9.4.

▲ Figure 9.3 A flight simulator pod

▲ Figure 9.4 A flight simulator flight deck

The cabin contains an aircraft cockpit just like the real thing. It includes screens, flight controls and instrumentation, an adjustable pilot seat, as well as a sound system and computer hardware. In a flight simulator, the control panel in the cockpit is identical to one in a real plane. Displays are generated by computers on the screens in the cockpit. In order to create realism, as the pilot 'takes off' in the simulator, he or she sees the airport as well as its surroundings. The simulation might show a baggage cart to the side of the runway and countryside in the distance. The sound system recreates the rush of air around the wings even though they are not there. In order to recreate the feeling of movement, the simulator is tilted and shaken by a number of hydraulic systems. These are designed to make the pilot feel that he or she is actually controlling a real plane.

Modern military and civilian aircraft are very expensive to build and because of the high production cost, airlines are reluctant to train beginners using real aircraft. Although a flight simulator is also expensive to buy, it is still regarded as the most cost-effective method for training pilots. The reason is that it can be operated without damaging any real aircraft. Trainee pilots just beginning their training obviously do not have much experience in flying aircraft, which means it is likely that they would crash and damage a real aircraft, putting their lives at risk and also creating a danger to the general public. For these reasons flight simulators are much more effective in training pilots to fly.

For a flight simulator to be realistic, it should make the pilot think that he or she is actually inside the cockpit of a real plane. Every instrument must function identically to a real airplane. For example, fuel gauges must show the rate at which the engine consumes fuel, taking into account the simulated thrust being applied as well as engine temperature. Flight simulators use powerful hydraulic rams to reproduce the effects of motion that pilots need to experience. The cockpits are constructed with an array of instruments such as joysticks, levers, switches, buttons, and sliders.

In order to recreate the sensation of a pilot manually landing the plane, it is important that what the pilot sees through the cockpit windows is life-like. To this end, sophisticated computer graphics are used in addition to the realistic cockpit instrumentation. The hardware involved is often a group of projectors, a back-projection screen and a panoramic spherical mirror. It is important for the display to have very high resolution.

In a flight simulator, the created environments are based on actual international airports, constructed from plans, maps, photographs and site visits. A database is needed which also contains information on the surrounding area of each airport for several miles. Building such a database requires special software tools.

9.2.2 Learning to drive a car

Simulators, for similar reasons to those for pilot training, are also used to teach people to learn to drive. It is safer to allow drivers to learn how to drive without causing damage to the car or themselves or, indeed, other drivers and the general public. The same type of graphics and animation software used for pilot training can be used to reproduce the situations that car drivers experience. Driving simulators can take several forms. One used by a driving school would probably be a full cab model, which looks like the driver's side of a car and includes real car parts, such as clutch, accelerator and brake pedals, signals, windshield wipers, and an ignition key.

Driving simulators replicate actual driving experiences. They allow you to experience driving at different times of day or night and driving in different weather conditions such as fog, heavy rain, snow, driving into a setting sun and so on. Driving simulators can also be used to monitor driver behaviour, performance, and attention span.

9.2.3 Natural-disaster planning

A number of natural disasters have occurred in the past decade. In 2018 there was a tsunami in Indonesia; also that year there were the Bangladesh floods and landslides as well as an earthquake in Papua New Guinea. Most natural disasters have a human cost, whether this is in terms of displaced persons, the spread of disease, disruption to transport, or damaged and destroyed homes. Natural disasters occur unexpectedly and usually result in huge loss of life and property. How to make effective contingency plans is an intriguing question constantly faced by governments and experts.

In order to plan for natural disasters, such as tsunamis, simulations are created to represent the flow of floodwater and the debris it carries. Computer modelling allows us to understand and analyse natural disasters. In recent years, the huge increase in computer power and speed, together with advances in the development of computer-based algorithms, has made it easier to produce computer models. This has made it possible to model the movement of billions of water particles to allow more accurate predictions of the effects of natural disasters such as tsunamis, floods and mudslides. This can help town planners to build structures which can withstand such disasters, as well as allowing the emergency services to co-ordinate an efficient response.

The major task for the creators of natural-disaster models is to provide information about disasters before they occur. The purpose of such modelling is to anticipate the likelihood and severity of natural disasters, from earthquakes and hurricanes to major crop failure. It can also be used to model what might happen in the event of a terrorist attack. The very first task is to create a catalogue of events, which is then used in the simulation. This is more difficult than it seems. It should be simple to create a catalogue from historical records. This does not work well, however, as only the most catastrophic events are likely to have been recorded; the smaller events might not have been deemed worthy of record or might not even have been witnessed by anyone at all.

This leads to a lack of reported small events, which leads to such events being underestimated. Equally troubling is that this can lead to the assumption that because a particular event has never been recorded as happening in the past, it is unlikely to happen in the future. In recent times, however, there has been an improvement in the collection of meaningful data and also in the methods of describing the collective behaviours of people in disaster situations.

In addition, there are mathematical models available. Mathematics is often used to describe natural phenomena. A tsunami is a high-speed sea wave usually caused by an underwater earthquake, landslide, or volcanic eruption. A tsunami when it reaches land is considered to be a shallow water wave. That means the height of the wave is smaller than the length of the wave. Mathematicians have produced equations relating the depth of water to the velocity of the wave, the length of the wave, the density of the sea water, the acceleration due to gravity and the energy of the wave. Similarly, mathematicians have produced equations and functions which represent the behaviour of volcanoes, earthquakes and tornadoes. These equations and functions can then be converted into computer models.

9.2.4 Nuclear science research

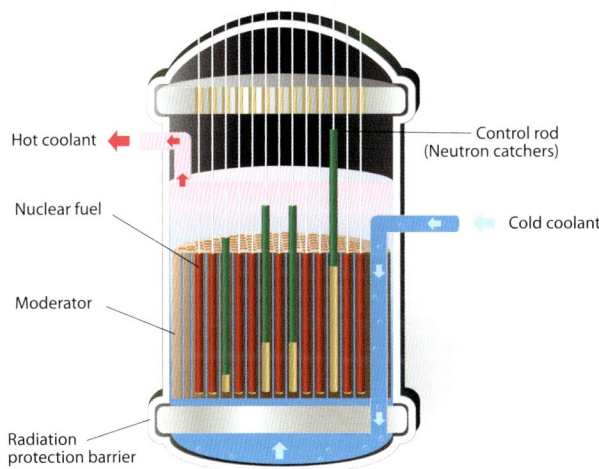

▲ Figure 9.5 A diagram of a nuclear reactor

Systems like nuclear reactors are very complex and very costly to construct and operate. If these systems are operated incorrectly, they can cause dangerous and even disastrous consequences. There is a great need to train the operators of nuclear research facilities. This is being done, increasingly, through the use of simulators. Nowadays, most reactors have a simulator. These simulators imitate all the devices in the reactor control room such as meters and sensors. Operators' skills are refined through the use of a simulated control-room environment. Computer simulations use mathematical algorithms to predict how different reactors will function.

Because scientists cannot forecast exactly how a nuclear reactor will respond to variations in its set-up or the conditions under which it is operated, every facility has to be designed and operated within large safety margins. This means that the efficiency of the forecast is reduced. The core of a fast-nuclear reactor may consist of an arrangement of stainless-steel pins which contain rods made of uranium or plutonium metallic alloys or oxides. Once the reactor is operating, these pins become extremely hot. In order to convert this heat to electricity and to prevent melting the stainless steel, thousands of gallons of liquid metal coolant such as liquid sodium are used. Rearranging these rods affects the neutron chain reaction and consequently the flow of coolant and the transfer of heat. In a real-life reactor, this can create extra expense and increase safety concerns. For these reasons, research scientists tend to use computer simulations. Reactors convert the uranium and plutonium inside each pin into energy through nuclear fission. The hope of nuclear scientists is to have more powerful computers in order to recreate experiments at a microscopic or even atomic level. High-speed supercomputers are increasingly being used to solve extremely difficult problems.

9.3 Using what-if analysis

Computer models all base their model upon a what-if analysis. This involves changing a model to find out what the results would be, if different input values are chosen. You need to know how to build, test and change data within models, or change the model itself to produce different sets of results.

Simple computer models can be created, tested and used in *Excel*. Some models only require a change in data.

> ### Task 9a
>
> Open the file **Widget.csv**. Enter appropriate formulas in cells B6, B7, B10, B12 and B17. Workers only work for eight hours a day. Calculate the cost to make each widget. Apply appropriate formatting to all cells (all currency values are to be shown in $ to two decimal places). Test your spreadsheet model.
>
> The cost of raw materials is $2, and the time taken to make each widget is 10 minutes. At the moment the company has two workers.
>
> Use the spreadsheet model to calculate:
> » how many workers are required to complete an order for 500 widgets placed on 7 January and required by 10 January
> » the sales price required per widget to make a $50 profit.
>
> Save the spreadsheet as **Task_9a**.

Open the file and enter into cell B6 the formula **=60/MINUTE(B5)** to calculate the number of widgets a worker makes in one hour. Enter into cell B7 the formula **=B3*B6** to calculate the number of widgets that are made per hour by all the workers. Enter in cell B9 the formula **=B4+B8/B6** to calculate how much it costs to make each widget. Enter in cell B12 the formula **=ROUNDUP(B11/(B7*8),0)** to calculate the time taken to manufacture, which is the number of widgets required by the customer divided by the number that can be made in the factory in one working day by all employees, rounded up to the nearest whole working day. Enter in cell B17 the formula **=B11*(B16-B9)** to calculate the number of widgets multiplied by the sales price of each widget minus the cost to make each widget.

9.3.1 Test the spreadsheet model

Before modelling the data, we must test the data model using a test plan and data that gives us (easy-to-calculate) known results. Check that the formulas you have entered give the correct results using data like this: one worker, $1 cost of raw materials, ½ hour to make one widget, and $10 per hour as the rate of pay. Each widget would cost $6 to make ($5 for ½ hour labour + $1 for raw materials). If 20 widgets were required it would take 1¼ working days (if each widget takes ½ hour to make, 16 can be made in one day), so the formula should round this up to two full working days. To test the profit/loss formula, values that make the calculations easy have been used (for example if a widget costs $4.50 to make, then to sell it for $5.50 makes the value 1, which is easy to multiply by). Cell B9 is tested in Table 9.4 so will not need testing again, but the results from B9 are used for this test. The test tables for this data and two other examples are shown in Tables 9.1 to 9.6.

▼ Table 9.1 Completed test plan for Task_9a: cell B6 – Version 1

Test type				Formula	Cell	B6
Data entry in cell:				Expected results	Actual result	Remedial action
B3	B4	B5	B11			
1	1	00:30:00	20	2	2	No
1	1	01:00:00	10	1	Division by zero error	Yes - error
2	2	00:15:00	200	4	4	No

Identify where the error has occurred. Correct the function in cell B6 so it becomes **=60/(60*HOUR(B5)+MINUTE(B5))**.

Attempt all the tests for cell B6 again. The resulting test plan for the retest should look like Table 9.2.

▼ Table 9.2 Completed test plan for Task_9a: cell B6 – Version 2

Test type				Formula	Cell	B6
Data entry in cell:				Expected results	Actual result	Remedial action
B3	B4	B5	B11			
1	1	00:30:00	20	2	2	No
1	1	01:00:00	10	1	1	No
2	2	00:15:00	200	4	4	No

Continue with the testing for the other formulas.

▼ Table 9.3 Completed test plan for Task_9a: cell B7 – Version 1

Test type				Formula	Cell	B7
Data entry in cell:				Expected results	Actual result	Remedial action
B3	B4	B5	B11			
1	1	00:30:00	20	2	2	No
1	1	01:00:00	10	1	1	No
2	2	00:15:00	200	8	8	No

▼ Table 9.4 Completed test plan for Task_9a: cell B9 – Version 1

Test type				Formula	Cell	B9
Data entry in cell:				Expected results	Actual result	Remedial action
B3	B4	B5	B11			
1	1	00:30:00	20	$6.00	$6.00	No
1	1	01:00:00	10	$11.00	$11.00	No
2	2	00:15:00	200	$4.50	$4.50	No

▼ Table 9.5 Completed test plan for Task_9a: cell B12 – Version 1

Test type				Formula	Cell	B12
Data entry in cell:				Expected results	Actual result	Remedial action
B3	B4	B5	B11			
1	1	00:30:00	20	2	2	No
1	1	01:00:00	10	2	2	No
2	2	00:15:00	200	4	4	No

▼ Table 9.6 Completed test plan for Task_9a: cell B17 – Version 1

Test type						Formula	Cell	B17
Data entry in cell:						Expected results	Actual result	Remedial action
B11	B16	B4	B8	B6	B9			
10	$5.50	2	10	4	$4.50	10	10	No
10	$6.50	2	10	4	$4.50	20	20	No
20	$6.50	2	10	4	$4.50	40	40	No

After testing, enter the data for two workers, with the cost of raw materials as $2 and the time taken to manufacture each widget as 10 minutes.

Now the model has been created, we will use it to calculate how many workers are required to complete an order for 500 widgets, placed on 7 January and required by 10 January.

Enter 500 in cell B11. Enter in cells B13 and B14 the dates 7 January and 10 January, and enter in cell B15 the formula **=B14-B13** and format this cell as a general number. To model the number of workers required you could just guess what number to place in cell B3 and keep trying until you get cell B12 to say 3, but this is not efficient.

Goal Seek

Instead of guessing the values (which may be reasonably easy in this simple model) we can use the Goal Seek feature. Place the cursor in cell B12. Use the **Data** tab and locate the **Forecast** group. From the **What-If Analysis** icon, use the drop-down menu to select **Goal Seek…**.

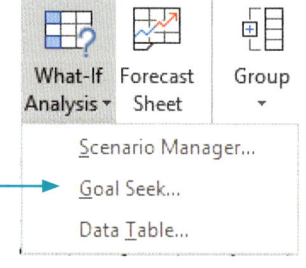

In the **Goal Seek** window, enter in the **To value:** box the value 3 (the number of days for the order taken from cell B15). Enter in the **By changing cell:** box the cell reference B3, then click OK.

Goal Seek will run, and after testing all possibilities will return a solution in the **Goal Seek Status** window.

The solution offered by *Excel*'s Goal Seek is to employ 4.08 workers.

This also suggests that these workers would make 24.48 widgets per hour.

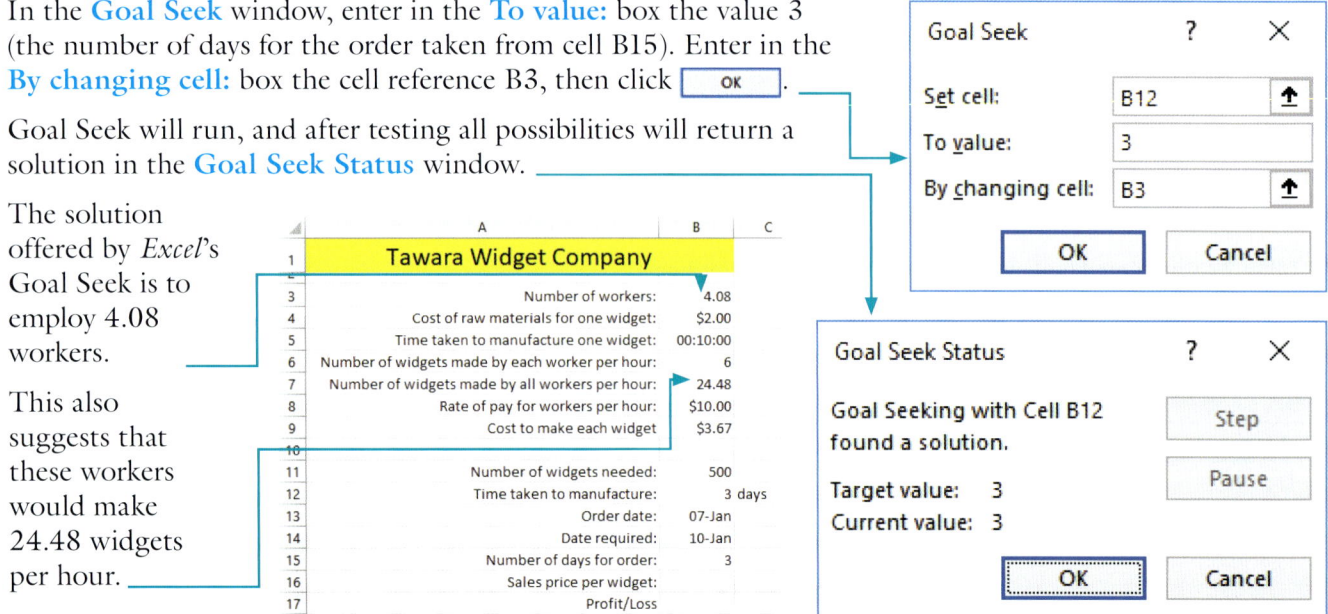

To produce a realistic working solution, the number of workers will need changing to 5 which will change the contents of cell B7 to 30. To calculate the sales price required per widget to make a $50 profit, place the cursor into cell B17, then from the **Data** tab select the **What-If Analysis** icon, then **Goal Seek…**. Edit the **Goal Seek** window to look like this.

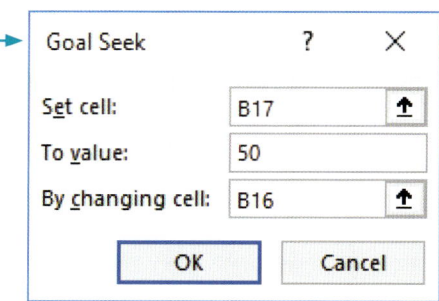

Click **OK** to run the Goal Seek. Goal Seek will run, and after testing all possibilities will return a solution in the **Goal Seek Status** window.

This returns in cell B16 a value of 3.76666666666667, which needs to be rounded up to $3.77 (it may appear to hold this already if you have formatted the cell to two decimal places). Change the value in cell B16 to $3.77 which will return a profit of $51.67.

> **Advice**
> Remember, formatting to a number of decimal places does not change the original value.

Activity 9a

Open the file **TuckShop.csv** and format it to look like this.

Enter appropriate formulas in rows 10 and 12. Enter formulas into row 13 to calculate the percentage profit/loss, which is the profit/loss divided by the wholesale cost per pack. Test your model.

Use the spreadsheet model to calculate the unit sales price for each product to return a percentage profit of (at least):
» 50%
» 20%.

Save the spreadsheet as **Activity_9a**.

	A	B	C	D
1	Healthy Eating Tuck Shop			
2				
3			Item	
4		Granola bar	Dried fruit	Carrot sticks
5	Outgoing			
6	Number in pack:	48	60	50
7	Wholesale cost per pack:	$24.00	$28.00	$16.00
8	Income			
9	Unit sales price:	$1.00	$1.00	$1.00
10	Total Sales price per pack:	$48.00	$60.00	$50.00
11	Profit/Loss			
12	Profit/Loss:	$0.00	$0.00	$0.00
13	Percentage profit/loss	0%	0%	0%

Practice questions

1. A company that sells cars uses a driving simulator so that customers and their teenage children can test drive the latest models of cars without going on the road.
   Describe the effects using this driving simulator could have on customers and the car sales company. [6]

2. An engineering company requires a new computer system to control a production process. Staff will need training before they can use the new system.
   Describe the benefits of using a computer simulation for this training. [3]

3. Climate change is another way of describing global warming. Many experts are using computer models to predict the changes which are occurring.
   Describe the drawbacks of using computer models to predict climate change. [6]

10 Database and file concepts

In this chapter you will learn about:
- flat file and relational databases
- relationship types in a relational database: one-to-one, one-to-many and many-to-many
- entity relationship diagrams
- key fields
- referential integrity
- the characteristics of data in unnormalised form
- the normalisation of data into First, Second and Third Normal Form
- the components of a data dictionary
- query selection using static and dynamic parameters
- different file types and their use
- proprietary, generic and open source file formats
- indexed sequential access
- direct file access
- hierarchical database management system
- management information systems.

In this chapter you will learn how to:
- create a relational database
- assign appropriate data types and field sizes
- create and use relationships
- assign key fields
- use referential integrity
- validate and verify data entry
- perform searches
- perform calculations within a database
- sort data
- design and create an appropriate data entry form
- design and create a switchboard menu within a database
- import and export data
- normalise a database to First, Second or Third Normal Form
- create a data dictionary and select appropriate data types.

Before starting this chapter you should:
- understand the terms 'file', 'record' and 'field'
- be able to create a flat-file database
- have studied validation and verification (Section 1.4)
- have studied testing a spreadsheet (Section 8.2)
- have studied the IF function in a spreadsheet (Section 8.1)
- have studied custom-written and off-the-shelf software (Section 2.4).

For this chapter you will need these source files:
- boat.png
- ellmau.csv
- FA-Cust1.csv
- FA-Cust2.txt
- FA-Invoice.csv
- FA-Product.csv
- freezer.png
- jobs.csv
- NP-Boat.csv
- NP-Customer.csv
- TB-Driver1.csv
- TB-Driver2.csv
- workers.csv

10.1 Database basics

A database is a program used to store data in a structured way. This includes the data that is stored and the links between the data items. All databases store data using a system of **files**, records and fields. There are two types of database: a **flat-file database** and a relational database.

10.1.1 Database types

Flat-file databases

A flat-file database stores its data in one table, which is organised using rows and columns. For example in this database about teachers, each record (row) in the table contains data about one person. Each column in the table contains a field which has been given a field name and each cell in that column has the same, predefined data type.

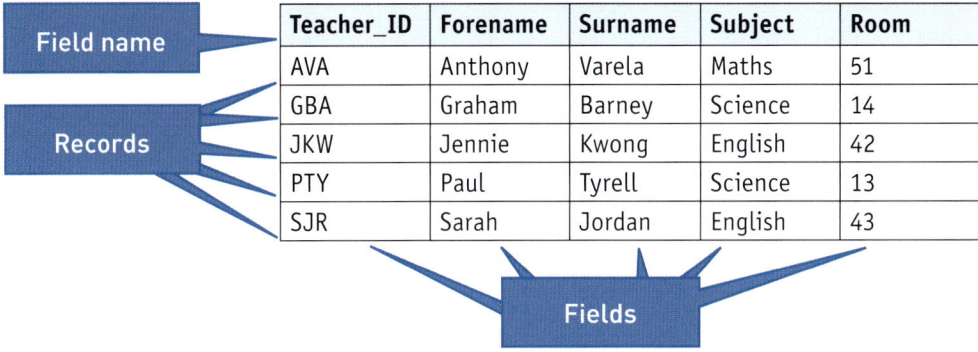

▲ Figure 10.1 A flat-file database table

Relational databases

A relational database stores the data in more than one linked table, within the file. The relational database is designed so that the same data is not stored many times. Each table within a relational database will have a key field. Most tables will have a **primary key** field which holds unique data (no two records are the same in this field) and is the field used to identify that record. Some tables will have one or more **foreign key** fields. A foreign key in one table will point to a primary key in another table.

Table 10.1 shows the structure of a flat-file database table if we add extra data such as the details of each student taught by each teacher, using the example seen above.

▼ Table 10.1 A flat-file database table

Teacher_ID	Forename	Surname	Subject	Room	Student_ID	Student_FName	Student_SName
AVA	Anthony	Varela	Maths	51	G12345	Jasmine	Hall
AVA	Anthony	Varela	Maths	51	G12346	James	Ling
AVA	Anthony	Varela	Maths	51	G12348	Addy	Paredes
AVA	Anthony	Varela	Maths	51	G12349	Hayley	Lemon
AVA	Anthony	Varela	Maths	51	G12351	Jennie	Campbell
GBA	Graham	Barney	Science	14	G12345	Jasmine	Hall
GBA	Graham	Barney	Science	14	G12348	Addy	Paredes
GBA	Graham	Barney	Science	14	G12349	Hayley	Lemon
JKW	Jennie	Kwong	English	42	G12345	Jasmine	Hall
JKW	Jennie	Kwong	English	42	G12349	Hayley	Lemon
JKW	Jennie	Kwong	English	42	G12351	Jennie	Campbell
PTY	Paul	Tyrell	Science	13	G12346	James	Ling
PTY	Paul	Tyrell	Science	13	G12351	Jennie	Campbell
SJR	Sarah	Jordan	English	43	G12346	James	Ling
SJR	Sarah	Jordan	English	43	G12348	Addy	Paredes

As it is possible for different teachers to teach in the same classroom, or for teachers or students to have the same forename and/or the same surname, these cannot be used as primary key fields. If the data is split into two tables, one for the teachers and one for the students, that are linked together, it can be stored and retrieved more efficiently, like this:

▼ Table 10.2 Teachers' table

Teacher_ID	Forename	Surname	Subject	Room
AVA	Anthony	Varela	Maths	51
GBA	Graham	Barney	Science	14
JKW	Jennie	Kwong	English	42
PTY	Paul	Tyrell	Science	13
SJR	Sarah	Jordan	English	43

▼ Table 10.3 Students' table

Student_ID	Student_FName	Student_SName	English	Maths	Science
G12345	Jasmine	Hall	JKW	AVA	GBA
G12346	James	Ling	SJR	AVA	PTY
G12348	Addy	Paredes	SJR	AVA	GBA
G12349	Hayley	Lemon	JKW	AVA	GBA
G12351	Jennie	Campbell	JKW	AVA	PTY

These two tables are linked with a one-to-many relationship, because one teacher's record is linked to many student records. The primary key fields (which *must* contain unique data) will be the Student_ID and Teacher_ID.

Flat-file database versus relational database

Using the example above, if only one of these sets of data (the teachers' data or the students' data) had to be stored, then a flat-file database would be ideal as long as there are no (or very few) repeated data items. If a database was to hold both of these sets of data, then a relational database would be best.

A relational database is designed so that the same data is not stored many times. If a data item is only stored once, then there is less chance of data entry errors and less time taken to enter new data or edit existing data in a database. If data is deleted from a database, it will also take less time to delete the data from one table, as the same data will not need deleting several times as it would in a flat-file database single table. It is also less likely that an item of data would be overlooked and not deleted. In addition, when searching a large database, it is generally quicker for the software to locate records as it will be searching through significantly less data. However, this is not always the case; where indexed fields are used it can be true but it is not always the case. It depends on the structure of each database and the quantity of data being searched.

You can see, comparing Table 10.1 with Tables 10.2 and 10.3 with just five teachers and five students, how much internal memory and external storage space is saved simply by not storing data more than once. Imagine the space saved for a school with 1500 students and 120 teachers, or in a national database with data on every vehicle and driver registered in a country. It is also much easier for users to produce reports when data is held using two or more tables in a relational database, rather than using two or more flat-file databases.

We will use *Microsoft Access*® to create our database. We can use this software to create a relational database if there are two or more tables of data. If we only have one table, we use the software to create a flat-file database.

10.1.2 Data types

When you create a new database table, you will set the data type for each field. This determines how *Access* will store, manipulate and display the data for each field. There are a number of data types that you can use and different packages may have different names for them. The list below shows the generic names for these data types, but depending upon the package used, you may have different names. For example in *Access*, an alphanumeric field is called a text field. The three main types of field are alphanumeric, numeric and **Boolean**:

» Alphanumeric data can store alpha characters (text) or numeric data (numbers) that will not be used for calculations. In *Access* this is called a text field.
» A numeric data type (as the name suggests) stores numeric values that may be used for calculations. It should not store numeric data that is not used for any calculations, like telephone numbers (data not used for calculations should be stored as an alphanumeric data type). In *Access*, this is called a number field. There are different types of numeric field including:
  – Integer fields store whole numbers. In *Access* you can select an integer field or a long integer field. It is wise to use a long integer field if it is going to contain three or more digits.
  – **Decimal** formats, which will allow a large number of decimal places or a specified restricted number, if this is set in the creation of the field properties when the database is set up.
  – Currency values, which allow currency formatting to be added to the display. These include currency symbols and regional symbols. The database does not store these as this would use up valuable storage space.
  – **Date and time** formats, which store a date and/or time as a number.
» A Boolean (or logical) data type stores data in a Yes/No (or True/False or 0/-1) format.

There are other data types as well, like **AutoNumber** which generates unique numbers. Some packages like *Access* have long and short versions of their data types, for example long text and short text (which are versions of alphanumeric data types) or long number and number.

Other data types can often be found in commercial databases. *Access* can store placeholders for media, such as images, documents, spreadsheets, presentations, sound bites and video clips. All of these are stored using an **attachment** data type or using an **OLE object** (**object linking and embedding**) data type. Media files tend to be used for web applications where a **back-end database** holds the media to be displayed in another application like a web page.

> **Advice**
>
> *Microsoft Excel*® is **not** suitable for database tasks because you cannot define data types.

10.1.3 Relationship types

Relationships are used to link tables together. There are three types of relationship, one-to-one (1–1), one-to-many (1–∞) and many-to-many.

One-to-one relationship

This relationship is where each record in one table relates to only one record in another table. For this type of relationship to exist, the linked fields in both data tables must contain the same unique values.

One-to-many relationship

This relationship is where each record in one table can relate to many records in another table. This is where the primary key field in one table relates to the foreign key field in the second table (see Section 10.1.4).

Many-to-many relationship

It is not possible to create a single many-to-many relationship in *Microsoft Access* as this is only a theoretical type of relationship. The many-to-many relationship is therefore created using two one-to-many relationships and a link table. For example if a company sells many different types of products and has many different customers, it is possible that one customer has bought different types of product from the company and a type of product has been sold to many different customers. In this case three tables and two relationships would be created. The primary key field in the Customers table relates to many records in the foreign key field within the Link table like this.

The primary key field in the Product table relates to many records in the foreign key field within the Link table like this.

An example of the Link table's contents can be seen here.

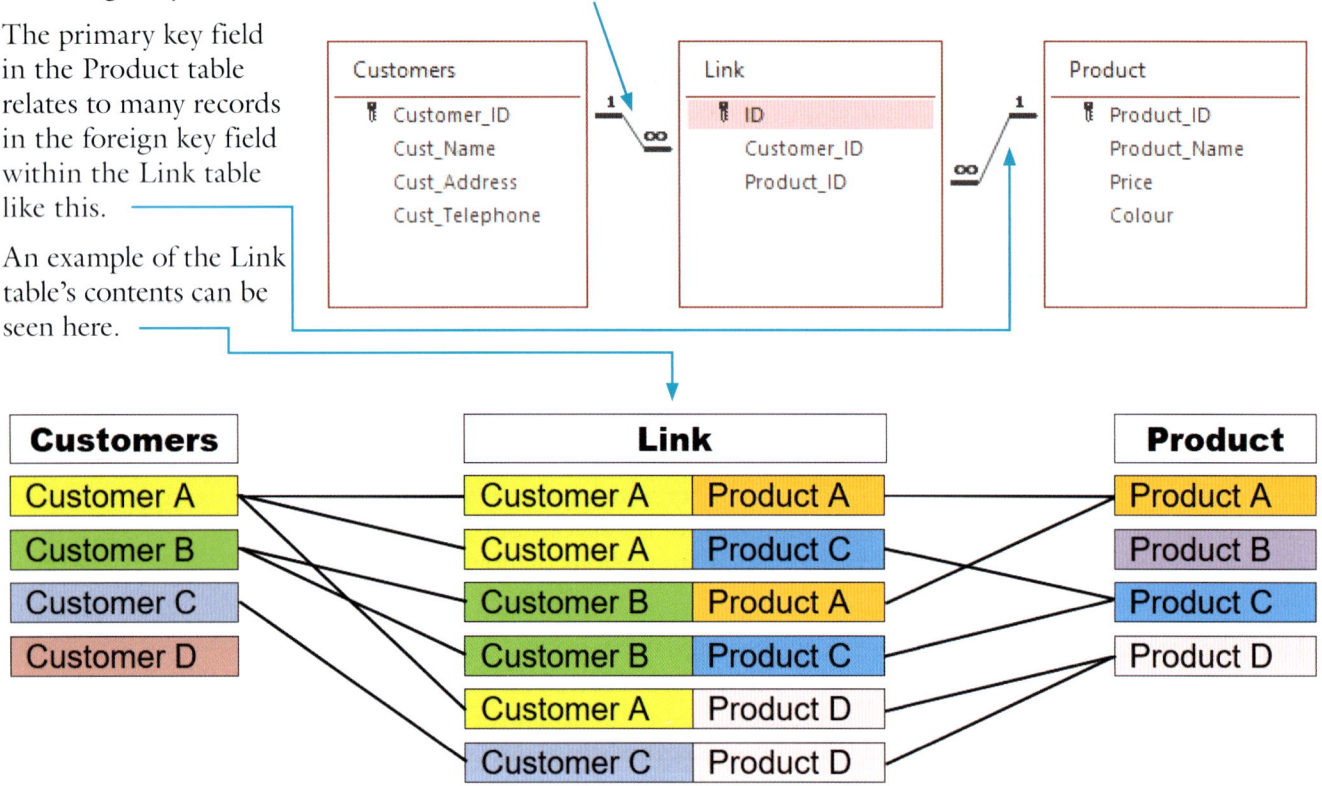

10.1.4 Key fields

Key fields are used to create the relationships between different tables in the database. All relationships between tables use key fields. There are three types of key field: primary key, foreign key and compound key.

Primary key

For a database table to be used as a relational table, it must have a primary key field. Primary key fields are single fields that contain unique data and cannot contain blank records. Because primary key fields are unique, they are also used in many databases as indexes. An index optimises the storage of table records within the database and helps to provide quicker searches.

Foreign key

A foreign key is a field in the table that is a primary key in another table. It links to that primary key to form a relationship between two tables.

Compound key

A **compound key** is a primary key created using a number of foreign key fields rather than a single foreign key, which together create unique data. In *Access*, no data in any of these foreign key fields can be blank.

10.1.5 Entity relationship diagrams

The three types of entity relationship diagram (ERD) are conceptual, logical and physical. Each ERD shows the design of the table structures and relationships between tables within a relational database. ERD designs include:

» entities, which are objects or items to be included (often the tables to be used)
» attributes, which are the facts about or properties of an entity (often the fields to be used)
» relationships, which are the links between the entities.

Entities are drawn using a rectangular box with rounded corners like this.

There are different ways of displaying the attributes for each entity; the easiest to use is as a list within the entity box like this:

Relationships are shown as lines between the entities.

One-to-one relationships in an ERD

One-to-one relationships have a single line at 90° placed at each end of the relationship line, for example one driver has one driver's licence.

One-to-many relationships in an ERD

One-to-many relationships have a single line at the 'one' end of the relationship line and at the 'many' end of the relationship line there are three lines.

For example one customer can make many purchases from a company. This one-to-many relationship looks like this.

Many-to-many relationships in an ERD

As we said above, although many-to-many relationships exist (theoretically), these must be designed as two one-to-many relationships. For example if a company sells many different types of products and has many different customers, it is possible that a customer has bought different types of product from the company and a type of product has been sold to many different customers. In this case the relationships in the entity relationship diagram look like this.

Note that relationship lines in the ERD do *not* identify the fields (called attributes in the ERD) used for the relationship, unlike a sketch of database relationships which would need primary and foreign keys identified.

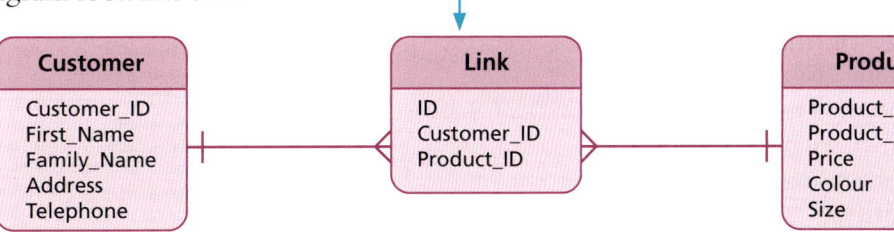

Conceptual design

Conceptual design is what you do when you create a database design/model which is not dependent upon the choice of **database management system (DBMS)**. A conceptual model uses information gathered from the business requirements and is the only type of ERD that allows generalisations. In other words, it allows 'a kind of' relationships, for example a rectangle is 'a kind of' shape, whereas logical and physical models have to have more precise relationship definitions. It is simpler than logical or physical design.

Logical design

Logical design is what you do when you create a database design/model where the DBMS has already been selected. Logical models also use information gathered from business requirements. They are more complex than a conceptual model in that they contain a map of rules that will be followed. Column types may be set to help the business analysis. Logical design is not about database creation.

Physical design

Physical design is what you do when you create the actual database based on the requirements gathered during logical database modelling so that it can be implemented within the DBMS. It contains more detail than the other types of ERD, for example column types from logical designs become data types, and field lengths are included in the design. Primary and foreign key fields are also included. Like conceptual and logical models, the relationships are not drawn from attribute to attribute (as you would find in a database relationship diagram).

> ### Task 10a
> The Tawara Widget Company has a number of different sites. The company is going to create a database to store details of its employees, their jobs and the site where they work. The company requires the site code, name of the site, address and telephone number. It also requires for each employee: their payroll number, name, address, job code, job description and rate of pay. All employees doing the same job are paid at the same rate of pay.
>
> Create conceptual, logical and physical entity relationship diagrams for this company.

Conceptual ERD

Identify the entities from this task. These will be: the employees, their jobs and the site where they work. Create the conceptual ERD (either hand drawn or on the computer) like this.

The attributes have been added to the diagram, using the details given in the task description. They are used to identify what information should be stored within each entity.

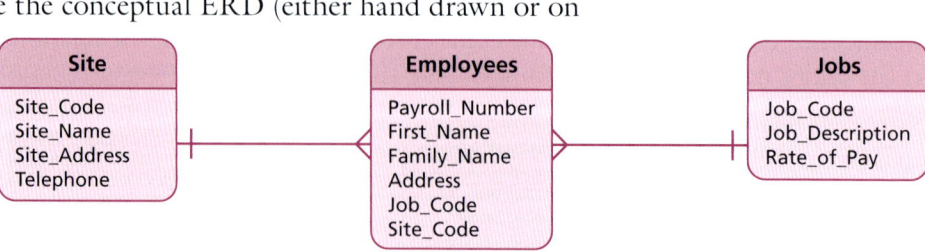

Logical ERD

Create the logical ERD in a similar way. Allow space within each entity box for another column to the right of the attributes. Complete the entities and attributes (as in the conceptual diagram). Add a new column to the right of the one holding the attributes. In this new column, place the type of data that the company holds for each entity. For each entity box, you now have the attribute and its associated data type.

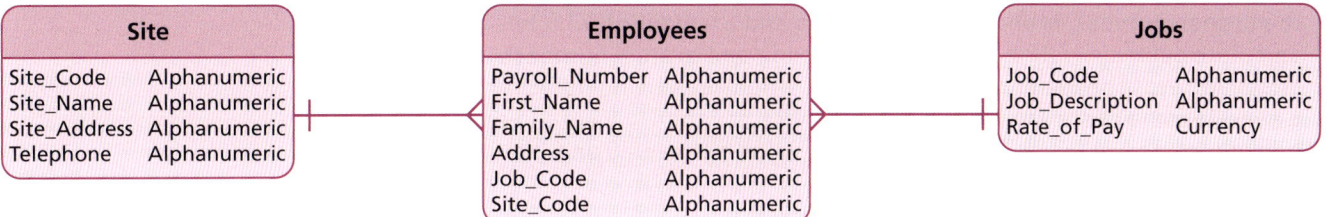

Physical ERD

The physical ERD is the one used to create the database, so must include the entities (tables) and attributes (fields) as well as the data types, field lengths and key fields. Both primary and foreign keys should be identified. Create this in a similar way to the conceptual and logical diagrams. The final diagram will look like this.

Site			
Site_Code	Alphanumeric	5	PK
Site_Name	Alphanumeric	34	
Site_Address	Alphanumeric	150	
Telephone	Alphanumeric	13	

Employees			
Payroll_Number	Alphanumeric	8	PK
First_Name	Alphanumeric	30	
Family_Name	Alphanumeric	36	
Address	Alphanumeric	150	
Job_Code	Alphanumeric	2	FK
Site_Code	Alphanumeric	5	FK

Jobs			
Job_Code	Alphanumeric	2	PK
Job_Description	Alphanumeric	60	
Rate_of_Pay	Currency	2 dp	

Activity 10a

The Tawara Bus Company uses 10 buses which it hires to customers. For each bus, it has its registration number (which is unique, seven characters long and consists of both numbers and letters), the make and model of the bus, as well as the number of seats on the bus. The company employs 26 drivers; each has a payroll number and the company stores personal data including their name, address, telephone number (stored as five numbers followed by a space then six numbers) and date of birth. All drivers are over 25 years of age. The company holds personal data about each customer, which includes a numeric customer number, their name, address, telephone number and email address. Each time a bus is hired for a journey it is given a unique booking number. The bus used, driver, date and time of the journey, the customer and number of miles are recorded.

Create conceptual, logical and physical entity relationship diagrams for this company.

Hint: Drawing each box side by side will not work for this activity.

Task 10b

The Tawara Motor Company is designing a database to store information on each employee, their job and the branch in which they work. For each employee, the database needs to hold information on their name, their payroll number (which is unique to them and has two letters followed by four digits, the first letter is always T or M), the branch in which they work and the code of the job that they perform. For each job, the job code and description must be stored along with the rate of pay for that job. For each branch of the company, the branch number, branch name, address and a weighting which is used to calculate salaries will be stored. The job code and branch numbers are both single letters. The weighting ensures that employees living in larger cities, where housing costs are higher, receive more pay than workers in areas where housing costs are lower. This is a decimal value between 0 and 2 stored to two decimal places.

Design and create this database.

10.1.6 Designing the database structure

Often the design of the database can be taken from the physical ERD. Before this, it is worth creating the conceptual ERD using three entities.

Entity and attribute names

Each entity name may be used as a table name, so it is important to keep them short and meaningful and for them to contain no spaces. In this task there are three distinct sets of information used: the information about the branches, the employees and their jobs. In this case we will use Branch, Employees and Jobs.

Each attribute name may be used as a field name, so again it is important to keep them short, meaningful and without spaces. They must be short enough to ensure that printouts fit on a page without wasted space but long enough to be meaningful. A good convention is to avoid using spaces within field names as some databases do not accept these. Assign the attributes to the entities like this.

Branch	Employees	Jobs
Code	PayNumber	JobCode
Branch	FirstName	Description
Address1	FamilyName	RateofPay
Address2	JobCode	
Address3	BranchCode	
Weighting		

Relationships

Each branch has many employees working there, so a one-to-many relationship will be needed between these entities. As some employees may perform the same job within the company, a one-to-many relationship will exist between the entities for jobs and employees. Add these changes to the ERD so it looks like this.

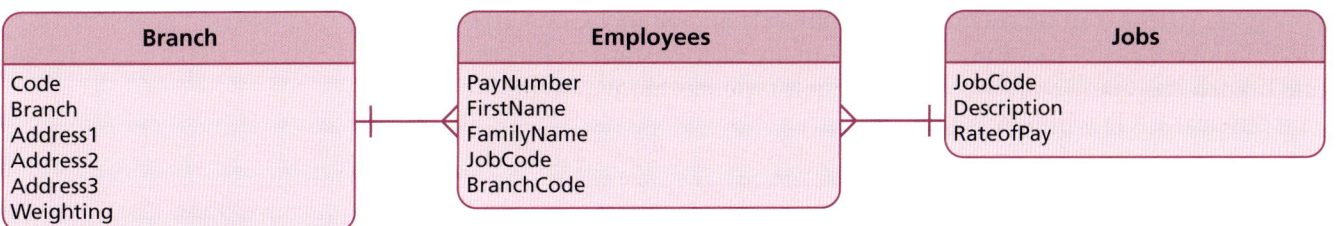

Data types

To turn the conceptual ERD into a logical ERD, the data types need to be considered and added. See the list of data types that could be chosen in Section 10.1.2 above.

> **Advice**
> Remember to only use a numeric field type if the data may be used for a calculation.

Add appropriate data types to create this logical ERD.

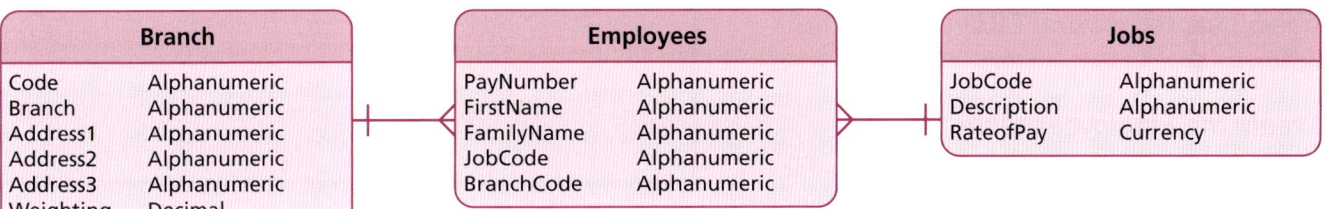

Field lengths

The length of each field needs to be considered and added to the ERD. If the data is available to examine, count the number of characters for the longest data item and set the field length to that number. For example for first names or family names, look for the longest possible name you can find (including hyphenated names) and set the field length to this. Field lengths only need to be considered for alphanumeric data types. Some of the information to help set the field lengths is in the task, for example the payroll number has two letters followed by four digits so will be six characters long. The branch code and job attribute in the job table are defined as a single alphanumeric character.

Key fields

To turn the ERD into a physical ERD, the key fields need to be considered and added. The primary key field for each entity/table will be the field that contains unique information. In the case of the Employees entity/table, you were told in the task that the employee's payroll number was 'unique to them' so this will be an obvious choice for primary key field. For the Jobs entity/table, the job code field will contain unique values and will be suitable to use as a primary key. For the Branch entity/table the branch code will also contain unique values and should become the primary key field for this table.

Foreign key fields can be added on the JobCode and BranchCode attributes of the Employees entity to allow the one-to-many relationships to be created, as one branch has many employees and one job type has many potential employees. The physical ERD will now look like this:

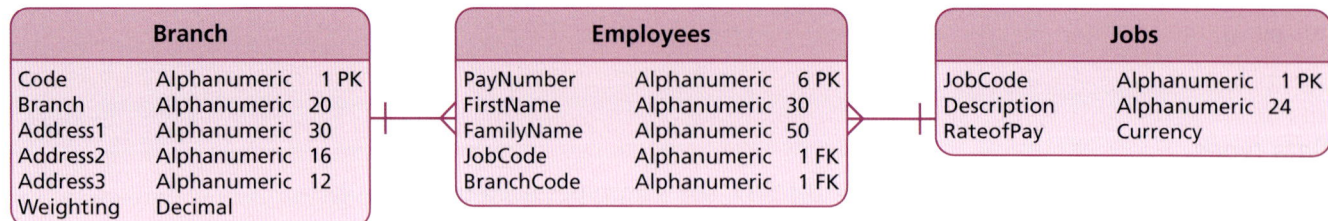

10.1.7 Creating the database structure

Open *Microsoft Access* and select the icon for **Blank database**. Select the folder you wish to store the database in and enter the database name as **Task_10b** before clicking on the **Create** button.

We will use the physical ERD to create this database.

Create the Employees table

Select the **Create** tab, then double-click on **Table**. Select the **Home** tab, then in the **Views** section, the icon for **Design View**.

When this is selected, the **Save As** window opens. Enter the entity name Employees and click on OK to name this table.

Access incorrectly assumes that you want an ID field as a primary key field, but the primary key for the Employees table will be the PayNumber. To delete the ID field, click the right mouse button in the ID box and from the drop-down menu select **Delete Rows**. This warning message appears.

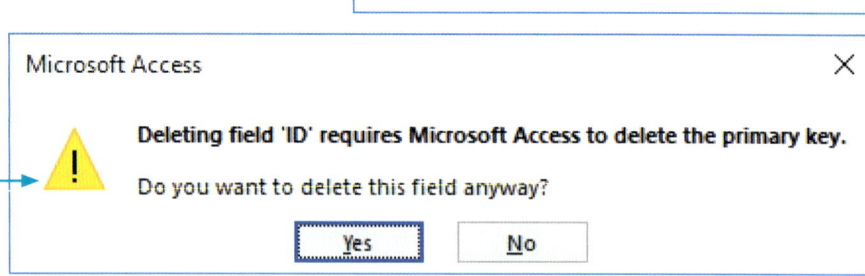

Click on Yes to delete the record.

Rename Field1 to be **PayNumber** and use the drop-down menu in the **Data Type** column to change the data type to **Short Text** if it is not already set as that.

> **Advice**
>
> In *Access*, for alphanumeric data, use a 'text' field type, either Short Text or Long Text.

Set a key field

To set a primary key, select the field that contains unique data, then click on the primary key icon. To set a compound key, hold down the <Ctrl> key and select all the fields that will comprise the compound key, then click on the primary key icon. A foreign key field is not set. However, it must contain the same data as the primary key to which it will be related.

For this Task, from the Design tab (which should be open) select the Primary Key button to set this as the primary key field for this table.

In the Field Size box change the size from 255 to 6 characters for the field length.

Move down to the row below PayNumber and enter the next field name, FirstName. Use this method to add all field names and data types to this table.

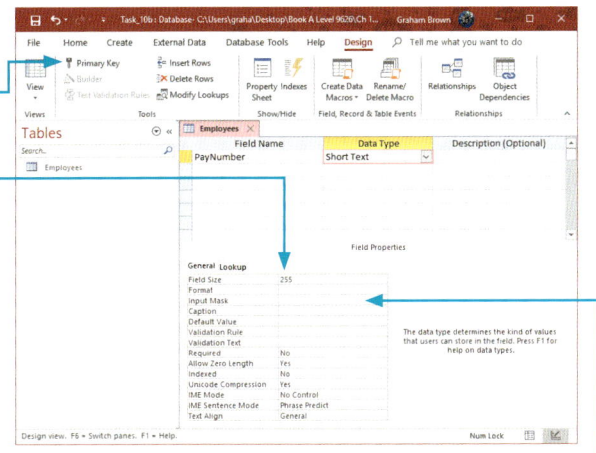

Input masks

An input mask will force a database user to restrict the data that they enter into a field. It is not a validation routine as it does not show an error message if incorrect data is added, but it does help to reduce data entry errors.

The Task stated that the payroll number 'has two letters followed by four digits'. We can use an input mask to restrict the data entry for each character within this field. Click the cursor in the PayNumber row. In the Input Mask section, enter 'LL0000' to restrict data entry to two alphanumeric characters followed by four numeric digits. Table 10.4 shows some of the characters that can be used within the input mask and the function of each.

▼ **Table 10.4** Input mask characters

Character	Function
0	User must enter a number
9	User can enter a number (optional)
#	User can enter a number, space, + or −
L	User must enter a letter
?	User may enter a letter
A	User must enter a letter or number
a	User can enter a letter or number
&	User must enter a letter, number or space
C	User can enter a letter, number or space

Validate data entry

The Task stated that for the payroll number, 'the first letter is always T or M'. We can use a validation rule to restrict the data entry for the first character within this field. Click the cursor in the PayNumber field. In the Validation Rule, enter Like "T?????" Or Like "M?????" to restrict data entry. In the Validation Text, enter an appropriate error message, such as The first character must be "T" or "M". Error messages like this should always tell the user how to correct the error.

A brief description can be added to each field by typing it in the **Description** area. This will help the user to enter data by giving them helpful instructions as to what is required in this field. This is not used by the database to check the data. Instead, it gives the user a little help when using the form to enter data. An example for the **PayNumber** field is: **Enter the employee's payroll number which is formatted as two letters followed by four numbers**. Save the table, which should look like this.

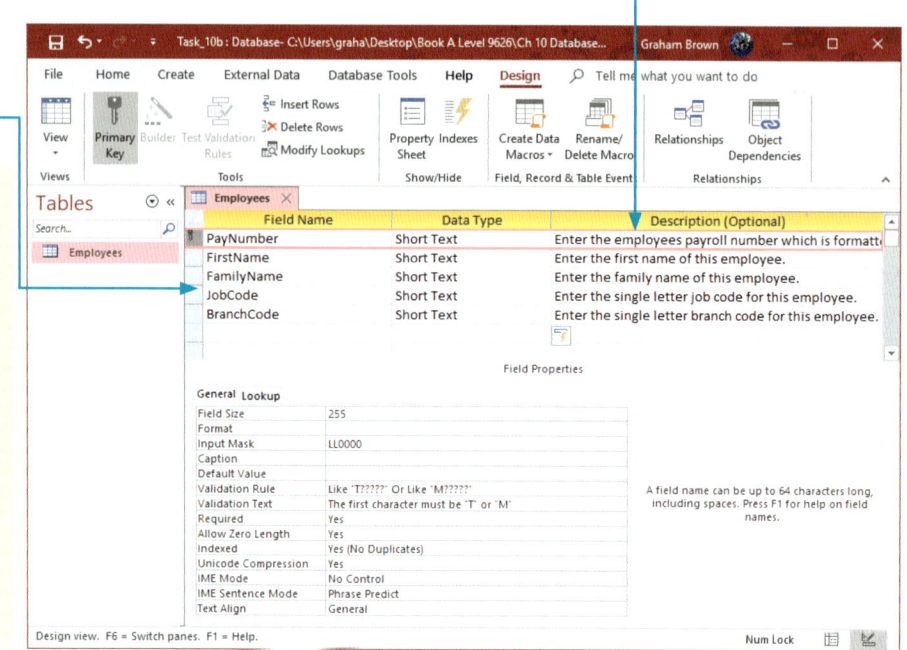

Advice

Here are some quick tips for setting other field types in *Access*.

» To set a percentage value, select a **Currency** data type, then in the **Format** section select **Percent**.
» To store a date or time, select a **Date/Time** data type, then select the **Format**.
» For a Boolean or logical field type select a **Yes/No** data type, then **Format** to choose from Yes/No, True/False or On/Off.

Create the Branch table

Create the Branch table using a similar method to that used for the Employees table. The **Weighting** field must be 'a decimal value between 0 and 2 stored to two decimal places'. For numeric fields, the **Field Size** box is used to display the type of numeric field. Use the drop-down menu to select **Double** from the list. The number of decimal places can be set using the **Decimal Places** section. For this Task, use the drop-down menu to select **2** decimal places. Add the validation rule and text as shown.

Set the **Code** field as the primary key. Save the table.

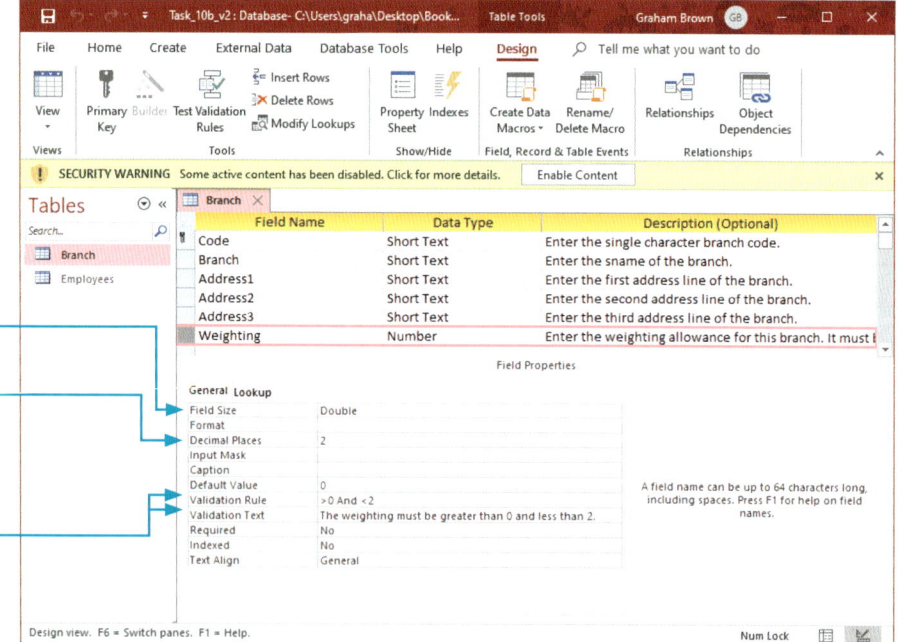

Advice

For integer field types in *Access*, use **Integer** for a whole number between 0 and 255 and **Long Integer** if larger than that.

Create the Jobs table

Create the Jobs table using the physical ERD and a similar method to the other tables. Set the **Job** field as the primary key. Save the table. The completed Jobs table should look similar to this.

Close all three tables before creating the relationships.

Create a relationship

Select the **Database Tools** tab, then in the **Relationships** section select the **Relationships** icon.

This will open the **Relationships** window in *Access*. All of the tables that have been created are visible in the **Show Table** window. Select each table in turn and click on **Add** to add each table to the **Relationships** window. When all of the tables have been added, close the **Show Table** window.

> **Advice**
>
> When table names and field names are referred to together, they are often shortened. This is done using notation in the form of TableName.FieldName. For example the notation Employees.PayNumber means the PayNumber field within the Employees table. This notation is used in the rest of this chapter.

The ERD shows there were two one-to-many relationships to create, one between **Branch.Code** and the Employees table. The two foreign keys in the Employees table are the **JobCode** and **BranchCode** fields, so the **BranchCode** field will be used. For the first FK, a relationship needs to be created between the **Branch.Code** and **Employees.BranchCode**. This will be a one-to-many relationship. Click on the **Code** field in the Branch table.

Drag and drop this field onto the **BranchCode** field in the Employee table.

The **Edit Relationships** window will now appear. Check that the correct table names and field names are present within this window.

You will notice that *Access* has created this as a one-to-many relationship type.

At this stage the relationship has not yet been created.

> **Advice**
>
> If the two fields joined by a relationship are both primary key fields, then *Access* will create a one-to-one relationship.

Referential integrity

Referential integrity forces table relationships to be consistent and avoids redundant data. This means that the data in the foreign key field must match the data in the primary key field. It prevents a user from accidentally changing or deleting data in one table, without the same action happening to the related data in another table.

To enforce referential integrity between these tables, tick the check box.

Tick the other two check boxes for Cascade Update Related Fields and Cascade Delete Related Records so that when you update or delete a primary key, *Access* automatically updates all records in the foreign key field/s that reference that primary key field. Click on Create to create the relationship. The relationship diagram will now show the relationship line. As referential integrity has been added to the relationship the diagram shows the join type on the relationship line.

> **Advice**
> To open the Edit Relationships window and edit the relationship at any time, double-click the left mouse button on the relationship line.

Create the relationship between Jobs.Job and Employees.JobCode using a similar method to that used for the Branch table. Make sure that you enforce referential integrity. Save the database, which will save the relationships you have created. The relationship diagram now looks like this.

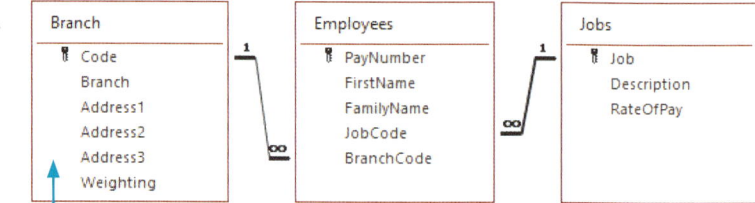

The table structure of the database has been created, so the validation rules and input masks need to be tested.

Testing the validation rules

Using the techniques learnt in Section 8.2, design your test plan to test the validation rule set in the Employees.PayNumber and Branch.Weighting fields. For the Employees.PayNumber field the rules set were Like "T?????" Or Like "M?????", so design your test plan as shown in Table 10.5.

▼ **Table 10.5** Sample test plan for Task 10b: Employees.PayNumber field

Test type	Validation	Table.Field	Employees.PayNumber	
Data entry	Data type	Expected result	Actual result	Remedial action
TX5443	Normal	Data accepted		
TA1234	Normal	Data accepted		
MX5443	Normal	Data accepted		
MA1234	Normal	Data accepted		
LX5443	Abnormal	Validation error message		
NA1234	Abnormal	Validation error message		
SX5443	Abnormal	Validation error message		
UA1234	Abnormal	Validation error message		
100400	Abnormal	Validation error message		

For the **Branch.Weighting** field the rules set were **>0 And <2**, so design your test plan as shown in Table 10.6.

▼ **Table 10.6** Sample test plan for Task 10b: Branch.Weighting field

Test type	Validation	Table.Field	Branch.Weighting	
Data entry	Data type	Expected result	Actual result	Remedial action
0.01	Normal	Data accepted		
1	Normal	Data accepted		
1.99	Normal	Data accepted		
0	Abnormal	Validation error message		
2	Abnormal	Validation error message		
Hello	Abnormal	Validation error message		
-67	Abnormal	Validation error message		
1000	Abnormal	Validation error message		

Use this test data to test both rules and complete the test plans.

Testing the input mask

Using the techniques learnt in Section 8.2, design your test plan to test the input mask set in the **Employees.PayNumber** field. The input mask set was 'LL0000', so design your test plan as shown in Table 10.7.

▼ **Table 10.7** Sample test plan for Task 10b: Employees.PayNumber field

Test type	Input mask	Table.Field	Employees.PayNumber	
Data entry	Data type	Expected result	Actual result	Remedial action
TX0001	Normal	Data accepted		
MA9999	Normal	Data accepted		
TTT443	Abnormal	Validation error message		
M31234	Abnormal	Validation error message		
TX544	Abnormal	Validation error message		
000001	Abnormal	Validation error message		

Use this data to test the input mask and complete the test plans.

Verify data entry

As well as using validation rules and input masks, it is also important to verify all data entry into your tables. As explained in Section 1.4, visual verification can greatly reduce the number of data entry errors. Carefully check that the data that has been entered into the tables matches exactly that provided in the source document.

Activity 10b

Create a word-processed Evidence Document. Place your name and school/college name in the header of all pages and save this as **Activity_10b**.

In Activity 10a, you designed a database structure for the Tawara Bus Company. Use your physical ERD to create this structure in your database package. Place in your Evidence Document evidence of:
» all the tables showing all of the data types
» all primary keys
» validation rules
» input masks
» relationship diagrams.

Save your database as **Activity_10b**. Save your Evidence Document.

10.1.8 Creating a relational database from existing files

Task 10c

Freeze-Air is a company that makes and sells fridges and freezers. It has supplied you with a physical ERD and its data files so that you can create a database. Each invoice contains only a single product item but may contain more than one of those items. The ERD is shown here.

Product		
Code	Alphanumeric	5 PK
Model	Alphanumeric	8
Price	Currency	
Capacity	Numeric/Integer	
Energy_Rating	Alphanumeric	5
Colour	Alphanumeric	10

Invoice		
Inv_Number	Numeric	PK
Cust_ID	Alphanumeric	6 FK
Product_Code	Alphanumeric	5 FK
Quantity	Numeric/Integer	
Order_Date	Date/Time	
Order_Time	Date/Time	
Collect_Date	Date/Time	
Collect_Time	Date/Time	

Customer		
Cust_ID	Alphanumeric	6 PK
Given_Name	Alphanumeric	16
Family_Name	Alphanumeric	24
Address1	Alphanumeric	30
Address2	Alphanumeric	20
Address3	Alphanumeric	20
Discount	Numeric/Percent	

The data to create the database is located in the files **FA-Product.csv**, **FA-Invoice.csv**, **FA-Cust1.csv** and **FA-Cust2.txt**. Create this database.

Sometimes you may have to create a database using given data files. In this Task, the database structure has been designed for you as an ERD but may be given as a data dictionary (see Section 10.11), or you may have to use normalisation to find the data structures (see Section 10.10).

Always start by closely examining any files given to you. When you study the files **FA-Cust1.csv** and **FA-Cust2.txt** you will see that the two files need merging and the order of the fields re-arranging before the data can be imported into a database. *Excel* would be the most appropriate package for this task. When examining these files, it is clear that FA-Cust1 will create the first part of the data file and FA-Cust2 the second. Open both files in *Excel* (drag the .txt file into *Excel*) and view them side-by-side.

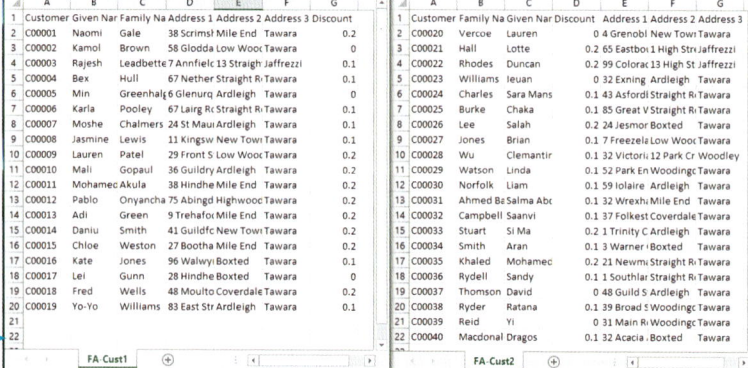

Copy and paste the columns in FA-Cust2 so that they are in the same order as those in the first file, like this.

Copy, from FA-Cust2, cells A2 to G22 and paste them into cells A21 to G41 in FA-Cust1. Make sure that you do not copy row 1 as this will cause import problems later. Save this file as **Cust.csv**.

Create a new blank database and save this as **Task_10c**.

Create the Product table

Select the **External Data** tab, then **New Data Source**.

Select **From File** from the drop-down menu, then **Text File** (as a .csv file format is a text format) like this.

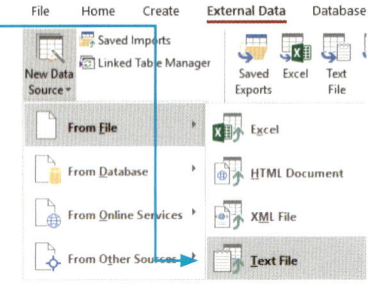

The **Get External Data – Text File** window opens. Select, in the **File name** box, the file FA-Product.csv. Make sure that the radio button for **Import the source data into a new table in the current database** is selected, then click OK.

This opens the **Import Text Wizard** window. As the data is a .csv file which has the data separated (delimited) by commas, select the **Delimited** radio button before clicking Next >.

In the next window, check that the radio button for **Comma** is used, then tick the check box for **First Row Contains Field Names**, before clicking Next >.

The next window is probably the most important. For each field, change the field name to match that shown in the physical ERD (or the data dictionary). In this instance, in the **Field Name:** box, change the **Product code** to **Code**, leave the **Data Type** as **Short Text**. As this field is to be the primary key field, change the **Indexed:** box from **Yes (Duplicates OK)** to **Yes (No Duplicates)**.

We will have to select the field lengths when the import is complete. Click on each field in turn, setting the field names and data types as shown in the ERD, then click Next >. *Access* will try and add its own primary key in an ID field, but we need to select the radio button for **Choose my own primary key**. Because we changed the index to no duplicates in the previous window, it will recommend the **Code** field. Click Next >. In the **Import to Table:** box, enter the table name **Product** as the entity name from the ERD and click Finish. You do not need to save the import steps so click **<Close>**.

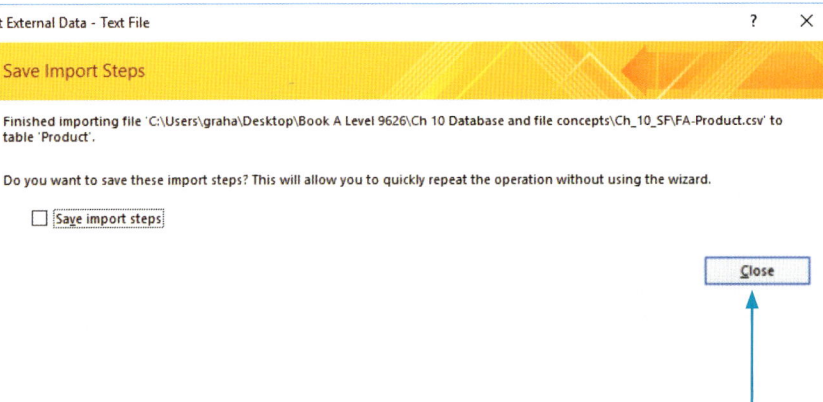

Select the **Home** tab and then open the **Product** table in **Design View**. For each alphanumeric field set the field size to match the physical ERD. Save the database.

Create the Invoice and Customer tables

Use the same techniques to import the files **FA-Invoice.csv** and **Cust.csv** as the Invoice table and Customer table respectively, so that the tables look like this:

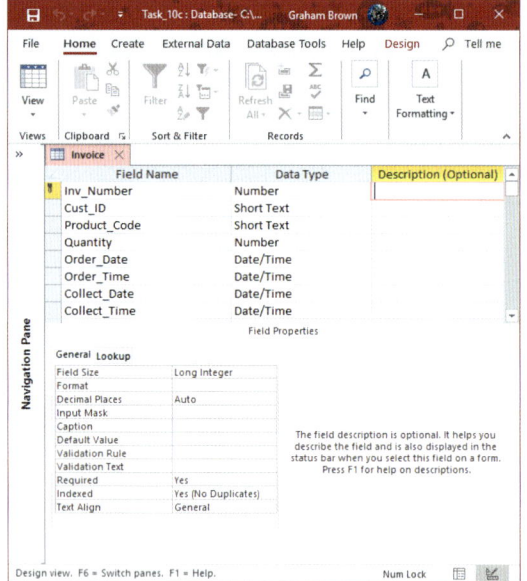
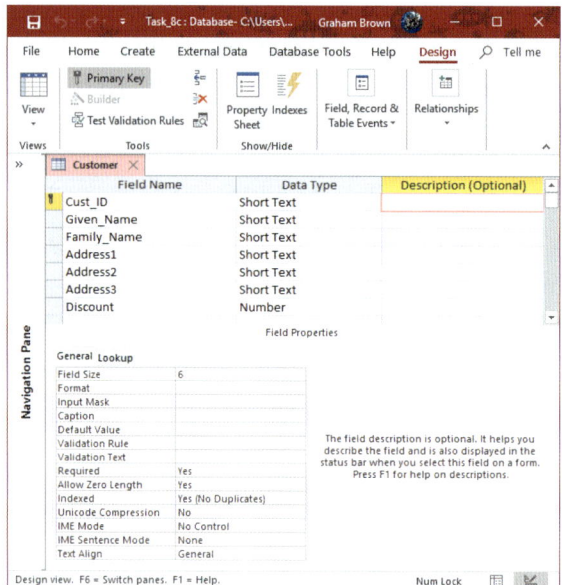

Create the relationships

Use the techniques learnt earlier in the chapter to create relationships that match the ERD, as shown here.

Save the database.

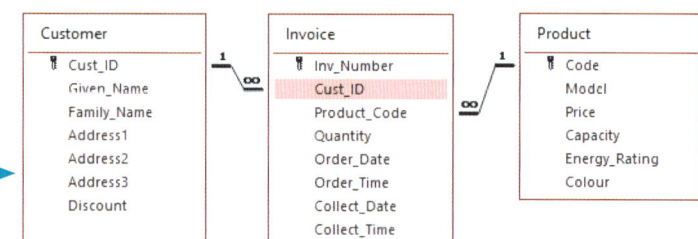

> ## Activity 10c
> You work for Nanna Pat's Boat Yard. It makes and sells small boats. You have been supplied with a physical ERD and the data files **NP-Boat.csv** and **NP-Customer.csv**. The ERD is shown below.
>
>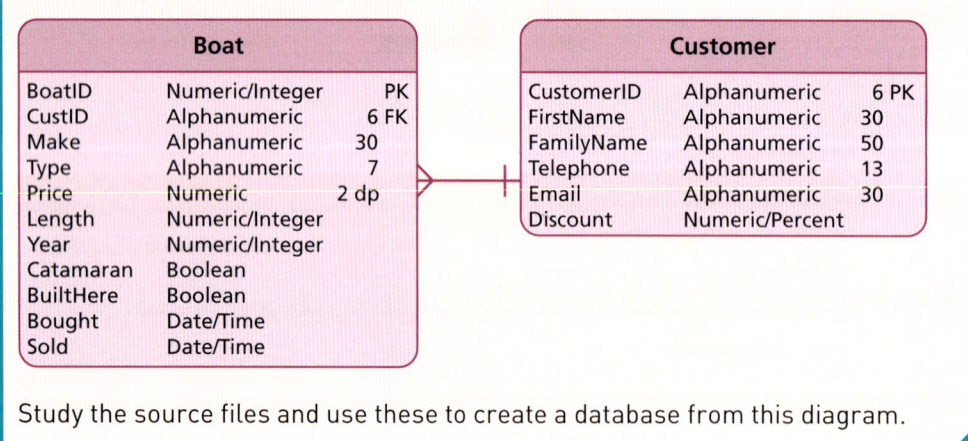
>
> Study the source files and use these to create a database from this diagram.

10.1.9 Querying the database

At this level you may be required to perform a query on a single criterion but are more likely to be required to perform them on multiple criteria and produce high-quality output to certain specifications. Performing searches in *Access* is carried out using queries and the high-quality output is done using reports.

Task 10d

Use a copy of the database saved in Task 10c to produce a report for the directors of the company. This will list the details of all the customers who are given no discount on their orders and, for each customer, list the model/s purchased and the price. Do not display the discount.

Copy the database saved in Task 10c and save this as **Task_10d**. Open this database.

As this report is to be presented to the directors of the company, it must have a professional appearance with an appropriate title and layout. As the task requires the details of the customers, all of the customer data should be displayed for those who meet the search criterion. Start the task by identifying which fields will be needed and which will not.

Look at each table and indicate which data will be required. One method is to list all the fields and tick or highlight those required. It is debatable as to whether the Cust_ID field is required, but if in doubt about a field, include it anyway.

▼ Table 10.8 Field selection for Task 10d

Customer table	
Cust_ID	✓
Given_Name	✓
Family_Name	✓
Address1	✓
Address2	✓
Address3	✓
Discount	✗

Invoice table	
Inv_Number	
Cust_ID	
Product_Code	
Quantity	
Order_Date	
Order_Time	
Collect_Date	
Collect_Time	

Product table	
Code	
Model	✓
Price	✓
Capacity	
Energy_Rating	
Colour	

Perform a simple numeric query on a single criterion

Select the **Create** tab, then in the **Queries** section, select the icon for **Query Wizard**. In the **New Query** window, choose the **Simple Query Wizard**, then click [OK]. Using Table 10.8, select from the **Tables/Queries** drop-down list each table in turn.

You will select the ticked fields from each table in Table 10.8. When you need to move to a new table, just select from the **Available Fields** box in turn. Select the **Discount** field too, as it will be used in the query although not displayed. Use the right arrow key to move the field name into the **Selected Fields** box.

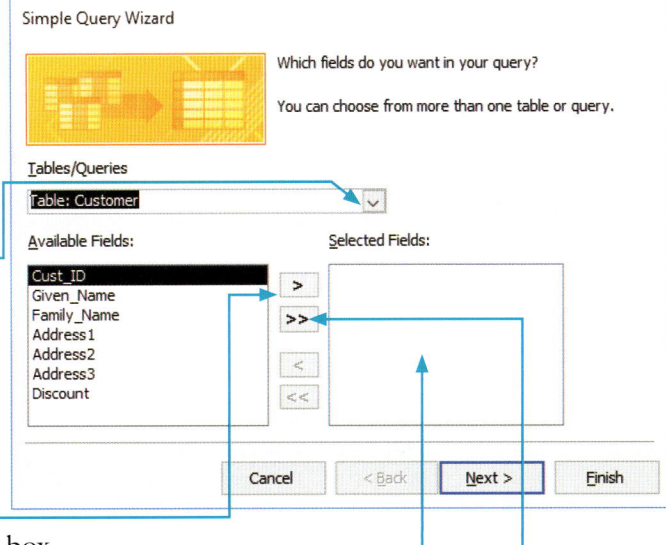

Advice

If all fields from a table are to be placed in the query then use this button.

When all the ticked (required) fields have been selected, click **Next >**. For this query we will select the radio button for **Detail** then click **Next >**.

Advice
We will look at summary queries later in the chapter.

For the final window, change the title of the query to a suitable title for the report.

Select the radio button for **Modify the query design**, then click on **Finish**.

The query will appear in **Design View**. The criterion for the query has not yet been added.

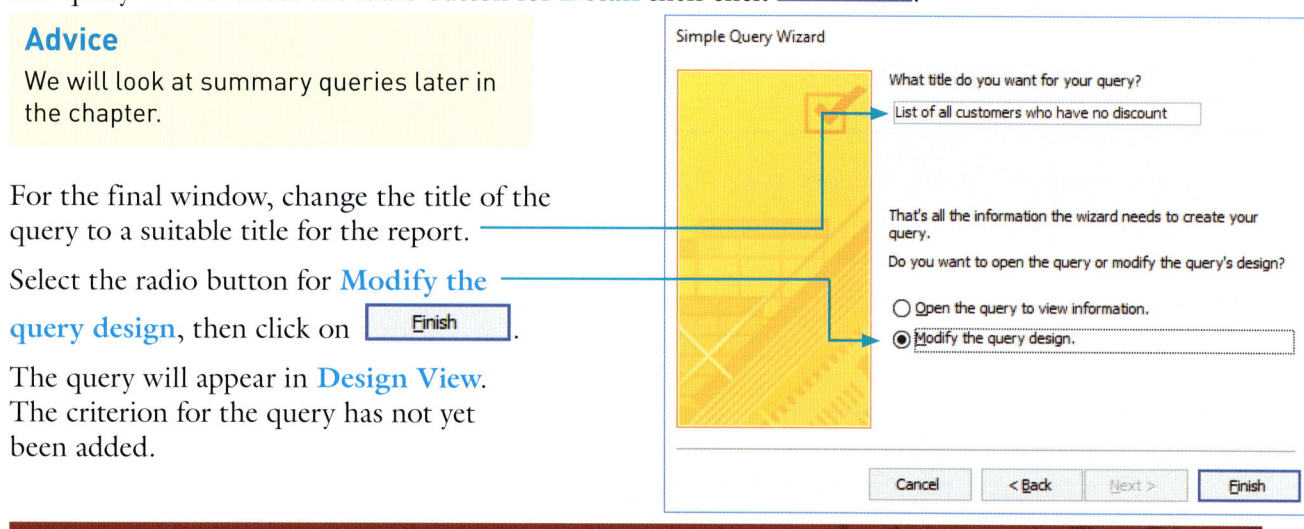

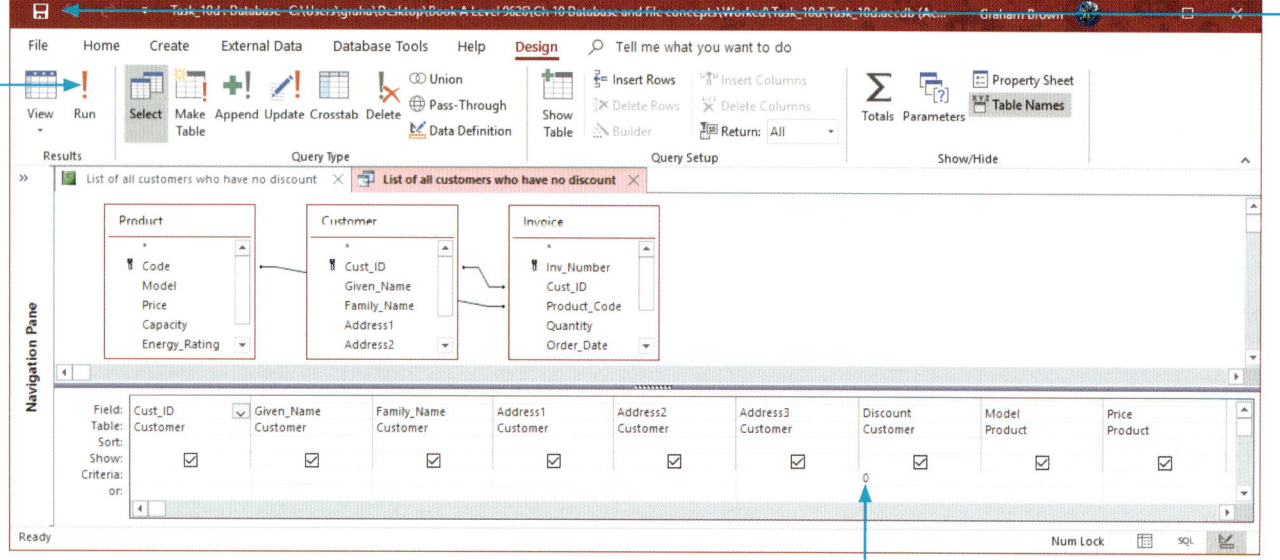

Move the cursor into the **Discount** column on the **Criteria:** row.

Enter the value **0** to indicate no discount. Save the query (using the save icon at the top left). Click on the **Run** icon to run the query.

The results of the query appear like this:

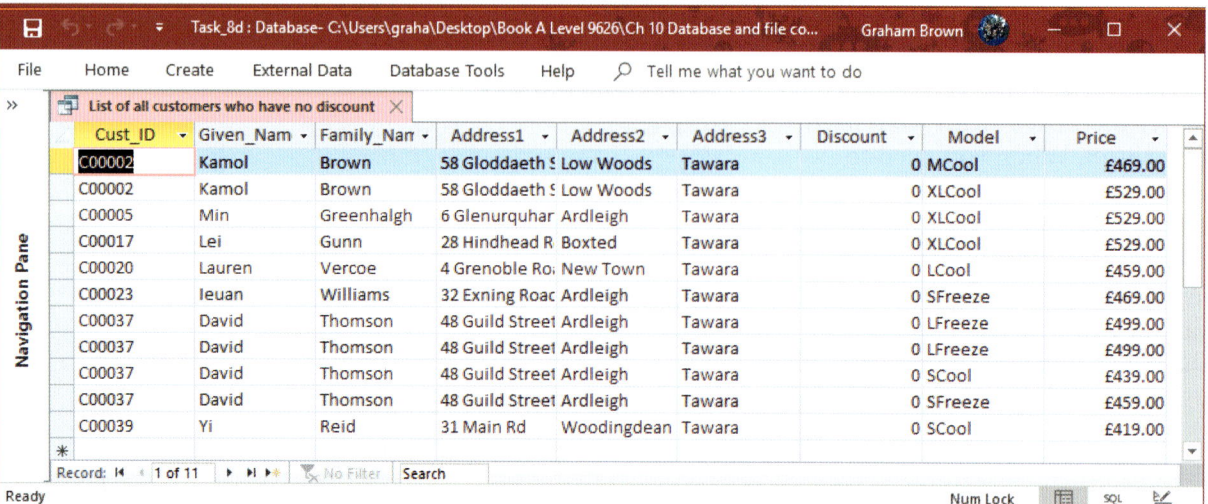

Advice

Use these numeric operators for queries involving numbers.

Operator	Meaning	Example	
=	Equal to	=6	Equal to 6
>	Greater than	>6	Greater than 6
<	Less than	<6	Less than 6
>=	Greater than or equal to	>=6	Greater than or equal to 6
<=	Less than or equal to	<=6	Less than or equal to 6
<>	Not equal to	<>6	Not equal to 6

Creating the report

The easiest way to create the report is to use the Report Wizard. Select the **Create** tab, and from the **Reports** section select the **Report Wizard** to open the Report Wizard window like this.

Select from the **Tables/Queries** drop-down list the query that we have just created.

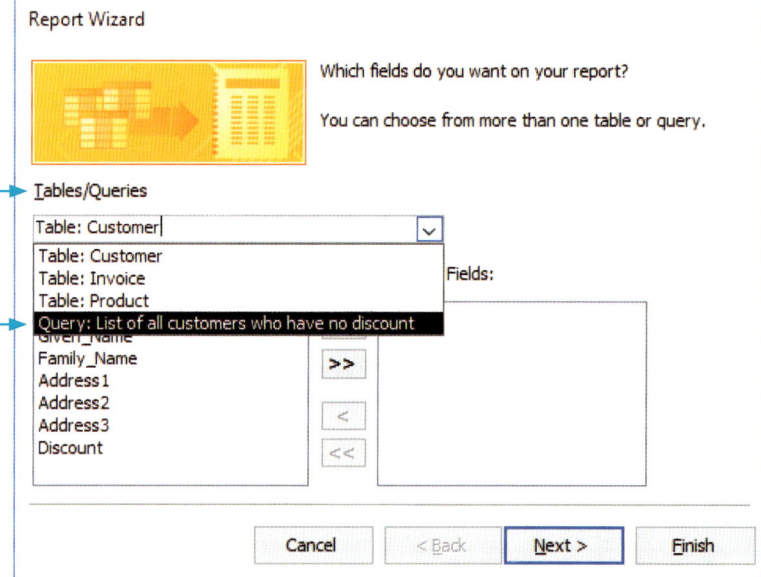

Move all fields except for **Discount** into the **Selected Fields:** box, then click Next >.

View the data **by Customer**, then click Next >. There is no need to select any groupings as *Access* has already added all the Customer details into the top group, so click Next >. There is no requirement in the question to sort or summarise the data so click Next >. Select the layout type that you require, but because there are a number of fields selected, it is easier to lay out the report with landscape orientation selected. Click in the **Orientation** section on the radio button for **Landscape**, then click on Next >. In the final Report Wizard window enter the title for the report; it should be meaningful to the intended audience (who are the directors of the company in this question).

Click on the radio button for **Modify the report's design**, then click Finish.

This opens the report in design view, so it looks like this.

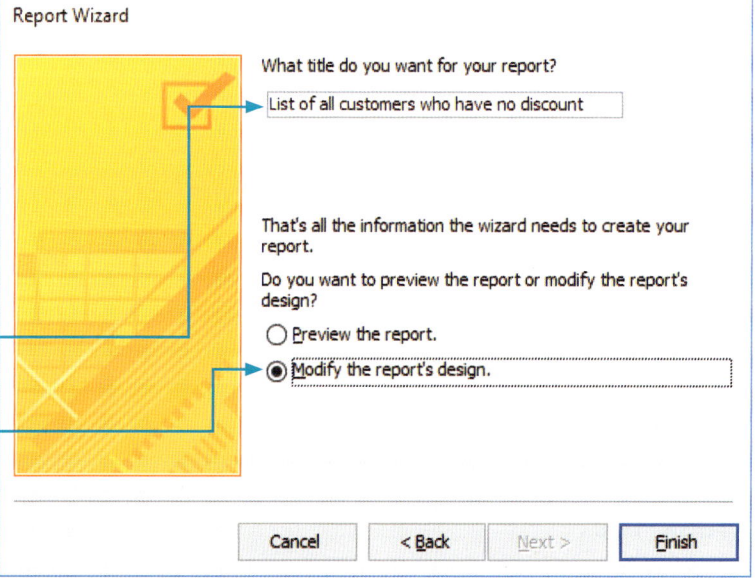

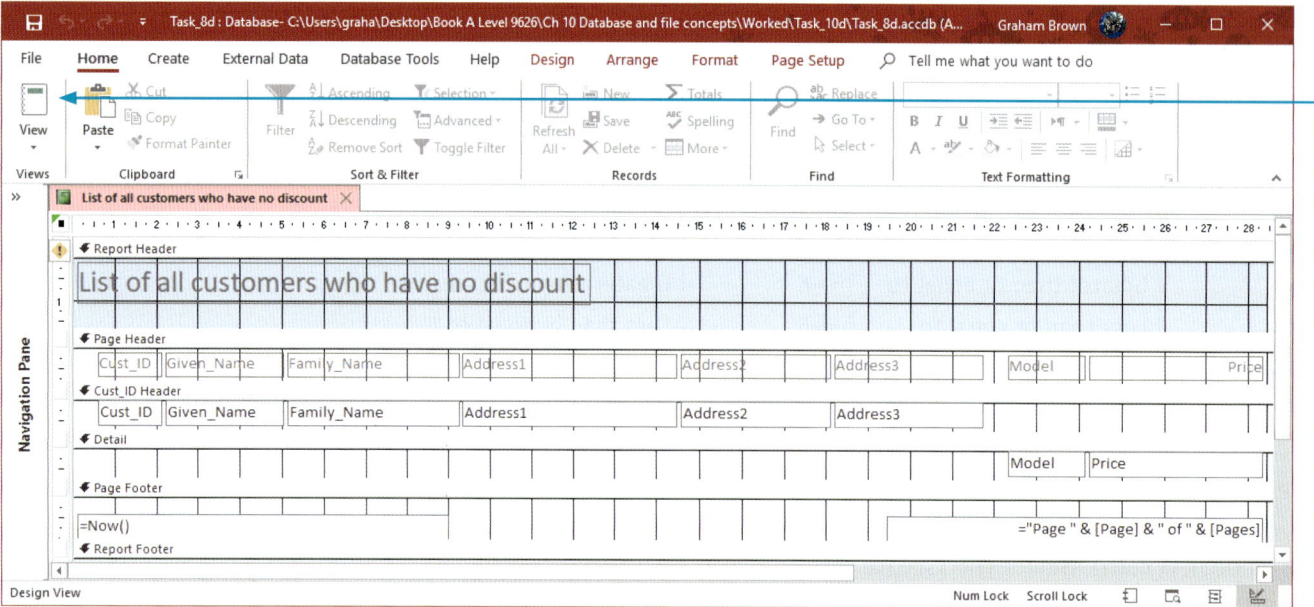

This is where each control can be adjusted if data does not fit within each control. To check if data fits within each control, select the Home tab, and in the View section select the icon for Report View.

Report View shows none of the data is longer than the control so save the report. It looks like this.

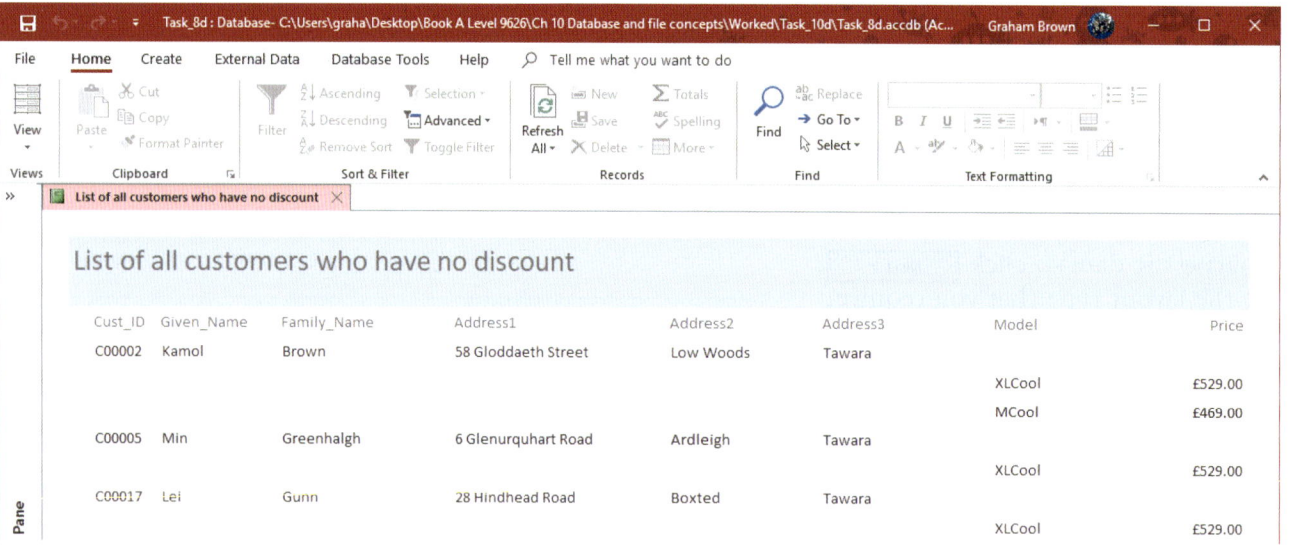

Activity 10d

Use a copy of the database saved in Activity 10c to produce a report for the shareholders of the company. Select and list all the details for the boats that have been sold and were made in the year 2006, as well as the full name of the customer who bought each boat.

Task 10e

Use a copy of the database saved in Task 10d to produce a report for the directors of the company. This will list the details of all the customers who are given between 5 and 15% discount (inclusive) on their orders. For each customer, list the model/s purchased, discount and price.

Perform a numeric query using multiple criteria

Create a query and report in a similar way to the one for Task 10d. Make sure that you select the fields from the tables and not the query from this Task. In the Design View of the query, in the Criteria: row of the Discount column, enter the formula (in this case inequality) >=0.05 AND <=0.15, like this.

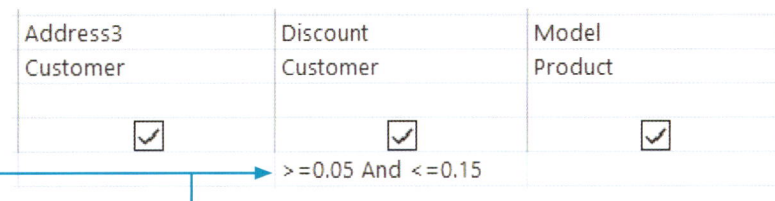

List of all customers who have between 5% and 15% discount (inclusive)

Cust_ID	Given_Name	Family_Name	Address1	Address2	Address3	Discount	Model	Price
C00003	Rajesh	Leadbetter	7 Annfield Rd	13 Straight Road	Jaffrezzi	10%		
							XLCool	£499.00
							MCool	£439.00
C00004	Bex	Hull	67 Netherpark Crescent	Straight Road	Tawara	10%		
							LCool	£459.00
C00006	Karla	Pooley	67 Lairg Road	Straight Road	Tawara	10%		
							SCool	£439.00
C00007	Moshe	Chalmers	24 St Maurices Road	Ardleigh	Tawara	10%		
							LCool	£489.00
							LCool	£479.00

This image only displays the first few matching records in this query.

Advice

This query used the **Boolean operator** AND to make sure that all the conditions were met before a record was selected. Other Boolean operators are OR which selects records if one of the conditions are met and NOT which selects records if a condition is not met. Combinations of these operators can allow us to perform complex queries, which we will meet later in the chapter.

Task 10f

Use a copy of the database saved in Task 10e to list the details of all products sold where the colour is Anthracite.

Perform a simple alphanumeric query on a single criterion

Create a query and report in a similar way to the one for Task 10e. Select all the fields from the Product table only. In the Design View of the query, in the Criteria: row of the Colour column, enter the text Anthracite. As you press the <Return/Enter> key, *Access* places speech marks around the text like this.

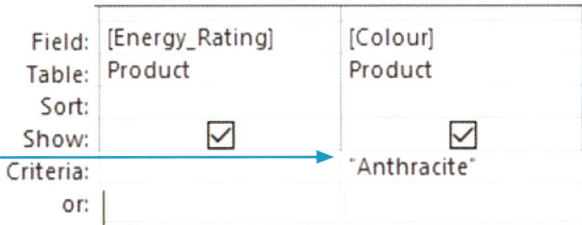

Products made in anthracite

Code	Model	Price	Capacity	Energy_Rating	Colour
X0003	XLCool	£529.00	400	A++	Anthracite
X0006	XLFreeze	£549.00	400	A++	Anthracite
X0009	LCool	£489.00	300	A+	Anthracite

Task 10g

Use a copy of the database saved in Task 10f to list the details of all products sold where the colour is Anthracite and the energy rating is A++.

Perform an alphanumeric query using multiple criteria

AND query

Create a query and report in a similar way to the one for Task 10f. Select all the fields from the **Product** table only. In the **Design View** of the query, in the **Criteria:** row of the **Colour** column, enter the text **Anthracite**. In the same **Criteria:** row of the **Energy_Rating** column, enter the text "A++" This time, type the double speech marks yourself as the plus signs confuse *Access*. It should look like this.

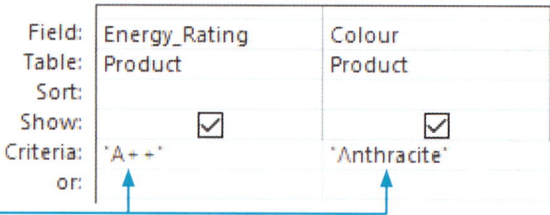

Products made in anthracite with A++ energy rating

Code	Model	Price	Capacity	Energy_Rating	Colour
X0003	XLCool	£529.00	400	A++	Anthracite
X0006	XLFreeze	£549.00	400	A++	Anthracite
X0012	LFreeze	£509.00	300	A++	Anthracite

Task 10h

Use a copy of the database saved in Task 10g to list the details of all products sold where the colour is Anthracite or Silver.

OR query

Create a query and report in a similar way to the one for Task 10g. Select all the fields from the **Product** table only. There are two methods for performing this query. The first uses the Boolean operator OR. In the **Design View** of the query, in the **Criteria:** row of the **Colour** column, enter the text **Anthracite OR Silver**. *Access* places the speech marks around the two colours like this.

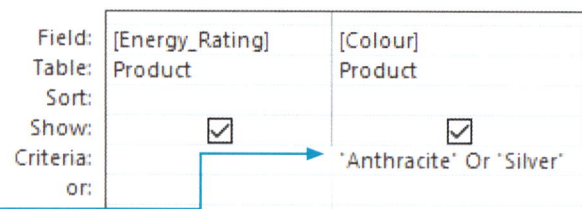

The second method places each part of the Boolean statement on a new row. In the Design View of the query, in the Criteria: row of the Colour column, enter the text Anthracite. Directly below this in the or: row of the Colour column, enter the text Silver, so that it looks like this.

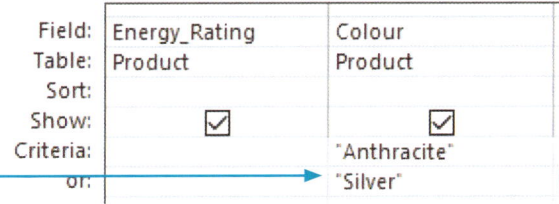

Both methods give the same result.

Models made in anthracite or silver

Code	Model	Price	Capacity	Energy_Rating	Colour
X0002	XLCool	£519.00	400	A++	Silver
X0003	XLCool	£529.00	400	A++	Anthracite
X0005	XLFreeze	£539.00	400	A++	Silver

NOT query

The same results can be achieved using a NOT query. As the company only makes models in white, silver and anthracite, selecting models where the colour is NOT white would also work. Create a new query and report in a similar way to the one for Task 10g. Select all the fields from the Product table only. In the Design View of the query, in the Criteria: row of the Colour column, enter the text Not White. *Access* adds the speech marks like this.

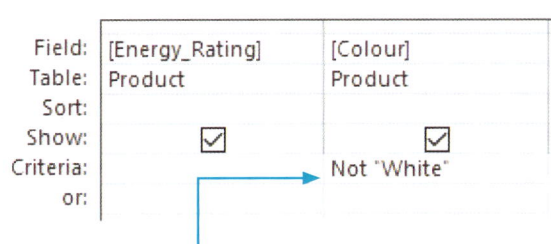

Wildcard searches allow us to search for part of a string. The * symbol is used to denote the wildcard. If the search criterion is like M* it will find all items starting with the letter M, if it was Med* it would find all items starting with the letters Med. For searches that 'ends in', the asterisk appears first, so *t would find all the items ending in t. If an item can appear anywhere in the field, then *X* would find the letter X anywhere within the field.

Task 10i

Use a copy of the database saved in Task 10h to list the details of all products sold where the model name:
» ends with Cool (to indicate it is a fridge).
» starts with M (to indicate it is a medium-sized item).
» contains L (to indicate it is a larger item).

Create three new queries and reports in a similar way to the one for Task 10h. Select all the fields from the Product table only. Open the query Design View for each query and move the cursor into the Criteria: row of the Model column. For the first query, enter the text *Cool. *Access* changes this to a 'Like' function so the criteria and result look like this.

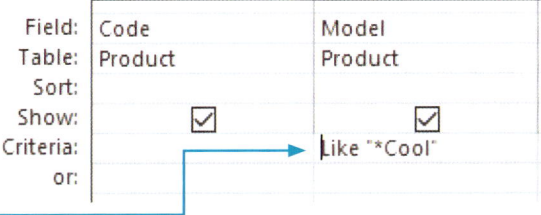

Models that are fridges

Code	Model	Price	Capacity	Energy_Rating	Colour
X0001	XLCool	£499.00	400	A++	White
X0002	XLCool	£519.00	400	A++	Silver
X0003	XLCool	£529.00	400	A++	Anthracite

For the second query, enter the text **M***, which again becomes a 'Like' function so the criteria and result look like this.

Model name starts with M

Code	Model	Price	Capacity	Energy_Rating	Colour
X0013	MCool	£439.00	250	A++	White
X0014	MCool	£459.00	250	A++	Silver
X0015	MCool	£469.00	250	A++	Anthracite

For the second query, it would appear sensible to enter the text ***L***, which again becomes a 'Like' function like this.

When this is tested it does not give the correct results.

Complex queries using multiple criteria

Because some of the fridges have the letter L at the end of their model name, these records are also included. To remove these, we must use more complex criteria that will allow the letter L in the **Model** field but will not output Models which start with the letter M or the letter S. Enter this text into the **Criteria:** row:
(Like "*L*" AND NOT Like "M*") AND (Like "*L*" AND NOT Like "S*"). This solution works but other criteria such as 'Like "*L*" AND NOT Like "*l"' would not work because records X0007–X0009 contain L at both the start and finish of each model name.

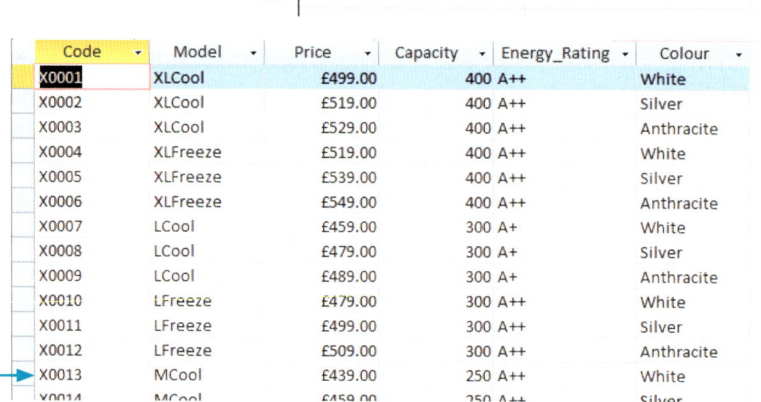

> ### Task 10j
> Use a copy of the database saved in Task 10i to produce a report for the directors of the company. This must contain the details of customers whose family name starts with the letter S or T, who live in Tawara and do not have 20% discount.

Use the same method as the previous tasks for selecting the data from the **Customer** table. Use an appropriate name for the query. In the **Design View** of the query, in the top **Criteria:** row of the **Family_Name** column, enter the text **S* OR T***. In the same row of the **Address3** column, enter the text **Tawara** and in the **Discount** column enter **NOT 0.2**, like this.

Field:	[Cust_ID]	[Given_Name]	[Family_Name]	[Address1]	[Address2]	[Address3]	[Discount]
Table:	Customer	Customer	Customer	Customer	Customer	Customer	Customer
Sort:							
Show:	✓	✓	✓	✓	✓	✓	✓
Criteria:			Like "S*" Or Like "T*"			"Tawara"	Not 0.2
or:							

Tawara customers with a family name that starts with S or T who do not get a 20% discount

Cust_ID	Given_Name	Family_Name	Address1	Address2	Address3	Discount
C00034	Aran	Smith	3 Warner Close	Boxted	Tawara	10%
C00037	David	Thomson	48 Guild Street	Ardleigh	Tawara	0%

To sum up, a simple query involves a search on one field only without the use of Boolean operators. A complex search involves a search on more than one field or the use of Boolean operators or a combination of both.

Date query

When dates are used within queries the dates must be placed within hash symbols so that *Access* treats the data as a date rather than as a calculation, for example #31/12/2020#.

> **Task 10k**
>
> Use a copy of the database saved in Task 10j to create a report for the manager, listing the invoice number, customer name, all product details, the order date and quantity ordered for all invoices where the order was placed in November or December 2019.

Use the same method as the previous tasks for selecting the data from the **Customer**, **Product** and **Invoice** tables. Use an appropriate name for the query. In the **Design View** of the query, in the top **Criteria:** row of the **Order_Date** column, enter the text **>#31/10/2019# AND <#01/01/2020#** like this.

Field:	Family_Name	Order_Date
Table:	Customer	Invoice
Sort:		
Show:	✓	✓
Criteria:		>#31/10/2019# And <#01/01/2020#
or:		

Orders placed in November or December 2019

Inv_Number	Given_Name	Family_Name	Order_Date	Quantity	Code	Model	Price	Capacity	Energy_Rating	Colour
33	Lauren	Patel	02/11/2019	1	X0013	MCool	£439.00	250	A++	White
34	Jasmine	Lewis	09/11/2019	1	X0002	XLCool	£519.00	400	A++	Silver
35	Linda	Watson	01/12/2019	1	X0001	XLCool	£499.00	400	A++	White
36	Aran	Smith	18/12/2019	1	X0009	LCool	£489.00	300	A+	Anthracite
37	Fred	Wells	23/12/2019	1	X0008	LCool	£479.00	300	A+	Silver

Task 10l

Use a copy of the database saved in Task 10k to create a report for the manager, listing the invoice number, the order date and time (in hh:mm format) of all orders placed in the afternoon.

Time query

Use the same method as the previous tasks for selecting the data from the **Invoice** table. Use an appropriate name for the query. In the **Design View** of the query, in the top **Criteria:** row of the **Order_Time** column, enter the text **>#12:00:00#**, like this.

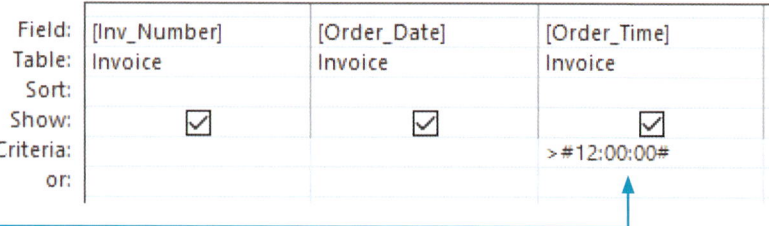

This lists the 31 orders that were placed during the afternoon. You will notice that the time is displayed in hh:mm:ss format both in the query and in the report as we did not specify the time format when we assigned data types to the field. To change the format of the displayed data in the report, open the report in **Design View**, select the control for **Order_Time** and in the **Format** tab select the drop-down menu for the **Format** box and choose the required format, in this case **Short Time** to display as hh:mm.

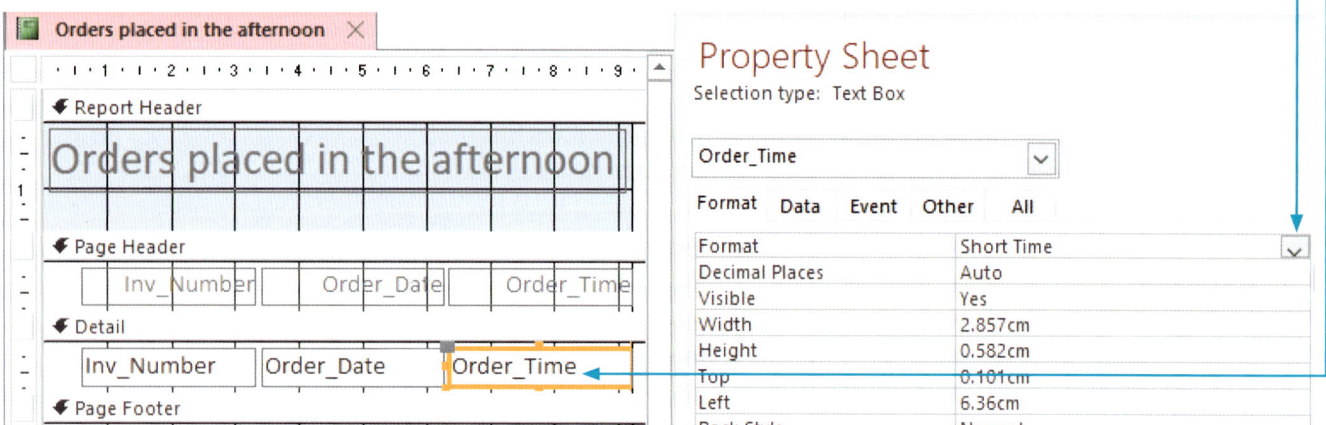

Advice

To locate records in a date field where the record is blank, set the criteria to **IS NULL**.

Activity 10e

Use a copy of the database saved in Activity 10d to produce the following reports for the shareholders of the company. Find the:
» names and email addresses of the customers with a phone number that starts with 001223.
» details of the unsold power boats that were built at this yard after 2012.
» make, type, price, year and date the boat was bought for all boats bought in the year 2020.

Static and dynamic parameter queries

Static parameter queries

The reports we have just created are based upon static parameter queries. This means that the parameter that the query uses (in Task 10d this was the value 0), has been set as a fixed value within the query. It can only be changed by recreating the query or editing its structure.

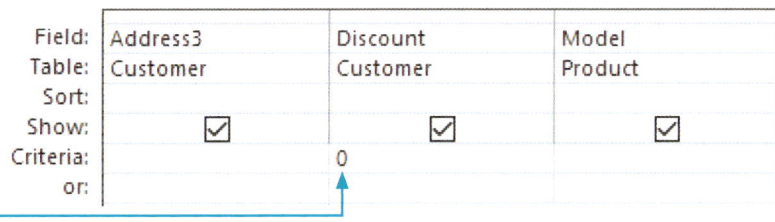

Dynamic parameter queries

Rather than creating a new query each time and changing the value/s within the query, we can create a dynamic parameter query. As the query is run, it asks the user to enter a value or formula which it then uses to select the appropriate records.

Task 10m

Use a copy of the database saved in Task 10l to produce a report for the directors of the company which will prompt the user to enter the discount criteria so as to select the customers who meet the given discount criteria, and for each of these customers to display their details, the model/s purchased, discount and price.

You will notice the similarity between this Task and Task 10d which required a discount set to 0.

Use the same method as the previous tasks, applying appropriate names to the query and the report. In the **Design View** of the query, in the **Criteria:** row of the **Discount** column, enter the text that will be used to instruct the user what to enter for the parameter query, for example: **[Enter the discount rate for this search]**. This text must be entered inside square brackets.

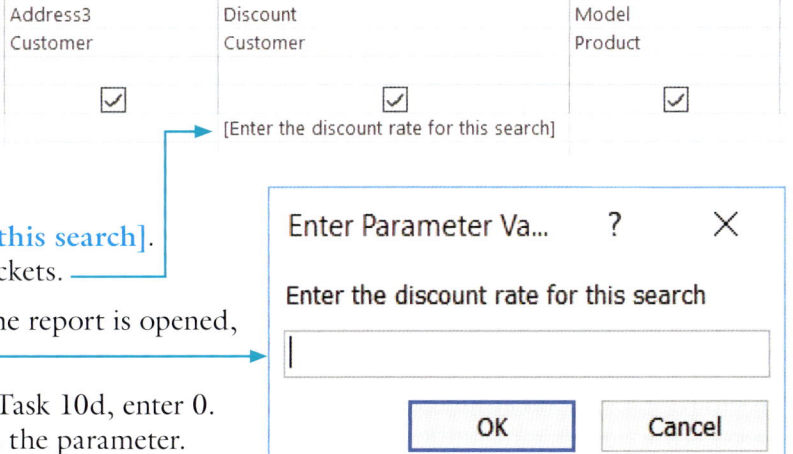

Create the report using this query. When the report is opened, the query is run and this window appears.

Enter the criteria into the box. To recreate Task 10d, enter 0. This query will only accept a single value as the parameter.

Task 10n

Use a copy of the database saved in Task 10m to produce a report for the directors of the company. The query will select the details of all the customers who are given between two (inclusive) discount values entered at run-time by the user. For each customer, list the model/s purchased, discount and price.

You will notice the similarity between this Task and Task 10e which required a discount of between 5% and 15%. This task requires two parameters, so the new parameter query will look like this.

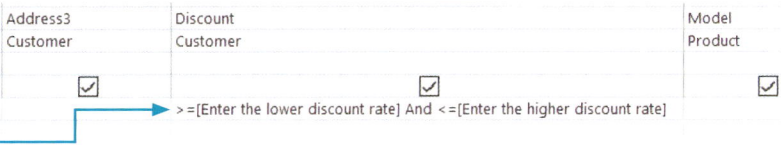

When this query is run, both parameter boxes will appear in turn. To recreate Task 10e, enter 0.05 for the lower discount rate and enter 0.15 for the higher discount rate.

The result should be the same data as Task 10e.

Task 10o

Use a copy of the database saved in Task 10n to produce a report for the directors of the company. The query will select the details of all the models in two colours selected by the user.

You will notice the similarity between this Task and Task 10h which required all products sold where the colour is Anthracite or Silver. This task requires two parameters so the new parameter query will look like this.

When this query is run, both parameter boxes will appear in turn. To recreate Task 10h, enter Anthracite for the first colour and Silver for the second colour.

Task 10p

Use a copy of the database saved in Task 10o to create a report for the manager, listing the invoice number, customer name, all product details, the order date and quantity ordered for all invoices where the order was placed between two dates input by the user.

You will notice the similarity between this Task and Task 10k which required orders placed in November or December 2019. This task requires two parameters so the new parameter query will look like this.

You will notice that the dates entered into the parameter query do not have to have # symbols around them like static date queries. When this query is run, both parameter boxes will appear in turn. To recreate Task 10k, enter the last day in October 2019 for the start date and enter the first of January 2020 for the end date.

To sum up, in a static query every time that the query is run it will search for the same data. If different data is to be searched for, the user would need to redesign the query to change the data in the criteria. Every time a parameter query is run a dialogue box appears asking the user to type in the data they are looking for. This would save the time of designing the query every time different data is required. In addition, a dynamic parameter query often requires the user to have less technical knowledge.

Activity 10f

Use a copy of the database saved in Activity 10e to create parameter queries that can be used to find:
- boats of a specified type
- boats of a specified type costing less or equal than a specified price
- boats of a specified type made after or on a specified year
- catamarans made in a specified year.

Use these parameter queries to list the make, type, price and year of the:
- yachts
- power boats costing less than £40 000
- yachts made after the year 2011
- catamarans made in the year 2015.

Nested queries

The term 'nested query' means running one or more queries based on the results of a previous query. This is often required when data is selected in a query and then used for a different type of query, for example producing a query to summarise data, based upon the results from a simple or complex query.

Task 10q

Use a copy of the database saved in Task 10p to compare the average prices of fridges and freezers made by this company.

Although this appears a relatively easy Task, there are several stages that need to be completed in order to succeed. To gather the correct data, we need to use the data from the **Product.Price** field together with the data from the **Product.Model** field to produce a field called **Type** which will only be calculated at run-time. Using this method saves you having an extra field stored permanently in the database and having duplicated data, as this information is held in the **Model** name. All model names with Cool in them are fridges and all model names with Freeze in them are freezers.

Fields calculated at run-time 1

Create a new query using the simple query wizard (as in previous tasks), selecting only the fields **Type** and **Price**. Name this query **Type and Price**. In the **Design View** of the query, in the **Field:** row of a new column, enter the text **Type:** to create a new field called **Type**. This field only appears when this query is run and is not stored in the database. Edit this text to become: **Type: Right([Model],6)** which adds a formula or calculation for this field. This extracts the last six characters of the model field like this.

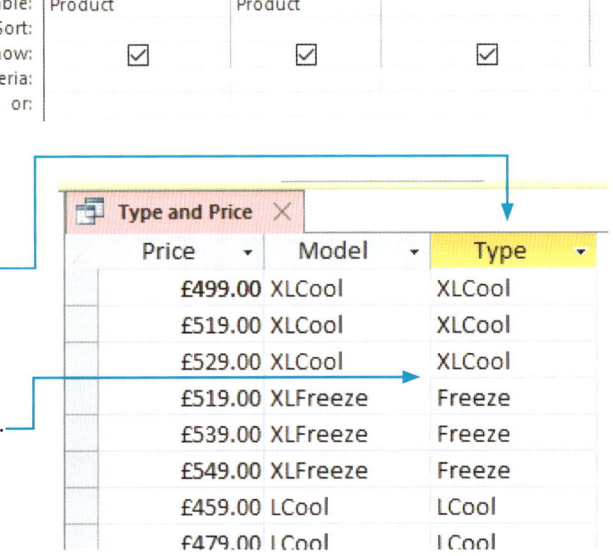

The results work well for all the freezers (although it would be better if it displayed 'Freezer' rather than 'Freeze'), but it leaves extra characters on the fridges (and uses 'Cool' rather than 'Fridge'), as can be seen here.

A conditional formula needs adding to this field so that if the result is 'Freeze' it displays 'Freezer', and if not it displays 'Fridge'.

IIF function

The IIF function works in the same way as the IF function in *Excel*. Edit the function in the Type field to:

Type: IIF(RIGHT([Model],6)="Freeze", "Freezer","Fridge") like this.

Test your results by clicking on the Run icon. If the results appear correct, then remove the tick from the Model field so that it is not displayed in the query (even though it will be used for the calculation). Run the query again to see the results, which should look like this.

Save this query, which will be the first stage of the nested query.

Summarise data in queries

Select the Create tab, then the Query Wizard icon, choose the Simple Query Wizard, then click OK. Select from the Tables/Queries drop-down list the Query: Type and Price, then move both the available fields into the Selected Fields box, then click Next >. From the next window select the radio button for Summary data.

The Summary Options button becomes visible so click on this.

This task requires 'the average prices of fridges and freezers' so select the tick box for Avg, then click OK to return to the previous window. Click Next >, apply an appropriate name for your query, for example Average prices of fridges and freezers, then click Finish. The average prices are shown like this.

This method can be used to summarise data with the sum, average, maximum and minimum values as well as counting the number of items in each category.

Crosstab queries

Data can also be summarised using a crosstab query which displays results in a similar format to a pivot table in a spreadsheet.

Task 10r

Use a copy of the database saved in Task 10q to count the number of people with each discount value in each area identified in the Address3 field. Use the Discount values as column headings.

Before starting a crosstab query, visualise how you want the results to appear. Work out what row headings, column headings and type of summary data you require and prepare a small hand-drawn sketch of how you would like the output to look. In this case the **Discount** field should be the column headings (as specified in the Task). The sketch should look similar to the one in Figure 10.2.

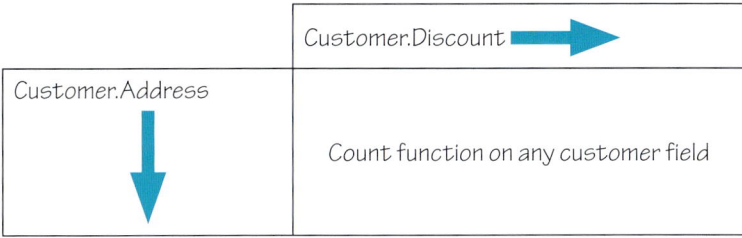

▲ Figure 10.2 Planning for a crosstab query

To create this table, select the **Create** tab, then the **Query Wizard** icon. From the **New Query** window, select the **Crosstab Query Wizard** then click [OK]. For this query we only need data from the Customer table so select **Table: Customer** in the top box, then click [Next >]. Figure 10.2 has only the **Address3** field as a row heading so move that field from the **Available Fields:** list into the **Selected Fields:** box, then click [Next >]. Select the **Discount** field for the column headings as shown in Figure 10.2 then click [Next >]. For the calculation, select any field in the **Fields:** box (in this case **Cust_ID** has been used) and in the **Functions:** box select the **Count** function. The tick in the check box for row sums can be removed if these are not required.

You can check that the **Sample:** diagram matches Figure 10.2 like this.

Click [Next >], give the query an appropriate name like **Number with each discount address areas**, then click [Finish]. The resulting crosstab may look similar to this.

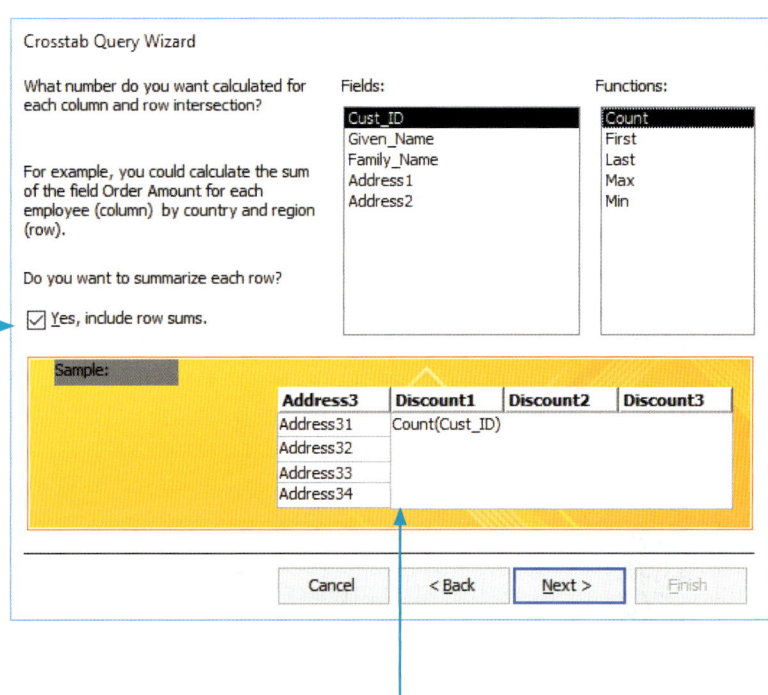

Activity 10g

Use a copy of the database saved in Activity 10f to create a two-dimensional grid showing, for all boats that have not been sold but were made after the year 2010, the number of each type of boat in stock and the year these were made. Do not show any row or column totals.

Task 10s

Create a new database for the Tawara Bus Company to hold information on its drivers using the files **TB-Driver 1.csv** and **TB-Driver2.csv**, which will be merged into a single table. Identify any duplicate records.

Start by examining the data files; they both have identical structures so can be imported as a single table. Open *Access* and click on the Blank database icon. Select the folder and add the filename **Task_10s** before clicking the Create button. Select the External Data tab, then the New Data Source icon, From File, then Text File. Locate the first source file **TB-Driver1.csv** so it appears in the File name: box.

Make sure the radio button for Import the source data into a new table in the current database is selected, then click OK . Import the data into the database but let *Access* select the primary key (if you select the primary key on the ID field this will automatically stop any duplicate records on the second import). Repeat this import for the file **TB-Driver2.csv**, this time selecting the Append radio button.

To test if there is duplicated data in this table, we will use a find duplicates query.

Find duplicates queries

Duplicate records can sometimes appear in a table, often where more than one user has input data into an *Access* database or where two or more external data files have been joined (appended) into a single table. To check for duplicate data in a table, first close the table if it is open. From the Create tab, select the Query Wizard icon. From the first New Query window select the Find Duplicates Query Wizard then click OK . Select the table (there is only one so it should already be selected) then click Next > . Ignore the ID field; as this is the primary key field, it cannot contain duplicate values. Move the other four fields from the Available fields: box into the Duplicate-value fields: box, then click Finish . The results look like this.

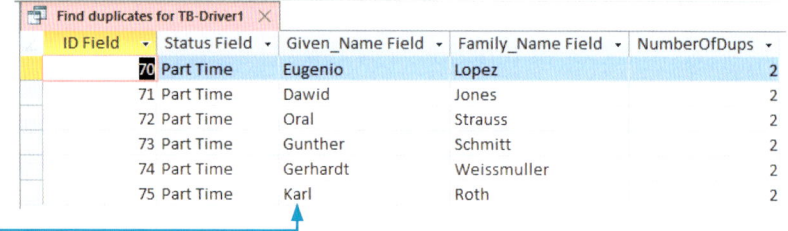

One copy of each record should be deleted (and this query run again) to make sure that the table contains no duplicate data. This is a useful tool for checking that we have normalised the database, as we shall see later in the chapter.

Fields calculated at run-time 2

> ### Task 10t
> Use a copy of the database saved in Task 10r to display details of fridges and freezers, as well as the price per cubic centimetre of each item.

Create a new query using the Simple Query Wizard which contains all the details of the Product table and open this query in Design View. Calculating the price per cubic centimetre involves taking the contents of the Price field and dividing this by the Capacity field. Add a new field called **Price_per_CC** that is calculated at run-time by entering the text **Price_per_CC: [Price]/[Capacity]**.

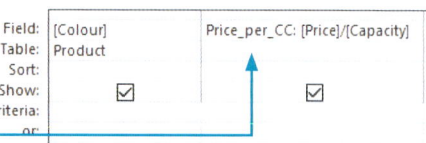

Create a report for this query. In the **Design View** of the report, select the **Price_per_CC** field, right mouse click and select **Properties** from the bottom of the drop-down list. In the **Format** tab change the **Format** to **Currency** and the **Decimal Places** to 2 so that the report looks similar to this:

Calculated field to obtain the price per cubic centimetre for each product

Code	Model	Price	Capacity	Energy_Rating	Colour	Price_per_CC
X0001	XLCool	£499.00	400	A++	White	£1.25
X0002	XLCool	£519.00	400	A++	Silver	£1.30
X0003	XLCool	£529.00	400	A++	Anthracite	£1.32
X0004	XLFreeze	£519.00	400	A++	White	£1.30

Activity 10h
Use a copy of the database saved in Activity 10g to create a report for your manager that displays the make, type, price, length, and price per foot for each yacht and power boat.

Task 10u
Use a copy of the database saved in Task 10t to display the invoice number, product model, quantity, order date and time, collection date and time, customer name, and number of days taken from placing the order until it was collected.

Create a new query using the **Simple Query Wizard** which contains all the fields specified in the Task. Open this query in **Design View**. Calculating the number of days between the order and collection involves subtracting the order date from the collection date. Add a new field called **Days** that is calculated at run-time by entering the text
Days: [Collect_Date]-[Order_Date] like this.

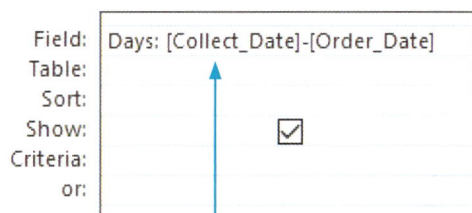

Create the report for this query which should look similar to this. When this report is examined it is clear that one order either took a very long time to process or there was an error in the data entry.

Time taken between order placed and collection

_Number	Model	Quantity	Order_Date	Order_Time	Collect_Date	llect_Time	Given_Name	Family_Name	Days
1	LCool	1	15/06/2018	14:35	23/06/2019	16:54:00	Clemantine	Wu	373
2	LFreeze	1	10/07/2018	14:03	12/07/2018	15:49:00	Sandy	Rydell	2
3	LFreeze	1	09/08/2018	11:23	10/08/2018	08:10:00	David	Thomson	1
4	XLFreeze	1	25/08/2018	15:13	04/09/2018	15:46:00	Naomi	Gale	10

Task 10v
Use a copy of the database saved in Task 10u to display the invoice number, product model, quantity, order date and time, collection date and time, customer name, and how many days have passed since the order was collected from the company.

Using similar methods to those described in Task 10u, create and open the query in **Design View**. Calculating the number of days that have passed since the item was collected requires *Access* to use today's date for the query (which is accessed using the NOW() function) and subtract the collection date from this. We only require the days part of this calculation so must select the integer part (using the INT function), as the decimal part holds the time values. Add a new

field called **Days** that is calculated at run-time by entering the text
Days: INT(Now()-[Collect_Date]) like this.

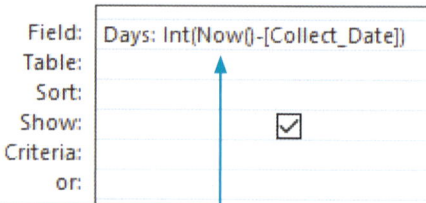

Create the report for this query which should look similar to this, although the days value will be different depending upon today's date.

Number of days since the item was collected

_Number	Model	Quantity	Order_Date	Order_Time	Collect_Date	llect_Time	Given_Name	Family_Name	Days
1	LCool	1	15/06/2018	14:35	23/06/2019	16:54:00	Clemantine	Wu	53
2	LFreeze	1	10/07/2018	14:03	12/07/2018	15:49:00	Sandy	Rydell	399
3	LFreeze	1	09/08/2018	11:23	10/08/2018	08:10:00	David	Thomson	370
4	XLFreeze	1	25/08/2018	15:13	04/09/2018	15:46:00	Naomi	Gale	345

> ### Activity 10i
> Use a copy of the database saved in Activity 10h to create a report for your manager that selects all the boats that were bought by the company and have not yet been sold. Display the make, type, price, the date bought, today's date, and the number of days that boat has been held in stock.

Calculated controls

Calculations can also be performed within a report, often to calculate the sum, average, maximum or minimum values of the selected data or to count the number of items selected within the report. All of these functions can be produced within the report in *Access*.

> ### Task 10w
> Open the file saved in Task 10v. Edit a copy of the report created for Task 10t to display, at the bottom of the report:
> » the average price per CC
> » the maximum and minimum price per CC
> » the number of items in this report.

Copy, paste and rename the report you created for Task 10v. Open this report in **Design View**.

Click the left mouse button on the bottom edge of the **Report Footer** and drag this down about two centimetres, so that this footer is now visible. Select the **Design** tab, move to the **Controls** section and select the **Text Box** icon.

Move down into the **Report Footer** and click the mouse button and drag to place a new control, in this case a text box directly below the **Price_per_CC** column. This positioning is important as this control will be used to calculate the average **Price_per_CC** for the data in this report.

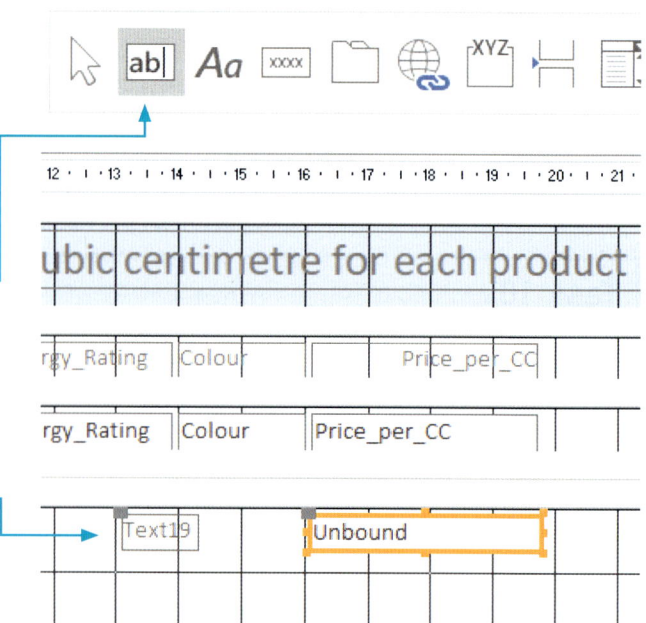

If the **Property Sheet** is not showing, right click the mouse button on the **Unbound** control that you have just created, then select **Properties** from the bottom of the drop-down menu. In the **Property Sheet**, select the **All** tab, find the **Control Source** section and type the formula **=AVG([Price_per_CC])** into this row. The round brackets are part of the AVG function; the square brackets tell *Access* that this is a field. **Format** this control to Fixed format with 2 decimal places like this.

The **Control** (in this case Text19) and **Property Sheet** will change to this.

Click in the label for this control (in this case called Label20) and change the **Caption** like this.

Stretch the label so that all the **Caption** text is fully visible, like this.

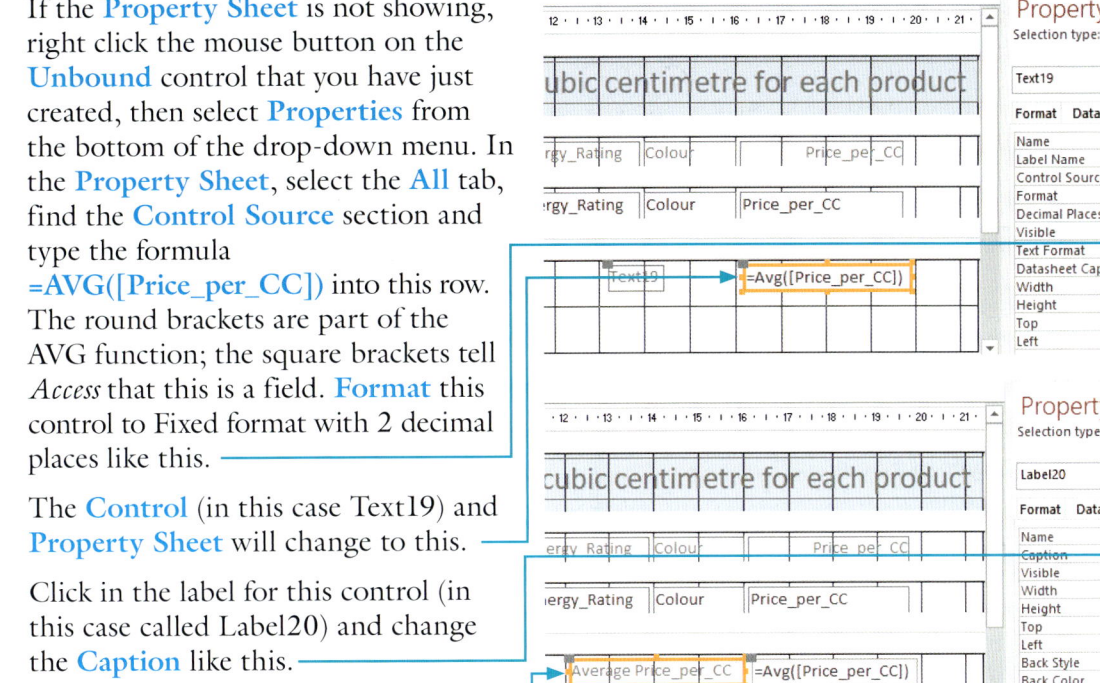

Change to the **Report View** and make sure that the control is in the correct place and appears to give the right answer (it is not too large or too small).

Rather than repeating this process three more times, it will be quicker to copy and paste these controls and edit each one to give the required results. Use the **lasso tool** to highlight both the **Text Box** and its **Label**. Use **<Ctrl C>** to copy, then use **<Ctrl V>** to paste the copies of these controls. Using **<Ctrl V>** pastes the new controls directly under the existing ones and you do not need to reorganise the controls. It also extends the bottom of the **Report Footer** as needed. If you right mouse click and use **Paste** from the drop-down menu, this pastes the controls in the top-left-hand corner of the **Report Footer** and you then have to drag and position each set of controls. Repeat **<Ctrl V>** until you have four sets of controls like this:

In the last three controls containing labels, change the **Captions** to **Maximum price_per_CC**, **Minimum price_per_CC** and **Number of items**. Select the second **Text Box** (for the maximum price per CC) and change the formula so that it becomes **=MAX([Price_per_CC])**. Change the formula for the minimum price per CC so that it becomes **=MIN([Price_per_CC])**. In the final control to count the number of items, change the formula so that it becomes **=COUNT([Price_per_CC])**. The controls should look like this:

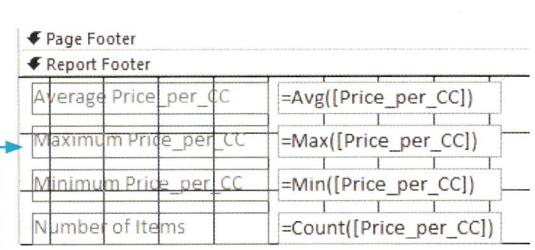

Advice

The **SUM** function can be used to calculate the total for any field, using the same method as **AVG**, **MAX**, **MIN** and so on.

In the **Property Sheet** pane for the final **Text Box**, change the **Format** back from Currency to a **General Number**. Set the **Decimal Places** for this control to 0.

Check the layout and calculations in *Report View*. The completed calculations look like this.

185	A+	Silver	£2.48
185	A+	Anthracite	£2.54

Average Price_per_CC	£1.79
Maximum Price_per_CC	£2.54
Minimum Price_per_CC	£1.25
Number of Items	24

Advice
Make sure that each control is wide enough to show the formula in full.

Activity 10j
Open a copy of the database saved in Activity 10i. Edit a copy of the report created for Activity 10h to display for the yachts and power boats at the bottom of the report the:
» average price of a boat
» average price per foot
» maximum and minimum price of a boat
» number of boats in this report.

Make sure that you use appropriate formatting for all data.

Query selection

Selection of the correct type of query from numeric, alphanumeric, date and type, simple, complex, nested, and crosstab to answer a specific question is critical. If the question requires summary data, consider if a two-dimensional table is required. If it is, then a crosstab query is the most appropriate to use; if it is not, then one (or more if nested queries are required) simple (or complex) query should enable any other questions to be answered. If only one field needs searching then it will be a simple query but if more than one field needs to be searched on it is called a complex query. It could be that the same criterion is to be used every time. If you are searching for the number of people who live in Bengaluru, a static query should be used. However, there will be times when the criterion needs to be changed every time the query is run. For example, you may need to find the number of people who live in different cities. In this case a dynamic query should be used. If data is to be entered by the user when a query is run, then a dynamic query is required but if the criteria need embedding into the query, so that the same query is run repeatedly even if the data in the tables change, then use a static query.

Data can be sorted into order, or collected into groups. Although *Access* has the ability to sort your data in both tables and queries, it is often easier to save the sorting until the report is created. Both sorting and grouping can be included in the report wizard, or they can be added to a report in design view.

10.1.10 Sorting and grouping data

Sorting data using the report wizard

Task 10x
Open the file saved in Task 10w. Create a new copy of the report for Task 10t, sorting the data into descending order of price, then into ascending order of capacity.

Use the query created in Task 10t. To produce this report, select the **Create** tab and click on the **Report Wizard** icon. In the **Tables/Queries** box select the query used for Task 10t. Select all fields using the double arrow key and then click on Next > twice to obtain this view:

Use the drop-down lists to select the **Price** field, then the **Capacity** field. For the **Price** field, click on `Ascending` to the right of this field and it will toggle (change) to `Descending`. When these fields have been set as shown, click on `Next >`. Run through the final stages of the wizard, giving this report a suitable name. This process is the same for other data types such as dates. You will notice that the report wizard changes the order of the fields to place the sorted fields first, like this:

Task 10x - Average price per cubic centimetre for each item						
Price	Capacity	Code	Model	Energy_Rating	Colour	Price_per_CC
£549.00	400	X0006	XLFreeze	A++	Anthracite	£1.37
£539.00	400	X0005	XLFreeze	A++	Silver	£1.35
£529.00	400	X0003	XLCool	A++	Anthracite	£1.32

Sorting data using design view in a report

Task 10y
Open the file saved in Task 10x. Edit the report created for Task 10w, sorting data into descending order of price, then into ascending order of capacity.

Open the report created in Task 10w in **Design View**. In the **Grouping & Totals** section select the **Group & Sort** icon.

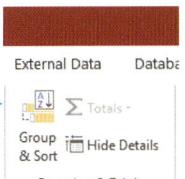

This opens the **Group, Sort and Total** pane at the bottom of the window. Select the button for **Add a sort**. Select the **Price** field from the drop-down list, then change the sort from 'from smallest to largest' to **from largest to smallest** using the second drop-down menu. Add a second sort using the **Add a sort** button. Select the **Capacity** field from the drop-down list; there is no need to change the sort type as it is already from smallest to largest (to match the ascending order required by the Task). The pane looks similar to this:

This has sorted the data as required but has not changed the order of the fields in the report like this:

Calculated field to obtain the price per cubic centimetre for each product						
Code	Model	Price	Capacity	Energy_Rating	Colour	Price_per_CC
X0006	XLFreeze	£549.00	400	A++	Anthracite	£1.37
X0005	XLFreeze	£539.00	400	A++	Silver	£1.35
X0003	XLCool	£529.00	400	A++	Anthracite	£1.32

Activity 10k
Open a copy of the database saved in Activity 10j. Edit a copy or recreate the report created for Task 10h sorted into ascending order of type and then descending order of price.

Grouping data

Grouping allows data to be collected and displayed together, not only as individual records but also with totals for each group of data.

Task 10z
Open the file saved in Task 10y. Create a report for the manager that displays all the details of all the products. Group this report by product type (whether the product is a fridge or freezer). Include the average price of all fridges and freezers.

Create a new query using the Simple Query Wizard which contains all the fields from the Product table and a calculated field called Type, as we did in Task 10q, using the formula Type: IIF(RIGHT([Model],6)="Freeze","Freezer","Fridge"). Save this query with an appropriate name. Select the Create tab, then Report Wizard to create a new report, then select the query you have just saved, moving all fields from the Available Fields: to the Selected Fields: box then click Next >. The Task requires us to group the report by product type, so double-click the left mouse button on Type to move it as a new group like this.

Click Next > three times, give the report an appropriate name, select Modify the report's design, then click Finish. Check all data within all fields is fully visible, then (if the Group, Sort and Total pane is not visible) select the Group & Sort icon. In the Group, Sort, and Total pane select the option for More alongside the grouping.

This extends the grouping options for Type like this.

Select the drop-down menu for with no totals like this.

In the Totals section, edit the Total On box to become Price and the Type to be Average. Tick the box where you would like these average values to appear, in this case in the group footer.

Select the Design tab, move to the Controls section and select the Text Box icon.

Move down into the Type Footer, click the mouse button and drag to place a new label to the left of the Average price control. Change the Caption of this label to Average price. Save this report which should look like this.

Activity 10l

Open a copy of the database saved in Activity 10k. Create a report for the manager that displays all the details of all the boats that are unsold, grouped by type then by catamaran, then sorted in ascending order of price, and calculate a grand total and the total price for each type.

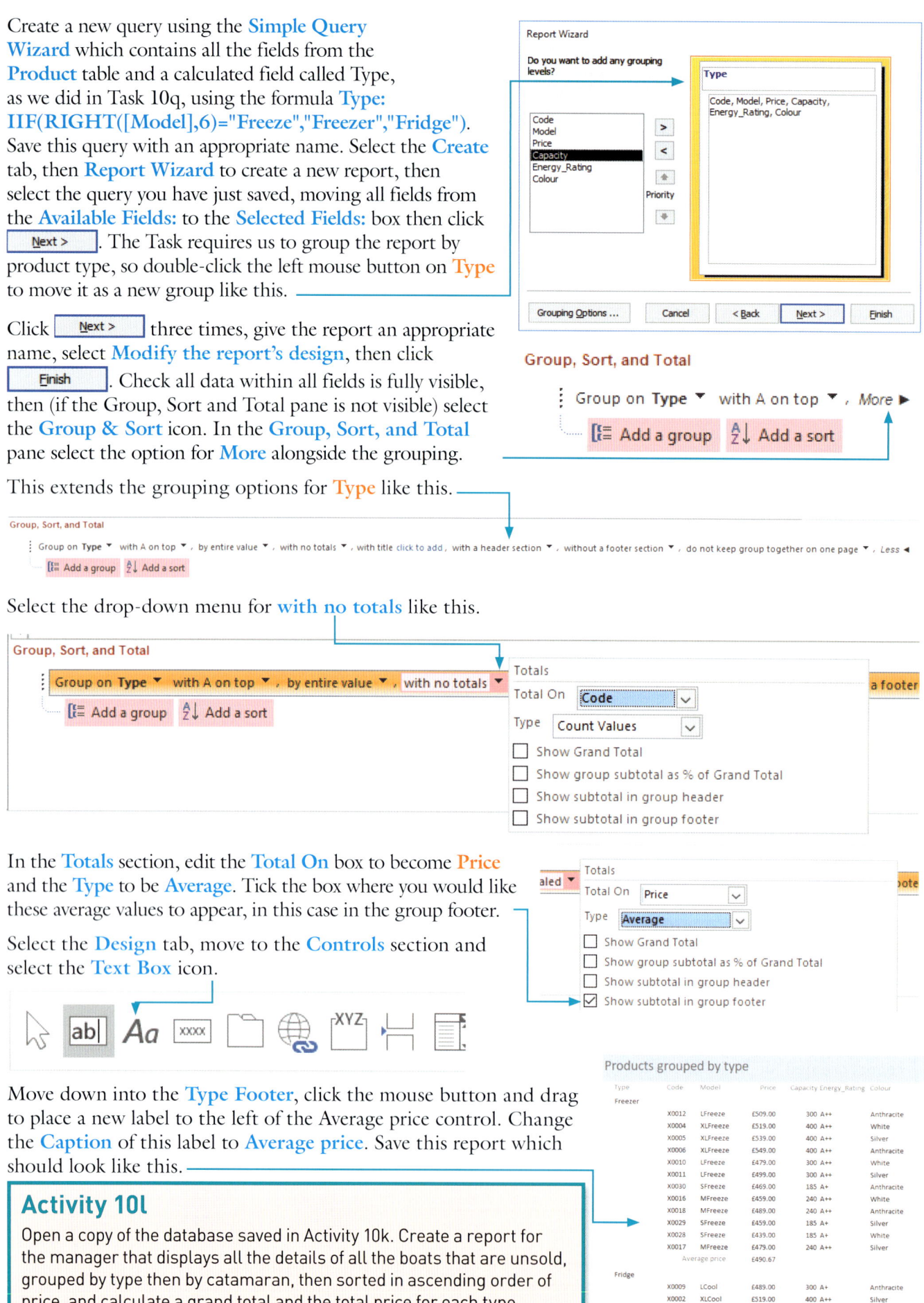

10.1.11 Creating a data entry form

> **Task 10aa**
> Open the file saved in Task 10z. Add new data entry forms to collect data for all fields in the Customer and Product tables. These will be used by employees of the company.

The best way to create a data entry form is to select the **Create** tab, then click on the **Form Wizard**.

The **Form Wizard** window opens. Select the table which holds the fields that you will include in the form. It is often sensible to have a new form for each table, but if a form needs fields from more than one table then place these fields in a query and select the query. For this Task we will select the **Customer** table first. Move all fields across from the **Available Fields:** to the **Selected Fields:** box like this.

Then click **Next >**. Choose that layout of the screen that you require (I chose **Columnar** for this Task) then click **Next >** again. Set the title of the form to **Customer** (*Access* may do this for you) then click **Finish** to open the form like this.

The bottom of the form has a navigation bar which can be used to move from record to record like this:

- Go to first record
- Current record
- Next record
- Last record
- New record

Consider the person who will use the form. This Task states that the form will be used by employees, who will almost certainly have had some training, so less detailed instruction about form completion will be required than for a form to be completed by, for example, the customer themselves. In the case of a customer entering data rather than an employee, instead of using labels containing field names, each label must describe the data to be entered and how to enter the data for each field. The form design must include appropriate font styles and sizes, with easy-to-read text, taking account of the user's age (for example a database for younger children will require simple language and large, simple font styles).

Script/cursive fonts which are more difficult to read should not be used unless specifically required by the company. Spacing between fields is vital if a form is to contain different blocks of information. For example in this form a block for the name, one for the address and one for the discount would be appropriate, with similar fields grouped together on the form. It is also important to assign suitable character spacing for each data item on the form. *Access* initially sets the length of all alphanumeric (text) fields to the field size (length) assigned when the table is created. However, we can see from this example that the Discount field control is far too large for the data that will be stored. It is important to use the white space on the form to separate these grouped areas but also to make sure that there are no large areas of white space on the finished form.

With all these considerations in mind we will edit this form to make it more suitable for the employees to use. The title of the form gives no instruction to the user as to its purpose. Change from Form View to Design View. Edit the Caption in the Form Header from Customer to an appropriate description like 'Form to enter customer details'.

Add some instructions for the user (this is more important if other people are using the form, rather than just employees). Drag the cursor so that it lassos all the controls together for each group on the form. Increase the white space between the groups. Increase the font size of all controls from 11 points to 14 points (as there is no shortage of space on the form). Reduce the size of the control for the data entry on the Discount field. All labels in the Detail section can have the text right-aligned (change this in the Properties window after lassoing all the labels). Edit the Caption for the label used for the Cust_ID field to make it more meaningful, for example 'Customer ID number'. Add a label to the right of this field and add appropriate text. The final stage must be to make sure the person entering the data knows that the Cust_ID field is the key field and cannot be left blank. Select the label icon and drag a new label onto the form to create a new control. Type the text into this control. With the control selected, right mouse click and select Properties; these can be used to change the font colour to red. The design view of the form has now been changed from this to this:

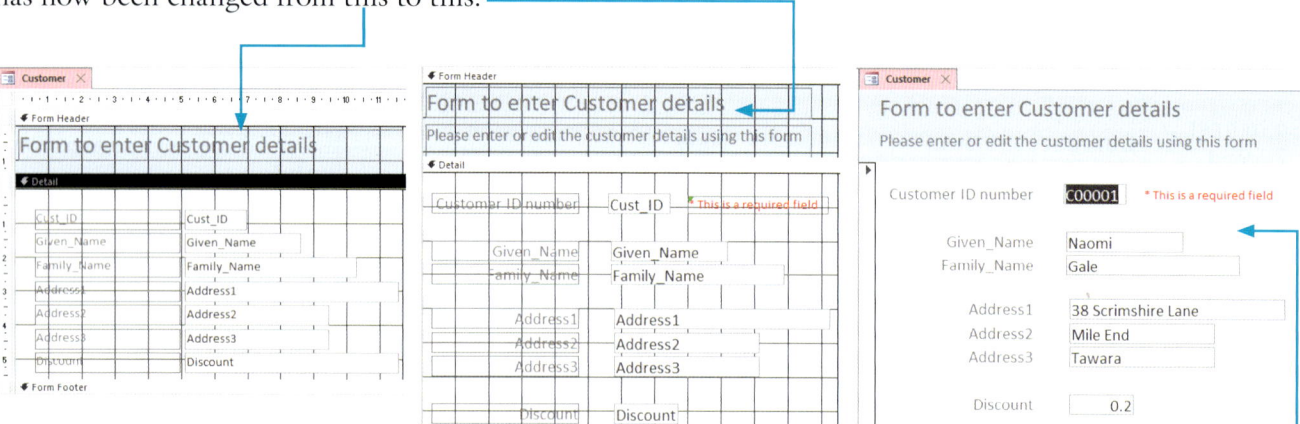

This diagram shows the edited form in Form View.

Repeat this process for the Product table to create the second data entry form. There are no obvious groups for this data although appropriate lengths for the controls will be needed. Again, because the data entry will be completed by an employee, there is no need to give details of data formats (for example the product codes are all the letter X followed by four digits). Apart from adding validation rules to the tables, there are two fields where data entry can be quicker, and data entry errors can be reduced, by developing the controls within the form. The Energy_Rating could be restricted to either A+ or A++, as the company only makes items with these ratings, and the Colour could be White, Silver or Anthracite (but extra colours may be added later). In order to do this, we will replace the controls for Energy_Rating and Colour with list boxes. Use the lasso method to move these two controls and their labels down; you may need to drag the form footer lower to create the space, like this.

From the Design tab, in the Controls section, select the drop-down list for More options.

Select from this drop-down list the icon for a List Box.

> **Advice**
> Using a list box will restrict data entry to only items from a given (or stored) list; if you require a user to be able to enter other data, choose a combo box.

Click and drag to draw the list box control onto the form. The List Box Wizard opens for this. For the energy rating there are only two options, which are unlikely to change, so we will select I will type in the values that I want, then click Next >. Enter the values like this.

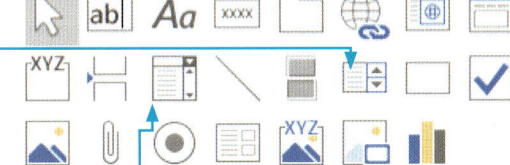

Click Next > and select the radio button for Store the value in this field: before selecting the Energy_Rating field from the drop-down list.

Click Next > and type Energy rating for the list box label, then click Finish. As the company may add extra colours to its range of products, we will create a new table called Colour and store this in the database. Create a single primary key field called Colour and enter each colour (Anthracite, Silver and White) as new records in this table.

Add a new list box control to the form and when the wizard opens, select **I want the list box to get the values from another table or query**, then click **Next >**. Select the option for **Table: Colour** then click **Next >**. Select this field, then click **Next >** three times. Select the radio button for **Store the value in this field:** before selecting the **Colour** field from the drop-down list. Click **Next >** and type **Colour** for the list box label, then click **Finish**. Test the form in **Form View** to make sure the correct data is being displayed for this record, return to **Design View** and remove the two original controls and their labels so that your form may look like this.

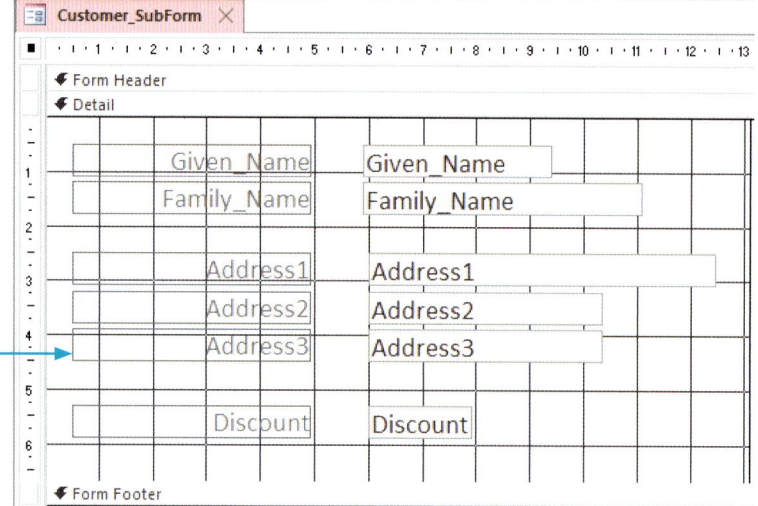

Creating a sub-form

Task 10ab

Open the file saved in Task 10aa. Add a new data entry form for the Invoice table which contains a sub-form for the Customer table. These will be used by employees of the company.

To complete this Task, we will place a customer's form inside a new invoice form. If we use the existing **Customer** form some labels would be inappropriate. Make a copy of the **Customer** form, rename it **Customer_SubForm** and edit it to look like this.

Create a new data entry form for the **Invoice** table using the method above. This time we will group the form into the order details, collection details and customer details. The customer details will be held in a sub-form. For each section of the form, add a new label to help the user find the data on the form easily. These labels will display **Order details**, **Collection details** and **Customer details**. With the form in **Design View**, to add the Customer sub-form select the **SubForm** icon.

Click and drag the form box below the **Customer details** label, allowing plenty of space for all fields within the Customer table to be visible.

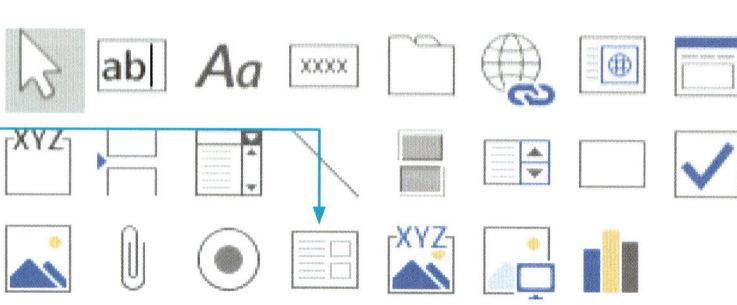

From the **SubForm Wizard** select the radio button for **Use an existing form** and select the **Customer_SubForm**, then click [Next >]. Select the radio button for **Choose from a list**, then **Show Customer for each record in Invoice using Cust_ID**, then click [Next >]. Name this sub-form **Customer_SubForm**, then click [Finish]. The **Invoice** form looks like this.

You will notice that the background colour of three labels has been changed (in the properties) to clearly identify the groups of data within the form.

Using radio buttons

Task 10ac

Open the file saved in Task 10ab. Edit the database so that the capacity of the product can be added using a radio button rather than typing the data. The capacity of all models is 185, 200, 240, 250, 300, or 400 cubic centimetres. Edit the data entry form for the Product table to use radio buttons to select one of these values.

Open the **Product** form in **Design View**. Delete the control and label for **Capacity**. Select the **Design** tab, then the **Option Group** icon from the **Controls** menu.

Click in the **Detail** section and drag the Option Group frame, making it large enough so that a number of radio buttons and their labels can be placed within it. The **Option Group Wizard** window opens. Enter the label names in the table like this.

Click [Next >], make sure the radio button for **No, I don't want a default** is selected, then click [Next >]. Edit the values for each button so that they are 185, 200, 240, 250, 300 and 400 respectively, then click [Next >]. Select the radio button for **Store the value in this field:** and select the **Capacity** field from the drop-down list, then click [Next >]. Leave the selection on **Option buttons**, then click [Next >]. Change the caption to **Capacity of the product**, then click [Finish]. The **Design View** of the form will look like this and the edited form in **Form View** will look similar to this.

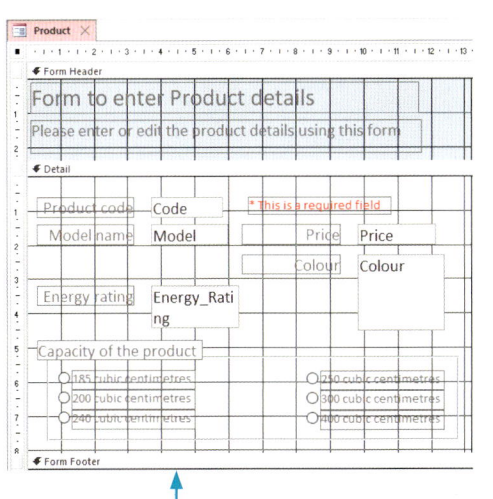
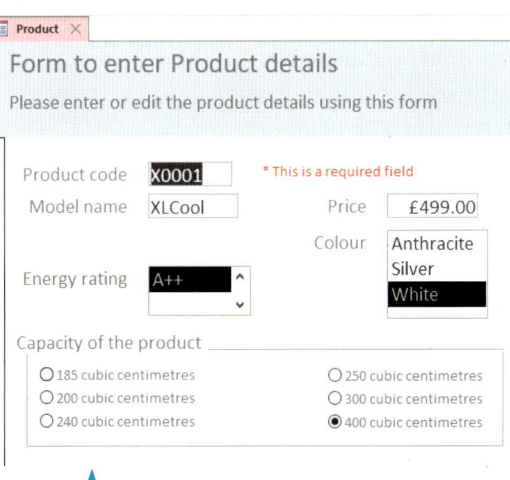

10.1 Database basics

321

Activity 10m

Open a copy of the database saved in Activity 10l. Create a single data entry form with a sub-form. Make sure that the forms include efficient methods of data entry that will help to reduce data entry errors. Discount values of 0, 5 and 10% are offered to customers; three types of boat are sold at the moment, but other types may be sold in the future.

10.1.12 Designing a switchboard

A switchboard (or menu) can be added to your database, so that it appears when the database is opened. This makes it easier for the user to add, edit or search for data. Once the switchboard has been designed and created, users with little database experience can usually use the database with ease. The switchboard will often include a title, instructions for the user and a number of navigation buttons to open forms and reports. It is important to design the switchboard so that it is easy to read, with well-spaced-out buttons and no large areas of white space. Company logos or other images can be included on the switchboard. Grouping similar items together is desirable.

Switchboards are usually designed on paper before they are created.

Task 10ad

Open the file saved in Task 10ac. Design and create a switchboard that allows users to enter or edit details of customers, invoices and products, as well as produce reports to display the models that are fridges, the products made in Anthracite and the price per cubic centimetre for each product. Display the image **freezer.png** on the switchboard. Save the database as **Task_10ad**.

This switchboard will be designed to group the title, instructions, forms, reports, and other items into different areas of the switchboard. As the image **freezer.png** is required, the aspect ratio of this image may determine the page layout, in this case it has portrait orientation so would ideally fit down the left or right side of the form. The form design may look like this.

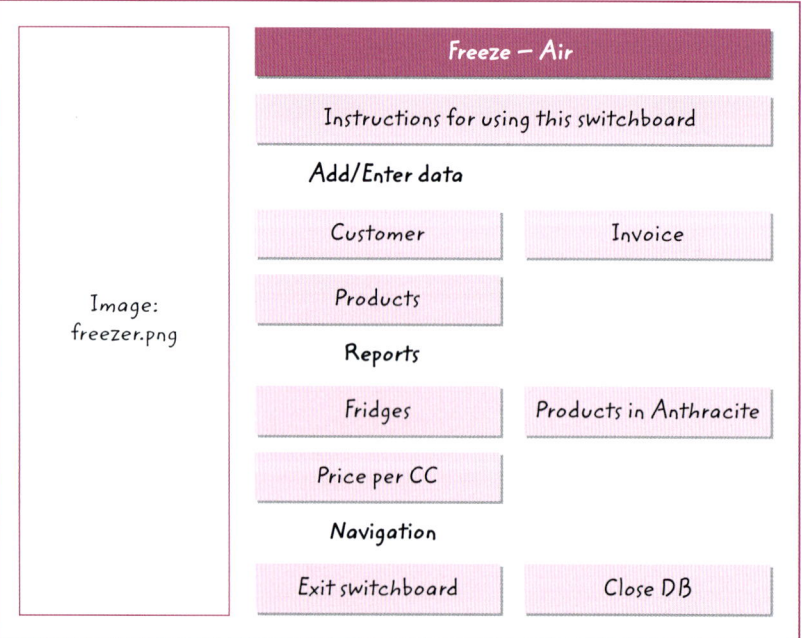

Even though this is only a very rough sketch, it allows the person creating the switchboard to plan the relative amounts of white space required for this design.

Creating a switchboard

Although *Access* has a **Switchboard Manager**, it is often quicker and easier to create your own switchboards/menus using *Access* forms. To create this form, select the **Create** tab, then the **Form Design** icon. The blank form is displayed and *Access* defaults to the **Design** tab. For the page title, a text box is required so select the **Label** control.

Drag the cursor over the page where this text is to be placed and enter the text 'Freeze-Air Main menu'. On the right in **Property Sheet** (click the right mouse button and select **Properties** if this is not visible) select the **Format** tab and enter an appropriate **Font Size**. Use the other property settings to adjust the text alignment, foreground and background colours, like this.

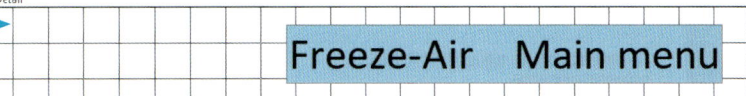

Use similar techniques to add the instructions and three group titles. From the **Controls** section, select the icon for a **Button** and drag out the rectangle to make the shape of a button. As you release the drag handle the **Command Button Wizard** opens.

This command button will link to the Customer form so in the **Categories:** box select **Form Operations** and in the **Actions:** box select **Open Form**, then click Next > .

Select the Customer form, then click Next > . Select the radio button for **Open the form and show all records**, then click Next > . Select the radio button for **Text** and enter the text **Customers** (there is no need to add a form as this button sits in the forms group of the menu page), then click Next > . Enter a meaningful name for this control like **Customers**, then click Finish . The button is created, but it may need editing to give it the appearance that you require. Resize the control using the drag handles. In the **Properties Sheet** add a border, change the foreground and background colours, font size and text alignment as required, like this.

Repeat this process for the other two form controls, selecting the Invoice and Product forms for the appropriate control. Use similar methods to add the three buttons for the reports, selecting **Report Operations** and **Open Report** before selecting the appropriate report for each button.

The **Exit menu** button is created using a **Form Operation** to **Close Form**. The final button used to close the database is created in the **Categories:** box using **Application** and **Quit Application**.

The image freezer.png is added to the form using the image icon.

Select this icon and drag the cursor to create the size of the image frame. As you release the drag handle the **Insert Picture** window opens. Select the image **freezer.png** then click on [Open]. The Design View and finished form will look like this.

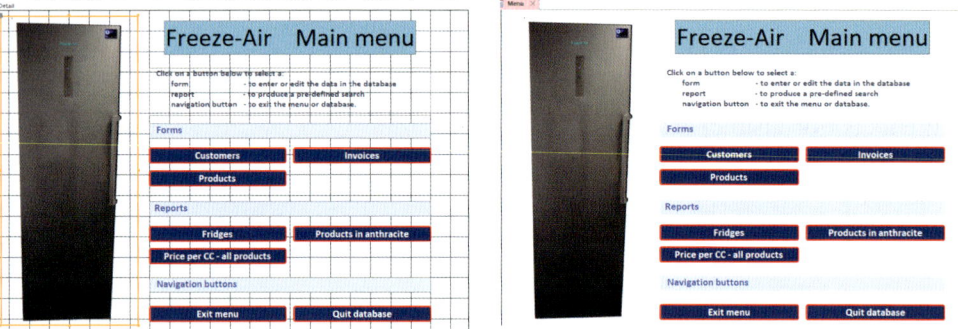

To improve the database, controls can be added to the other three forms to close the form and allow a user to return to this main menu from each form. Save the database as **Task_10ad**.

> ### Activity 10n
> Open a copy of the database saved in Activity 10m. Design and create a switchboard that allows users to enter or edit details of boats and customers, as well as produce reports to display the boats bought in the year 2020 and the grouped report for all unsold boats. Apply an appropriate colour scheme and display the image **boat.jpg** on the switchboard.

10.1.13 Importing and exporting data

Importing data

We learnt how to import data at the start of the chapter. The only input formats that you are likely to experience are comma separated values (.csv) and text (.txt) files; both have similar properties. Should source data be presented in **rich text format** (**.rtf**), it will need to have all formatting removed so that it can be saved in text (.txt) format before it can be imported.

Exporting data

Tables, queries and reports can all be exported from *Access* for use in other applications. Make sure that the object to be exported (it is the same for tables, forms, reports and so on) is not open within the database. Click the right mouse button on the object to be exported, then select the **Export** option from the drop-down menu like this.

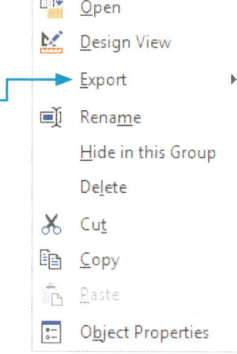

When you click on **Export**, a sub-menu with the type of export appears. Select the file type you require from this list, which is likely to be .pdf or .rtf format for a completed document. If the exported data is to be used for another purpose (for example as a source data for a mail merge) it may be exported in .txt or even *Excel* format. If the data exported is required in .csv format, select the option for .txt and change the file extension from .txt into .csv. In each case, you must enter an appropriate file name then click on [OK].

10.2 Normalising data

Normalisation is a technique used to reduce the duplication of data in a relational database. It helps to:
» organise data in an efficient way
» remove redundant (duplicated) data
» make sure that only related data is stored in a table.

This technique is a multi-step process, where each step has a rule that improves the efficiency of the database. These rules are called the Normal Forms, numbered from 0 to 5. We will only study un-normalised data (0NF) to Third Normal Form (3NF). Fourth and Fifth Normal Form (4NF and 5NF) are beyond the scope of this book and exam.

10.2.1 Un-normalised Form (0NF or UNF)

If a database is not normalised it is called an un-normalised database, often shortened to 0NF or sometimes UNF. This is often a flat-file database (a single table) that contains duplicated data (which is called redundant data) and complex data structures (more than one item of data, otherwise known as non-atomic data) stored within a single field.

> ### Task 10ae
> Change the following data on students, their house and their courses, into First, Second and Third Normal Form.
>
> | Student name | House | Course |
> | --- | --- | --- |
> | Jane Smith | York | Computing, Mr Brown, Room 53, Maths, Miss White, Room 104, Statistics, Miss White, Room 104 |
> | Ruksana Patel | Lancaster | Business, Mrs Grey, Room 3, Maths, Miss White, Room 104, Science, Mr Green, Room 24 |
> | Jamal Aziz | York | Computing, Mr Brown, Room 53, Maths, Miss Black, Room 102 |
> | Jet Li Xa | Lancaster | Maths, Miss Black, Room 102, Science, Mr Green, Room 24 |
> | Jane Smith | Edinburgh | Statistics, Miss White, Room 104 |
>
> Each teacher teaches in a single classroom.
>
> Save your work at each stage as **Task 10ae_1**, **Task 10ae_2** and **Task 10ae_3**.

Create a new word-processed document which contains the data structures for this task. Save this as **Task 10ae_1**.

10.2.2 First Normal Form (1NF)

The rules for a database normalised to 1NF are:
» All data is stored in a database table.
» A unique key must exist in each table (often a primary or compound key).
» Only **atomic data** is stored (which is data stored to the lowest level of data and cannot be broken down any further).
» Each field has a unique name.
» Each record is unique (so there are no repeated rows in any table and a primary/compound key exists in each table).
» There are no repeating groups of columns.

To convert the data in Task 10ae into 1NF we must first ensure that a unique key exists in each table. There are two students with the same name. Looking at this data it is possible (although unlikely) that two students with identical names, houses and courses could exist, so a new field is required as a primary key field. We will call this field **St_ID**. Each **St_ID** will be a unique number.

The data given contains non-atomic data. The student's name can be split into their **Forename** and their **Surname**. The course details can be split into the **Subject**, **Teacher** and **Room**. Each course should also have its own unique ID (the data presented is not unique, for example Mr Brown may take two different Computing classes in the same classroom at different times). We will call this field **Course_ID**. For the key to be unique we have a compound key with both **St_ID** and **Course_ID**.

This data would therefore satisfy First Normal Form as all data is in a table, there are no repeating groups of columns, each record is unique, each field has a unique name, and all data is atomic. There are two key fields, one for the students and one for the courses.

St_ID	Forename	Surname	House	Course_ID	Subject	Teacher	Room
0001	Jane	Smith	York	1042	Computing	Mr Brown	53
0001	Jane	Smith	York	1400	Maths	Miss White	104
0001	Jane	Smith	York	9961	Statistics	Miss White	104
0002	Ruksana	Patel	Lancaster	9606	Business	Mrs Grey	3
0002	Ruksana	Patel	Lancaster	1400	Maths	Miss White	104
0002	Ruksana	Patel	Lancaster	5000	Science	Mr Green	24
0003	Jamal	Aziz	York	1043	Computing	Mr Brown	53
0003	Jamal	Aziz	York	1402	Maths	Miss Black	102
0004	Li Xa	Jet	Lancaster	1401	Maths	Miss Black	102
0004	Li Xa	Jet	Lancaster	5000	Science	Mr Green	24
0005	Jane	Smith	Edinburgh	9961	Statistics	Miss White	104

Save your document.

Why is this data in First Normal Form (1NF) and not in Second Normal Form (2NF)?

This data is not in 2NF because the:
- **Subject**, **Teacher** or **Room** fields are not dependent on the **St_ID** field
- **Forename**, **Surname** or **House** fields are not dependent on the **Course_ID** field.

10.2.3 Second Normal Form (2NF)

The rule for a database normalised to 2NF is that the table must be in 1NF and any non-key attributes that only depend on part of the table key are placed in a new table.

Looking at this data, it is grouped into two types of data, the data relating to each course and the data relating to each student. The data must be split into three tables: courses, students and a table to link the two to meet 2NF. The first table will relate to the courses.

Course_ID	Subject	Teacher	Room
1042	Computing	Mr Brown	53
1043	Computing	Mr Brown	53
1400	Maths	Miss White	104
1401	Maths	Miss Black	102
1402	Maths	Miss Black	102
5000	Science	Mr Green	24
9606	Business	Mrs Grey	3
9961	Statistics	Miss White	104

You will notice that we have sorted the data to help ensure there are no duplicated records. The second table will consist of the students.

St_ID	Forename	Surname	House
0001	Jane	Smith	York
0002	Ruksana	Patel	Lancaster
0003	Jamal	Aziz	York
0004	Li Xa	Jet	Lancaster
0005	Jane	Smith	Edinburgh

The third table will link each student to each course.

St_ID	Course_ID
0001	1042
0001	1400
0001	9961
0002	1400
0002	5000
0002	9606
0003	1043
0003	1402
0004	1401
0004	5000
0005	9961

Save your document as **Task 10ae_2**.

Why is this data in Second Normal Form (2NF) and not Third Normal Form (3NF)?

This data is not in 3NF, because the **Teacher** and **Room** fields are dependent on each other, as each teacher always teaches in the same room.

10.2.4 Third Normal Form (3NF)

The rule for a database normalised to 3NF is to make sure that any non-key attributes that are more dependent on other non-key attributes than the table key, are removed to a new table.

Looking at this data, the course table must be split into two tables: one for the courses and one for the teachers to meet 3NF.

The first table will relate to the courses.

Course_ID	Subject	Teach_ID (Foreign)
1042	Computing	T2
1043	Computing	T2
1400	Maths	T5
1401	Maths	T1
1402	Maths	T1
5000	Science	T3
9606	Business	T4
9961	Statistics	T5

The second table will relate to the teachers. The third table will relate to the students.

Teach_ID 🔑	Teacher	Room
T1	Miss Black	102
T2	Mr Brown	53
T3	Mr Green	24
T4	Mrs Grey	3
T5	Miss White	104

St_ID 🔑	Forename	Surname	House
0001	Jane	Smith	York
0002	Ruksana	Patel	Lancaster
0003	Jamal	Aziz	York
0004	Li Xa	Jet	Lancaster
0005	Jane	Smith	Edinburgh

The fourth table will link each student to each course.

St_ID 🔑	Course_ID 🔑
0001	1042
0001	1400
0001	9961
0002	1400
0002	5000
0002	9606
0003	1043
0003	1402
0004	1401
0004	5000
0005	9961

Save your document as **Task 10ae_3**.

Sometimes when you change the data into one type of form, it is also set at the next level. For example, if data was changed into 2NF, it may also set as 3NF without any other changes required, if it also meets the rules for 3NF.

Activity 10o

Change the following data on customers, the company they work for and their contact details, into First, Second and Third Normal Form.

Customer name	Company	Contact details (3 x address, zip code, 2 x telephone numbers)
Surjan Patel	Binaccount	Manley Hall, Manley, Tawara, 4303, 05551 275236, 05551 175394
Jay Murray	Tawara Bus Company	14 Main Street, Grovecourt, Tawara, 4303, 05551 245975, 05551 175401
Lucy Murray	Easy Doors	41 Dilbridge Road, Dockside, Port Peppard, 4302, 05553 128835, 05553 245046
Surjan Patel	DBM Systems	56 Green Lane, , Littleton, 4254, 05554 168431, 05554 179139
Joseph Norris	Tawara Bus Company	14 Main Street, Grovecourt, Tawara, 4303, 05551 245975, 05551 175401
Victoria Wilkins	Tawara Bus Company	14 Main Street, Grovecourt, Tawara, 4303, 05551 245975, 05551 175401

Save your work at each stage as **Activity 10o_1NF**, **Activity 10o_2NF** and **Activity 10o_3NF**.

10.2.5 Advantages and disadvantages of normalising data

Less money needs to be spent on storage as the file size is smaller due to there being no redundant data, but the resulting larger numbers of tables require more relationships to be designed. This takes more time, when designing a large database, as well as requiring workers to have greater knowledge. This, in turn, may lead to workers having to be brought in from other companies and adds to the cost.

It groups data logically as well as reducing inconsistent data in tables by enforcing referential integrity, but having more tables than an unnormalised database makes it difficult to monitor where particular data is.

With the larger number of tables, setting up complex queries can be more difficult, although searching on one table will be much faster as there is less data to go through. The processing of data can be slower with a greater number of tables and links to navigate.

Any changes which are needed in certain records can be made automatically to any related records.

With no duplicated data there will be fewer errors in the data and making changes to a table is easier as there is less data to alter.

Removing non-atomic data may not always be a good idea if putting data into separate fields serves no useful purpose.

10.2.6 Normalise a database to a specified Normal Form

While this can be completed to 1NF or 2NF, most tasks are likely to require a solution to 3NF.

> ### Task 10af
>
> All employees who work in the ski school, a restaurant or hotel in Ellmau, Austria are to have their details stored in a database. All database field names must be short, meaningful, consistent in style, and contain no spaces or underscores.
>
> Using suitable software, open and examine the data in the files **ellmau.csv** and **jobs.csv**.
>
> Use this data to create a relational database normalised to 3NF.
>
> Save your database as **Task 10af**.

Open and examine the files **ellmau.csv** and **jobs.csv** in *Excel*. In the jobs data file, the column headings that will become field names are long and contain spaces and so do not meet the requirements of the question. These can be replaced with shorter text like **JobCode** and **Description** before resaving the data file in a new folder (never overwrite the original source files in case you need to refer to them again). The data in this file is all atomic and does not need to be split into other columns. In the ellmau data file, the employee's names are not atomic data, as they contain both forename and surname. This needs splitting into two columns. Insert two new columns before column A. In cell L2, place the formula **=LEFT(C2,FIND(":",C2)-1)** to extract the forename of this employee. In cell M2, place the formula **=RIGHT(C2,(LEN(C2)-FIND(":",C2)-1))** to extract the surname for this employee. Replicate these formulas down to row 171. Copy the contents of cells L2 to M171, move the cursor into cell A2, right click the mouse button and from the drop-down menu, select the icon for **Paste Special** (values).

> **Advice**
> A different way of splitting the names into two columns is to insert a new column between A and B. Highlight column A and select from the **Data** section, **Text to Columns**. Select **Delimited**, then a colon as the delimiter.

As there is no key field on this table (because theoretically there could be two people with the same names, dates of birth and jobs), insert a new column A and add an ID column. Add a unique number for each employee in this column. Enter in cell B1 the text **Forename** and in cell C1 the text **Surname**. Delete columns D, M and N. Edit cell E1 to **JobCode**, cell F1 to **Employer**, cell G1 to **EmpAdd1**, cell H1 to **EmpAdd2**, cell I1 to **EmpType** and cell J1 to **EmpCode**. This data is now structured in 1NF. Save the file.

To change this into 2NF, it needs splitting into a table for each employee and for each employer. Save the file as two different files, one with the name **Employer** and one with the name **Employee**. Open the Employee file and delete columns F to I inclusive. Save this file. Open the Employer file and delete columns A to E inclusive. Move the new column E so that it becomes column A, as this is the code for each employer. Sort the data into ascending order on the EmpCode column (do not include row 1 in your sorted data). Keep one row for each EmpCode and delete any other rows containing that code so that no duplicated rows are present. Save the file.

To change this into 3NF, the data in the Employer file must be split into two tables, as the EmpType column contains only three values, which would be better placed in a separate table with a simplified key. Save a copy of this file as **Type**. Open the Type file, delete columns A to D and insert a new column A. Delete all duplicated rows of data. In cell A1, add the text 'EmpTypeCode' and copy the first letter from the data in column B into the cells in column A, like this.

Save this file. In the Employer file, replace in column E the text Hotel with the letter 'H', Restaurant with the letter 'R' and Ski school with the letter 'S'. Save this file. Use these data files to create a new relational database with four tables like this.

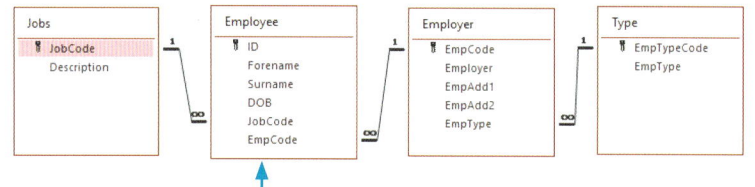

10.3 Creating a data dictionary

A data dictionary is a file (often a document) containing descriptions of and information about the structure of data in a database. It is often designed before a database is created as an alternative to an entity relationship diagram.

10.3.1 Components of a data dictionary

The data dictionary should always contain the table names, field names, data types (and sub-types where appropriate), field size (for alphanumeric fields), the primary key and foreign key fields, and other metadata which could include input masks, validation rules or default values, if applicable. Some of this data, for example data types and field lengths, can be worked out from data files that have been given to you.

10.3.2 Create a data dictionary and select appropriate data types

> **Task 10ag**
> Using the database you created in Task_10af, create a word-processed data dictionary for this database. Save your database and your data dictionary with the filename **Task_10ag**.

To create this database to 3NF, four data tables are required. These relate to the employee, the employer, jobs, and type of employer. For each database table, create a table in your word-processing software. Add the table name to the top of the table and list all the field names in the left column like this:

Table name: Jobs				
Field name	Data type	Field size	Key field	Metadata (including: input mask, validation rules, default value, etc.)
JobCode	Alphanumeric/Text			
Description	Alphanumeric/Text			

Examine the data for this table and it can be seen that the data for both fields contains only letters, so an alphanumeric data type will be needed. For each field, identify the number of characters for the longest item of data. For the **JobCode** field, there are two data items which are three characters in length; these are 'TBI' and 'TSI', so record the field size as three characters and set the field size to this length in the database. For the **Description** field, the longest data item is 'Trainee Snowboard Instructor' which is 28 characters in length, so record the field size as 28 and set the field size to this length in the database.

> **Advice**
> You will need to remove any relationships to other tables in your database before editing the field size within this table. These can be recreated when the field sizes have all been set.

The primary key field for this table will be the JobCode, because it contains unique data, so indicate in the key field column that this is the primary key. As there is no other metadata for this table (input masks and validation rules would not be appropriate and no field requires default values), complete the table like this:

Table name: Jobs				
Field name	Data type	Field size	Key field	Metadata
JobCode	Alphanumeric/Text	3	PK	
Description	Alphanumeric/Text	28		

10.4 File and data management

Program and data files are stored with different file types and each file is stored with a file extension, which informs the operating system which program to use to open the file. File formats fit into different categories, including proprietary, open source and generic file types.

10.4.1 Different file types and their uses

Program files are normally saved with .exe file extensions. These are often opened by clicking on a program icon; examples include *Microsoft Word*®, *Access* and *Excel*.

Proprietary file formats are formats belonging to a company, organisation or individual and are often specific to a program created and copyrighted by that company, for example current versions of *Microsoft Word* save files with a .docx file extension (earlier versions of the program were saved with a .doc extension). These files contain your data (in this case documents) but they are stored in the formats of the particular package and contain more information than just the contents that you can see when the document is opened. The exact details of the data stored, the encoding and its structure within the file are often kept secret and/or patented by the company or organisation which created the software. Proprietary software allows only people with licences to use it.

Open source file formats are formats used for storing data that are published for anyone to use. They can be opened by both proprietary and open source software. The details of the way the data is stored are available for all users and software developers, rather than just the organisation which created the structure. Open source file formats tend to be free, because not everyone can afford proprietary software, and the code is open to inspection and sometimes to amendment without breach of copyright legislation. Users can transfer files from a work computer to a home computer using open source software more easily, particularly if they do not have proprietary software on their home computer.

Generic file formats allow you to save files so that they can be opened on any platform, for example fields created on a PC can be read/imported on an Apple Mac®, mobile phone and so on, and vice versa. However, the files may not contain all the formatting that can be saved in a package-specific format and it is not always possible to open proprietary file formats on other platforms.

Common generic text files include:

- Comma separated values: These files have a .csv file extension. This file type takes data in the form of tables (that could be used with a spreadsheet or database) and saves it in text format, separating data items with commas.
- Text: These files have a .txt file extension. A text file is not formatted and can be opened in any word-processor software.
- Rich text format: These files have an .rtf file extension. This is a text file type that saves some of the formatting within the text.

Common generic image files include:

- **Graphics interchange format (GIF)**: These files have a **.gif** file extension. This format stores still or moving images and is an efficient method of storing images using a smaller file size, particularly where there are large areas of solid colour. It is widely used in web pages.
- **Joint photographic expert group (JPEG)**: These files have a **.jpg** (or sometimes a .jpeg) file extension. This format stores still images but not moving images. It is an efficient method of storing images using a smaller file size and is widely used in web pages.

- **Portable document format (PDF)**: These files have a .pdf file extension. This is a document which has been converted into an image format. It allows documents to be seen as an image so they can be read on most computers. The pages look just like they would if they were printed but can contain clickable links and buttons, form fields, video and audio. In pdf format, you can protect a document to stop others from editing it.
- Portable network graphics (PNG): These files have a **.png** file extension. It is a file format that compresses graphics (image) files without any loss of image quality. It was created to replace graphics interchange format and is now the most used lossless image compression format on the internet.
- **Moving Pictures Experts Group layer 4 (MPEG-4)**: These files have an **.mp4** file extension. It is not a single file format, but is a multimedia **container** which is used for storing video files, still images, audio files, subtitles and so on. This container is often used to transfer video files on the internet.

Common generic audio files include:

- **Moving Pictures Experts Group layer 3 (MPEG-3)**: These files have an **.mp3** file extension. It is a compressed file format used for storing audio files. This format cannot store still or moving images. The file sizes are relatively small but have high quality, which makes it suitable for use on the internet.

Common generic files used for website authoring include:

- **Cascading style sheet**: These files have a **.css** file extension. This is a stylesheet which is saved in cascading stylesheet format and is attached to one or more web pages (often written in HTML) to define the page's colour scheme, fonts and so on.
- **Hypertext Markup Language (HTML)**: These files have an **.htm** (or sometimes an .html) file extension. This is a text-based language used to create markup that a web browser will be able to interpret to display information on a web page.

Common generic compressed files include:

- **Roshal archive (RAR)**: These files have a **.rar** file extension. This is a container which can hold almost any file type in a compressed format. It is used to reduce the number of bytes needed to save a file, either to save storage space or to reduce transmission time. It was developed for *Windows* by a Russian software engineer named Eugene Roshal and takes its acronym from **R**oshal **AR**chive.
- **Zip**: These files have a **.zip** file extension. This is a container which can hold almost any file type in a compressed format. It is used to reduce the number of bytes needed to save a file, either to save storage space or to reduce transmission time.

10.4.2 Sequential access

Before studying the indexed sequential method of accessing data, it is important that we understand what is meant by sequential access. In a sequential file, records are stored one after the other, in the order in which they were added to the storage medium, usually magnetic tape. To read data from or write data to tape, sequential files must be used.

There are two ways that records can be arranged in a sequential file. One way is to have the records in some sort of order using a key field. A key field is one which is unique to every record, that is every record has a different value in that field. This is called ordered sequential. Alternatively, the records might be arranged with no thought given to their order, so they appear to be unordered. Whether the file is ordered or unordered affects the way in which the data is processed, as well as the type of processing that can be used. An unordered sequential file is often referred to as a serial file, as the only method for retrieving information is to go through each record one by one.

▼ Table 10.9 Unordered file

ID	Customer name	Country
7	Andreza Bistene	USA
3	Joseph Pinkerton	USA
8	Hannah Joseph	USA
10	Tyler Moncrieff	USA
6	Javier Hernandes	Chile
4	Nabil Mohammed	Brazil
5	Luis Nova	Venezuela
9	Addy Addu	Chile
1	Pingu Morales	Brazil
2	João Sousa	Brazil

▼ Table 10.10 Ordered file

ID	Customer name	Country
1	Pingu Morales	Brazil
2	João Sousa	Brazil
3	Joseph Pinkerton	USA
4	Nabil Mohammed	Brazil
5	Luis Nova	Venezuela
6	Javier Hernandes	Chile
7	Andreza Bistene	USA
8	Hannah Joseph	USA
9	Addy Addu	Chile
10	Tyler Moncrieff	USA

In an ordered file, the records are put in the order of a key field such as customer ID, as shown above. In an unordered file, the records are not in any particular order.

There are a number of disadvantages to using sequential files. For example, the only way to add new records to a sequential file is to store them at the end of the file. A record can only be replaced if the new record is exactly the same length as the original. Records can only be updated if the data item used to replace the existing data is exactly the same length.

The processing of records in a sequential file is slower than with other types of file. In order to process a particular record, all the records before the one you want have to be read in sequence until you get to the one you want. The use of sequential files is recommended only for those types of application where most or all the records have to be processed at one time. (See Chapters 1 and 4 for further details on the use of master and transaction files, which would use sequential access.)

10.4.3 Indexed sequential access

Indexed sequential files are stored in order, and unlike sequential access files which are stored on tape, these are stored on disk to allow some form of direct access. Each record consists of fixed-length fields. This is a leftover from the use of magnetic tapes where records had to be stored in the order they were written to the file. The use of ordering facilitates faster access, because a table of indexes is used to allow the search to jump to a particular place on the disk rather than going through all the other records to get to the right point.

With an indexed sequential system, the records are in order. For example if a database of employees was stored in order of surname (although this is unlikely unless it was part of a compound key field) and you wanted to find a record starting with the letter 'F', such as Fullman, then the place for 'F' is looked up in the table of indexes and the disk head is moved to the index point which is the start of the 'F's. It makes the searching faster. Although all the Fs would still need to be searched to get to Fullman, the computer does not need to search through any of the surnames beginning 'A' to 'E'.

Banks use sequential access systems for batch-processing cheques. This system would have to be at least indexed sequentially for faster access to records for online banking. Indexed sequential files are used with hybrid batch-processing systems, such as employee records. The index allows for direct access when individual records are required for human resource/personnel use. The records are held sequentially to allow for serial access when producing a payroll, since all records are processed one after the other.

10.4.4 Random access

Random access (which is sometimes called direct data access) is the quickest form of access. It does not matter where in the file the desired record is; it will take the same amount of time to access any particular record. Each record is fixed length and each has a key. The computer looks up the key and goes to the appropriate place on the disk to access it.

10.4.5 Using database management systems (DBMS)

Hierarchical database management systems

Hierarchical database management systems were developed in the 1960s and have a tree-like structure, where data is stored as records connected together through links. This allows fast access to data, because large amounts of data are bypassed as you go down the levels of its structure. Consider a family tree structure: A woman can have many children but the children can only have one mother. Records can have several records below them in this structure but only one record at the level immediately above their level. In other words, a 'parent' record can be linked to many 'child' records. In such a structure, records can only have one-to-one or one-to-many relationships. They are still used by some organisations, mainly in legacy systems (systems which although out of date are still in use). Their other major use is the storage of file names in Windows registries to hold details of files.

Network database management systems

Network database management systems were developed to overcome many of the faults associated with the hierarchical type. Many existing databases still rely on this form of DBMS. Each record can be linked to several other records. Many-to-many relationships are possible. Each 'child' record can be linked to several 'parent' records and each 'parent' record can be linked to several 'child' records. Again, many of the uses of this type of DBMS are in legacy systems but they can still be found in banking, airline reservations and government census and tax records.

Object-oriented database management systems

As the name suggests, this is a DBMS that involves the creation and modelling of data as objects. Object-oriented database management systems (OODBMS) consist of fields that contain objects such as images, video and audio files. Object fields are larger than fields found in other DBMS. Data is managed using object-oriented programming languages such as Python, Java and Visual Basic or a specialised query language. These systems are often used in applications that are required to handle complex data structures and relationships. Examples are CAD/CAM systems, geographic information systems and document management systems. Applications that require the integration of different data types and sources, such as multimedia data or data from multiple sources, are suitable for OODBMS.

Relational database management systems

As you have already discovered, a relational database consists of a number of separate tables that are related in some way. Data from one table can then be combined with data from another table when producing reports. For example, relational tables could be used to represent data from a payroll application and from a human resources application. The standard programming language in large applications to deal with relational tables is structured query language (SQL), which is used for queries and producing reports. The uses of relational databases are many and varied. A hospital may use a relational database to store and retrieve patient records. A company delivering packages could use one to track their packages and deliveries. Relational databases are used to process e-commerce and banking transactions.

Advantages and disadvantages of the different types of DBMS

One advantage of hierarchical databases is that they allow for fast data retrieval but they can be difficult to maintain and update, since changes to the data structure can affect the entire database. Hierarchical databases make it easier to work with large amounts of data as complex information is broken down into smaller, more manageable pieces. Hierarchical databases are easy to edit and add or delete information because all the data is stored in a single table. The data structure makes it easier to understand and navigate. They can use less storage compared to other types of DBMS. Data retrieval is easier than retrieving data from separate tables as with a relational database.

One disadvantage of hierarchical databases is that they are not very suitable for handling complex data relationships or changes in data structures. Another is hierarchical databases do not support complex data manipulation operations, such as data aggregation or data mining. They are not compatible with other DBMS, making it difficult for them to interact with other systems and applications. There is also little standardisation of their use and management. They lack the flexibility of relational and network DBMS as they only have one-to-many relationships. Hierarchical databases often contain duplicate copies of data leading to wasted storage space. With relational databases this is less of a problem.

One of the advantages of relational databases is they can have complex relationships between different data sets, unlike a hierarchical database. They also allow the creation of more complex queries using data from several tables. If two tables are not linked, however, adding a new record to one table is difficult because extra data must be entered in the other table first.

Relational DBMS can be set up to prevent duplicate entries or creation of data types that don't match. Tables can be set up so that only certain users can access that data, ensuring sensitive data is kept secure. A disadvantage of using a relational database is that adding a new field to a table may mean that other tables have to be changed. This takes time and may require expert knowledge. When a database contains a large amount of data the processing of it can slow down.

A network database is easy to design just as a hierarchical database is, whereas relational databases are more difficult to set up. A network database provides faster access to data than a hierarchical database as the data is better linked up with many relationships. Network DBMS are capable of handling several types of relationships unlike hierarchical databases which cannot handle many-to-many relationships. More intricate queries can be developed than those using a hierarchical database.

A disadvantage of a network DBMS is its structure is very complex, making it harder to make changes to the data structure without affecting the overall structure of the database. In addition, the design and the structure of such a database is not user-friendly. Like a hierarchical DBMS, it suffers from not having any database standard. Another disadvantage of a network database is that it has less scalability than other databases because the relationships between data elements can become more complex as the database grows. Creating a query for a network database can be difficult because the relationships between data items are more complex and difficult to understand.

An advantage of OODBMS is that they support complex data types such as images, videos, and documents more efficiently than other DBMSs. OODBMS are more secure than relational databases due to their authentication and authorisation features. They can more easily handle the data storage and retrieval of large-scale applications. The disadvantages of OODBMS are that support is limited compared to that provided for relational databases. It can take time to learn how to develop this type of database, as it requires expertise in both the object-oriented programming language and the database management system.

10.4.6 Management information systems

A management information system (often shortened to MIS) is a computer-based tool that organises and evaluates data for an organisation. It can be a single software program or in a larger organisation is likely to be a number of products used together. It often includes the management of web content, documents, records, learning and the content used for learning. The features of a good MIS are that it will be:

» flexible in the analysis and evaluation of data from many sources and in many different ways
» easy to use without computer expertise
» versatile enough to support different skills and knowledge
» collaborative to generate communication between managers and other staff within a company.

Many organisations use MIS for:

» process control
» human resource management
» accounting and finance
» sales and marketing
» inventory control
» office automation
» enterprise resource planning
» management reporting.

Management information systems can help a company to run more efficiently, because they provide details about the past performance of the company as well as how it is performing at the moment and, most importantly, they help to predict how well the company may do in the future. An MIS will help managers make decisions about actions they may need to take. It is a computer-based system that provides managers with the tools to manage their departments. An MIS is made up of a number of components, such as the hardware resources used within the system, software in the form of decision support systems, people management applications, and also project management systems. Within the MIS are systems designed specifically for managing the marketing aspects of the business as well as accounting information systems which have a variety of accounting functions. There are also systems used for human resources aspects of an organisation which store information about employees. A company using an MIS tends to have a management information system manager. Their job involves running the company's information and technology systems. Their role typically involves analysing business problems as well as designing and maintaining the computer applications which are needed to solve the company's problems.

An MIS helps with all aspects of project management and database retrieval applications. Information retrieval is the ability to take different types of data in the storage media and to produce information in a meaningful format. A properly designed storage and retrieval system performs searches efficiently and accurately. In some cases, it even suggests alternative courses of action for management to take. Department managers will use management information systems to gather and analyse information about various aspects of the organisation such as personnel, sales, revenue, and production. Management information systems are used to produce reports and charts about the different aspects of the company at regular intervals for department managers and directors. Reports are produced using a reporting tool which provides quick access to summarised reports coming from all departments. These reports help them evaluate their company's performance. The comparison of regular reports to previous reports, as well as using graphs, helps managers to detect trends in profits and sales.

Practice questions

All workers in one town in Austria are to have their data stored in a central database. This will be used to create a number of reports, which must be produced to a professional standard.

All reports must be designed to fit on an A4 page in portrait orientation and be a single page wide. All text must be at least 12 points high. Display all currency values in euros with two decimal places.

All field names must be short, meaningful, consistent in style, and may contain neither underscores nor spaces.

1 Using suitable software, open and examine the data in the file **workers.csv**.

  Use this data to create a small efficient relational database normalised to the 3NF. [52]

2 A New Year Awards Ceremony for 'Unsung Heroes' will take place on 1 January next year where awards will be presented to workers who, on today's date, will have been at the same place of employment for more than 25 years, are not any type of chef and have not previously been given an award.

Create this report so that address lines 1 and 2 are omitted and so data is grouped and formatted like this:

Job	Surname	Forename	Start Date
Chambermaid			
AktivHotel Hochfilzer			
Dorf 34			
6352 Ellmau			
	Kern	Luca	13/08/1979
	Kern	Katharina	26/08/1981
Hotel Der Bär			
Kirchbichl 9			
6352 Ellmau			
	Hauser	Luca	27/03/1991
Cleaner			

The title has been blurred out but should be a white sans-serif font on a black background. [22]

3 Create a report to display only the workers who are trainees but do not work at the Skischule. Identify, for each job described, how many trainees work in each place of employment. The report must have an appropriate title, be in a tabular format and display the place of work as row headings. Do not include totals for each place of work. [7]

4 Save your database with the filename 'Austria' and export this report in portable document format with the filename 'Trainee'. [1]

11 Video and audio editing

In this chapter you will learn about:
- methods of video compression
- the features found in video editing software
- the features found in sound editing software
- sampling rate and sampling resolution
- methods of sound compression
- different editing techniques and the reasons why they are needed.

In this chapter you will learn how to:
- edit a video clip
- edit a sound clip.

For this chapter you will need these source files:
- Activity_11c_audio.mp3
- Activity_11e_audio_1.mp3
- Activity_11e_audio_2.mp3
- Balloon.mp4
- Blue_Sky_Blues.mp3
- EQ11-Bali1.mp4
- EQ11-Bali2.mp4
- EQ11-Bali3.mp4
- EQ11-Bamboo.mp3
- EQ11-Voice.mp3
- Fish_Eagle.jpg
- Fish_Eagle.mp4
- Lake_Naivasha.mp4
- Nurse_Shark.png
- Shark.mov
- Swim_through.mov
- Task_11e_audio.mp3
- Task_11j_audio_1.mp3
- Task_11j_audio_2.mp3
- Task_11n_audio.mp3
- Task_11o_audio.mp3
- Task_11p_audio.mp3
- Voiceover_1.mp3
- Voiceover_2.mp3
- Voiceover_3.mp3
- Voiceover_4.mp3
- 11i.wav

Sound and video editing can be carried out using a number of different packages. In this book, we will be using *Windows Movie Maker®* for video editing and *Audacity®* for audio editing, but the principles of these exercises can be applied using similar techniques in most editing packages. With *Movie Maker*, make sure that you save the project at each stage throughout the chapter, as well as saving/exporting in the required format. Throughout this chapter you will be learning different editing techniques and the reasons why they are needed.

> **Advice**
> Although *Windows Movie Maker* has been replaced by the Video Editor section in the *Microsoft Photos®* application, Video Editor does not have all the required features you need to know and use.

11.1 Video editing

11.1.1 Frame size and aspect ratio

Before a video can be created or edited, the physical size of the completed frames and their aspect ratio need to be determined. This will often depend upon the audience and device/s that will view the completed video. The aspect ratio of each video frame is the relationship between the width and height. There are two common ratios: 16:9 which is sometimes called widescreen, and 4:3 which was originally called standard, although in recent years 16:9 video has become more common than 4:3. These aspect ratios determine the relationship between the number of pixels in the width compared to the number of pixels in the height, so for a 4:3 ratio, the video could be 400 pixels wide by 300 high, 1200 wide by 900 high, or 4000 wide by 3000 wide and so on. The frame size, along with its aspect ratio, is set when the video published/exported is saved.

Task 11a

Create a new video and set the image ratio to 16:9. Create title frames using the image **Fish_Eagle.jpg**.

Add the text 'Lake Naivasha, Kenya' in a dark blue, 60 point, sans-serif font, so that it looks like this:

Add an appropriate animation effect to place this text. Display this title for 8 seconds.

Save the video in 1920 x 1080 format as **Task_11a.mp4**.

Open *Movie Maker* and select the **File** tab, then from the top of the drop-down menu, select the **New project** option. Select the **Project** tab, then in the **Aspect ratio** section the option for 16:9 using the **Widescreen** icon.

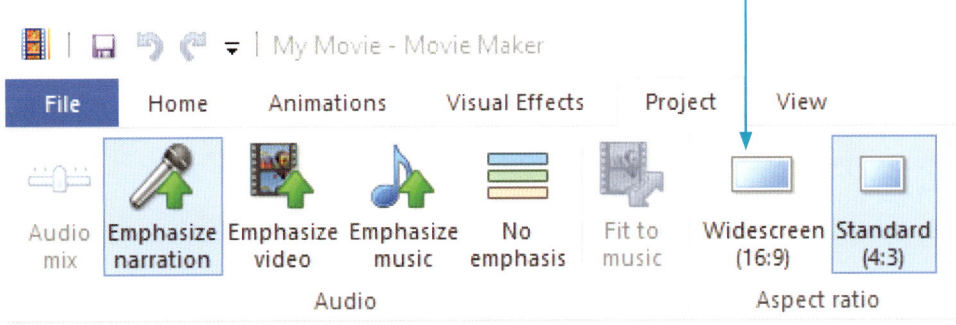

11 Resize and crop a still image to an aspect ratio

Before creating the title frames, read the Task to find out what size the final video will be. In this task the finished movie will be 1920 wide by 1080 pixels high. Compare this to the image **Fish_Eagle.jpg**, which is 3764 pixels wide by 3092 pixels high. This image will need to be both resized and cropped to fit an aspect ratio of 16:9.

Open your graphics/image editing package (*Adobe Photoshop* has been used here, but most packages will perform similar functions for this Task). Select the image size using the **Image** tab, then from the drop-down menu select **Image Size**.

In the **Image Size** window, tick the box for **Constrain Proportions**.

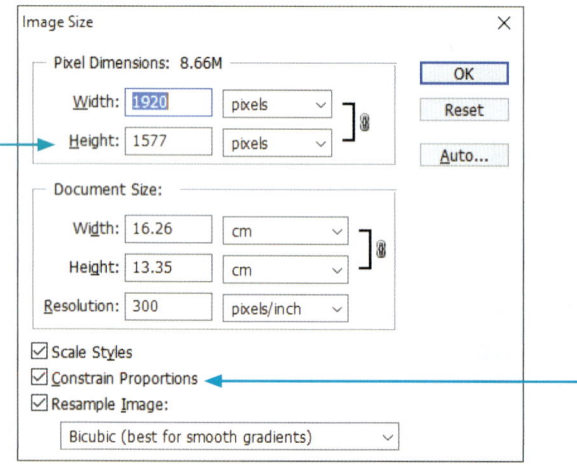

This **Constrain Proportions** check box maintains the aspect ratio of the image so that it does not become distorted. Now change the **Pixel Dimensions:**, **Width:** to **1920**. You will see that the image height now becomes 1577 rather than the 1080 that we require. Click OK.

To crop this image with accuracy, select the **Image** tab, then the **Canvas Size** option. Change the units from centimetres into **pixels** using the drop-down menu.

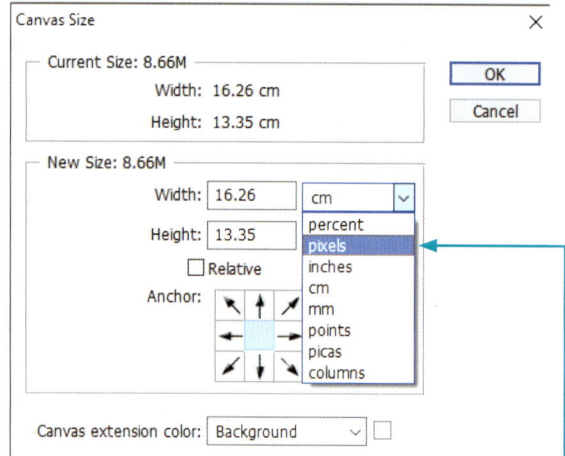

Looking at the image provided we need to keep the eagle which is placed in the bottom-right corner. In the **Anchor** box, click the mouse in the bottom-right corner. The anchor will change to this.

Enter the new height of **1080** into the height box, then click OK. A warning will be given, click on Proceed and the image will be cropped. Resave the image using **File** and **Save As…** and name the image **Fish_eagle1.jpg**, so you do not overwrite the original source file.

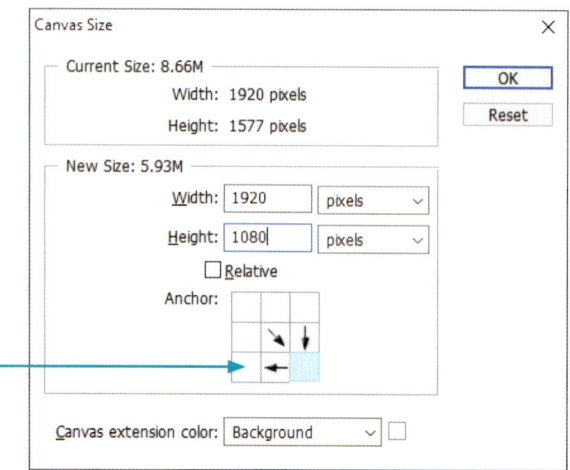

11.1.2 Insert a still image

The reasons for inserting still images are similar to those for extracting still images. You may have a particularly nice photograph which, if included in the video, would enhance it greatly. You may have extracted a frame to edit it and now you want to put it back. Open *Movie Maker* and using *File Explorer* open the folder containing the **Fish_Eagle1.jpg** image so that both packages are visible on the screen at the same time. Drag the image **Fish_Eagle1.jpg** from *File Explorer* into the right half of the *Movie Maker* window like this.

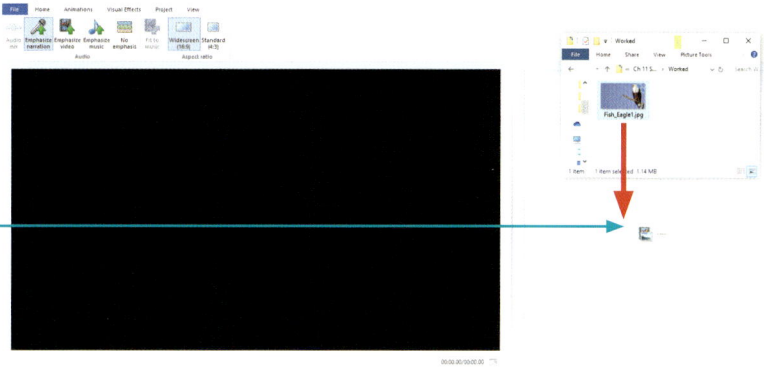

To change the duration of this background, click on the image, then on the **Edit** tab. In the **Adjust** section, find **Duration** and use the drop-down menu to set this to **8** seconds.

> **Advice**
> Always use the **Duration** to set timings as this is more accurate than manually dragging the timer bar.

11.1.3 Create text-based slides

It is sometimes necessary to add text to videos. It is generally accepted that adding text to videos engages the audience and helps them retain the video content better. To add the title text, click on the **Home** tab, then in the **Add** section select the icon for **Title**. This will add a black title background in front of the eagle image. Ignore this for now, and add the text **Lake Naivasha, Kenya** into the centre of the window. It may wrap onto two lines like this.

As you edit this the tab at the top changes to the **Format** tab. You can use this to set the **Text duration** to **8** seconds to match the eagle image. Set the **Start time** to **7:00s** so that the title text now overlays the eagle image. You will not be able to see this text until you move the slider button to about 10 seconds.

Click back on the text box, and in the **Font** section, click on the drop-down menu for the **Text Colour** to select a dark blue colour (you may need to use **More Colours** to find one dark enough). Change the **Font** size to **60** and make sure a sans-serif font is selected. In the **Paragraph** section, choose the icon to left-align the text. Click on and drag the text box into the top-left corner so that it looks like the image in the Task box. On the right side of the package, click the mouse on the old title (it is the black/grey rectangle below the eagle image). With this black/grey area selected, press the **<Delete>** key on the keyboard. The project looks similar to this.

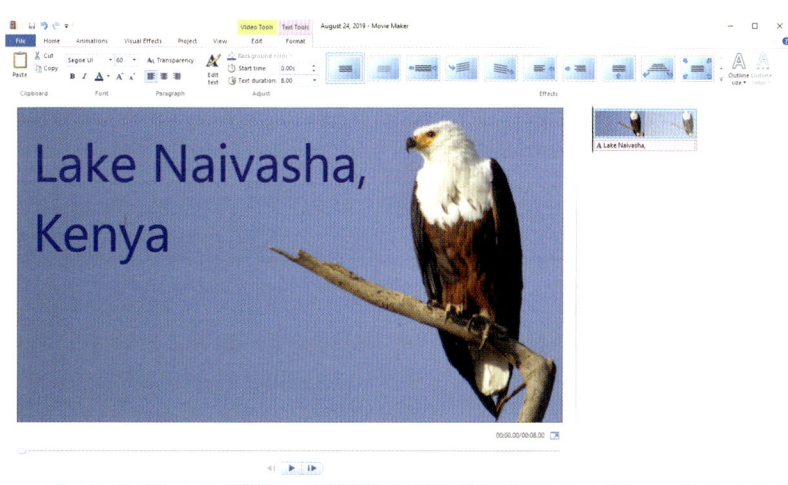

11.1.4 Add animation effects

The final instruction in the Task before saving the movie is to 'Add an appropriate animation effect to place this text'. This instruction means that only the text is animated, not the background image as well. Click on the text so that it is selected, then from the Format tab, in the Effects section, select any of the effects (except the first one). Choose one that does not place the text over the eagle part of the image and means that the text can be easily read by the audience in the time it will be displayed. If details of your target audience are given, make sure the font and animation are suitable for them.

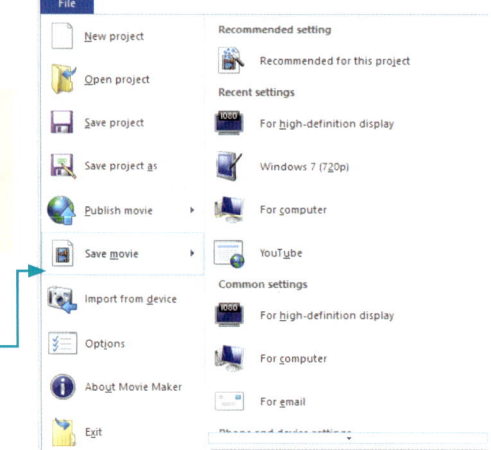

Advice

Please note that some of the effects work when the text enters the frame and some when it leaves. Make sure that the effect you choose matches the question.

Play the video clip to make sure it matches the specifications in the Task. The video needs to be saved in the required format, so from the File menu, select Save movie, then for the required resolution (1920 × 1080) select For high-definition display. In the Save Movie window enter Task_11a.mp4 then click Save.

Activity 11a

Create a new video and set the image ratio to 4:3. Create title frames using the image **Nurse_Shark.png**.

Add the text 'Diving in Cozumel, Mexico' in a yellow, 48 point, serif font, so that it looks like this:

Add an appropriate animation effect to place this text. Display this title for 6 seconds.

Save the video in 960 × 720 format as **Activity_11a.mp4**.

11.1.5 Trim a video clip

Video clips can have the start or end of them removed. This process is called trimming and can be completed with accuracy using the trim tool. Trimming and cropping a video can serve similar purposes. They are both used to remove parts of the video that are unwanted. The introduction to a video may be too long or boring so needs to be shortened or trimmed. There may be some unwanted material such as a person or something that has appeared in the background without you realising. This will need to be cropped. Trimming can be used to bring video and audio into synch if they are not already.

Task 11b

Open and examine the file **Lake_Naivasha.mp4** in a video editing package. Trim this clip so that it starts after 8 seconds and finishes 4.5 seconds later. Remove all sound from this video clip.

Save the video in 1920 × 1080 format as **Naivasha2.mp4**.

Open *Movie Maker* and drag the file **Lake_Naivasha.mp4** into the right-hand side of the *Movie Maker* window. Select the Edit tab. In the Editing section left-click on the Trim tool icon.

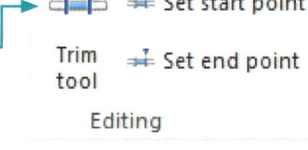

This opens the Trim tab where precise values can be entered for the start point and end point. Place in the Start point box the value 8 and in the End point box enter the value 12.5 (which is the start point value plus the 4.5 seconds for the clip). Click on the Save trim icon.

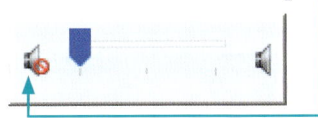

11.1.6 Remove audio from a video clip

Open the Edit tab and select the Video volume button. This will open a small volume window. Drag the slider handle to the left end of the volume bar to mute all sound within the clip.

Play the video clip to make sure it matches the specifications in the original question. From the File menu select Save movie, then for the required resolution (1920 × 1080) select For high-definition display. In the Save Movie window enter **Naivasha2.mp4** then click Save.

> ### Task 11c
> Open and examine the file **Lake_Naivasha.mp4** in a video editing package. Trim this clip so that only the first 6.3 seconds remain. Remove all sound from this video clip. Speed up this clip so that it runs at 1.5 times the playing speed.
>
> Save the video in 1920 × 1080 format as **Naivasha1.mp4**.

Reopen the file **Lake_Naivasha.mp4** in *Movie Maker*. Trim the clip using the method described above, with a Start point at 0 seconds and an End point at 6.3 seconds, then remove the audio from the clip.

It may be helpful to note that other video software use separate tracks for video and audio, so sound can be removed by deleting the audio track in alternative software.

11.1.7 Alter the speed of a video clip

To speed up this clip so that it runs at 1.5 times the playing speed, click on the Edit tab, then in the Adjust section use the drop-down menu for Speed, changing the speed from 1× to 1.5×. The length of the video clip will change from 6.3 seconds to 4.2 seconds. Use the method described in Task 11b to save the clip with the filename **Naivasha1.mp4** and a resolution of 1920 × 1080.

11.1.8 Splice/join together video clips

To splice means to join two pieces of video together; it was a term used for physically cutting and joining traditional film footage stored on a reel. To join two elements of video in *Movie Maker* means to place one element after another within your project.

Task 11d

The purpose of the video is to advertise the spectacular bird species that can be found at Lake Naivasha in Kenya. All stills and video clips were filmed by GBRvideo at Lake Naivasha in Kenya in August 2019.

Create a new video clip saved in 1920 x 1080 format as **Task_11d.mp4**.

Place the title frames stored in **Task_11a.mp4** at the start, followed by **Naivasha1.mp4** then **Naivasha2.mp4**. Take an extract of the final frame of Naivasha2.mp4 and use this as a background image for the caption frames, which should appear for 10 seconds. Add caption frames to indicate that there are over 400 species of birds seen in the Great Rift Valley and that Lake Naivasha offers some of the best bird-viewing opportunities. Place, after the caption frames, the video clip **Fish_Eagle.mp4**. Complete the video with appropriate credits lasting 6 seconds.

To start this Task, we need to create a new project and splice together the files **Task_11a.mp4**, **Naivasha1.mp4** and then **Naivasha2.mp4**, which is achieved by dragging each file into the right portion of the screen using the methods from Task 11a. When these clips have been placed in the correct order it should look similar to this.

Your editor may show more intermediate frames similar to this.

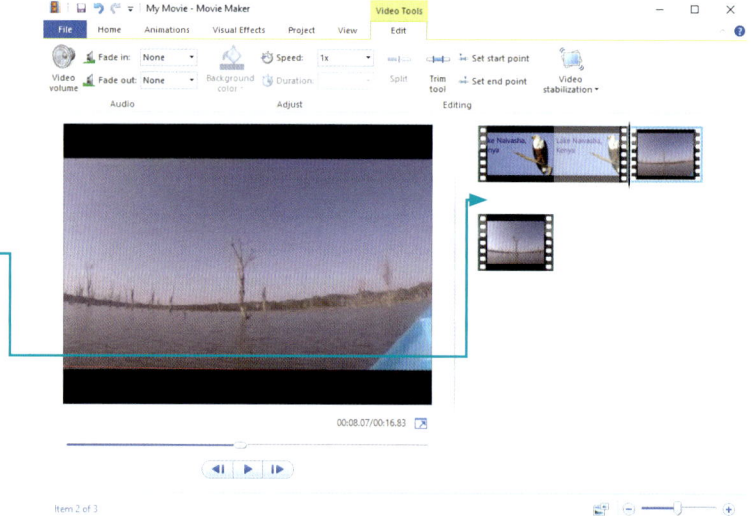

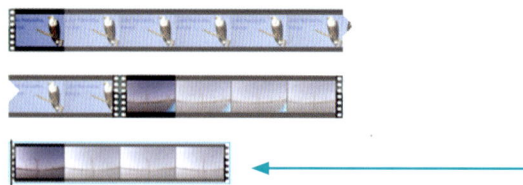

11.1.9 Extract a still image from a video clip

One reason for extracting a still image from a video clip is that a particular frame of a video might make a wonderful photograph to put in a photograph album. Another reason is that you might want to edit or add other graphics to the image using a graphics package before reinserting it back into the video.

Slide the timing bar so that it is placed at the end of the **Task11a.mp4** video like this.

Select the **Home** tab, then the **Snapshot** icon. Give the snapshot a meaningful filename, such as **Snapshot_n2end.png**. As this snapshot is saved, it is automatically added to the end of the project with a duration of 7 seconds. Click on the additional frames that represent this image to select it, then from the **Edit** tab, change the duration to **10** seconds.

11.1.10 Add a caption/subtitle

Captions/subtitles are needed for many reasons. The people watching the video may have a hearing impairment and so subtitles help them understand what is being said. It may be that they are watching the video in a public place where there is lots of background noise making it difficult to hear, such as a shopping mall. They could be in a quiet public place, such as a doctor's waiting room, where if they switched on the audio it could be disruptive, annoying other people. The viewers might not speak the same language as the people in the video and so subtitles in different languages could be provided. People tend to be more engaged with a video if the caption is reinforcing what the video is saying.

The text for the caption must give all of the information required by the Task, which is that 'there are over 400 species of birds seen in the Great Rift Valley and that Lake Naivasha offers some of the best bird-viewing opportunities'. All of this information must be included but do not copy it directly. Consider the information and how you would like to present it. This is a long piece of text, so perhaps two different sets of caption text can be used. Select caption text that is easy to read and appropriate for the audience, using short easy-to-read sentences without complex punctuation or words which may be difficult to read or understand, for example:

Lake Naivasha sits in the Great Rift Valley. It is home to over 400 species of birds.

Lake Naivasha offers some of the best bird-watching in the world.

To add a caption over the top of the still image, slide the timing bar so that it is at the start of the caption frames like this.

From the **Home** tab select the **Caption** icon. Replace the text 'Enter text here' with the first part of the caption. You are now in the **Format** tab. In the **Adjust** section, use the lower drop-down menu to set a duration of 5 seconds. Unless animation effects are required by the Task, do not use them. Check that the text is appropriately placed within the frames so that the background image is still clearly visible. Select a font style that is easily read by the audience and if no styles are specified, select an appropriate font for the content of the presentation and the audience, for example if it is for young children, a clear font (not cursive or with too many/long serifs) in a large easily read font size would be appropriate. For this caption, dragging the text box up the frame gives better contrast and is more easily readable. Move the timing bar to the end of this caption and add a second 5-second caption using the second example line above.

If subtitles are required, use caption text to create a subtitle. You can have text that appears to be a subtitle and other text that appears to be a caption on the screen at the same time. To do this change the font style/size/enhancements of the subtitle text so that it is clear that it is different from the caption text.

11.1.11 Create credits

Credits are used to inform viewers about the people involved in the creation of the video. They are a way of acknowledging the help, support and contributions made by people in the making of the video. Feature-length productions usually have both opening and closing credits. Opening credits are shown at the beginning and closing credits at the end. In short videos, all credits are placed at the end of the video. Their purpose is to inform viewers who was involved in the production of the video, for example filming, editing, sound, and so on, and usually include production details like filming location, date and so on.

For this video, the information has been embedded into the Task and must also include your details as you have edited the video. The credits need to include:

Filmed by: GBRvideo

Location: Lake Naivasha, Kenya

Date filmed: August 2019

Edited by: <Your name>

> **Advice**
> Ensure that your name, centre number and candidate number are included in the credits.

To add these credits to the video, place the timing bar at the end of the video and from the Home tab, in the Add section, click on the Credits icon. The tab will change to the Format tab. Replace the text [Enter name here] with the text from the list above including replacing the text <Your name> with your name. The credits will, as the default setting, scroll up the screen and be set on a black background like this, but other types of animation effect can be chosen from the Effects section. If an image or clip is also used behind the background credits, ensure that the credits are formatted and positioned so that they have good contrast and can be easily read. From the Format tab, in the Adjust section, use the drop-down menu to adjust the duration to 6 seconds.

Play the video clip to make sure it matches the specifications in the original question. Use the method described in Task 11b to save the clip with the filename **Task_11d.mp4** and a resolution of 1920 × 1080. Save the project as a .wlmp file.

> **Advice**
> If projects are to be edited at a later date, save the project in .wlmp format and then export in the format (such as MP4) that is instructed in the Task. Often you must submit files in the specified format. Files submitted as *Movie Maker* projects (.wlmp files) only contain links to the files, not the files themselves, so anyone viewing the file cannot see the content.

Activity 11b

The purpose of the video is to advertise the spectacular diving that can be found in Cozumel, Mexico. All stills and video clips were filmed by GBRvideo at Palancar Reef, Cozumel, Mexico, August 2004.

Create a new video clip saved in 960 × 720 format as **Activity_11b.mp4** using this storyboard.

0 secs: Title frames from **Activity_11a.mp4**.

6 secs: **Swim_through.mov** trimmed so that it starts at 15 seconds and the clip is 51 seconds long.

57 secs: Take an extract of the final frame of your clip from **Swim_through.mov** and use this as a background image for the caption frames, which should be placed for 8 seconds. Add caption frames to indicate Jacques Cousteau once called Cozumel 'the most spectacular diving site in the world'.

65 secs: **Shark.mov** trimmed so that only the first 6 seconds remain.

71 secs: Take an extract of the final frame of your clip from **Shark.mov** and use this as a background image for the credits, which should be placed for 10 seconds. Add appropriate credits to match the brief to this background.

Export your video in .mp4 format with the filename **Activity_11b.mp4**.

Task 11e

Open your video clip saved as **Task_11d.mp4**.

Add the soundtrack Task_11e_audio.mp3 to this video so that it starts after 5 seconds.

Save the video clip in 1920 x 1080 format as **Task_11e.mp4**.

Export this movie in the same format into .wmv, .mov and .avi file formats.

11.1.12 Add audio to a video clip

Adding audio may be necessary if the creator of a video wishes to get an emotional response from the viewer as well as explaining the details of the video sequence. A human voice used in narration can convey the emotion or mood of the video's contents. Music can be added, whether it be the strings of a violin suggesting sadness, the rousing sound of a song for a happy ending, or the calming effect required when observing sea life. Sound effects are needed when action needs emphasising, for example, explosions, speeding cars, and people walking, to name just a few.

Open the file **Task_11d.mp4** in *Movie Maker*. As the audio clip is to be added after 5 seconds, move the timing bar to exactly 5 seconds from the start. From the Home tab, select the Add music icon which opens a drop-down menu like this.

This icon is used to add any audio clip, including music or narration. Select the option for Add music at the current point....

Locate the folder containing the file **Task_11e_audio.mp3** and double-click on this file. This will now be included as part of your project and will show as a green ribbon like this.

Test the audio to ensure that it plays when the video is played.

Use the method described in Task 11b to save the clip with the filename **Task_11e.mp4** and a resolution of 1920 x 1080. Save the project in .wlmp format.

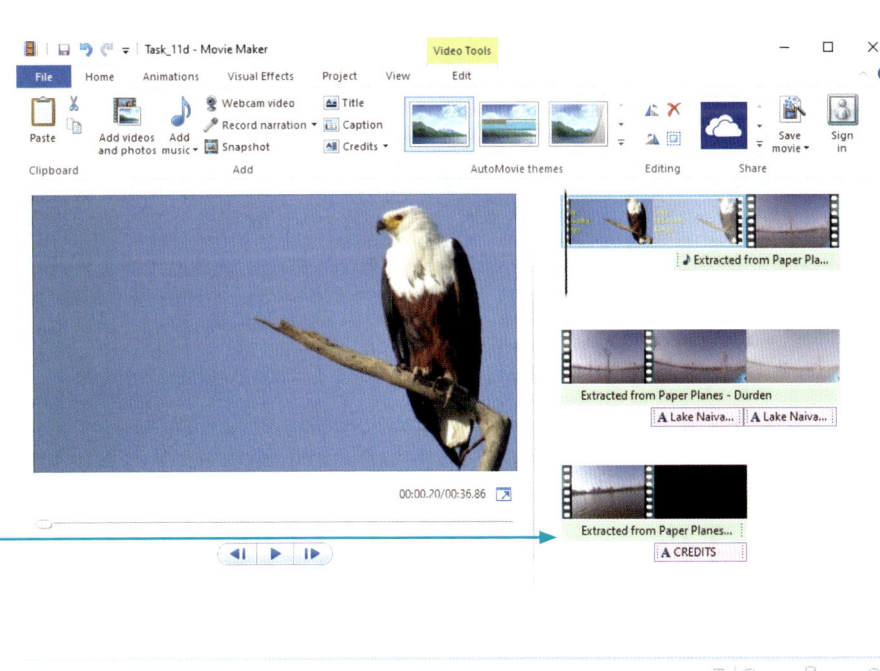

11.1.13 Export in different file formats

Not all devices and platforms are able to play videos in all formats, which means some videos are incompatible with particular devices or platforms. Videos may also need to be converted to reduce their file size to allow them to be attached to an email or to reduce storage space. The compression algorithm that particular formats use can be different and so the file format may need changing.

To export the movie into .wmv (Windows Media Video file) format, from the **File** tab, select **Save movie**, then **For high definition display**. Enter in the **File name:** box the text **Task_11e.wmv** and use the drop-down menu in the **Save as type:** box to select **Windows Media Video File (*.wmv)**. Click **Save**.

To change the movie into .mov and .avi formats requires the use of a video conversion program.

> **Advice**
>
> There are many programs for video conversion available, but make sure that you are familiar with the use of one that is downloaded to your computer, for when you are unable to access an online converter.

Prism® has been chosen for this Task, but there are many similar free-to-use products available. Open this application and select the **Add File(s)** icon. Double-click on the file **Task_11e.mp4** and the file is added to the **List of Files to Convert**. Select in the **Output Format:** box **.avi**, then click **Convert**.

To export this file into .mov format, return to the **Output Format:** drop-down list and change the setting to **.mov**, then click the **Convert** button.

> ### Activity 11c
>
> Open your video clip saved in **Activity 11b**.
>
> Add the soundtrack **Activity_11c_audio.mp3** to this video so that it starts after 3 seconds. Save the video clip in 960×720 format as **Activity_11c.mp4**.
>
> Export this movie in the same format into .wmv, .mov and .avi file formats.

11.1.14 Compress video to different resolutions

So far, we have created videos with high resolution in 16:9 and 4:3 format. We can use both *Movie Maker* and the video conversion software to save our projects in different resolutions for different media.

> ### Task 11f
>
> Open your video clip saved as **Task_11e.mp4**.
>
> Publish/save this movie with the filename **Task_11f** so that it can be viewed on YouTube, on an Android™ phone with medium resolution, and burnt on a DVD.

Open the project for Task 11e in *Movie Maker*. Publishing the movie to web-specific requirements for video-sharing services on the internet like YouTube can be completed by using the File tab, Publish movie and selecting YouTube. Select a resolution that will allow your movie to be uploaded to the video-sharing service. For this example, to create a smaller file size, we will choose a smaller image resolution, for example 1280 × 720. You will then need to sign into your account to use this service. This element of the process will not be assessed.

To save it in a format suitable for an Android phone, use the File tab, then Save movie and scroll down the drop-down list to the right to find and select the option for Android phone (medium) which will save the video in 960 × 720 format. Enter the filename Task_11f_android.mp4 then click Save.

To burn this video onto a DVD, we need a copy of Express Burn® software installed which runs alongside *Prism*. Open the video conversion software, in this case *Prism*, and use the method described in Task 11e to add the file Task_11e.mp4. Select File from the toolbar and Burn files as DVD Video Disk with Express Burn. Type in the text Lake Naivasha for the Disc label: then click on Burn. A DVD writer with a blank DVD is required for this to function.

11.1.15 Add fading effects: part 1

Sometimes it is necessary to indicate that a period of time has passed between two scenes in a video. This can be achieved by fading out the first scene and fading in to the next scene. It is sometimes used because the video creator wishes to ease the audience's emotions after a troubling scene. A more significant end or beginning requires a slower fade with more time spent on a blank/black frame.

If the creator wishes to show that only a few hours or minutes have passed between scenes, a fairly quick fade can be used. A longer fade can be used when the time interval between scenes is days/months/years.

Fading effects in *Movie Maker* can be created using transitions which are used to link two parts of a video together with a smooth change, rather than one frame moving to a very different frame and giving a very jerky effect as the video plays. They can also be created using pre-set fades too, and from black or white (as we will see later in this chapter).

Task 11g

Open the project saved in Task 11e. Add a 2-second transition between the clips Naivasha1 and Naivasha2. Save the video clip in 1920 × 1080 format as **Task_11g.mp4**.

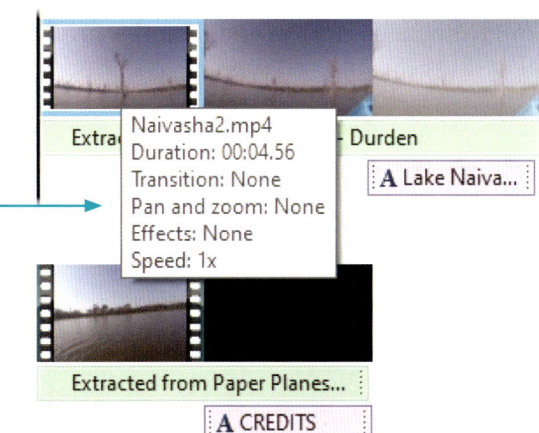

Advice

If you hover the mouse over any part of the right side of the *Movie Maker* window, the information about the element is shown like this.

Open the project saved in Task 11e in *Movie Maker*. To add the transition, click on the video Naivasha2.mp4 (between 12.4 and 16.8 seconds) to select it within the package.

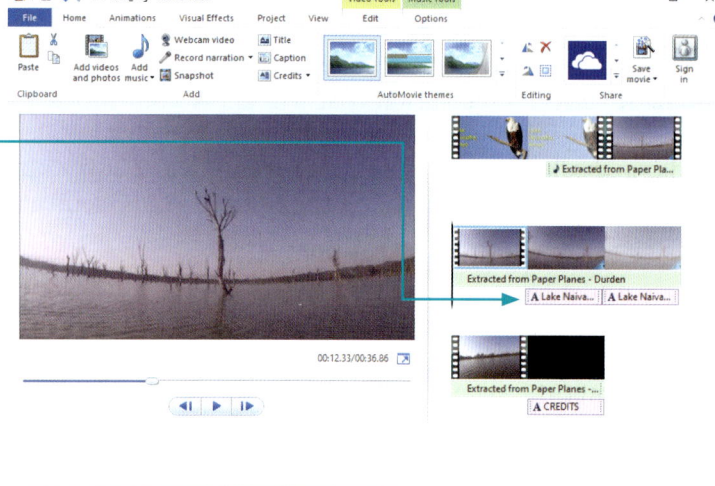

Select the **Animations** tab. In the **Transitions** section are a number of different transitions that can be applied, although only the first few transitions can be seen at the moment. There are many more different transitions listed in the scrolling menu, including dissolves, wipes, reveals and so on. Select the type of transition that would be appropriate; a very small dissolve for this transition has been chosen here.

The question requires a 2-second transition; the default for this transition is 1.5 seconds.

Use the drop-down menu for **Duration:** to change this value to **2.00**. Play the video and check that it gives a smooth transition between the two video clips. Use the method described in Task 11b to save the clip with the filename **Task_11g.mp4** and a resolution of 1920×1080. Save the project in .wlmp format.

Activity 11d

Open the project that you saved in Activity 11b.

Add 1-second transitions between the title frames and swim_through.mov and between the caption and shark.mov. Save the video clip in 960×720 format as **Activity_11d.mp4**.

11.1.16 Add pan and zoom effects

The pan and zoom effects are applied to a video or image to make it slowly zoom in on a focal point of an image or pan across an image. In *Movie Maker* there are three effects: pan only, which moves from one focal point to another; zoom in, which removes some or all of the outside of an image; and zoom out, which moves from a focal point to a wider angle view.

Task 11h

Create a new video and set the image ratio to 16:9. Place the video **balloon.mp4** at the start. Use pan and zoom to remove the hand and camera from the video clip. Take a snapshot of the final frame and zoom in on the balloons.

Save the video in 1920×1080 format as **Task_11h.mp4**.

Create a new *Movie Maker* project with a 16:9 aspect ratio. Drag the file **balloon.mp4** into the project. Watch the video clip to see where the hand and camera appear. Select the **Animations** tab. In the **Pan and zoom** section, use this button to show all pan and zoom options.

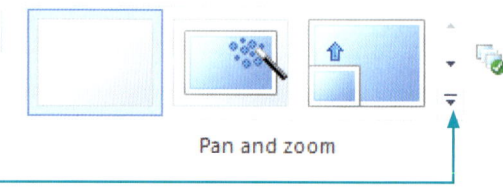

We need to zoom in to remove the left edge of the image. This means we need to select an option from the **Zoom in** section. Select the option for **Zoom in right** to remove the left edge of the clip. Several other zoom in options would also remove this. Move the timing bar to the end of the clip and select the **Home** tab followed by the **Snapshot** button to make a still image of the final frame of the zoomed video clip. Select this clip and again select the **Animations** tab, then in the **Pan and zoom** section, select the option for **Zoom in left**. Play the video again to see this effect. Use the method described in Task 11b to save the clip with the filename **Task_11h.mp4** and a resolution of 1920 × 1080. Save the project in .wlmp format.

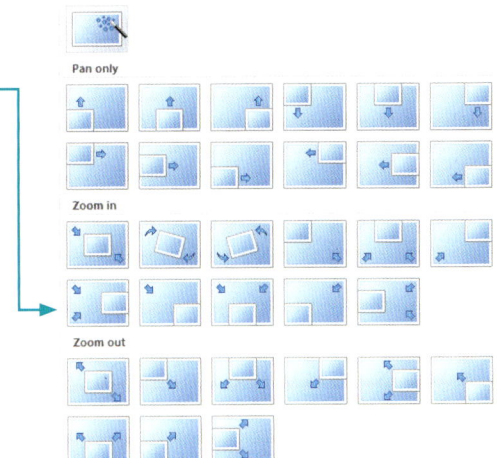

11.1.17 Use filters and colour correction

Movie Maker has limited facilities for colour filters and colour correction, but these are available from the **Visual Effects** tab which contains a range of different filters to apply to the image/clip. Different effects can be applied to the same still image or clip by cutting and splicing these into smaller segments and applying the effect to one segment. For example a 1-second portion of a video can be blurred by splitting the video clip into three and applying the blur filter to the 1-second segment. Multiple effects can be applied to the same clip using the **Multiple effects...** button at the bottom. The black and white effects can be used to give the film clip an old-fashioned look (using sepia) or by turning the video into a black and white version, or to make something look colder using the cyan tone.

> ### Task 11i
> Open the project saved in Task 11h. Give the savannah the appearance of being a cold place using a filter. Add an effect to ensure that the frames fade to black for the credits. Add credits to the clip on a black background and display this for 7 seconds. Save the video in 1920 x 1080 format as **Task_11i.mp4**.

Open the project saved in Task 11e in *Movie Maker*. Hold down <Ctrl> and press <A> to select all elements of the video. From the **Visual Effects** tab, in the **Effects** section for **Black and white**, select the option for **Cyan tone**.

Click on the still image and again from the **Effects** section select the drop-down menu, then the **Multiple effects** button. This opens the **Add or Remove Effects** window.

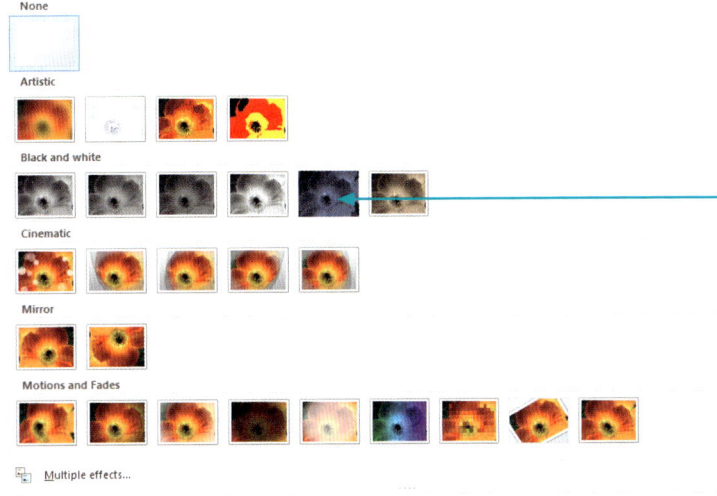

11.1.18 Add fading effects: part 2

From the **Available effects:** list, double-click on the option for **Fade out to black** to add this to the **Displayed effects:** list, then click Apply. Place the timing bar at the end of the snapshot and from the **Home** tab select the **Credits** button. Add your name to these credits. Use the method described in Task 11b to save the clip with the filename **Task_11i.mp4** and a resolution of 1920 × 1080. Save the project in .wlmp format. It now looks like this.

11.1.19 Video compression

There are two types of file compression used for both video and audio files. These are lossless compression and lossy compression. Lossless compression does not lose any data when the compression takes place. Lossy compression loses data.

There are many factors that affect the file size of a finished video. One is the number of pixels used for the product. The size of a Standard Definition file (SD) is often 720 × 480 pixels, which is 345 600 pixels per frame, whereas for a High Definition file (HD) it is 1920 × 1080 pixels, which is 2 073 600 pixels per frame, and a 4K Ultra High Definition file (UHD) is 3840 × 2160 pixels, which is 8 294 400 pixels per frame. The higher the definition of the file, the higher the number of pixels per frame required to store and stream it. A second factor is the length of the video. Generally, the shorter the video, the less storage space and streaming time are required. Another factor is the number of frames per second. Much of the video produced is set at 30 frames per second (fps) and if the number of frames per second is reduced from 30 fps to 15 fps the video will require half of the original storage capacity and time taken to stream it. Other factors can also affect the compression; these include whether progressive frames or interlaced frames are used, how frequently the key frames occur, and the audio sample rate and render quality, but these factors are beyond the scope of this book.

The video is stored in a container, like .mp4, .wmv, .mov, .avi and so on, and each of these holds the compressed data in a different form. The software used to display the video (it may be video-playing software, a browser or even software within a games console) must recognise the container holding the video and understand the **codec** used to encode the video within the container. The same codec must be used in the software to decode the video. If the codec is not present in the software (for example it has not been added to your browser as an add-in), then the video within the container will not play.

The word 'codec' is short for coder/decoder and is a program that knows how to handle video (and/or audio) when stored in a particular format. It is used to compress and decompress the data, often for/from data transmission.

Data might be compressed so that it can be shown on a phone with a small display, such as 426 × 240 pixels. If this compressed video was displayed on a UHD monitor or television, it would be very pixelated and almost unviewable. It is therefore critical that the correct container and compression method is selected when data is saved. Unless storage space and bandwidth for streaming are very limited, always select a lossless format.

11.2 Audio editing

> **Task 11j**
> Open the track **Task_11j_audio_1.mp3** in your audio editing package. Normalise this track to remove any DC offset. Add **Task_11j_audio_2.mp3** as a new track, clipped so that only the first 19.1 seconds remain, to start before the first clip. Trim the whole clip so that only the first 45 seconds remain.
>
> Add a 3-second fade-in to the clip and a 5-second fade-out. Change the clip from stereo into mono. Save the audio clip as **Task_11j.mp3**.

11.2.1 Import a new track

Open *Audacity*. There are two methods to open a file: the first is to open *File Explorer* in the folder showing the source files. Drag the file **Task_11j_audio_1.mp3** from *Explorer* into *Audacity*. This track has now been imported into the audio editor. The second method is to use the **File** tab, then **Open** before selecting this file and clicking ⟦Open⟧.

11.2.2 Normalise an audio clip including removing any DC offset

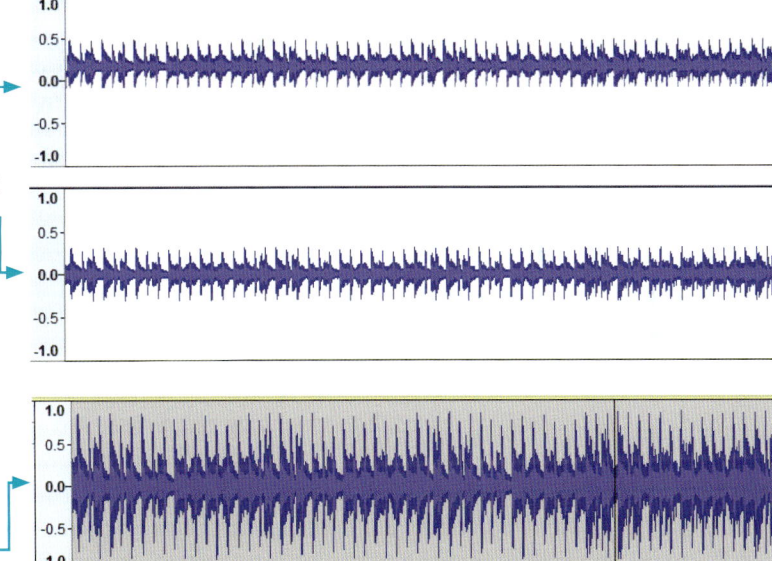

If an audio file has been converted from analogue to digital values using a faulty audio device, it may appear to have an offset waveform like this ⎯⎯ rather than like this, which is centred along the 0 line. ⎯⎯

To remove the **DC offset** (which can cause clicks, distortion and loss of audio volume), select the **Effect** tab, then from the drop-down menu select **Normalize…**. The **Normalize** window will open; make sure the **Remove DC offset** box is ticked, then click ⟦OK⟧. In this case it will change the waveforms of both left and right audio channels to this. ⎯⎯

This also has the effect of amplifying both channels.

11.2.3 Add a track to an audio clip

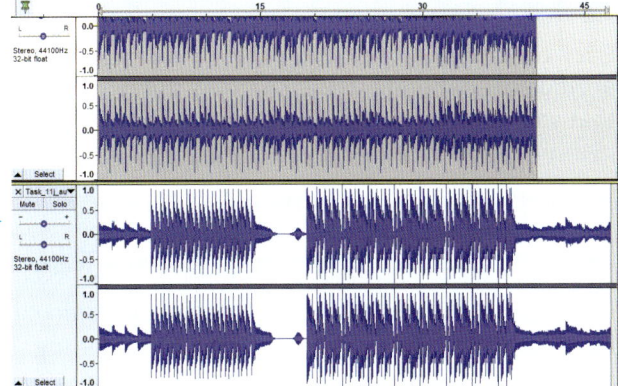

To add a new track to an existing clip, either drag the new clip into the editor or from the **File** tab select **Import**, then from the drop-down menu select **Audio**. Select the file **Task_11j_audio_2.mp3**, then click ⟦Open⟧. Both audio tracks are sitting alongside each other like this. ⎯⎯

11.2.4 Trim to remove unwanted material

Trimming an audio clip is necessary at times to remove unwanted material such as a sudden noise accidentally recorded. It is also needed if the length of a track has to be reduced. This could be due to only a certain length being required or to save storage space on a medium. Having added the second audio clip to the editor we now need to remove the end of the clip so that only 19.1 seconds remain. Click on the lower audio file to make sure it is selected. Set the timer using the Audio Position counter to 19.100 seconds.

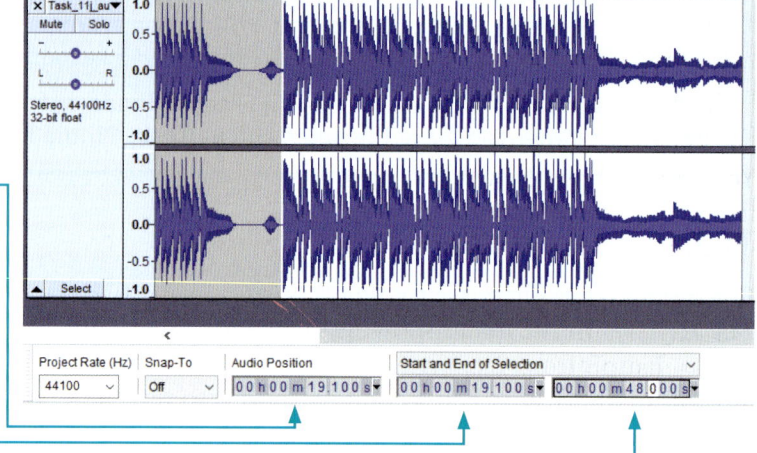

This is much more precise than using the timing bar. Set the Start of Selection counter to the same value.

Set the End of Selection counter to just after this audio clip finishes; I have chosen 48 seconds.

To remove the end of these tracks, click on the Cut icon.

11.2.5 Splice/join together

There are a number of reasons why we might want to splice/join together two audio clips. One could be to make two recordings flow uninterrupted into one another. Another could be that we want to combine the best parts from several recorded performances of the same act.

As the audio clip from the previous section will be the first clip in our completed clip, we need to change the order of the clips. Move to the lower clip and select the button for Open menu…, like this.

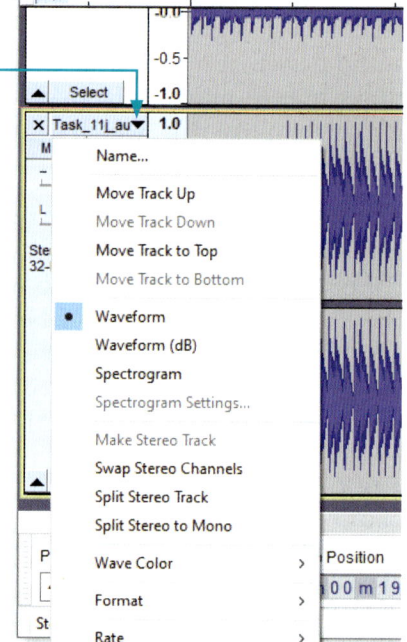

From the drop-down menu select Move Track Up. The clips are now reordered.

To splice the clips together, select all tracks using the Select tab followed by All (or <CtrlA>). Move to the Tracks tab to open the drop-down menu. Select Align Tracks to open another menu, then Align End to End. Select the Tracks tab again followed by Mix, then from the sub-menu select Mix and Render. This will merge the two clips into a single stereo clip.

Use the methods shown in Section 11.2.4 to trim the clip so that only the first 45 seconds remain.

Advice

Each time you perform any action in the audio editing section, listen on your headphones to the changes that you have made, to ensure that you have completed each task without distortion.

11.2.6 Fade in and fade out

Fade-in and fade-out are two very necessary features of audio editing.

We may need to fade in and/or fade out an audio clip. This could be because a piece of music does not contain an obvious ending. The beginning/end of an audio clip may not be smooth so a fade-in/fade-out is necessary. There may need to be a clear section of silence before the audio starts. A fade-in may be needed if it is necessary to soften the sound of the drum and/or percussion instruments.

Use the **Start of Selection** and **End of Selection** counters to highlight the first 3 seconds of the clip from the previous section to create the fade-in. Select the **Effect** tab, then from the drop-down menu select **Fade In**. The waveform at the start of the clip changes from this to this.

You can see how the waveforms have been compressed (making the audio quieter) in the second version.

Use the **Start of Selection** and **End of Selection** counters to highlight the last 5 seconds of the clip to create the fade-out. Select the **Effect** tab, then from the drop-down menu select **Fade Out**. The waveform at the end of the clip changes from this to this.

11.2.7 Change from stereo to mono

This compresses all the audio data held in the two tracks (stereophonic with left and right channels) within the clip into a single monophonic track. Select the clip, then from the **Tracks** tab drop-down menu select **Mix** and from the sub-menu select **Mix Stereo Down to Mono**. The two tracks have now been mixed down into a single mono track.

11.2.8 Amplify

Because one track was originally quieter than the other, mixing down the two tracks leaves an imbalance in the final volumes. The section from 19.1 seconds to the end of the track is quieter than the first, as can be seen by the much shorter waveforms, so this section is highlighted and adjusted from the **Effect** tab using **Amplify**. *Audacity* calculates the amount of amplification required (to match the loudest portion of the clip) so from the **Amplify** window, make sure that the **Allow clipping** box is unchecked, then click [OK]. This amplify effect will be used later in the chapter to add voiceovers to a clip.

Save the project using the **File** tab, then **Save Project**, then from the sub-menu select **Save Project As** which gives a warning message that this will not generate a playable audio file but an editable project that can only be opened in *Audacity*.

> **Advice**
> Ensure you use the correct formats when you export audio files.

11.2.9 Export in .mp3 format

You must export this clip into .mp3 format so that it can be played on a range of devices. From the **File** tab select **Export**, then from the sub-menu select **Export as MP3**. This opens the **Export Audio** window.

Enter the **File name:**, in this case **Task_11j.mp3**, and ensure that the **Save as type:** box contains **MP3 Files**. In the **Format Options** section, tick the check box for **Force export to mono**.

Do not edit the other sections in the **Format Options**. We will study some of these later in the chapter. Click [Save]. Enter in the **Edit Metadata Tags** window details of the originator(s) and track(s), such as title, year and so on; original works are being used to create this clip, but both artists have allowed their work to be used for commercial purposes as long as credit is given to them. Do not use other artists' material unless it is copyright-free or you have permission to use it from the originator.

Click [OK] to save this work.

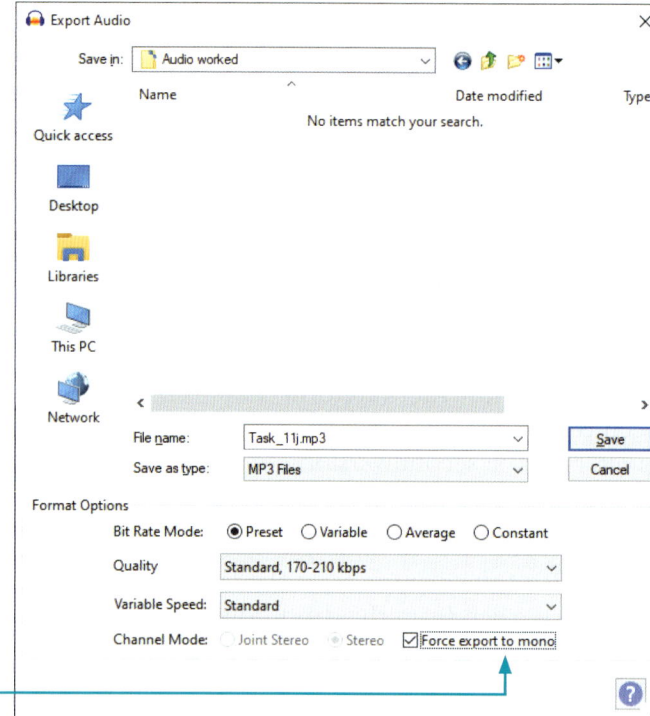

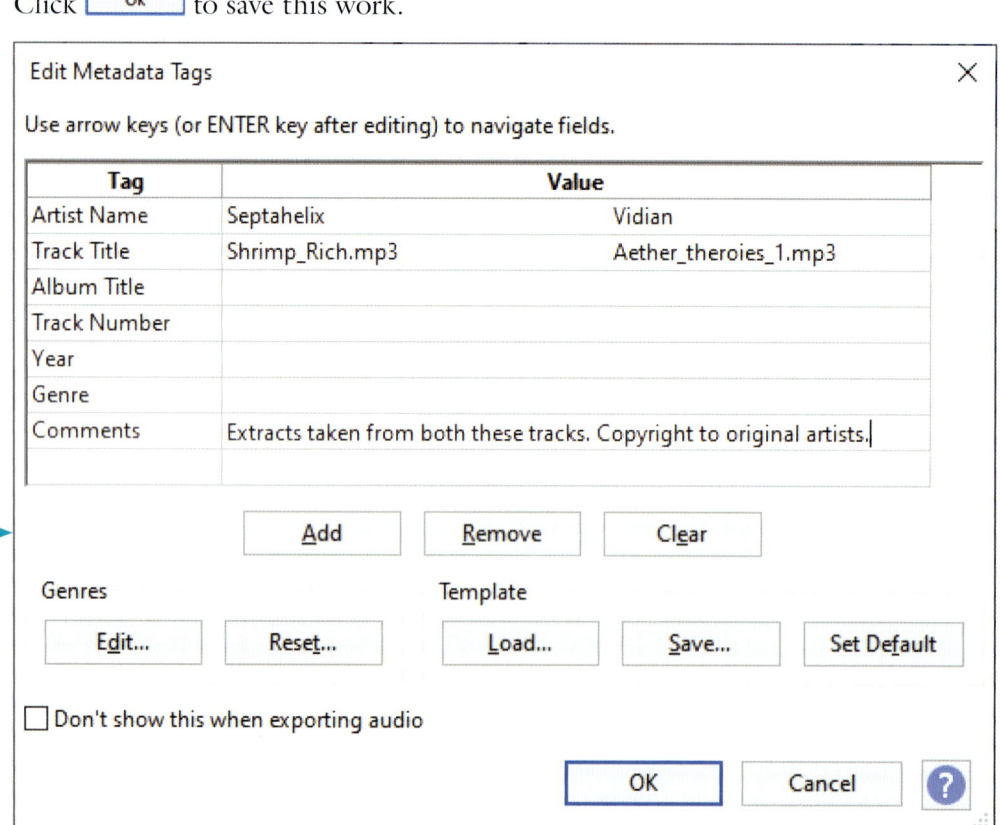

Activity 11e

Open the track **Activity_11e_audio_1.mp3** in your audio editing package. Clip this track so that only the section between 1 minute 17.2 seconds and the end of the clip remains. Normalise this track to remove any DC offset. Add **Activity_11e_audio_2.mp3** as a new track, clipped so that only the first 30.6 seconds remain, to start before the existing clip. Save the audio clip as **Activity_11e.mp3**.

Task 11k

Open the track **Task_11j.mp3** in your audio editing package. Change the pitch of the first 19 seconds of the clip from the key of C into the key of F. Speed up the whole clip so that it plays at 1.4 times the speed. Save the audio clip as **Task_11k.mp3**.

11.2.10 Change the pitch

Open the file **Task_11j.mp3** in *Audacity*. This clip is made up of two separate clips which were created in different keys, so we are going to change the pitch of one part of the clip from the key of C into the key of F to match the other part of the clip. Highlight the first 19 seconds of the clip. From the **Effect** tab select **Change Pitch…**. This opens the **Change Pitch** window. In this instance, *Audacity* has suggested that we make the changes that we want to make to satisfy the Task.

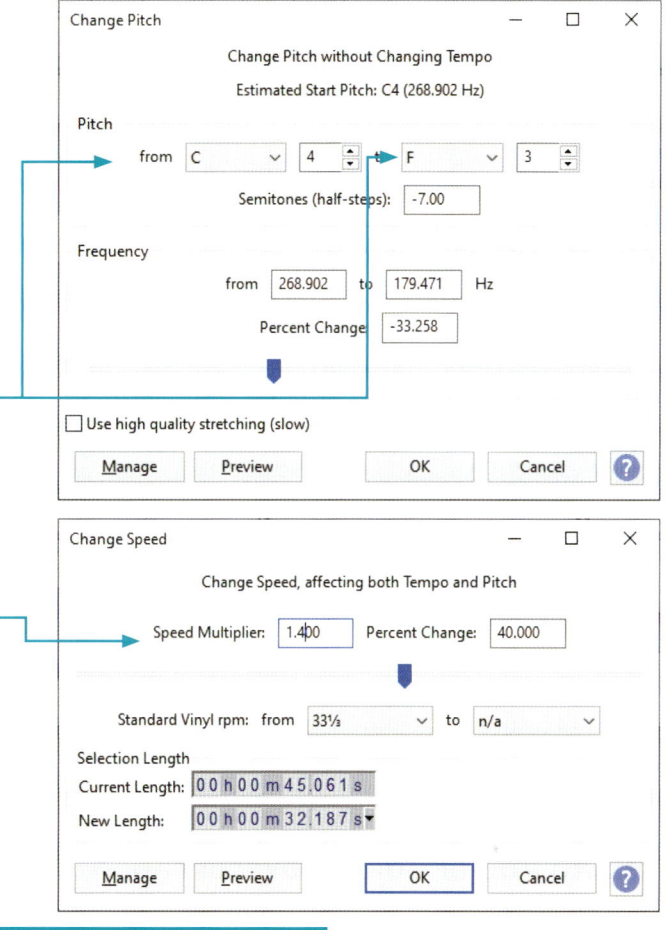

11.2.11 Alter the speed

Select all of the audio clip using **<CtrlA>**. From the **Effect** tab, select **Change Speed…**. This opens the **Change Speed** window. Set the **Speed Multiplier** box to 1.400.

Click **OK** to change the clip speed. The length of the audio clip changes from about 45 seconds to about 32 seconds. Save the *Audacity* project file and export the audio clip as **Task_11k.mp3**.

Activity 11f

Open the track **Activity_11e.mp3** in your audio editing package. Change the pitch of the first 30.6 seconds of the clip from the key of C sharp/D flat into the key of F sharp/G flat. Slow down the whole clip so that it plays at 0.9 times the speed. Save the audio clip as **Activity_11f.mp3**.

11.2.12 Apply noise reduction

Noise reduction helps to reduce background sounds that are constantly present like hiss, hum, buzzing or fan noise, but cannot remove irregular background noise such as clicks, noise from traffic or from an audience.

Task 11l

Open the track **Voiceover_1.mp3** in your audio editing package. Remove the background noise from this clip. Trim the clip so that it is 4 seconds long.

Save the audio clip as **Task_11l.mp3**.

Open the file **Voiceover_1.mp3** in *Audacity*. To reduce the background noise in the clip, we must select an area of the clip that contains only the background noise, for example between 3.3 seconds and 8.09 seconds like this.

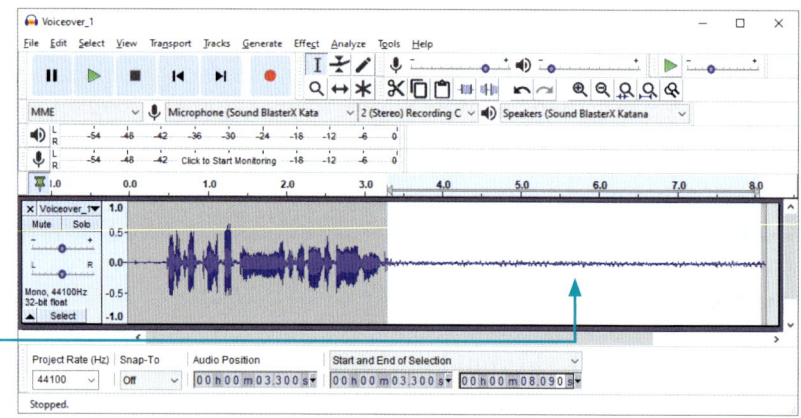

Then from the Effect tab, select Noise Reduction…. The Noise Reduction window opens. In the Step 1 section, click on Get Noise Profile. Select all of the audio file with <CtrlA>, then from the Effect tab, select Noise Reduction… again. In the Step 2 section, there are settings that you can change, but the default settings are almost always good enough for cleaning up the types of background noise we mentioned earlier (like hiss, hum, buzzing or fan noise). These settings have probably been set to around a 12-decibel reduction with a sensitivity of 6 and 3 bands of frequency smoothing like this.

Click OK to include the noise reduction. Trim the clip so that it is 4 seconds long, save the project and export the clip as **Task_11l.mp3**.

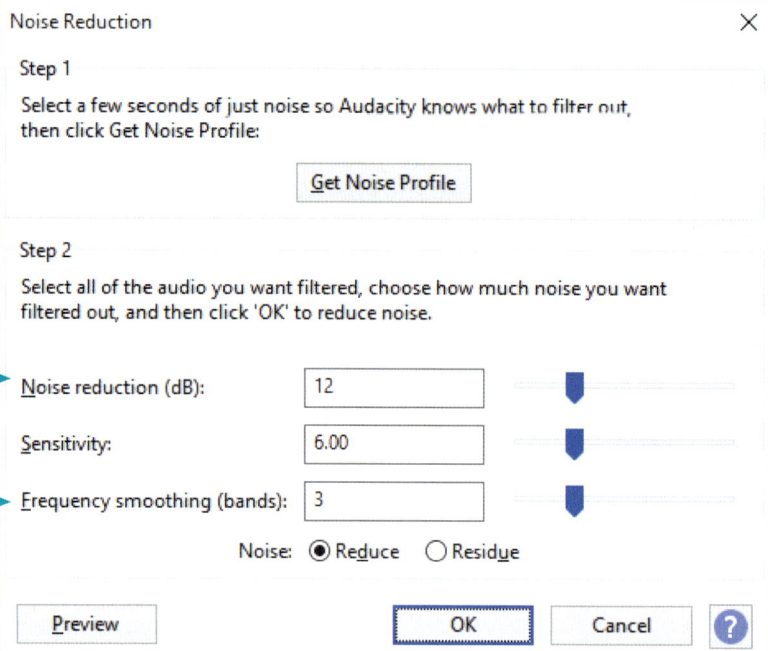

11.2.13 Add or adjust reverberation

Reverb is an effect whereby the sound produced by a musical instrument or **amplifier** is made to echo slightly (reverberate). The effect sounds like an echo that you would get when speaking or playing an instrument in a cavern or an empty room. By altering the size of the cavern or room, the amount of echo changes. Reverb can be used to give extra 'depth' to a voice.

Task 11m

Open the track **Voiceover_2.mp3** in your audio editing package. Add a small amount of reverb to the track between 4 and 6 seconds. Add increased treble to the track between 8 and 11.8 seconds. Add increased bass to the track from 11.8 to 15 seconds. Save the audio clip as **Task_11m.mp3**.

Open the file **Voiceover_2.mp3** in *Audacity*. To add reverb to only the part of the track between 4 and 6 seconds, we must highlight only that section of the track. From the **Effect** tab, select **Reverb...** to open the **Reverb** window which appears like this.

Click [Preview] to listen to the results of the reverb using your headphones.

To edit the reverb so that only a small amount is added (to make the voice sound slightly richer/deeper) but with not as much echo, you can change the settings for the **Room Size** to 60% and the **Reverberance** to 40% (they may already be set to this). Listen again using **Preview**. You can edit each setting a small amount until you get the effect that you require.

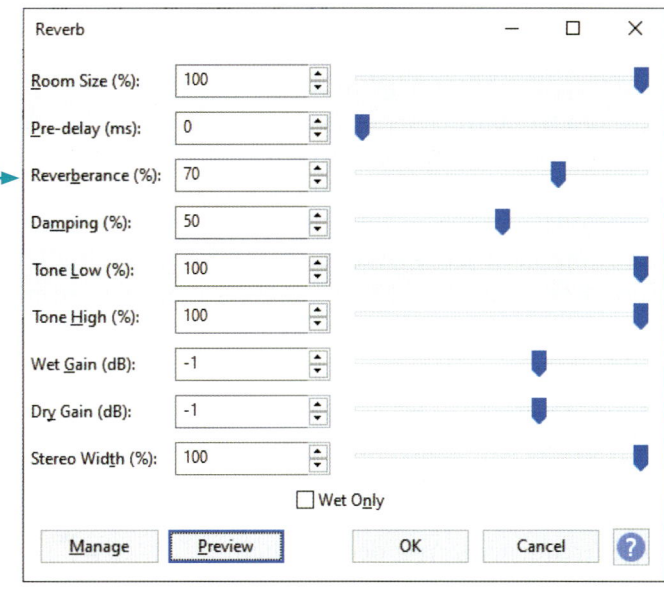

11.2.14 Apply equalisation

Equalisation, often referred to as EQ, means boosting or attenuating (reducing) the levels of different frequencies in a signal. In its simplest form, it is a process whereby you can boost the bass (low frequencies) or treble (high frequencies) in an audio clip, but the sounds are equalised to get exactly the overall effect that is required. Equalisation can also be used to place filters to reduce certain noise frequencies (either high or low frequencies) on audio clips.

To add increased treble to the track between 8 and 11.8 seconds, first select this section of the audio clip. Versions of *Audacity* from 2.3.3 onwards do not have an equalisation section, but this can be completed using the **Effect** tab, then **Graphic EQ...**. You are presented with a series of sliders, the bass (low) frequencies are on the left and the treble (high frequencies) are on the right. We can increase all the treble frequencies by 10 decibels and for the upper mid-range frequencies we can show a gradual increase like this.

Click [OK] to set this equalisation. Listen to the clip and you can adjust the volume of each frequency range to make the voice sound as you want it.

To increase the bass to the track from 11.8 to 15 seconds, first select this section of the audio clip, then move the sliders to set the EQ like this.

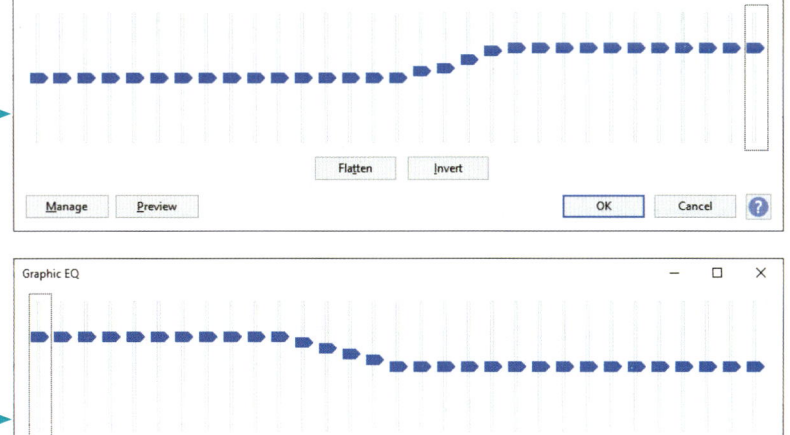

Again, listen to the clip and you can adjust the volume of each frequency range to make the voice sound as you want it. Save the project and export the clip as **Task_11m.mp3**.

> ### Activity 11g
> Open the track **Blue_Sky_Blues.mp3** in your audio editing package.
>
> Add lots of reverb to the track between 20 and 30 seconds. Add increased bass to the track from 40 to 50 seconds. Add increased treble to the track between 1 minute and 1 minute 10 seconds. Save the audio clip as **Activity_11g.mp3**.

11.2.15 Apply a high-pass filter

A high-pass filter is used in sound systems to allow high frequencies to pass through while filtering or cutting low frequencies. A physical high-pass filter (device) was originally connected to smaller speakers (tweeters) to remove the bass elements being sent to these speakers. High-pass filters are also included as optional settings in some top-end microphones when they are used for recording in offices or warehouses. They can filter out the low frequency noises from devices like fans or air-conditioning units. In terms of editing existing audio, we can use software to replace the hardware filters when dealing with audio clips.

Always apply the noise reduction filter before applying a high-pass filter. In *Audacity* the high-pass filter attenuates (reduces the effect of) the lower frequencies. You decide where the attenuation starts and the number of decibels per octave used for that attenuation. It is similar to setting a graphic equaliser like this.

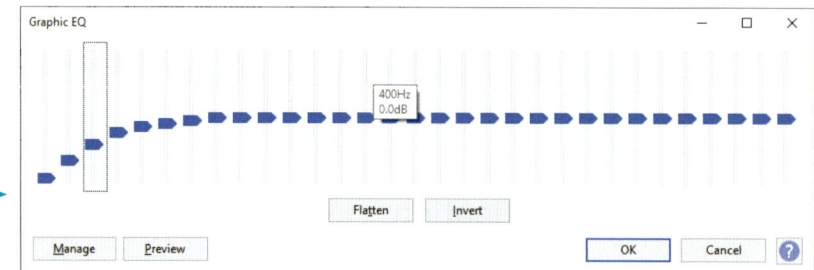

> ### Task 11n
>
> Open the track **Task_11n_audio.mp3** in your audio editing package. Apply a high-pass filter to the clip between 10 and 20 seconds. Save the audio clip as **Task_11n_high.mp3**.
>
> Reopen the track **Task_11n_audio.mp3** in your audio editing package. Apply a low-pass filter to the clip between 10 and 20 seconds. Save the audio clip as **Task_11n_low.mp3**.

Open the file **Task_11n_audio.mp3** in *Audacity*. Select an area of background noise from 40 to 50 seconds (even though the sound in this part of the clip may be barely audible) and from the **Effect** tab select **Noise Reduction...** then click `Get Noise Profile`. Select all of the audio file with **<CtrlA>**, then from the **Effect** tab, select **Noise Reduction...** again in the **Step 2** section and click `OK` to set the noise reduction.

Highlight the clip between 10 and 20 seconds and from the **Effect** tab select **High-Pass Filter...**, then click `OK`. You can see from the waveform and hear on this clip that this has a dramatic effect on the audio clip, because much of the music was previously low-frequency bass. Save the project and export the clip as **Task11n_high.mp3**.

11.2.16 Apply a low-pass filter

We do not need to reapply noise reduction when using a low-pass filter, because these functions remove some of the same frequencies. Start a new project and reopen the file **Task_11n_audio.mp3** in *Audacity*. Highlight the clip between 10 and 20 seconds and from the **Effect** tab select **Low-Pass Filter...** and click `OK`. You can see from the waveform and hear on this clip that this has had a lesser effect than the high-pass filter, because there are fewer high-frequency notes in this clip, although the audible effect is still noticeable. Save the project and export the clip as **Task11n_low.mp3**.

> ### Activity 11h
> Open the file you saved as **Activity_11g.mp3** in your audio editing package. Apply a high-pass filter to the clip between 1 minute 20 seconds and 1 minute 30 seconds and apply a low-pass filter to the clip between 1 minute 40 seconds and 1 minute and 50 seconds. Save the audio clip as **Activity_11h.mp3**.

11.2.17 Apply echo and decay

Echo copies the same sound and replays it after a given delay time, usually with fewer decibels than the original sound. The process of losing decibels each time the sound is repeated is called 'decay'. Creative use of echo can make a single sound/note into much more.

> ### Task 11o
> Open the track **Task_11o_audio.mp3** in your audio editing package. Apply echo with a 0.5 second time delay and a decay factor of 0.8 to the clip between 10 and 20 seconds. Save the audio clip as **Task_11o.mp3**.

Open the file **Task_11o_audio.mp3** in *Audacity*. Highlight the clip between 10 and 20 seconds. From the **Effect** tab, select **Echo…** to open the **Echo** window. Set the **Delay time:** at **0.5** seconds and the **Decay factor:** at **0.8** like this.

Click **OK**. You can see from the waveform and hear on this clip that each echo has added to the original sound in this section, like this.

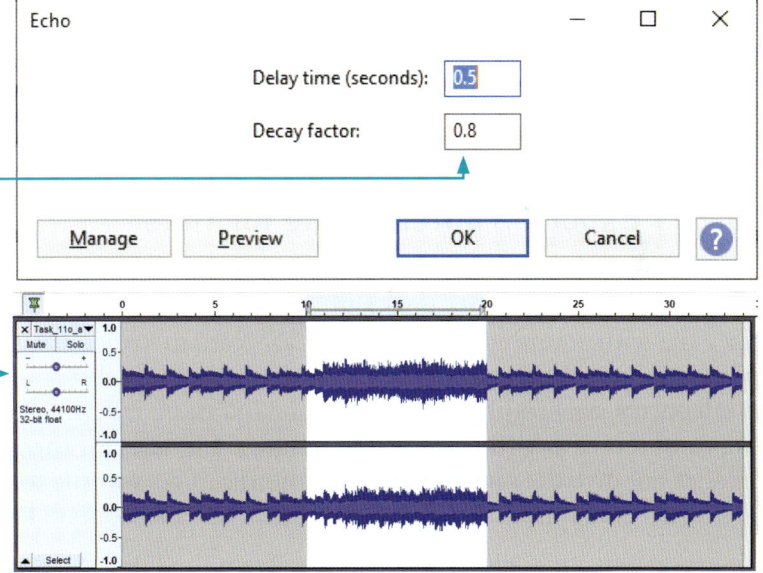

> ### Task 11p
> Open the track **Task_11p_audio.mp3** in your audio editing package. Add the track **Voiceover_3.mp3** so that it starts at 10 seconds and is clearly audible. Save the audio clip as **Task_11p.mp3**.
>
> Export this clip into **.wav**, **.m4a**, and **.aac** formats.
>
> Export this clip as **Task11p_Blu-ray.mp3** for use in a *Blu-ray Disc*™ player.

11.2.18 Overdub to include a voiceover

Overdubbing is usually carried out by adding extra instruments to a music track to give it a richer, fuller sound. It means putting other tracks on to existing tracks. It is often needed when a voiceover is needed to explain the contents of the existing audio track.

Open the file **Task_11p_audio.mp3** in *Audacity*. Add the second clip **Voiceover3.mp3** so that both clips are present. Click on the track **Voiceover_3** and move the timer bar to exactly **11** seconds. From the Tracks tab, select Align Tracks, then from the sub-menu select Start to Cursor/Selection Start. Although the voiceover can be heard, it would be better to reduce the volume (number of decibels) of the instrumental clip during this period. Highlight the instrumental (track without the voiceover) track between 10 seconds and 21 seconds. From the Effect tab, select Amplify. In the Amplify window, change the Amplification (dB): setting to -6 (which will reduce the volume of this section by 6 decibels) like this.

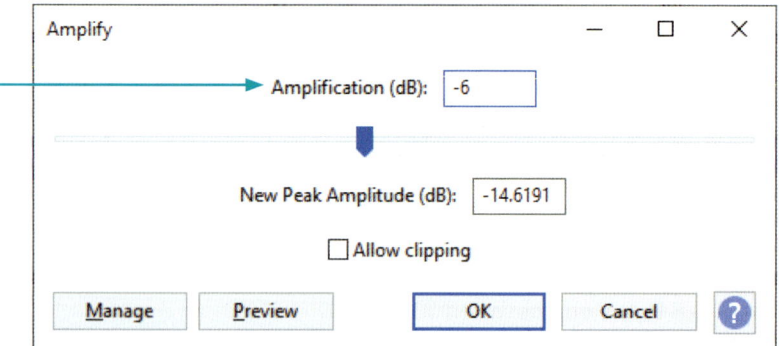

Click [OK] to reduce the volume of this section of the track. To mix the tracks together, use <CtrlA> to select all tracks. Move to the Tracks tab to open the drop-down menu. Select Mix, then from the sub-menu select Mix and Render to New Track. Save your project. Now remove the original instrumental and voiceover clips from the project and save again with a new filename. Export the clip as **Task_11p.mp3**.

11.2.19 Export in different file formats

We have already exported this clip into .mp3 format (which is the most commonly used digital format). Both .mp3 and .m4a (which is an unprotected file extension for a file with advanced audio coding [.aac]) are in lossy format, where sound quality is not always as good as the original recording. Both .aac and .m4a (which is effectively .aac) coding is a more recent and more efficient coding format than .mp3. In most cases (though not always), the file sizes for these formats are smaller but give the same quality of audio file. The .wav file format is a container that often contains lossless files, although it can also contain lossy data.

To export this audio file into .wav format, from the File tab, select Export, then from the sub-menu select Export as WAV, which will allow you to save the file with different sampling rates. These can be found in the Save as type: box, like this.

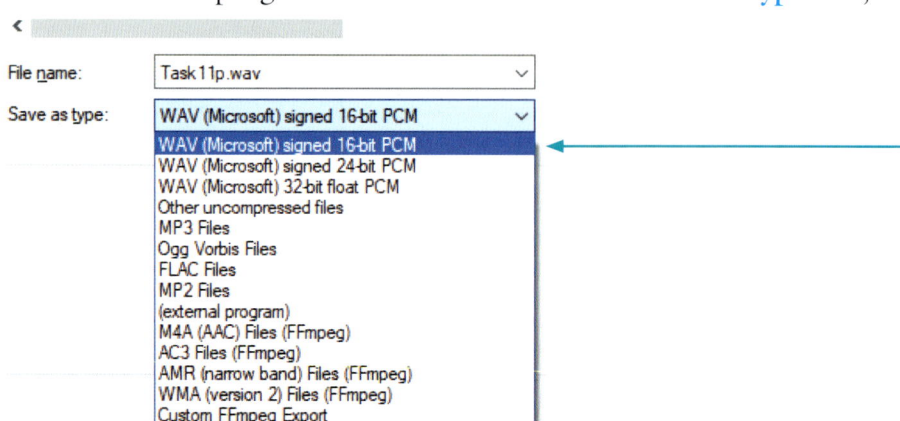

We will export this into .m4a and .aac formats later in the chapter. The reasons for using different formats are the same as for videos.

11.2.20 Sampling rate and resolution

Sampling rate

Digital audio is a series of discrete bursts called samples. Because the frequency of these samples can be very fast it makes the sound seem continuous. A sample rate is the number of samples of audio carried out within a given time, usually within 1 second. It can be measured in hertz (Hz) or kilohertz (kHz). The higher the sample rate, the more accurately the original sound can be represented and the better the recording sounds. The most common sample rate is 44 100 Hz (44.1 kHz), which is the default in *Audacity* and for audio CDs. The sample rate for DVDs is 48 000 Hz and for *Blu-ray Discs* it is 96 000 Hz.

Some lower sample rates such as 22 050 Hz (used for AM radio and older, poor-quality .mp3 recordings for floppy disk and original MP3 players) and 11 025 Hz are not really suitable for music, because a great deal of sound quality, particularly high frequency sound, is lost. Similarly, 32 000 Hz is suitable for speech but not really for music, and 8 000 Hz is the lowest sample rate and is only used for VoIP.

The human ear is limited in regard to the frequencies it can actually hear and, as you get older, the range of frequencies that you can hear decreases. The human ear has a hearing range of around 20 000 Hz (20 kHz), meaning that for high-quality audio, the sample rate should not be below 40 000 Hz (40 kHz) samples per second. If sampling rates much higher than this are used, most people cannot hear the difference. Each sample occupies a certain amount of storage space so the lower the sampling rate, the smaller the file size is and consequently less storage space is required. A higher sampling rate results in more storage space being required and a greater file size.

Sampling resolution

This is the number of bits per sound sample. The higher the sampling resolution, the more accurately the wave form of the sound will be converted from analogue to digital. The higher the sampling resolution, the greater the file size. The earliest video games used 8-bit resolution. This means each sound sample takes one of 256 different values, which is not really enough to accurately represent music audio. The standard for compact disc audio and many sound cards is 16-bit resolution, but modern-day digital audio is normally found in 16-bit or 24-bit resolutions.

The frequency for each sample is stored in a number of bits, called sampling resolution or bit depth. Each additional bit doubles the number of values that can be stored. If data is stored in 16-bit format, each sampled data item can have 65 536 possible values; 24-bit audio stores a possible 16 777 216 values. These values are all per channel and for stereo audio two channels are needed which doubles the sampling and memory requirements. New 32-bit standards are also available, but many experts argue that the difference between 24-bit and 32-bit is beyond the capacity of human hearing. Data is compressed when the audio is exported.

11.2.21 Compress to different sample rates to suit different media

As WAV is an uncompressed format, only the bit rate needs to be selected in the **Save as type:** box.

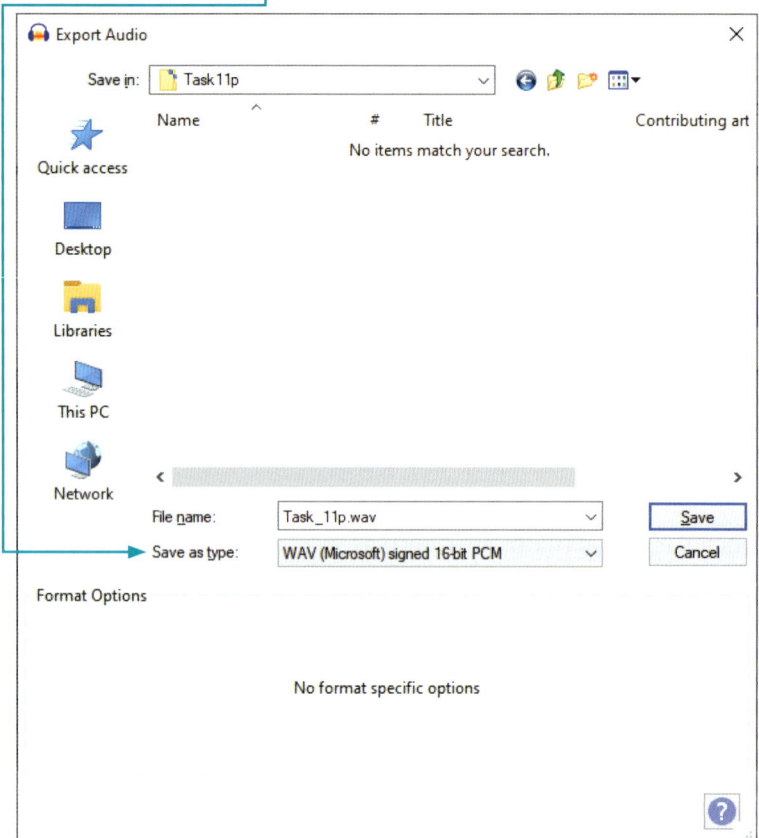

If a file size is going to be too large to stream/upload/download/store, then 16-bit would be the most suitable, although some audio quality is lost. It is a balance between quality and file size and the needs of the audience (which have not been specified in the question). To complete the saving for Task 11p, select the option for 24-bit then click **Save**.

To complete Task 11p we must export this audio file into .m4a format. From the **File** tab, select **Export** then from the sub-menu select **Export Audio....** From the **Export Audio** window select, in the **Save as type:** box, the option for **M4A (AAC) Files (FFmpeg)**. You may need to download an additional dll file in order to do this.

To export this audio file into .mp3 format for use in a *Blu-ray Disc* player, we need a sample rate of 96 000 Hz (or 96 kHz). From the **File** tab select **Export**, then from the sub-menu select **Export as MP3**. Enter the filename **Task_11p_Blu-ray**. From the **Export Audio** window select, in the **Bit Rate Mode:**, the radio button for **Constant** and using the drop-down menu for **Quality** select **96 kbps** like this.

Click **Save**.

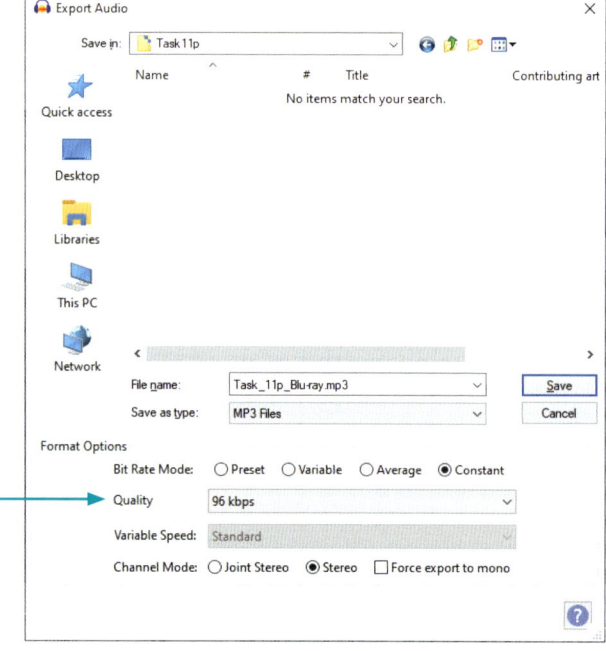

> **Activity 11i**
>
> Open the track **11i.wav** in your audio editing package. Clip this track so that it finishes after 33 seconds. Make a second copy of this clipped track. Then on this clipped track:
> » amplify it by 10 decibels
> » set a 3-second fade-in after 0 seconds
> » set a 3-second fade-in after 24 seconds
> » set a 3-second fade-out after 6 seconds
> » set a 3-second fade-out at the end
> » remove the audio between 9 and 24 seconds.
>
> Add the track **Voiceover_4.mp3** so that it starts at 10 seconds and is clearly audible. Apply noise reduction to this track. Save your project. Mix and render the tracks into a single stereo clip and export this clip into .wav and .mp4a formats. Export this clip as **Activity11i_DVD.mp3** for use in a DVD player.

11.2.22 Audio compression

As we have already seen earlier in the chapter, sample rate and sampling resolution are two key elements in determining file size. Some audio compression, called dynamic compression, works by reducing a signal's dynamic range, this is the range between the loudest and quietest parts of an audio signal. Compression works by attenuating (quietening) the loudest parts of the signal and then boosting the results so that all the quieter parts become slightly louder and the loudest parts slightly quieter. The waveform of a compressed signal appears to be more compressed than the original, although this gives, argue some audio engineers, better sound quality.

As stated earlier in this chapter, the two main types of compression are lossy and lossless, although there are many types of these compressions which go beyond the scope of this book, like multiband compression (sometimes called limiting), lookahead compression, brickwall limiting, sidechain compression (sometimes called ducking), mid-slide compression and parallel compression. Some of these compression methods are used in live mixing of music during concerts, as well as audio files.

Lossless compression does not reduce the quality of the audio file at all. Data is not lost, so an audio file can be recreated exactly as it was when originally created. There are a number of algorithms used to perform this and most look for repeating patterns of data.

Lossy compression removes some of the audio file's data and so reduces the quality of the audio file, but does create smaller file sizes than lossless compression. Once this data has been removed from the files it can never be replaced. Some of the sounds in the low and high frequencies within a track are outside the range of human hearing. For this reason, some lossy compression software will remove these frequencies.

Most audio files are stored and streamed in containers. Even lossless formats, like .wav files, are containers that store the audio data, sample rate, sampling resolution (bit rate) and track numbers. Many lossy containers contain the audio files and meta tags, which indicate which codec or codecs can be used to decode the audio as well as the name of the artist, track, track number and so on.

Practice questions

You have been supplied with the following source files:
» EQ11-Bali1.mp4
» EQ11-Bali2.mp4

- » EQ11-Bali3.mp4
- » EQ11-Bamboo.mp3
- » EQ11-Voice.mp3

You will develop a short video clip to advertise tourism in Bali. The original video clip was filmed in Bali in August 2018 by GBRvideo and the audio was recorded by Karla Marie.

All video and audio clips produced must be of a professional standard.

1. Open the file **EQ11-Bamboo.mp3** in a suitable audio editing package. Use noise reduction facilities to remove all background noise. Amplify the audio from 8 seconds to the end of the clip to be the maximum volume without clipping. Remove the first 2 seconds of the clip. Insert a silent portion into the clip starting at 17.7 seconds and finishing at 25.4 seconds. Fade out the clip between 16.7 and 17.7 seconds. Fade in the clip between 25.4 and 26.4 seconds. Fade out the clip between 35.3 seconds and 37.3 seconds and remove the end of this clip at this point. Save your audio clip as **Bali_sound.mp3** in stereo with a sample rate of 44 kHz. [14]

2. Open the file **EQ11-Bali1.mp4** in your video editing software. Set the aspect ratio to 16:9. Remove all audio from the clip. Save your video clip as **Bali1.mp4** with a resolution of 1920 x 1080 pixels. [4]

3. Take a still image from the first frame of this clip. Save it with the filename **Bali1** in a format suitable for importing into your video editing software. [2]

4. Take a still image from the last frame of this clip. Save it with the filename **Bali2** in a format suitable for importing into your video editing software. [1]

5. Open the file **EQ11-Bali3.mp4** in your video editing software. Set the aspect ratio to 16:9. Remove all audio from the clip. Edit the speed of the clip so that it is 0.25 x the original speed. Save your video clip as **Bali3.mp4** with a resolution of 1920 x 1080 pixels. [4]

6. Use the files saved in steps 1 to 5 to create a video clip to match this storyboard.
   All text should be displayed in a consistent sans-serif font.
   Do *not* use transitions unless instructed to do so. [27]
   - » Time 0 seconds:
     – Background image **Bali1** placed for 6 seconds.
     – The text 'Bali' placed as a title in the top-right corner for 6 seconds.
     – Audio clip **Bali_sound.mp3** saved in step 1 plays.
   - » Time 6 seconds:
     – Video clip **Bali1.mp3** plays.
   - » Time 11.74 seconds:
     – Background image **Bali2** placed for 7 seconds.
     – Caption text: 'for forested volcanic mountains, iconic rice paddies, beaches and coral reefs' broken down into phrases.
     – Added for 6 seconds.
   - » Time 17.74 seconds:
     – Video clip **EQ11-Bali2.mp3** plays.
   - » Time 25.38 seconds:
     – Video clip **Bali3.mp3** plays.
     – Display appropriate credits for 12 seconds to scroll up the black area on the left of the screen. Include the text 'Edited by:' followed by your name in the credits.

7. Export your video clip as **Bali_video.mp4** for use in a Blu-ray Disc player. [1]

Glossary

Abnormal data Data that is of the wrong type (for example text in a numeric field) or is outside the accepted range (for example an exam mark percentage of 110).

Absolute reference Fixes a cell reference within a spreadsheet so that when the formula is copied this cell reference never changes.

Accuracy of information A measure of how free information is from errors and mistakes. It often depends on the accuracy of the original data which was collected.

Actor A group of actuators which combine collaboratively to perform actions and make decisions, such as a robot.

Actual result Results obtained as a result of live testing.

Actuator A device used for controlling a device using the output from a computer.

Adware There are two types where the malicious type is generally regarded as being more of a nuisance than anything else. It automatically generates advertisements on a user's computer.

Aggregated information When the personal details of a number of people are combined to provide information without individually identifying anybody.

Algorithm A list of well-defined instructions which can be followed in order to produce a solution to a problem. Often written in pseudocode, the instructions can be rewritten as a computer program.

Alphanumeric A field type where any character is acceptable input; sometimes called text.

Ambient temperature The temperature of the air surrounding the device being used to measure the temperature.

Amplifier A part of a sound system that makes the sound louder so that it can be easily heard.

AND Used to search with two (or more) conditions, where both (or all) conditions must be true for the result to be true.

Anonymised information When the personal details of people are stored without identifying anybody by name.

Anti-virus software Its function is to detect and remove viruses.

ASCII (American Standard Code for Information Interchange) This is the standard set of codes (numbers) used by the computer to identify each letter and symbol of the character set (on the keyboard).

Asymmetric encryption Sometimes referred to as public-key encryption, this uses two different keys, a public key used to encrypt a message and a private key used to decrypt the message.

ATM (active traffic management) Forms the basis of smart motorways, using variable speed limits and controlling the use of the (hard) shoulder for occasional use by moving traffic.

ATM (automated teller machine) A machine, operating as part of a computer system, that dispenses cash and performs other banking services when a customer inserts a bank card.

Atomic data Data that cannot be broken down into smaller parts/greater detail.

Attachment A data type used within *Microsoft Access* to add one or more files (including documents, presentations, images, sound bites and so on) to the records in your database.

Authentication In IT terms, this is the process of verifying the identity of a person or device, such as when using a user ID and password when logging on to a computer system.

AutoNumber A data type used within *Microsoft Access* to generate a unique integer value for each record created. If a record is deleted its unique number is not reused.

AutoSum Function in *Microsoft Excel* that adds together the contents of a number of cells.

AVERAGE Function in *Microsoft Excel* that calculates the mean (average) of a list of numbers.

AVERAGEIF Function in *Microsoft Excel* that looks at the cells within a given range and calculates the mean (average) from those cells that meet a given condition.

AVERAGEIFS Function in *Microsoft Excel* that looks at the cells within a given range and calculates the mean (average) from those cells that meet a number of given conditions.

Back-end database A database that is accessed through an application program rather than the database software. It does not store database elements like queries, forms or reports.

Back-up software Used to keep copies of files in case the originals should get corrupted.

Backward chaining Used within an expert system and often referred to as goal-driven because it starts with a goal or set of goals that basically establish which rules are to be followed.

Batch processing Occurs when several transactions are collected over time and then processed all at once at a later time.

Behavioural-based detection A method of detecting viruses by looking out for abnormal behaviour.

Biometric data Computer data relating to an individual's physical characteristics, including iris and retina recognition, fingerprints, voice recognition, among others. It uniquely identifies a user when used for accessing a computer system.

Bit A shortened form of binary digit, this is the smallest unit of measurement of computer data. It contains a single binary value of 0 or 1.

Block A number of sectors on a hard disk smaller than a track.

Boolean A logical thought process developed by the English mathematician and computer pioneer George Boole. See Boolean operator.

Boolean operator The logical operators AND, OR and NOT, which are used in searches (queries) within data handling applications and return the values True or False.

Bot Short for internet robot, this performs tasks that are normally undertaken by a human. When used to gather information over the internet it is called a web crawler.

Botnet A server that takes control of a network of infected computers.

Calibration Process of trying to ensure the accuracy of sensors.

Cascading style sheet A style sheet saved in CSS format with a .css file extension. This allows the stylesheet to be attached to each web page to define how to display the content layer of a web page, without the page needing to contain the style tags. If a number of cascading style sheets are used, the multiple style definitions will cascade into one.

CASE...ENDCASE A type of condition used within an algorithm that results in one of a number of alternative actions being carried out.

Certificate authority (CA) Sometimes referred to as a certification authority, this is an organisation that issues digital certificates enabling transparent, trusted transactions to take place.

Character check See Type check.

CHAR Function in *Microsoft Excel* that returns the character specified by a code number.

Check digit A validation check carried out on numeric data consisting of long strings of digits. A calculation is performed on the individual digits to produce a check digit. After data input or transfer, the check digit is recalculated to make sure it produces the original result.

Checksum A method of verification used to check that a whole file has been transferred accurately, unlike parity checks which check individual bytes.

Ciphertext Encrypted information produced as a result of encryption performed on plaintext.

Client The name given to a computer or workstation that runs programs and accesses data stored on a server when part of a client–server network.

Client–server network A computer network which has a central computer called a server to which many other computers or workstations are connected. These computers or workstations are called clients.

CODE Function in *Microsoft Excel* that returns the ASCII value of a given character.

Codec A circuit capable of converting audio and video data into a digital form and then compressing it for transmission across a network.

Comma separated values A file format (.csv) used to hold tabular data, such as from a spreadsheet or database table. It stores it as a text-based file using commas (or semi-colons in some locations) to separate the data items.

Command line interface (CLI) A means of interacting with a computer using commands in the form of successive lines of text.

Compiler Converts statements written in a high-level language program into machine code for the computer to execute.

Compound key Multiple fields within a relational database table that together contain unique data (no two records within these combined fields can contain the same data). No field within the compound key can contain a blank record.

Computer fraud Involves using a computer to take or alter electronic data or to gain unlawful use of a computer or system illegally to benefit financially.

Computer sabotage Making deliberate attacks which are intended to cause computers or networks to cease to function properly.

CONCATENATE Function in *Microsoft Excel* that joins several text strings into one string.

Conditional branching The direction from one statement in an algorithm to another part of the algorithm depending on the result of a comparison being made.

Confidential data Data which is personal and possibly sensitive which should not be revealed to other workers.

Consistency check Sometimes referred to as an integrity check, this is a validation check which checks that data across two fields is consistent.

Container A file format which holds data.

Contiguous data Data that is placed together and can be selected in a single range.

Control system The use of microprocessors or computers together with sensors to control a process.

Control total Similar to a hash total, in that the total value of a numeric field is calculated; unlike hash totals they can be useful when calculating the total amount in a profits field.

COUNT Function in *Microsoft Excel* that looks at the cells within a given range and counts the number of cells that contain numeric values.

COUNTA Function in *Microsoft Excel* that looks at the cells within a given range and counts the number of non-blank (not empty) cells.

COUNTIF Function in *Microsoft Excel* that looks at the cells within a given range and counts the number of these cells that meet a given condition.

COUNTIFS Function in *Microsoft Excel* that looks at the cells within a given range and counts the number of these cells that meet more than one given condition.

Cray XC40 Supercomputer Produced for the UK Meteorological Office in 2017, with 172 000 processor cores.

.css See Cascading style sheet.

.csv See Comma separated values.

Currency The unit of money used by countries, for example Egyptian pounds, Kenyan shillings, and rupees in India and Pakistan.

Custom-written software Specially written software for a particular task and developed for a specific company or business.

Cyber-fraud The illegal use of the internet to deceive people in order to obtain money, goods or services from them.

Data Often used as a shortened form of the term 'raw data', this comprises characters, symbols, images, audio clips and video clips, which on their own have no meaning.

Database A collection of data items and the links between them which is organised and stored so that it can interrogated.

Database management system A complex software system used to control, maintain and manage a database.

Database of facts Contains all the known facts of information relating to a specific scenario which an expert system has been created for.

Data compression See Zip.

Data logging Using an electronic device to record data over time with a built-in sensor to monitor or measure physical variables.

Date and time A field type within *Microsoft Access* that is used to store dates and times.

DAY Function in *Microsoft Excel* that returns a number between 1 and 31 as the day value from a given date.

DBMS See Database management system.

DC offset Unwanted DC output voltage which appears at the output of an operational amplifier (op-amp) as well as the desired signal. It is the average offset of a waveform from zero. It can be a cause of clicks, distortion and loss of audio volume.

Decibel A unit used to measure the intensity of a sound (how loud a sound is).

Decimal A data type used within numeric fields in *Microsoft Access* to store the data in decimal format (including storing numbers like 5.67).

Decryption key The key used to convert ciphertext back into plaintext. In many systems the decryption key is the same as the encryption key.

Defragmentation Process of organising the data on a disk to reduce the number of blank sectors.

De-identification Removing items of personal data from a record, such as an individual's name.

Delete utility A type of utility software that deletes the pointers telling the operating system where to find a file.

Device driver A small program that enables the operating system and application software to communicate with a hardware device.

Dialogue interface Allows a user to communicate with a computer or device using their voice.

Digital certificate Used to verify the identity of a computer or server and is issued by a certificate authority.

Digital divide A term used to describe the gap between those people who are able to access modern technology and information and those with restricted or no access.

Direct data Data that is collected for a specific purpose or task and is to be used for that purpose and that purpose only.

Direct data access See Random access.

Direct data sources Sources that provide the data gatherer with data that is first-hand.

Disk formatting The configuration of a storage medium ready for initial use.

Disk surface The surface of a platter within a disk drive which stores the data.

Double data entry Involves the entry of data twice; both versions are compared by the computer and the person entering the data is alerted by the computer to any differences.

Double-entry verification A form of verification. See Double data entry.

Duty of confidence The responsibility of an employee to the company they work for and the customers whose personal details they are dealing with to maintain the confidentiality of the data.

Electoral register Often referred to as the electoral roll, this is an example of an indirect data source. It is a list of adults who are entitled to vote in an election, containing information such as name, address, age and other personal details.

Electronic funds transfer (EFT) The electronic transfer of money from one bank account to another using computer-based systems.

Email account An arrangement with an ISP (see Internet service provider) which allows users to send and receive emails using an address provided by the ISP.

Encryption key Used for scrambling and unscrambling data; it is used to convert plaintext into ciphertext and back again.

EXACT Function in *Microsoft Excel* that compares two strings/cells and looks for an exact match including case.

Expected result The list of expected results of tests that are going to be performed on a new system.

Expert system A computer-based system used to solve problems usually of a diagnostic nature using databases of expert knowledge.

Explanation system The part of an expert system which explains the logical reasoning that the expert system has used to come to its conclusions.

Extended ASCII An extended version of the ASCII system which provides extra characters with codes from 128 to 255.

Extreme data A type of data used to test a system. Where data must be within a certain range, extreme data is the data on either boundary of the range.

False positive Happens when a virus detection algorithm is so general that even though only a small number of suspicious bytes are identified, the virus-free file is still considered to be containing a virus.

Fault tolerance Ability of a mainframe to remain in use even when one or more of its individual components fails.

Field A place used to hold a single data item within a database.

File A logically organised collection of records, usually where all the records are organised so that they can be stored together within a database.

File compression See Zip.

File copying A type of utility software used to create a new copy of a file.

FIND Function in *Microsoft Excel* that returns the starting position of one text string within another text string.

Firewall The hardware or software designed to prevent unauthorised access to a computer network.

Flat-file database A database structure using a simple two-dimensional table.

FLOPS A measure of how many floating-point operations per second a supercomputer can perform.

Flowchart A data or program flowchart which represents an algorithm in diagrammatic format.

Footer The area at the bottom of a document between the bottom of the page and the bottom margin, sometimes used to enter page numbers or the same text consistently over a number of pages.

Foreign key The name given to a field in a table within a relational database, which is linked by a relationship to a primary key field in another table.

Format check Sometimes referred to as a picture check, is a validation check used to check that data is in a particular format, usually a combination of numbers and letters in a specific order.

Formula A calculation containing either mathematical operators that can be used to add, subtract, multiply, divide and calculate indices (powers) of a number or containing functions to perform a variety of tasks.

Formulas Sets of instructions used to perform a calculation in a spreadsheet; can include cell references, numbers, mathematical symbols and, in some cases, predefined functions.

Forward chaining Used within an expert system and often referred to as data-driven, because the data entered into the system determines which rules are selected and used.

Frame One still image that is part of a moving image.

General Data Protection Regulation (GDPR) The set of rules used by the European Union to govern the protection of data.

Gesture-based interface A user interface designed to interpret human gestures and convert these into commands a computer can understand.

.gif See GIF.

GIF (graphics interchange format) A format used to save still or moving images. This is an efficient method of storing images using a smaller file size, particularly where there are large areas of solid colour. It is widely used in web pages.

Global digital divide The term used to describe the digital divide between countries.

Goal Seek A feature in *Microsoft Excel* to give a required output by repeatedly changing one item of input data using trial and error until it finds the required output.

GPU (graphics processing unit) Similar to a computer's CPU but designed specifically for performing mathematical calculations that are necessary for rendering all images on the computer's screen.

Graphical user interface (GUI) A user interface which uses windows, icons, menus and pointers, to carry out commands.

Greenhouse Referred to as a glasshouse in some countries; used in cooler countries to grow plants which normally only grow in warmer climates.

Hacker A person who uses their programming skills to gain unauthorised access to computers or computer networks.

Hash total A verification method where totals are calculated in numeric fields in order to check data has been transferred accurately, but unlike control totals, the results of the calculations are usually meaningless, such as adding up a number of student IDs.

Header The area at the top of a document between the top of the page and the top margin, often used to enter the same text consistently over a number of pages.

Heat maintenance Often referred to as heat management, this is the controlling of heat produced by large computers to manageable levels.

Heuristic-based detection A method of detecting viruses by looking at source code to detect a pattern.

High-level formatting A type of disk formatting usually carried out by a user.

HLOOKUP Function in *Microsoft Excel* that looks up data using data organised in horizontal rows.

Hosts file A basic text file which contains the name and the IP address corresponding to URLs.

HOUR Function in *Microsoft Excel* that returns a number between 0 and 23 as the number of hours shown on a 24-hour clock from a given time.

.htm See HTML.

HTML (Hypertext Markup Language) A text-based language used to create markup, so that web browser software can display information in different ways.

HTTPS (Hypertext Transfer Protocol Secure) The protocol whereby HTTP data is securely transferred over a network, usually the internet.

Identity theft Illegally using another person's name and personal information in order to obtain credit, loans and state benefits.

IFS A function in Excel that can be used in place of nested IF functions. It looks at a number of given conditions and performs a defined operation when the first True condition is met. If none of the conditions are met it can return an error message.

IF Function in *Microsoft Excel* that looks at a given condition and performs an operation if the condition is met, or a different operation if the condition is not met.

IF...THEN...ELSE A comparison or condition made within an algorithm which leads to another action or statement being carried out or an alternative (ELSE).

INDEX Function in *Microsoft Excel* that returns a value or reference of a cell, at the intersection of a given row and column in a given range.

Indirect data Data that is obtained from a third party and used for a purpose different to that for which it was originally collected.

Indirect data sources Third-party sources that the data gatherer can obtain data from. One such source is the electoral register.

Industrial espionage Spying in order to discover the secrets of a rival manufacturer or other industrial company.

Inference engine Performs the reasoning in an expert system using a rules base and interrogates the database of facts.

Information Data which has been put in context or given a meaning, usually after some form of processing, sometimes by computer.

INPUT A statement used in pseudocode requiring a value to be entered.

INT Function in *Microsoft Excel* that calculates the integer (whole number) part of a number and ignores all digits after the decimal point.

Integer A whole number with no decimal places.

Internet protocol (IP) Provides the protocol, or set of rules, which governs how data is sent from one computer to another on the internet. Each computer on the internet has an IP address that uniquely identifies it from all other computers on the internet.

Internet service provider (ISP) A company or organisation that provides services for accessing and using the internet.

Interpreter Converts statements written in a high-level language program into an intermediate form for the computer to execute.

Invalid character check See Type check.

IPsec (internet protocol security) The authentication of computers and encryption of data so as to provide secure encrypted communication between two computers over an IP network.

JPEG (joint photographic expert group) A format used to save still images, originally named by the Joint Photographic Expert Group. This is an efficient method of storing images using a smaller file size and is widely used in web pages.

.jpg See JPEG.

Knowledge base A component of an expert system usually comprising a database of facts and a rules base.

Knowledge base editor Software that allows a knowledge engineer to edit the rules and facts within the knowledge base of an expert system.

Knowledge engineer Person responsible for designing an expert system after initially collecting information and data to do with the system.

Label A value entered as text (or alphanumeric), often in a spreadsheet cell or text control box displayed within elements of a database package. This term is also used within programming but is not within the scope of this book.

Lasso tool A tool used to select objects by holding down the left mouse button and dragging the mouse around it, then letting go.

LEFT Function in *Microsoft Excel* which returns a given number of characters from the start of a text string.

LEN Function in *Microsoft Excel* which counts the number of characters within a string.

Length check A validation check which is performed on fields of an alphanumeric type to check whether the required number of characters has been entered for that field, no more and no fewer.

Levels of access The sets of permissions or restrictions granted to particular groups of users when accessing a computer network. Some users will have access to parts of the system that others do not have. It provides an effective part of network security.

Limit check A validation check applied to one boundary in a numeric range either at the upper or lower limit, unlike a range check which tests both boundaries.

Linear progression The writing of algorithmic statements or instructions in a sequence that uses the result of one statement in subsequent statements.

Link editor See Linker.

Linker Also known as a link editor, this combines object files or modules that have been created using a compiler into a single executable file.

Lookup A generic term for looking up specified results from a table of data.

LOOKUP Function in *Microsoft Excel* which looks up data using the first row or the first column of a range of cells and returns a relative value.

Loop A sequence of instructions within an algorithm or flowchart which is repeated a certain number of times.

Lossless compression (images) This allows the original images to be recreated with no loss of quality. (sound) This preserves all the information from the original sound sample.

Lossy compression (images) This involves the loss of resolution of parts of the image (often in parts of the image where it will be least noticed). (sound) This does not preserve all the information from the original sound sample and involves some degradation of the signal.

LOWER Function in *Microsoft Excel* that returns a string as lower-case characters.

Low-level formatting The basic form of disk formatting which permanently erases data on a disk and is usually only carried out by manufacturers.

Mainframe computer Used mainly by large organisations for bulk data processing applications such as the census, industry and consumer statistics, and transaction processing.

Malicious bot Behaves like a worm as it can replicate itself and is designed to feed back to a server called a botnet.

Malware Short for malicious software, the overall term for software which is a threat to computer security.

Massively parallel processing The type of parallel processing employed by a supercomputer's operating system making use of the very large number of processor cores.

Master file A relatively permanent file that contains information which is only periodically updated. An existing master file is used together with a transaction file to produce a new updated master file.

MATCH Function in *Microsoft Excel* that returns the relative position of an item within an array, that matches a value from a given location.

MAX Function in *Microsoft Excel* that calculates the maximum value in a list of numbers.

Mean time between failures (MTBF) The average amount of time a system is available between each time the system fails.

Measurement See Monitoring.

MEDIAN Function in *Microsoft Excel* that places a list of numbers into sequential order and returns the middle number from that list. If there are two middle numbers, the mean of those two middle numbers is calculated and displayed.

Microprocessor A central processing unit on a single integrated circuit chip.

MID Function in *Microsoft Excel* which returns, from a given starting position, a given number of characters from the middle of a text string.

MIN Function in *Microsoft Excel* that calculates the minimum value in a list of numbers.

MINUTE Function in *Microsoft Excel* that returns a number between 0 and 59 as the number of minutes shown on a clock from a given time.

MIPS (millions of instructions per second) The measure of how fast a mainframe's CPU is dealing with computer program instructions.

MODE Function in *Microsoft Excel* that extracts the most frequently occurring number from a list.

Model A computer representation of a real-life process.

Modelling Creating a programmed simulation of a situation or process that will allow data to be changed and examining the effect that this has on other data. This is often done with a spreadsheet and can be used to predict future trends.

Monitoring The observing and measuring of physical variables using microprocessors or computers together with sensors.

MONTH Function in *Microsoft Excel* that returns a number between 1 and 12 as the month value from a given date.

.mp3 See MPEG-3.

.mp4 See MPEG-4.

MPEG-3 (Moving Pictures Experts Group layer 3) A file format used for storing audio files. It compresses the original audio (sound) file to about 1/12 of the original file size while keeping the original sound quality.

MPEG-4 (Moving Pictures Experts Group layer 4) A file format (multimedia container) used for storing video files. It is also used to store audio (the soundtrack to the video) and other data like subtitles and still images.

Network policies Sets of rules that allow companies to choose who is allowed to access their computer network. They control an individual's use of the network once they have gained access.

Non-contiguous data Data that is not placed together (often in a spreadsheet) and cannot be selected using a single range.

Normal data Data that is within an acceptable range and is usual for the situation.

NOT Used to search with two (or more) conditions, where if the condition is True the result is False and if the condition is False the result is True.

Null string Often referred to as the null character, this has an ASCII value of 0 and when sent to a printer causes no action to be taken.

Number A value entered in numerical format; for example '5' is a number, whereas 'five' represents a number but is in text format.

Numeric Consisting of numbers: real, integer or fixed decimal.

Object linking and embedding (OLE) A Microsoft technology that allows a file to be embedded and linked at the same time to other documents/files and objects. Data that is available in other applications can be accessed and manipulated from the current document/file.

Off-the-shelf software Software which already exists and is available straight away, ready for use.

OLE object A data type used within *Microsoft Access* to store files created in other programs such as graphics, spreadsheets, or documents.

OR Used to search with two (or more) conditions, where one (or more) of the conditions must be true for the result to be true.

Padlock An icon used by most websites. When closed it indicates a secure connection has been established.

Parallel processing The way a mainframe operating system splits a large task into several sub-tasks then recombines the results once processed.

Parity bit The bit added to a byte when parity checks are used.

Parity check A type of verification used to check data has been transmitted accurately; involves the use of a parity bit.

Password A string of characters, often used in conjunction with a user ID as part of the authentication process when logging on to a computer or other digital device or software.

Payment Card Industry Data Security Standard (PCI DSS) A set of rules helping company websites to process bank card payments securely and to help reduce card fraud.

Payroll The payment of workers by a company. Payrolls are usually produced using batch processing at the end of each week or month.

.pdf See PDF.

PDF (portable document format) A method of allowing documents to be read/used on most computers. The pages look just like they would when they are printed but can contain clickable links and buttons, form fields, video, and audio. You can protect a document to stop others editing it.

Penetration test Carried out by companies by employing somebody to deliberately attack their computer network.

Performance metrics Measure of how well a computer processor deals with data.

Pharming Malware that when downloaded unknowingly, corrupts the Hosts file causing the user to be taken to a fake website when they enter what they regard as an authentic URL.

Phishing When fraudsters try to obtain personal banking details using email.

Physical variable A measurable property of a material or system such as temperature, pressure, light and so on.

Pi The number Pi (π) is a mathematical constant. It is approximately equal to 3.14159 and is used in mathematical calculations involving circles.

Picture check See Format check.

Plaintext Information that is readable by humans or a machine; the state that text exists in before it is encrypted.

Platter An individual disk within several disks which constitute a disk drive.

.png See PNG.

PNG (portable network graphics) A format used to save still images. This is an efficient method of storing images and was designed to replace .jpg (JPEG). It is widely used in web pages.

Presence check A validation check used to check that data has been entered into a field and, if not, an error message is produced saying data must be entered.

Primary key A single field within a relational database table that contains unique data (no two records within this field can contain the same data). A primary key field cannot contain a blank record.

PRINT A statement used in pseudocode to output data; usually the result of some processing.

Private key A key used in the encryption process to decrypt a message when using asymmetric encryption but is used for both encryption and decryption when using symmetric encryption.

Program flowchart See Flowchart.

Proprietary and open source software Proprietary file formats are formats belonging to a company, organisation or individual whose use is controlled by them. The structure of these file formats is often hidden from others. Open source file formats are formats used for storing data that are freely available for anyone to use. The structure of these file formats is published and available for others to use.

Pseudocode A way of writing algorithms in a language independent of any programming languages.

Pseudonymised data Like de-identification but instead of removing the personal items of data they are instead replaced with a temporary ID.

Public key A key used in the encryption process available to many users for the purpose of encrypting a message using asymmetric encryption.

Public key encryption See Asymmetric encryption.

Quality of information A judgement on how good or reliable information is.

Random access Process of directly accessing data stored on a non-sequential storage system like HDD or SSD, without searching through other records to find it.

Range check A validation check performed on fields containing numeric data to check that the value entered is within an upper and lower boundary range.

Ransomware A type of malware hackers use to encrypt a user's computer data, which will only be decrypted by the hacker on receiving payment of a ransom.

.rar See RAR.

RAR (Roshal archive) A file format for a container (like a .zip file) that can hold other files and folders. Its name comes from Roshal archive and it was developed by a Russian software engineer, Eugene Roshal.

RAS Measure of reliability, availability and serviceability of mainframes.

READ A statement used in pseudocode to obtain data from a file.

Real-time processing The processing of data immediately with no delay, usually involving the use of sensors.

Record A collection of fields containing information about one data subject (usually a person) or one object within a database.

Relational database A database structure where data items are linked together with relational tables. It maintains a set of separate, related files (tables), but combines data elements from the files for queries and reports when required.

Relative reference Automatically adjusts a cell reference within a spreadsheet to refer to different cells relative to the position of the formula during the replication process.

REPEAT...UNTIL A type of loop found in algorithms where the instructions are executed at least once and then repeated until the condition is true.

Reverb A shortened form of reverberation, which is an electronically produced echo effect in recorded music.

Rich text format A file format (.rtf) used for text-based files that saves the formatting within the document, so allowing some formatting to be passed from one application package to another.

RIGHT Function in *Microsoft Excel* which returns a given number of characters from the end of a text string.

Rootkit A type of malware that allows an attacker to have remote access to a victim's computer continuously.

ROUND Function in *Microsoft Excel* that rounds a number to a specified number of decimal places.

ROUNDDOWN Function in *Microsoft Excel* that rounds a number down to a specified number of decimal places.

ROUNDUP Function in *Microsoft Excel* that rounds a number up to a specified number of decimal places.

.rtf See Rich text format.

Rules base The set of rules applied by an inference engine of an expert system, usually consisting of series of IF...THEN statements.

Sandbox A virtual environment within which virus codes are allowed to be executed to observe their behaviour.

SEARCH Function in *Microsoft Excel* that returns a number to represent the position of the first character or text string that matches a given character or text string.

SECOND Function in *Microsoft Excel* that returns a number between 0 and 59 as the number of seconds shown on a clock from a given time.

Sector Smallest unit of data storage within a disk drive and a subdivision of a track.

Secure sockets layer (SSL) The predecessor of transport layer security (TLS); found to have many security flaws and replaced by TLS which provided greater security when transmitting data.

Sensor Converts a physical characteristic, such as temperature, light or pressure, into a signal which can be measured electrically by a computer.

Sequential access A method of accessing data on a medium, by having to read through all the preceding data first.

Server A computer that provides data to other computers connected to it on a network. There are many types of server including application servers, file servers, mail servers, and web servers.

Shell The major part of an expert system; usually consists of the user interface, explanation system, inference engine, and knowledge base editor.

Signature-based detection A method of detecting viruses based on recognising existing viruses.

Smart home Uses computers and a router to enable control of devices within the home.

Smart motorway Section of motorway that uses active traffic management.

Smishing A variation of phishing whereby an SMS is used rather than an email to try to obtain personal banking details.

Spyware A type of malware designed to collect information about a computer user's activities without their knowledge.

SSH (secure shell) An encryption protocol used to enable remote logging on to a computer network, securely.

String A type of data that contains characters and may include text, punctuation, numbers. The term is often interchanged with the term 'alphanumeric data'.

SUBTOTAL Function in *Microsoft Excel* that calculates the function for a range of value or cells. The parameters passed to this function determine how it will work; for example it could be used to calculate the SUM, the AVERAGE and so on.

SUM Function in *Microsoft Excel* that adds up a list of numbers or specified cells.

SUMIF Function in *Microsoft Excel* that looks at the cells within a given range and adds the total from those cells that meet a given condition.

SUMIFS Function in *Microsoft Excel* that looks at the cells within a given range and adds the total from those cells that meet a number of given conditions.

Summit A supercomputer which, as of the beginning of 2020, was the world's fastest supercomputer.

Supercomputers The largest and fastest computers available.

Symmetric encryption Encryption which uses the same key to encrypt and decrypt a message, a private key. Often regarded as a faster process, it is less secure than asymmetric encryption.

System software Programs that run and control a computer's hardware and application software.

Table A two-dimensional grid of data organised by rows and columns within a database. Each row of the table contains a record. Each column in the table represents a field and each cell in that column has the same (predefined) field type.

TEXT Function in *Microsoft Excel* that changes the formatting of the current cell from a numeric or date/time value into text format.

Text file A file format (.txt) used for text-based files that contain an unformatted ASCII file, although there are file format variations depending upon the operating system. These files can be opened in any word processor.

Track Part of the surface of a disk platter which stores data, with each track containing several sectors.

Transaction In IT terms, when a computer makes an amendment to a database, be it adding, deleting or changing values in a record, a transaction is said to have been carried out.

Transaction file A file which contains all ongoing transactions in a batch-processing system.

Transaction processing A form of processing which can consist of more than one computer processing operation, but these operations must combine to form a single transaction.

Transport layer security (TLS) An improved version of the secure sockets layer (SSL) protocol and which has now superseded it. The term SSL/TLS is still sometimes used to bracket the two protocols together. It is used to provide secure transfer of data when accessing web pages.

Trojan horse A type of malware usually obtained as a result of the user thinking they have downloaded authentic software. It enables the author of the software to take control of the user's computer.

Type check Sometimes referred to as a character check or even invalid character check, is a validation check which checks that data is of a specific data type so that, for example, when digits are required to be entered, a letter is not entered by mistake.

.txt See Text file.

UNIVAC 1 First computer produced for general sale in 1951.

UPPER Function in *Microsoft Excel* that returns a string as only upper-case characters (capitals).

User ID A string of characters identifying a user when accessing a computer; nearly always used in association with a password.

User interface The means by which the user interacts with a computer system, involving the use of input devices and software. It is a key component of any computer system but particularly as part of an expert system.

Utility software Often supplied with the operating system, this mainly helps to manage files and their associated storage devices.

Validation A process where the software checks that the data entered into it is reasonable, often in a database or spreadsheet.

Verification A process that checks the accuracy of data entry or that the data has not been corrupted during transmission.

Virus A type of malware that is designed to spread from one computer to another causing changes in the way each computer operates.

Vishing A phone call made in order to get an individual to divulge their personal or banking details.

VLOOKUP Function in *Microsoft Excel* that looks up data using data organised in vertical columns.

VPN (virtual private network) Enables a network belonging to a company or organisation to be extended using a public network, usually the internet, via tunneling protocols to encrypt and decrypt data.

Weather station A system used to monitor the weather in terms of physical variables such as temperature, rainfall, hours of sunlight and so on.

WEEKDAY Function in *Microsoft Excel* that returns a number between 1 and 7 from a given date. If the day is a Sunday, 1 is returned, Monday is 2, and so on.

WHILE...ENDWHILE A type of loop found in algorithms where the instructions are repeated while the condition is true.

Worm Self-replicating malware whose sole purpose appears to be occupying as much disk space and bandwidth as possible.

WRITE A statement used in pseudocode to output data to a file.

WSAN (wireless sensor and actuator network) A network of devices using combined actuators which are called actors to automatically help the user, such as in a smart parking system.

YEAR Function in *Microsoft Excel* that returns the year from a given date.

XLOOKUP A function in *Excel* that performs either a horizontal or a vertical lookup of data from a list within the same worksheet or workbook or from an external data file. XLOOKUP can perform lookups with backwards referencing.

.zip See Zip.

Zip To reduce the number of bytes needed to save a file, either to save storage space or to reduce transmission time.

Index

A

.aac (advanced audio coding) files 364
absolute cell referencing 183–4
Access see databases
access rights 63, 124
accessibility barriers 145–6, 152
accuracy of information 8
active traffic management (ATM) 92–3
actors 95–6
actuators 83–7
adware 131
aggregated information 122
air-conditioning systems 41–2, 91, 97
algorithms 98–9
    CASE…ENDCASE 103–5
    counts 105–6, 108, 110
    file handling 113
    flowcharts 114–17
    IF…THEN…ELSE 100–3
    input and output 99–100, 113–14
    loops 105–9
    nested loops 110–11
    FOR…NEXT 106–7
    FOR…NEXT…STEP 107
    REPEAT…UNTIL 29–30, 108
    subroutines (procedures) 111–13
    variables 99
    WHILE…ENDWHILE 107–8, 109
alphanumeric data 279
amplification, audio clips 357
AND NOT query 302
AND query 300
animal identification, expert systems 161–2
animation effects 344
anonymised data 121, 122
anti-malware software 136
anti-virus software 68–9, 129, 135–6
artificial intelligence (AI) 165–6
ASCII (American Standard Code for Information Exchange) 24–5
    converting strings to ASCII values 211–12
aspect ratio 340–1
assigning to variables 99
asymmetric (public key) encryption 12, 17
atomic data 325
audio
    addition to a video clip 349
    removal from a video clip 345
audio editing
    adding a new track 355
    amplification 357
    changing from stereo to mono 357
    changing the pitch 359
    changing the speed 359
    compression 366–7
    echo and decay 363
    equalisation 361
    exporting 358, 364–5, 366
    fade-in and fade-out 356–7
    high-pass filters 362
    importing a new track 355
    low-pass filters 362–3
    noise reduction 359–60
    normalising 355
    overdubbing 363–4
    reverb 360–1
    sampling rate and resolution 365
    splicing 356
    trimming 356
audio file types 333
authentication techniques 123–4
autonomous vehicles 43–6
AutoSum 186
AVERAGE function 187
AVERAGEIF function 197
AVERAGEIFS function 197–8

B

back-up software 67
backward (goal-driven) chaining 162–3, 163–4
bank transfers 34–5, 56
banking
    investment analysis 157–8
    online 143–4
bar charts 244, 245–6
batch processing 27–8, 55
    advantages and disadvantages 47
    credit card and debit card accounts 31–2
    customer orders and accounts 33
    payroll 32–3
    utility bills 31
behavioural-based malware detection 69
BIN2DEC function 212
BIN2HEX function 212
biometrics 123–4
BIOS (Basic Input/Output System) 62
Boolean (logical) data 279
botnets 132
bots 131–2, 166
building information modelling (BIM) 267–8
burglar alarm systems 42, 92, 96
business-to-business (B2B) transactions 37–8

C

calibration 80–2
captions, addition to a video clip 346–7
car engine fault diagnosis 159–60
carbon dioxide (CO_2) sensors 77, 78
car-park barrier systems 94–5, 96
cascading sheet style (.css) files 333
CASE…ENDCASE 103–5
cell referencing 183–4
cells 169
    formatting 220–8
    merging 178
    naming 184
    protecting 179–80
censuses 5, 6, 55
central processing unit (CPU) 49
    multiple processor cores 54
    speed 52–3
central-heating systems 40–1, 90–1
certificate authorities (CAs) 13
chaining
    applications 164–6
    backward (goal-driven) 162–3
    forward (data-driven) 163–4
CHAR function 212
character checks 20
charts 244
    automatic updating 258
    bar charts 245–6
    combo charts 249–51
    data selection 244–5
    exporting 252
    formatting 245–6, 250–1
    line graphs 248–9
    pie charts 246–8
    secondary axes 250–1
    selecting chart type 245
check digits 21
checksums 25–6
ciphertext 11
climate research 58, 263–4
Cloud-based services 51, 59
COBOL (Common Business-Oriented Language) 51, 60
CODE function 212
codec (coder/decoder) 354
combo charts 249–51
comma separated values (.csv) files 170, 332
command line interfaces (CLIs) 67–8, 72–3
    advantages and disadvantages 74–5
comments, addition to spreadsheets 226
Community Earth System Model (CESM) 58
compilers 59–60
    comparison with interpreters 61
    cross compilers 64
compound keys 281, 287
compression 66–7
    audio files 366–7
    file types 333
    video files 350–1, 354

computer-aided design (CAD) 266–8
computer availability 138–9
computer fraud 132–3
computer modelling
    climate change 263–4
    in construction industry 266–8
    financial forecasting 260–1
    population growth 261–2
    queue management 264–6
    traffic flow 266
    weather forecasting 262–3
computer sabotage 134–5
CONCATENATE function 208
concatenating strings 207–8
conceptual design, databases 282
conditional branching (selection)
    CASE...ENDCASE 103–5
    IF...THEN...ELSE 100–3
conditional formatting 226–8
conditional functions 192–200
confidentiality, personal data 121–2
connector symbols, flowcharts 115, 116–17
consistency checks 22
consumer statistics 56
control systems 43, 44, 76, 82–3
    actuators 83–7
    advantages and disadvantages 96–7
    air-conditioning systems 41–2, 91
    burglar alarm systems 42, 92
    car-park barriers 94–5
    central-heating systems 40–1, 90–1
    greenhouses 40, 88–90
    sensors 83
    smart homes 96
    traffic lights 95
    traffic/ pedestrian flow systems 42, 92–4
    wireless sensor and actuator networks 95–6
control totals 26–7
cooling systems, mainframes and supercomputers 54
count-controlled loops 105–6, 108, 110
COUNT function 191
COUNTA function 191
COUNTBLANK function 191
COUNTIF function 195
COUNTIFs function 195–6
Cray supercomputers 50, 57
credits, addition to videos 347–8
cropping an image 342
cross compilers 64
crosstab queries 308–9
.css (cascading sheet style) files 333
.csv (comma separated variables) files 170, 332
currency formatting 221
custom-written software 70, 71, 259
cyber espionage 133–4

D

Dark Web 134
data 1–2
    checking accuracy of 18–27
    direct and indirect 2–7
    encryption 10–18
    personal *see* personal data
    quality of information 7–10
data compression *see* compression
data dictionaries 330–1
data entry forms 317–19
    list boxes 319
    radio buttons 321
    sub-forms 320–1
data logging 3, 4
data processing 2, 27
    advantages and disadvantages of different methods 47
    batch processing 27–33
    online processing 34–9
    real-time processing 40–6
data protection 14–15, 121–2
data theft 133
data types 279
database management systems (DBMSs) 335–7
database queries 294–5
    alphanumeric queries on multiple criteria 300–2
    alphanumeric queries on one criterion 299–300
    complex queries using multiple criteria 302
    crosstab queries 308–9
    data summaries 308
    dates 303
    find duplicates 310
    IIF function 308
    nested 307–8
    numeric queries on multiple criteria 299
    numeric queries on one criterion 295–7
    performing calculations 309–14
    reports 297
    selection of query type 314
    static and dynamic parameters 305–6
    time 304
databases 276
    creation from existing files 292–4
    entity relationship diagrams 281–3
    exporting data 324
    grouping data 315–16
    importing data 292–4, 324
    key fields 280–1
    normalising data 325–30
    relationship types 279–80
    sorting data 314–15
    structure creation 286–91
    structure design 284–6
    switchboards 322–4
    types of 277–8
databases of facts 155
date formats 222
DATE function 213
date queries 303
DATEDIF function 215
DAY function 214
DC offset 355
DEC2BIN function 212
DEC2HEX function 212
decay, audio files 363
decimal places 221
defragmentation 66
de-identification 120
delete utility 68
delivery vehicle scheduling 160–1
developing countries, digital divide 153
device drivers 62, 63
diagnoses, expert systems 159–60, 164–5
dialogue interfaces 73–5
digital certificates 13
digital divide 137–8
    causes 138–40
    and disability 145–6, 152
    effects 140–4
    global 153
    groups affected by 146–9
    and level of education 150
    older people 146–8
    reducing the effects of 145–6
    in rural areas 148–50
    socioeconomic factors 150–2
digital skills 138
    training courses 145
direct data 2
    advantages and disadvantages 7
    sources 2–3
    uses 3–4
disability
    adaptation of technologies 145
    digital divide 145–6, 152
    smart-home technology 97
    and user interfaces 74
disk defragmentation 66
disk formatting 65–6
disk management systems 69–70
disk partition 66, 69–70
double data entry 23–4
driving simulators 270
drones 45–6
drug research 57
duplicate records, location in databases 310
duty of confidence 121–2
dynamic parameter queries 305–6

E

echo, audio files 363
economic factors, digital divide 140

education, digital divide   141–2, 146, 150
electoral register (electoral roll)   5, 6
electric actuators   86
electromagnetic field sensors   83
electronic data interchange (EDI)   36–7
electronic funds transfer (EFT)   34–5, 56
ELSE   101–3
email
    encryption   16
    malware prevention   136
    phishing   126–7, 129, 132
employment, effects of digital divide   142–3
encryption   10–11, 121
    advantages and disadvantages   17–18
    need for   10
    symmetric and asymmetric   11–12
    uses   14–17
encryption protocols   12–14
ENDCASE   103–5
ENDIF   101–3
ENDWHILE   107–8
entity relationship diagrams (ERDs)   281–3
    database structure design   284–6
environmental monitoring   78
equalisation (EQ), audio files   361
error messages   63
error trapping   194
EXACT function   211
Excel see spreadsheets
.exe files   332
expert systems   155–7
    advantages and disadvantages   166–7
    artificial intelligence   165–6
    car engine fault diagnosis   159–60
    chaining   162–6
    in computer gaming   165
    delivery vehicle scheduling   160–1
    financial planning   158
    insurance planning   158–9
    investment analysis   157–8
    medical diagnosis   160
    mineral prospecting   157
    plant and animal identification   161–2
explanation systems   156
exponential growth   261–2

F

fading effects
    audio clips   356–7
    video clips   351–2, 354
fault tolerance   53
fear of IT   140, 146
file copying   67–8
file deletion   68
file handling   113
file management   63
file management systems   69
file types   332–4

fileless malware   132
files
    indexed sequential access   334–5
    random access   335
    sequential   28, 333–4
filters, video editing   353
financial forecasting, modelling   255–6, 260–1
financial planning expert systems   158
FIND function   209–10
firewalls   122–3
flat-file databases   277, 278
    see also databases
flight simulators   268–70
floating point operations per second (FLOPS)   53
flowchart symbols   114
flowcharts   98, 114–17
font colour   224
font size   224
font style   224
foreign key fields   277, 280, 286
format (picture) checks   21
formatting charts   245–6, 250–1
formatting disks   65–6
formatting rules   226–8
formatting spreadsheets   220–8
formulas   182–3
    automatic recalculation   258
FOR…NEXT   106–7, 111
FOR…NEXT…STEP   107
FORTRAN (Formula Translation)   59
forward (data-driven) chaining   163–6
fractions, spreadsheets   222
fraud   132–3
freeze and unfreeze panes, spreadsheets   181–2
functions
    averages   187, 197–9
    conditional   192–200
    counting   191, 195–6
    INDEX   204, 205–6
    INT   188
    ISERROR   194
    lookup   200–3
    MATCH   204–6
    maxima and minima   187, 198
    names   185
    nested   192, 199–200
    number base conversions   212–13
    rounding   188–90
    strings, working with   207–12
    SUBTOTAL   200
    summing   186–7, 196–7
    time and date   213–17
    TRANSPOSE   207

G

gaming, use of expert systems   165
gas sensors   77
General Data Protection Regulation (GDPR)   121

generic file formats   332
genetic analysis   57
geography, and digital divide   139–40, 148
gesture-based interfaces   74
    advantages and disadvantages   74–5
GIF (graphics interchange format) files   332
global digital divide   153
Goal Seek   256, 261, 274–5
government websites   144
graphical user interfaces (GUIs)   68, 73
    advantages and disadvantages   74–5
graphics interchange format (GIF) files   332
graphs *see* charts
greenhouse control systems   40, 88–90, 97
gridlines, displaying in spreadsheets   174, 176
grouping data   315–16
guidance systems   43

H

hackers   10
    cyber espionage   134
    penetration testing   123
hard disks   65
    defragmentation   66
    encryption   15–16
    formatting   65–6
    management systems   69–70
    partition   66, 69–70
hardware firewalls   123
hash totals   26
headers and footers   173, 175–6
headings, displaying in spreadsheets   174, 176
healthcare
    effects of digital divide   141
    medical diagnosis expert systems   160
    monitoring technologies   80
heat maintenance, mainframes and supercomputers   54
heuristic-based malware detection   68–9
HEX2BIN function   212
HEX2DEC function   213
hierarchical databases   335, 336
high-pass filters   362
high performance computers, availability of   138–9
HLOOKUP function   201
hosts file   126
HOUR function   216
HTML (Hypertext Markup Language) files   333
human errors   259, 260
humidity sensors   77, 79
hydraulic actuators   86
Hypertext Transfer Protocol Secure (HTTPs)   16–17

INDEX

I

IBM z15 mainframe computer 50
identity theft 121
IF function 192–3
   nesting 199–200
IFERROR function 194
IFS function 194
IF...THEN...ELSE 100–3
IIF function 308
income, and digital divide 150–2
INDEX function 204, 205–6
indexed sequential files 334–5
indices (powers) 182, 183
indirect data 2
   advantages and disadvantages 7
   sources 4–5
   uses 6
industrial espionage 133–4
industry statistics 55
inference engines 156
   chaining 162–6
information 1–2
   accuracy 8
   quality of 7–10
infrared sensors 78
infrastructure 138
input 99–100
   writing algorithms 113–14
input masks 287
   testing 291
insurance planning expert systems 158–9
INT function 188
intellectual property theft 134
internet access 139, 145
Internet of Things (IoT) 96
Internet Protocol Security (IPsec) 12, 13, 14, 18
interpreters 60–1
interviews 3, 4
invalid character checks 20
investment analysis expert systems 157–8
ISERROR function 194
ISNOTEXT function 210
ISNUMBER function 210
ISTEXT function 210

J

job opportunities, effects of digital divide 142–3
JPEG (joint photographic expert group) files 67, 332

K

key fields 277, 280–1, 285, 287
keyloggers 131
knowledge base editors 156
knowledge bases 155, 156
knowledge engineers 156, 157

L

languages, high-level 59–60
LEFT function 208
LEN function 209
length checks 20–1
levels of access 124
lifespan, mainframes and supercomputers 51
light level control systems 89–90
light sensors 77, 79
limit checks 22
line graphs 244, 248–9
linear solenoid actuator 84
link tables 280
linkers (link editors) 61–2
Linux 54
LISP (List Processor) 60
list boxes 319–20
logical design, databases 282, 283
lookup checks 21–2
lookup functions 200–6
loops 105
   nested 110–11
   FOR...NEXT 106–7
   FOR...NEXT...STEP 107
   REPEAT...UNTIL 29–30, 108
   WHILE...ENDWHILE 107–8, 109
lossless and lossy compression 66–7, 367
low performance computers 139
LOWER function 211
low-pass filters 362–3

M

macros 256, 261
magnetic actuators 87
mainframe computers (mainframes) 49–50
   advantages and disadvantages 58–9
   characteristics 51–4
   uses 55–6
malware
   consequences 135
   prevention of 135–6
   types of 130–2
   uses 132–5
management information systems (MIS) 337–8
many-to-many relationships 280, 281
margins, spreadsheets 172–3, 176–7
massively parallel processing 54
master files 28–31
MATCH function 204–6
MAX function 187
MAXA function 187
MAXIFS function 198
mean time between failures (MTBF) 51–2
measurement technologies see monitoring technologies
mechanical actuators 87
MEDIAN function 199
medical diagnosis expert systems 160
memory management 63

MID function 209
millions of instructions per second (MIPS) 52–3
MIN function 187
mineral prospecting expert systems 157
MINUTE function 216
MODE function 199
modelling 254
   climate change 263–4
   in construction industry 266–8
   financial forecasting 255–6, 260–1
   population growth 261–2
   queue management 264–6
   reasons for 257
   spreadsheets and drawbacks 257–60
   traffic flow 266
   weather forecasting 262–3
   what-if analysis 255–6, 272–5
modelling software 256–7
moisture control systems 89
monitoring technologies 76
   advantages and disadvantages 82
   sensors 77–8
   uses 78–80
MONTH function 214
Moving Pictures Experts Group (MPEG) files 333
.mp3, .mp4 files 333, 364, 366
   exporting an audio file in .mp3 format 358
multipoint calibration 81–2
multitasking 63

N

natural-disaster planning 270–1
natural language processing (NLP) 166
nested functions 192, 199–200
nested loops 110–11
nested queries 307–8
network databases 335, 337
network policies 124–5
news, online access to 144
NEXT 106–7
noise reduction, audio files 359–60
normalising audio clips 355
normalising databases 325–30, 329–30
   advantages and disadvantages 329
   first normal form (1NF) 325–6
   second normal form (2NF) 326–7
   third normal form (3NF) 327–9
NOT query 301
nuclear science research 271–2
numeric data 279

O

object-oriented database management systems 335–6, 337
observation 3, 4
Office for National Statistics (ONS) survey on internet use 137–8, 147
off-the-shelf software 70–1

older people, digital divide   146–8
one-to-many relationships   280, 281
one-to-one relationships   279, 281
online banking   143–4
online processing   34
    advantages and disadvantages   47
    automatic stock control   35–6
    business-to-business transactions   37–8
    electronic data interchange   36–7
    electronic funds transfer   34–5
online shopping   38–9, 144
open source file formats   332
operating systems   62–4
    mainframes and supercomputers   54
OR query   300–1
OTHERWISE   105
output   99–100
    writing algorithms   113–14
overdubbing   363–4
oxygen (O_2) sensors   77, 78

P

page specifications, spreadsheets   171–7
pan effects   352–3
parallel processing   49, 54
parity checks   24–5
passwords   63
    protecting spreadsheet contents   179–80
Payment Card Industry Data Security Standard (PCI DSS)   12–13
payroll processing   32–3
PDF (portable document format) files   243, 333
penetration testing   123
percentage displays, spreadsheets   222
performance metrics   52–3
personal data   120–1
    confidentiality   121
    preventing misuse of   125–9
    security precautions   122–5
pH control systems   90
pH sensors   77, 78, 81
pharming   126, 129, 133
phishing   126–7, 129, 132, 133
phone calls, vishing   128–9
physical design, databases   282, 283
picture (format) checks   21
pie charts   244, 246–8
pilot training   268–70
pivot charts   242–3
pivot tables   240–2
plant identification, expert systems   161–2
platters   65
pneumatic actuators   86
PNG (portable network graphics) files   333
population growth models   258, 261–2
portable document format (PDF) files   243, 333
post-condition loops   105

power-line communication (PLC)   145
pre-condition loops   105
presence checks   19
pressure sensors   77, 79
primary key fields   277, 280, 285, 287
procedures (subroutines)   111–13
program files   332
proprietary file formats   332
proximity sensors   83
pseudocode   98–9
pseudonymised data   120
public key (asymmetric) encryption   12, 17

Q

quality of information   7–8
    influencing factors   8–10
quantum mechanics   57
querying a database *see* database queries
questionnaires   3, 4
queue management   264–6

R

rack and pinion rotary actuator   84–5
radio buttons   321
random access (direct data access)   335
range checks   19–20
ransomware   17, 132, 133
RAR (Roshal archive) files   333
RAS (reliability, availability and serviceability)   51–2
real-time processing   40
    advantages and disadvantages   47
    air-conditioning systems   41–2
    autonomous vehicles   43–6
    burglar alarm systems   42
    central-heating systems   40–1
    greenhouses   40
    guidance systems   43
    traffic flow control systems   42
    wireless sensor and actuator networks   42–3
reed switches   78, 79
referential integrity   289–90
relational databases   277–8, 336–7
    see also databases
relationship creation, databases   289–90
relationship types   279–80
relative cell referencing   183–4
relevance of data   8–9
remote deletion   125
REPEAT...UNTIL   29–30, 108
    nested loops   110–11
reports   297–8
resizing an image   342
reverb (reverberation), audio files   360–1
rich text format (.rtf) files   332
RIGHT function   208
robotics   166
rootkit   131
Roshal archive (RAR) files   333
rotary actuators   84–5
rounding functions   188–90

rural areas, digital divide   138, 143–4, 148–50

S

sabotage   134–5
sampling rate   365
sampling resolution   365
sandboxes   69
scareware   132
searching spreadsheets
    using date and time   236–8
    using numeric data   235–6
    using substrings   238
    using text filters   234–5
SECOND function   216
Secure Shell (SSH)   12
Secure Sockets Layer (SSL)   12–13, 14, 18
security
    confidentiality of personal data   121–2
    mainframes and supercomputers   52
security precautions   122–5
security threats   126–9
    see also malware
sensors   40, 76–8
    autonomous aircraft   44
    in burglar alarm systems   92
    calibration   80–2
    central-heating systems   41
    in control systems   83
    in a greenhouse   88
    wireless sensor networks   96
sequential files   28, 333–4
serial processing   49
shell, expert systems   156
shopping, online   38–9, 144
signature-based malware detection   68
simulations   268
    advantages and disadvantages   268
    driving simulators   270
    flight simulators   268–70
    natural-disaster planning   270–1
    nuclear science research   271–2
smart homes   96–7
smart motorways   92–3
smartphones
    internet access   139
    remote deletion   125
    smishing   127–8, 129, 133
social interaction, effects of digital divide   143
social media
    confidentiality issues   121, 122
    manipulation of   166
socioeconomic group, and digital divide   150–2
soft actuators   85–6
software
    anti-malware   68–9, 129, 135–6
    computer-aided design (CAD) packages   266–8
    custom-written versus off-the-shelf   70–1

for modelling 256–7
see also system software; utility software
software firewalls 123
software updates 125, 136
sorting data
    databases 314–15
    spreadsheets 239–40
sound sensors 77
speech recognition 73–4
speed 52–3
splicing
    audio clips 356
    video clips 345–6
split windows, spreadsheets 180–1
spreadsheet creation 228–9
spreadsheet models 255–7
    drawbacks 259–60
    effectiveness 257–9
    financial forecasting 255–6, 260–1
    Goal Seek 256, 261, 274–5
    what-if analysis 272–5
spreadsheet searches
    using date and time 236–8
    using numeric data 235–6
    using substrings 238
    using text filters 234–5
spreadsheet testing 230–33, 272–4
spreadsheets
    basics 169
    cell referencing 183–4
    charts *see* charts
    comments 226
    conditional formatting 226–8
    counting functions 190–1
    displaying gridlines and headings 174, 176
    editing cells, rows and columns 174, 177–8
    error trapping 194
    exporting data 243, 252
    fonts 224
    formatting 220–8
    formulas 182–3
    freeze and unfreeze panes 181–2
    functions *see* functions
    headers and footers 173, 175–6
    hide and unhide rows and columns 178
    importing data 170–1
    merging cells 178
    naming cells and ranges 184–5
    page layout 171–7
    pivot charts 242–3
    pivot tables 240–2
    protecting cells and contents 179–80
    rounding functions 188–90
    sorting data 239–40
    split windows 180–1
    text orientation 223–4
    validation rules 218–20
    wrapping text 178–9

spyware 131, 132–3
SSH (Secure Shell) 12
SSL (Secure Sockets Layer) 12–13, 14, 18
static parameter queries 305
STEP 107
stock control 35–6
strings
    changing case 211
    comparing 211
    concatenating 207–8
    conversion to ASCII values 211–12
    extracting data from 208–9
    LEN function 209
    searching within 209–10
sub-forms 320–1
subroutines (procedures) 111–13
subtitles, addition to a video clip 346–7
SUBTOTAL function 200
SUM function 186–7
SUMIF function 196–7
SUMIFS function 197
supercomputers 50, 263
    advantages and disadvantages 59
    characteristics 51–4
    uses 56–8
switchboards 322–4
symmetric encryption 11–12
system software 59
    compilers 59–60, 61
    cross compilers 64
    device drivers 62
    interpreters 60–1
    linkers 61–2
    operating systems 54, 62–4
    uses 64
systems encryption 15

T

temperature control systems 88, 90–1
temperature sensors 77, 79
    calibration 80–2
test plans, spreadsheets 230–33, 272–4
text (.txt) files 332
TEXT function 216–17
text orientation 223–4
thermal actuators 86–7
throughput, mainframes and supercomputers 53
time and date, spreadsheet searches 236–8
time and date functions 213–17
time and date formats 222
time queries, databases 304
touch sensors 83
tourism, effects of digital divide 143
traffic flow modelling 266
traffic light control systems 95, 97
traffic/pedestrian flow control systems 42, 92–4
transaction files 28
    updating a master file 28–31
transaction processing 56
transactions 27

Transport Layer Security (TLS) 12–13, 14, 18
TRANSPOSE function 207
travelling salesman problem 161
trimming
    audio clips 356
    videos 344–5
Trojan horses (Trojans) 130
truncating a number (ROUNDDOWN) 189–90
tsunamis 270, 271
turbidity sensors 77
two (twin)-factor authentication 124
two-point calibration 81
type checks 20

U

ultrasonic sensors 83
un-normalised databases (0NF, UNF) 325
updates 125, 136
UPPER function 211
user interfaces 67–8, 72–5
    expert systems 156
utility software 62, 64–5
    disk management systems 69–70
    file management systems 69
    uses 65–9
UV sensors 77

V

validation 18, 18–22, 258–9
    advantages and disadvantages 27
    difference from verification 22–3
validation rules
    databases 287–8
    spreadsheets 218–20
    testing 230–33, 290–1
vane rotary actuator 85
variables 99, 259
verification 18
    advantages and disadvantages 27
    databases 291
    difference from validation 22–3
verification methods 23–7
video compression 350–1, 354
video editing
    adding audio 349
    adding text 343
    altering the speed 345
    animation effects 344
    captions/subtitles 346–7
    credits 347–8
    exporting in different formats 350
    extracting a still image 346
    fading effects 351–2, 354
    filters and colour correction 353
    frame size and aspect ratio 340–1
    inserting a still image 343
    pan and zoom effects 352–3
    removing audio 345
    resizing and cropping an image 342
    splicing 345–6
    trimming 344–5

viruses   130
vishing   128–9, 133
visual checking   23
VLOOKUP function   201–2
voiceovers   363–4

W

waiting line management   264–6
water pollution monitoring   78–9
.wav files   364
weather forecasting   4–5, 57, 262–3
weather stations   79–80

web crawlers   131
WEEKDAY function   213
what-if analysis   255–6, 272–5
WHILE…ENDWHILE   107–8, 109
    nested loops   110
wi-fi   139
wildcard searches   301–2
wireless connections   139
wireless sensor and actuator networks (WSANs)   42–3, 95–6
wireless sensor networks (WSNs)   96
worms   130–1
wrapping text   198

X

XLOOKUP function   202–3

Y

YEAR function   214, 215

Z

zip files   333
zoom effects   352–3